W0245910

APPLICATIONS OF SYNCHROTRON RADIATION

Editors

C.R.A. CATLOW
Wolfson Professor of Natural Philosophy
Royal Institution
London

and

G.N. GREAVES
Head of Materials Science
SERC Daresbury Laboratory
Warrington
and
Visiting Professor of Chemistry
University of Keele

Blackie
Glasgow and London

Published in the USA by
Chapman and Hall
New York

Blackie and Son Ltd
Bishopbriggs, Glasgow G64 2NZ
and
7 Leicester Place, London WC2H 7BP

Published in the USA by
Chapman and Hall
a division of Routledge, Chapman and Hall, Inc.
29 West 35th Street, New York, NY 10001-2291

© 1990 Blackie and Son Ltd
First published 1990

Softcover reprint of the hardcover 1st edition 1990

All rights reserved
No part of this publication may be reproduced,
stored in a retrieval system, or transmitted,
in any form or by any means,
electronic, mechanical, recording or otherwise,
without prior permission of the Publishers.

British Library Cataloguing in Publication Data

Applications of synchrotron radiation.
1. Synchrotrons. Radiation
I. Catlow, C. R. A. (Charles Richard Arthur), *1947–* II.
Greaves, G. N.
539.735

ISBN-13: 978-94-010-6664-8   e-ISBN-13: 978-94-009-0395-1
DOI: 10.1007/978-94-009-0395-1

Library of Congress Cataloging-in-Publication Data

Applications of synchrotron radiation / edited by C.R.A. Catlow and
G.N. Greaves.
  p.   cm.

  1. Synchrotron radiation.   2. Synchrotron radiation sources.
I. Catlow, C.R.A. (Charles Richard Arthur), 1947-
II. Greaves, G.N.
QC793.5.E627A66   1989
543—dc20   89-15910
                                                CIP

Preface

Synchrotron radiation became available in a routine and regular manner to the scientific community in the early 1980s. Since that time the use of techniques employing synchrotron radiation has proliferated, so that the unique properties of this form of electromagnetic radiation are now having a major impact on several areas of physical and biological sciences. Not only have several new techniques become available but new opportunities with existing methodologies, e.g. diffraction, have been opened up.

In this book we provide a survey of some of the most important applications of synchrotron radiation, with a strong emphasis on the fields of chemistry and materials science. An introduction to the properties of the radiation and its instrumentation is given in chapter 1. The following chapters describe the use of synchrotron radiation in high resolution powder diffraction for structural studies of crystalline materials and in diffraction topography for imaging defects in single crystals. The role of EXAFS in investigations of amorphous and disordered crystalline solids and of biological systems is highlighted. The important enhancements to surface science techniques offered by synchrotron radiation are then reviewed. Later chapters describe more specialist applications, including trace-element analysis, protein crystallography, X-ray microscopy, and atomic and molecular spectroscopy.

It is not possible in a book of this length to survey the whole of this field, but we aim to have pointed the reader towards some of the most exciting and topical areas. Synchrotron radiation studies are developing rapidly; and the horizons are broadening, with the availability of new advanced sources both in Europe and the USA in the 1990s. We hope that this book will indicate some of the future directions of the field.

C.R.A. Catlow
G.N. Greaves

Contributors

C.R.A. Catlow Royal Institution, 21, Albemarle Street, London, WIX 4BS, UK

A.V. Chadwick University Chemical Laboratory, University of Kent, Canterbury, Kent, CT2 7NH, UK

M. Dudley Department of Materials Science and Engineering, State University of New York at Stony Brook, NY11794-2275, USA

P.J. Duke SERC Daresbury Laboratory, Warrington, WA4 4AD, UK and Kings College, London, UK

J. Evans Department of Chemistry, University of Southampton, Southampton, SO9 5NH, UK

C.D. Garner Department of Chemistry, University of Manchester, Manchester, M13 9PL, UK

H.C. Gerritsen Department of Molecular Biophysics, University of Utrecht, Buys Ballot Laboratory, PO Box 80.000, 3508 TA Utrecht, The Netherlands

I.D. Glover Department of Physics, University of Keele, Keele, Staffordshire, ST5 5BG, UK

G.N. Greaves SERC Daresbury Laboratory, Warrington, WA4 4AD, UK

S.J. Gurman Department of Physics, University of Leicester, Leicester, LE1 7RH, UK and SERC Daresbury Laboratory, Warrington WA4 4AD, UK

J.R. Helliwell SERC Daresbury Laboratory, Warrington, WA4 4AD, UK and Department of Chemistry, University of Manchester, Manchester M13 9PL, UK

D.A. King Department of Physical Chemistry, University of Cambridge, Cambridge, UK

Y.K. Levine Department of Molecular Biophysics, Buys Ballot Laboratory, PO Box 80.000, 3508 TA Utrecht, The Netherlands

J. Miltat   Laboratoire de Physique des Solides Bât 510, Université de Paris-Sud, 91405 Cedex, France

D. Norman   SERC Daresbury Laboratory, Warrington, WA4 4AD, UK

C. Robertus   University of Utrecht, Buys Ballot Laboratory, PO Box 80.000, 3508 TA Utrecht, The Netherlands

G. Ungar   School of Materials, University of Sheffield, Sheffield, S10 2PZ, UK

H. Van Langen   Department of Molecular Biophysics, Buys Ballot Laboratory, PO Box 80.000, 3508 TA Utrecht, The Netherlands

R.D. Vis   Department of Physics, Free University of Amsterdam, De Boelelaan 1081, Amsterdam

J.B. West   SERC Daresbury Laboratory, Warrington, WA4 4AD, UK

Contents

7   EXAFS studies of ionically conducting solids   171
A.V. CHADWICK

8   Applications of EXAFS to the study of metal catalysts   201
J. EVANS

9   Looking at solid surfaces with synchrotron radiation   221
D. NORMAN and D.A. KING

1 Synchrotron radiation instrumentation

G.N. GREAVES and C.R.A. CATLOW

1.1 Introduction

Almost all modern synchrotron radiation sources employ a dedicated *storage ring*. The basic principles of the operation of such sources are simple. Electrons are accelerated and injected into the storage ring; they are kept at a fixed energy within the orbit of the accelerator by application of a constant magnetic field. The synchrotron radiation is emitted tangentially from the beam orbit into a very small angle called the *photon opening angle*. The energy lost from the beam due to emission of the electromagnetic radiation is replenished by radio frequency power provided by a klystron. Thus a storage ring may be envisaged as a device for transforming radio frequency radiation into the broad electromagnetic spectrum characteristic of synchrotron radiation. This encompasses X-ray to microwave radiation and is strongly polarised in the plane of the storage ring—particularly at X-ray energies.

Two key characteristics of a source are the energy of the electrons in the beam, and the beam current. The former range from 0.5 to 5 giga electron volts (GeV) and the latter from 50 to 500 milliamps (mA). Characteristics for some and planned storage rings around the world are given in Table 1.1. Electrons at these energies travel close to the speed of light.

A schematic illustration of a storage ring is shown in Figure 1.1. The beam circulates around a tube, which is operated at high vacuum (less than 10^{-9} torr). The whole ring has an overall diameter typically of 30 m; a view of the ring at the SRS Daresbury is given in Figure 1.2(a). The beam is bent by a succession of powerful magnets separated by straight sections. At the SRS sixteen 1.2 T magnets bend the 2 GeV beam through a radius of 5 m. Additional magnetic devices known as *wigglers* and *undulators* can be inserted in the straight sections between the bending magnets to provide alternative radii of curvature for the electron beam. By altering the curvature of the electron orbit, the peak in the white continuum of synchrotron radiation can be shifted to shorter or longer wavelengths. At the SRS Daresbury a superconducting 5 T wiggler magnet produces a hard X-ray beam with energies up to 100 kilo electron volts (keV). An undulator made from 0.3 T permanent magnets emits strongly in the range 100 eV to 1 keV.

The distinction between dipole radiation, emitted when the beam is bent by

Table 1.1 Synchrotron radiation facilities worldwide.

Name	Location	Electron energy (GeV)	Typical beam current (mA)
SRS	Daresbury, UK	2	300
DCI	Orsay, France	1.8	300
NSLS	Brookhaven, USA	2.5	500
		0.75	400
DORIS II	Hamburg, FRG	3.7–5.3	100
BESSY I	W. Berlin, FRG	0.805	300
SPEAR	Stanford, USA	3.5	100
ALS	Lawrence Berkley, USA	1.5–1.9	400
under construction			
ADONE	Frascati, Italy	1.5	100
ALADDIN	Wisconsin, USA	1	100
PHOTON FACTORY	Tsukuba, Japan	2.5	200
BESSY	Berlin, FRG	0.8	300
ESRF	Grenoble, France	6	100
under construction			
APS	Argonne, USA	7	200
under construction			
SUPERACO	Orsay, France	0.8	400
ELETTRA	Trieste, Italy	1.5–2	200
under construction			
MAX	Lund, Sweden	0.5	50
UVSOR	Okazaki, Japan	0.75	60
VEPP-3	Novosibirsk, USSR	2.2	100
BEPC	Beijing, China	2.6	100
HESYRL	Hefei, China	0.8	150

conventional magnets, and the radiation emitted by insertion devices (wigglers and undulators) can be seen in Figure 1.3. Dipole radiation emanates from a single element of the electron beam trajectory. For insertion devices, many elements of the oscillating electron beam contribute to the emitted radiation. The curvature of the electron beam in a wiggler magnet is severe and the total intensity from the multiple sources is additive, i.e. proportional to N where N is the number of oscillations or periods of the electron trajectory. Undulator magnets deflect the electron beam by less than the photon opening angle giving rise to interference effects between the travelling electron and the photon it has emitted. As a result the continuum of synchrotron radiation emission is broken up into a harmonic series. Thus undulator radiation can be quasi-monochromatic and the intensity can approach N^2 at the harmonics of the device.

The electron beam is injected in to the ring as a series of *bunches* to match the klystron frequency. Each bunch is typically a few centimetres in length and so synchrotron radiation has a natural pulse width of 100 ps. Accordingly a single electron bunch may be used for time-resolved experiments. In general, during

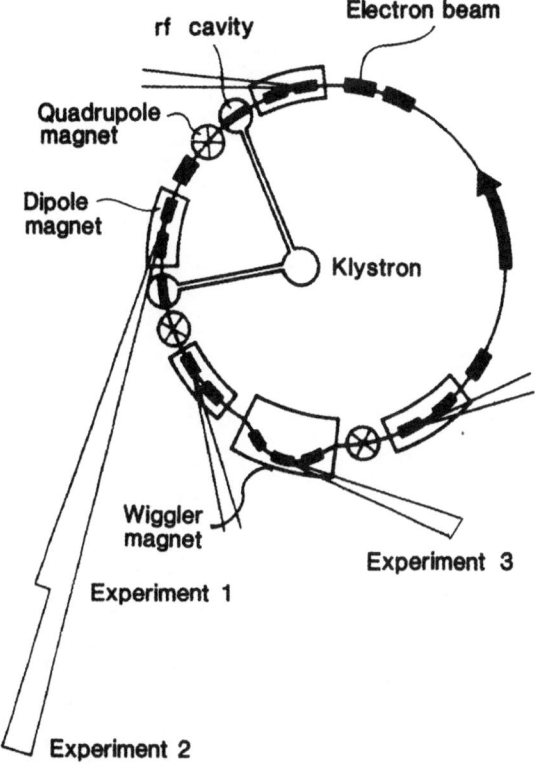

Figure 1.1. A schematic diagram of a storage ring with experimental beam lines.

operation, electrons will slowly be scattered out of the beam orbit partly due to electron–electron interaction (important in low energy machines) but mainly due to scattering from the gaseous molecules which are present in significant concentrations even under high vacuum conditions. As a consequence the electron current and hence the intensity of the synchrotron radiation decays exponentially. It will normally be necessary to refill the storage ring every 12–24 h.

Synchrotron radiation sources will generally support numerous experimental stations. These are constructed along beam lines which, as shown in Figure 1.2(a), collect the radiation emitted tangentially as the beam passes through the bending magnets or insertion devices. The synchrotron radiation source is usually referred to as the *tangent point*. In order to achieve a satisfactory geometry for the experiment it may be necessary to deflect the beam using a mirror; focussing mirrors may also be employed to increase the intensity of the sample as discussed later in this chapter. For experiments requiring X-rays, it is possible to conduct experiments in air; such experiments will, however, normally be enclosed in hutches to provide radiation protection, because they offer a direct line of sight to the GeV electron beam. In

(a)

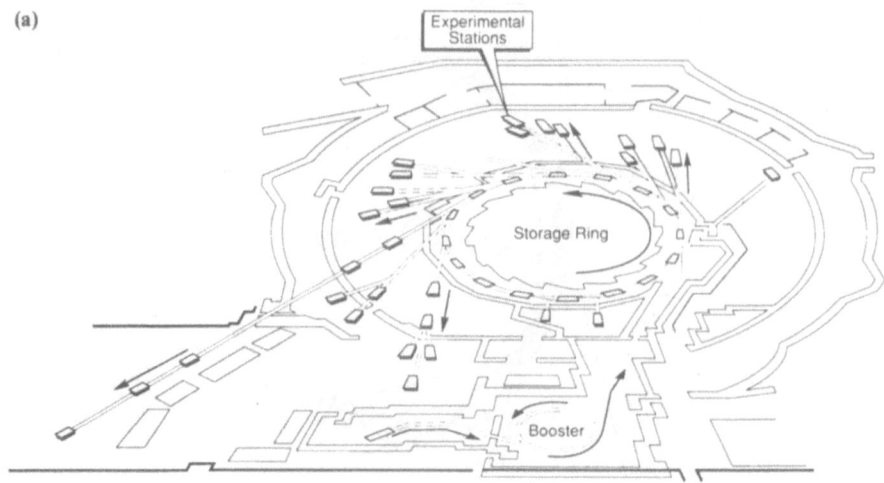

(b)

Figure 1.2.   The SRS at Daresbury Laboratory. (a) An overview of the ring and the beam lines;
(b) a VUV beam line, close up.

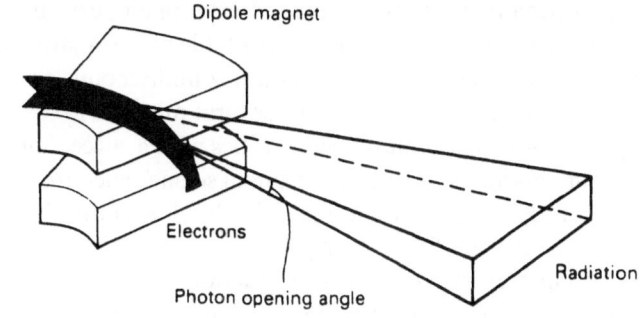

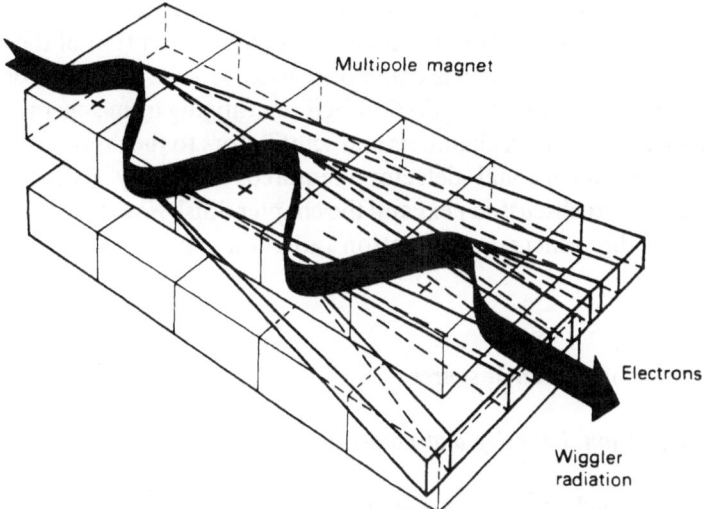

Figure 1.3. Schematic representation of an electron beam and radiation emitted by dipole and wiggler magnets.

contrast, for the soft X-ray and vacuum ultraviolet (VUV) regions of the spectrum, experiments are usually out of sight of the machine tangent point, and heavy shielding is not necessary. Nevertheless, evacuated chambers are needed, owing to the strong absorption of such radiation by air. In both cases experiments necessarily run remotely. Computer control is a prerequisite and the operator will seldom be able to see the experiment from the terminal area. TV monitors are often used to reveal the specimen area. Laser systems can be useful to survey samples into position and align optical elements.

A synchrotron radiation facility will therefore comprise a large group of experimental stations (typically 10–40 in number) clustered along the beam lines which radiate tangentially from the storage ring. Figure 1.2(b) shows a typical location on beam line 6 at the SRS. Each station will use a certain region of the synchrotron radiation spectrum (IR, VUV, X-ray or hard X-ray).

A station will normally perform one type of experiment, e.g. spectroscopy or diffraction, but multipurpose stations are available at many storage rings. The duration of a measurement may be from a few milliseconds to many hours. Experiments will last from a few days to several months. Most synchrotron radiation sources operate as multi-user facilities with a succession of scientists running experiments across a wide spectrum of science encompassing physics, chemistry, materials science and biology. This book is about the syndrome of multidisciplinary science offered by modern synchrotron radiation sources.

The present chapter provides an introduction to the instrumentation of synchrotron radiation experiments. We begin by reviewing the fundamental physics of the source, be it a dipole magnet or an insertion device. X-ray mirrors are discussed next. These are essential optical elements in many experimental stations, both for focussing and for deflecting the radiation emitted from the storage ring. We follow this by a description of the various monochromator configurations commonly used. A wide variety of detectors are employed at synchrotron radiation facilities ranging from general purpose simple devices such as ion chambers and scintillators to the more complicated one- and two-dimensional detectors required for X-ray diffraction and imaging. These are discussed next and the chapter finishes with descriptions of the layout of the principal synchrotron radiation experiments.

1.2 Synchrotron radiation sources

1.2.1 *Collimation, intensity and polarisation*

A classical charged particle travelling in a circular orbit through a magnetic field B emits electromagnetic radiation isotropically at the Larmor frequency. In storage rings, electrons or positrons travel close to the speed of light. In the stationary frame of the experimenter, synchrotron radiation is relativistically collimated in the forward direction and the emission is Doppler spread with a high energy cut-off in the X-ray region. Figure 1.3 shows schematically how synchrotron radiation from a dipole magnet is emitted tangentially to the particle orbit. The photon opening angle which defines this collimation is simply given by

$$\sigma_{ph} \sim \frac{1}{\beta\gamma} \tag{1.1}$$

where $\beta = v/c$ and $\gamma = E/m_0c^2 - v$ being the particle velocity, E its energy and m_0 its rest mass. For GeV electrons, $v \sim c$, $m_0c^2 = 0.511$ MeV and so σ_{ph} is typically 0.5 mrad or less.

The Larmor frequency, $eB/2\pi mv$, is 'blue shifted' as v approaches c to give the broad white spectra shown in Figure 1.4. The *critical energy* which

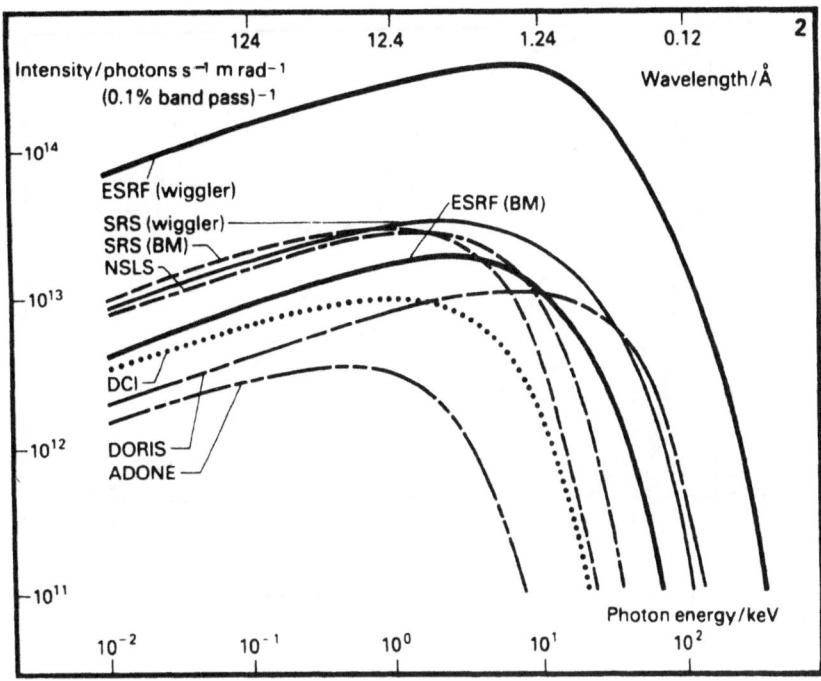

Figure 1.4. Spectral intensity distributions for various synchrotron radiation sources. Some of these are listed in Table 1.1.

characterises the X-ray cut-off is given by

$$E_c = 0.667 \, B \, E^2 \qquad (1.2)$$

For a typical dipole magnet field of ~ 1 T, an electron beam of a few GeV will emit X-rays with a characteristic energy of several keV. The spectral distributions for different storage rings are compared in Figure 1.4. The X-ray intensity integrated over σ_{ph} for a 0.1% band pass (bp) is given by

$$I \sim 5 \times 10^{10} \sigma_{ph} E^2 I_b f(E/E_c) \, \text{ps}^{-1} \, \text{mrad}^{-1} (0.1\% \, \text{bp})^{-1} \qquad (1.3)$$

where I_b is the particle beam current in mA. The 'universal function' $f(E/E_c)$, which can be deciphered amongst the spectra in Figure 1.4, has a maximum value of 1. Taking the SRS as an example, this runs at 2 GeV with a beam current of around 300 mA. Accordingly $\sigma_{ph} = 0.26$ mrad and the maximum intensity is 1.5×10^{13} ps^{-1} in a 0.1% band pass for a 1 mrad horizontal aperture. This should be compared with the bremsstrahlung intensity from a rotating anode—the most powerful laboratory source. This is limited to 10^5–10^6 ps^{-1} for this same band pass and solid angle aperture. Another important difference relates to the non-isotropic character of $f(E/E_c)$. An X-ray tube emits unpolarised radiation; whereas synchrotron radiation, particularly for energies close to E_c, is very strongly horizontally polarised in the orbit plane. Out of plane, the vertical component is increasingly important

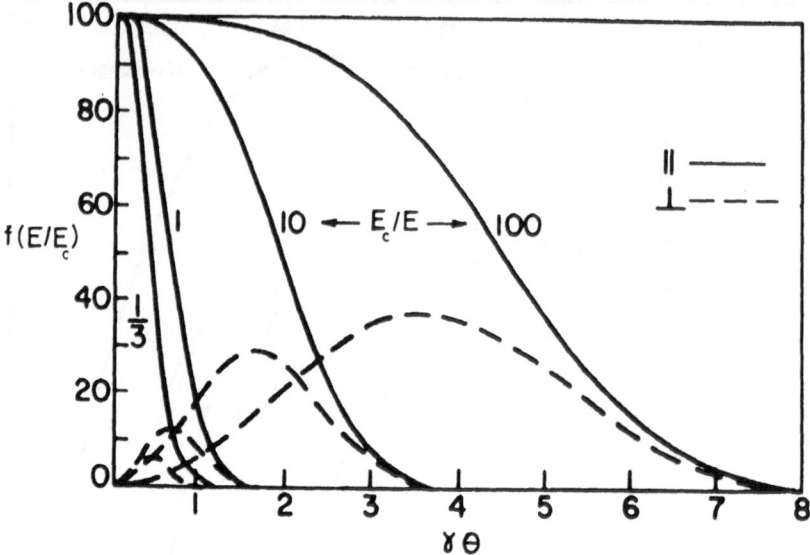

Figure 1.5. Parallel and perpendicular components of f (E/E_c) plotted as a function of vertical angle θ for various energies E.

leading to radiation having different degrees of circular polarisation. The vertical and horizontal components of $f(E/E_c)$ are plotted in Figure 1.5 as a function of angle, θ, with respect to the orbit plane. Figure 1.5 also shows how the photon opening angle, which is given by Eqn. 1.1 at $E = E_c$, broadens as the photon energy decreases. The collimation of synchrotron radiation is energy dependent.

A more physical indication of the X-ray intensity of synchrotron radiation can be obtained by expressing this in terms of the total power, P, emitted into a horizontal milliradian fan. This is given by

$$P = 4.22 \times 10^{-3} E^3 IB \quad \text{W mrad}^{-1} \tag{1.4}$$

For a conventional bending magnet (Figure 1.3) with $B = 1$ T and a 2 GeV beam, this amounts to tens of watts per milliradian. For a 4 GeV machine like DORIS, the emitted power per milliradian is hundreds of watts. However, for the insertion devices to be described shortly, the synchrotron radiation power delivered into a milliradian aperture can be as high as several kilowatts making the cooling of the first optical element in a beam line mandatory.

1.2.2 Brilliance and time structure

The geometry of synchrotron radiation beam lines and the heavy shielding around the storage ring result in large source to experiment distances of 10 m or more. This fact can be used to advantage in some experiments where high angular resolution is needed. For instance a 1 mm slit at 20 m subtends 50 μrad or 10 arc s at the source, which is comparable to the rocking curve width or angular acceptance of perfect crystals. In other cases where the photon flux at the sample needs to be maximised, focussing geometries must be used to image the source as we will describe later. Particle beams in present storage rings have finite cross-sections of between 0.5 and 10 mm^2 which result from the oscillations inherent in the magnetic lattice of the storage ring, called *betatron oscillations*. Future machines like the ESRF and the APS will have smaller cross-sections approaching 0.1 mm^2. It is the particle beam cross-section which primarily limits the brilliance of the imaged X-rays. If σ_V and σ_H are the vertical and horizontal root mean square (RMS) deviations of the particle beam with respect to the orbit plane, the spectral brilliance of a synchrotron radiation source is defined to be

$$B = \frac{I}{2.36^3 \times \sigma_V \sigma_H \sigma_{ph}} \quad \text{ps}^{-1}\,\text{mrad}^{-2}\,\text{mm}^{-2}(0.1\%\,\text{pb})^{-1} \qquad (1.5)$$

Thus for a 2 GeV, 300 mA, 0.5 mm^2 source like the SRS, the brilliance is $8 \times 10^{12}\,\text{ps}^{-1}\,\text{mrad}^{-2}\,\text{mm}^{-2}\,(0.1\%\,\text{bp})^{-1}$. Given perfect optics, this figure represents the largest number of photons per second in a given band pass that can be focussed onto unit area at the experiment. If the image is reduced in size by defocussing, the photon flux will increase, but the angular dispersion of the X-ray will also increase leaving the brilliance the same. This is an example of the application of *Liouville's theorems:* the brilliance of the image cannot exceed that of the source. In the case of monochromitising optics, the imaged brilliance may actually be reduced because the band pass will increase if the curvature of the diffracting element does not match the angular dispersion of the incident X-rays.

The same interdependence between spatial and angular distributions is also found in the phase space of the particle beam itself. For each spatial dimension, σ, there is also an angular divergence, σ', for particles in the beam. The product, $\sigma\sigma'$ is referred to as the *machine emittance* and is a fundamental design parameter of a storage ring. Typical values are 10^{-2}–10^{-4} mm mrad. Because the beam is continuously focussed and defocussed round the machine separate values vary but the product remains constant. In formal terms the problem of optimising an experiment to a synchrotron source reduces to that of matching the respective emittances. However in the design of storage rings at least one of the angular divergences of the particle beam (usually σ'_V) is much smaller than σ_{ph}. If X-rays are monochromatised vertically, the instrumentation problem simplifies to minimising optical aberrations to maximise the inherent brilliance of the source in this direction.

The particle beam in a storage ring also has a longitudinal dimension. This is because the radio frequency (rf) power which is coupled to the beam to replenish the power, P, lost through synchrotron radiation emission, breaks up the beam into a series of bunches. Synchrotron radiation is therefore pulsed rather than continuous. The longitudinal source size or bunch length is governed by the frequency of the rf supply and also the machine geometry and the beam current. Typical values are a few centimetres which, with relativistic particles, will result in pulse lengths for the emitted radiation of fractions of a nanosecond. Machines can often operate in various configurations ranging from a single bunch circulating with an average current of tens of milliamps to 100 or more bunches circulating with an average current of hundreds of milliamps. For example in multibunch mode the SRS accommodates 160 bunches which produce a light pulse with an RMS of 120 ps every 2 ns. This quasi-continuous time structure is the preferred mode of operation for most experiments. *Single bunch mode* is advantageous for timing experiments (see Chapter 15). A single bunch circulating round the SRS generates a 250 ps light pulse every 320 ns. The beam current I_b is about an order of magnitude less than in multibunch mode: typically 10–20 mA. Nevertheless, this results in an instantaneous brilliance of $2 \times 10^{15} \, \text{ps}^{-1} \, \text{mrad}^{-2} \, \text{mm}^{-2} \, (0.1\% \, \text{bp})^{-1}$, which should be compared with the peak brilliance of X-ray laser plasmas. These are the brightest 'single shot' X-ray sources available to date and have instantaneous brilliances in the range $10^{18}–10^{19} \, \text{ps}^{-1} \, \text{mrad}^{-2} \, \text{mm}^{-2} \, (0.1\% \, \text{pb})^{-1}$. We will see in the next section that insertion devices offer the prospect of much higher brilliance than bending magnets or even 'single shot' X-ray laser plasmas.

1.2.3   *Insertion devices*

The radius R of the electron orbit in a storage ring is given by

$$R = 3.34E/B \, \text{m} \tag{1.6}$$

Thus for a GeV beam travelling through a magnetic field of around a tesla the radius of orbit, R, will be several metres. Storage rings are built much larger with overall radii of 30 m or more. The extra circumference is made up from substantial straight sections between the dipole magnets. These accommodate focussing and defocussing magnets, rf cavities and so on but usually have spare capacity for insertion devices. Next generation storage rings are being designed around the concept of one insertion device for each straight section.

Insertion devices are multipole magnets which present the particle beam with a periodic magnetic field constraining it to oscillate but without any net deflection. A whole range of orbit radii, R, are possible, enabling the critical energy, E_c, to be tailored to particular experiments. More importantly, the synchrotron-radiation is emitted from a multiplicity of sources as illustrated in Figure 1.3 which results in improved intensity, I, and brilliance, B, compared to the source from a standard bending magnet.

Various types of multipole magnet are possible. The simplest, called a *wavelength shifter*, has only three poles but uses superconducting magnets to generate a pronounced 'blip' in the electron orbit giving rise to a source with high E_c. Multipole wigglers generally utilise conventional electromagnets. As these allow a smaller pole spacing, $\lambda_0/2$, and hence more oscillations in the same length, the total intensity is increased, although the electron beam curvature and the corresponding E_c of the synchrotron radiation will be similar to that for a standard dipole magnet. If permanent magnets are used λ_0 can be smaller still. Undulators are constructed in this way. By creating an electron beam deflection α comparable to the photon opening angle σ_{ph}, interference effects occur between the emitting particle and the emitted photon. This leads to further improvements in the intensity and brilliance of the emitted photon beam.

If $B_0(T)$ is the peak magnetic field of an insertion device, the deflection angle of the particle beam, α, is given by

$$\alpha = 0.477 \frac{B_0 \lambda_0}{E} \quad \text{mrad}$$

and the lateral displacement, s, by

$$s = \alpha \frac{\lambda_0}{\pi} \times 10^{-3} \text{mm}$$

where λ_0 is in millimetres. To take two extreme examples at the SRS, the 5 T wavelength shifter has a deflection angle of 60 mrad and a displacement of 9.5 mm, whereas for the 0.3 T permanent magnet undulator $\alpha = 0.3$ mrad and $s = 0.57$ mm. It is evident from Figure 1.3 that the intensity of the synchrotron radiation emitted from an insertion device scales with the number of magnetic poles, N. For a multipole wiggler, the brilliance is modified by the lateral extent of the multipole source, s

$$B_W \sim \frac{N B 2.35 \sigma_H}{(s + 2.36 \sigma_H)} \tag{1.7}$$

where B is given by Eqn. 1.5.

The brilliance spectrum of an undulator source is illustrated in Figure 1.6. The continuous spectrum that would be obtained from an equivalent multipole wiggler is decorated by a sequence of modes whose wavelengths, λ_i, are given by

$$\lambda_i = \frac{\lambda_0}{2i\gamma^2} \left(1 + \frac{\alpha^2 \gamma^2}{2} + \gamma^2 \theta^2 \right)$$

where θ is the observation angle with respect to the undulator axis-interference effects are two-dimensional. For geometrical and engineering reasons λ_0, the pole period, cannot be much smaller than about 50 mm. Accordingly for a photon opening angle $1/\gamma$ of 0.25 mrad, if the deflection angle α is of similar size (i.e. $\alpha \gamma = K \approx 1$) the undulator fundamental occurs around 20 Å in the soft X-ray region. For a 6 GeV machine on the other hand, for which $\alpha = 0.09$ mrad, an undulator with a deflection parameter, K, of unity will have a fundamental

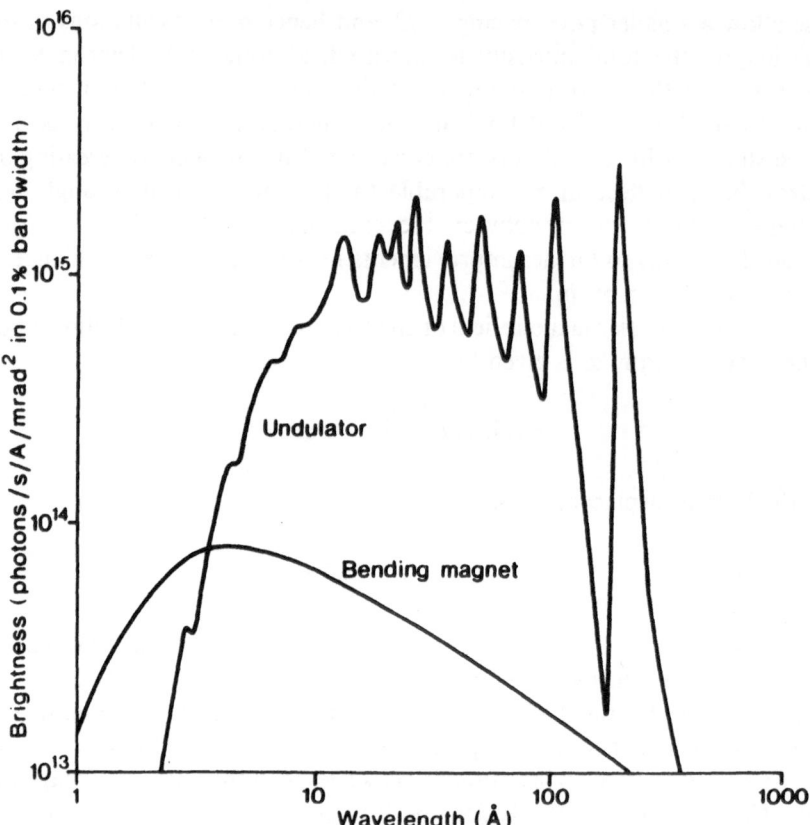

Figure 1.6. Spectral intensity of the SRS undulator compared to a bending magnet.

$\lambda_1 = 3\,\text{Å}$, with harmonics at hard X-ray wavelengths. The brilliance of the undulator modes is not seriously affected by the horizontal displacement s and at best scale geometrically with the number of poles, i.e.

$$B_U \leqslant N^2 B \qquad (1.8)$$

With the low emittance sources planned for the next decade, the undulator modes distinguishable in Figure 1.6 will be far better defined. By using pin-hole collimation, it will be possible to aperture quasi-monochromatic radiation without recourse to a diffracting element.

For present insertion devices with 10 or more poles Eqns. 1.7 and 1.8 indicate that the synchrotron radiation brilliance is improved compared to dipole radiation by one or two orders of magnitude, i.e. 10^{14}–$10^{16}\,\text{ps}^{-1}\,\text{mrad}^{-2}\,\text{mm}^{-2}$ (0.1% bp). The next generation synchrotron radiation sources however will have increases in beam energy, 6–7 GeV, and decreases in source size ($\ll 1\,\text{mm}^2$). This will lead to bending magnet source brilliances of $10^{14}\,\text{ps}^{-1}\,\text{mrad}^{-2}\,\text{mm}^{-2}$ (0.1% bp)$^{-1}$. The larger straight sections will accommodate bigger insertion devices comprising 100 poles or more with source brilliances in the range 10^{16}–10^{18} ps^{-1} mrad^{-2} mm^{-2} $(0.1\%\text{bp})^{-1}$ (see Table 12.2). Instantaneous or peak source brilliances per pulse are likely to be as high as 10^{23} ps^{-1} mrad^{-2} mm^{-2} $(0.1\%\,\text{bp})^{-1}$. In every respect these machines will furnish tunable pulsed X-rays with intensity and brilliance many decades higher than any alternative X-ray source.

For a fuller discussion of the theory of synchrotron radiation and accelerator physics, the reader is referred to recent reviews by Hofmann [1] and Suller [2] and references therein.

1.3   Mirror optics

A synchrotron radiation experiment generally utilises 1 or 2 mrad of the horizontal fan of radiation emitted by the storage ring. Because of the large source to experiment distance, which may be 20–30 m, focussing optics are often employed to exploit the extreme brilliance of the source. Diffraction and scattering experiments at hard X-ray energies require a focussed spot either at the sample or a little further downstream at the detector. For VUV experiments like UPS or soft X-ray spectroscopy, the standard arrangement is to focus twice, at the entrance slits and exit slits of the grating mono-chromator. Whilst focussing of X-ray and UV radiation can in principle be achieved by employing curved diffracting elements (crystals or gratings), in practice prefocussing mirrors are invariably employed. For VUV beam lines they offer the additional advantage that the photon beam can be deflected by substantial amounts so that the different experimental stations can be conveniently juxtaposed. Optimum optical arrangements for the principal synchrotron radiation experiments are described later in this chapter. Specific details can be found elsewhere in the book alongside the relevant chapters.

1.3.1   *Total external reflection*

The phenomenon of total external reflection is embodied in classical electromagnetism [3]. At X-ray energies the refractive index of materials is slightly less than unity and at shallow angles of incidence, refraction is reduced in favour of *total external reflection*. This occurs at a critical angle which, in practical units, is given by

$$\theta_c \sim \frac{20}{E}\sqrt{\rho}\ \text{mrad} \tag{1.9}$$

where E is the photon energy in keV and ρ the physical density at the surface in g cm^{-3}. Below θ_c the refracted wave is evanescent in the surface of the material, penetrating only a few tens of angstroms for keV X-rays. Above θ_c, the reflectivity approaches zero and the penetration depth approaches $\mu/\sin\theta$,

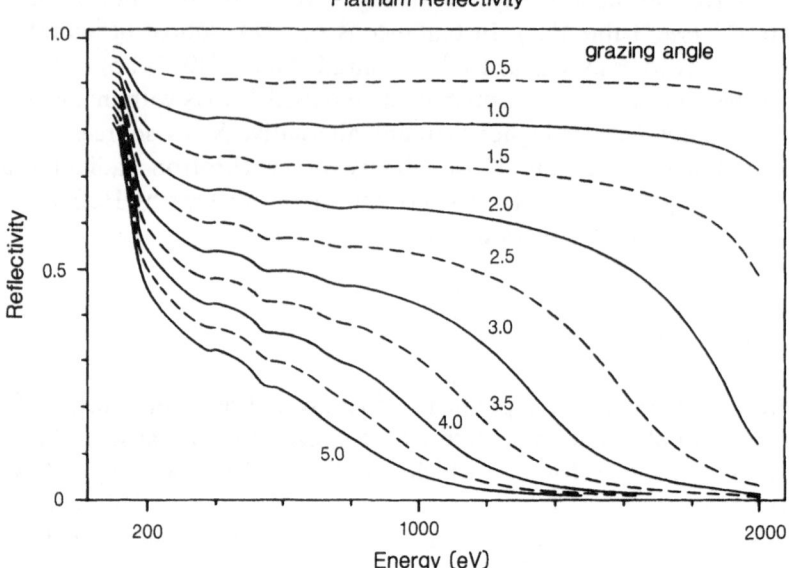

Figure 1.7.   Spectral reflectivity of Pt for grazing angles θ from 0.5° to 5° [4].

where μ is the absorption coefficient. For heavy metals like Pt and Au at 10 keV, θ_c is 9 mrad or 0.5°, increasing to 180 mrad or 10° at 500 eV. Corresponding values for silica, on the other hand are 3 mrad and 60 mrad at 10 keV and 500 eV, respectively.

Reflectivity profiles for Pt in the 200 eV to 2 keV energy range are plotted in Figure 1.7 for a variety of angles of incidence [4]. The high energy cut-off can be seen shifting upwards as the grazing angle is lowered. Mirrors are useful not just to focus with, but also to remove unwanted high energy radiation. The following practical points are worth stressing. At X-ray energies θ_cs are typically a fraction of a degree so mirrors need to be long, e.g. 70–100 cm, in order to intersect a beam which may only measure a few millimetres vertically. By contrast VUV mirrors can be much shorter but their heat loading is much greater. X-ray mirrors seldom absorb more than 10% of the total power, P, whereas the opposite is true of mirrors designed to reflect in the ultraviolet. Accordingly X-ray mirrors are usually fabricated from silica in contrast to VUV mirrors which are often made from SiC or alternatively from metal. The latter can be water-cooled although this can often introduce strain and distortion: in many ways uncooled refractory materials like SiC are more practical. Whatever the mirror material an X-ray surface finish is essential. A surface roughness of 5 Å (RMS) with a figure of 5 arc s (24 μrad) is readily achieved and generally considered adequate.

Metal coatings like Pt or Cr are applied to polished substrates by standard vacuum deposition techniques. Total thicknesses need not be greater than a few hundred angstroms on account of the shallow penetration of X-rays in the condition of total external reflection. Metal coatings enable the mirror to be inclined at a larger critical angle than the uncoated substrate. However, such thin films operate best in a high vacuum environment which minimises degradations from the deposition of contaminants from the beam line vacuum in the intense ambient radiation. This situation can also result sometimes in blistering of the coating and sometimes the opposite where the metal diffuses into the substrate. The deterioration of mirror surfaces exposed to synchrotron radiation is far from understood. Effects are considerably reduced of course when the X-ray mirror is not the first optical element in the beam line.

More practical details can be found in the comprehensive reviews by West and Padmore [5] and Howells [6].

1.3.2   Toroids and cylinders

Figure 1.8(a) illustrates the elliptical geometry which is the appropriate mirror configuration for focussing a point source. Although ellipsoidal mirrors can be figured for shallow angles of incidence and similar source and image distances, curvatures in the longitudinal (or meridian) and transverse (or sagittal) directions are vastly different. The required shape, however, is closely approximated by a *toroid* which can be readily generated by bending a long cylinder along its length as shown in Figure 1.8(b).

If the source to mirror and mirror to image distances are u and v, respectively, the longitudinal and transverse radii, R_M and R_S, are given by

$$\frac{1}{R_M} = \frac{\sin\theta}{2}\left(\frac{1}{u} + \frac{1}{v}\right), \quad \frac{1}{R_S} = \frac{1}{2\sin\theta}\left(\frac{1}{u} + \frac{1}{v}\right) \quad (1.10)$$

θ being the glancing angle of incidence. For a stigmatic image, $R_S = R_M\sin^2\theta$ and so with θ being usually 10 mrad or less R_S is four or more decades smaller than R_M. Suppose we take a toroidal mirror for focussing hard X-rays as an example. For a source to mirror distance of 15 m typical at synchrotron sources, 2:1 defocussing (i.e. $u/v = 2$) and a glancing angle of 6 mrad (0.34°), $R_S = 6$ cm and $R_M = 1.7$ km. Whilst the maximum horizontal aperture, $2R_S/u$, is 8 mrad and clearly adequate, the mirror will need to be more than 60 cm long in order to intercept a vertical fan of 0.25 mrad (i.e. σ_{ph}). By comparison, a VUV mirror might be only 20 cm in length. Because of the huge difference between tansverse and longitudinal radii, double focussing with a single mirror is often replaced by single focussing with two orthogonally placed cylindrical mirrors or with a cylindrical mirror and a cylindrical crystal. The two most commonly used bending mechanisms to create cylinders are shown in Figure 1.8(c, d). The ability to alter the transverse and longitudinal focus separately is particularly useful in achieving a stigmatic image at the entrance slits of a grating monochromator, at the specimen in a protein crystallography experiment or at the detector in a SAXS measurement.

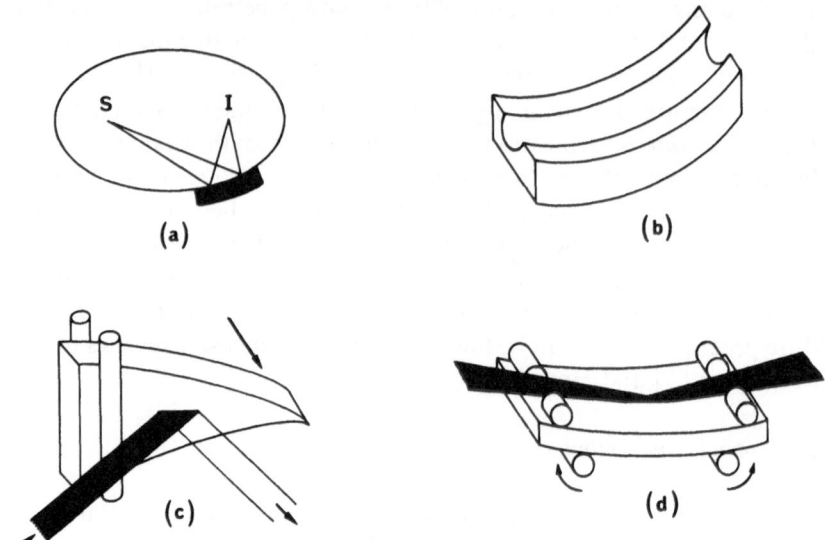

Figure 1.8. Focussing shapes for X-ray and VUV optics. (a) Ideal ellipsoid to produce aberration free image, I, of source S. (b) Toroid, produced by bending a cylinder, closely approximating to an ellipsoid. A cylinder profile can be produced by bending a triangular plate at its apex (c), or by exerting a rectangular plate to opposite couples at either end (d).

1.3.3 *Filters and windows*

Mirrors operating at X-ray wavelengths will often be preceded by filters or windows of C or Be. These have several uses. They remove the UV component which would otherwise be reflected or scattered by subsequent diffracting surfaces thereby contaminating the monochromatic beam. Such filters will naturally protect a mirror from the power associated with the unwanted UV radiation. Finally filters can be used to separate the beam line vacuum from the machine vacuum. For soft X-rays carbon foils 1 μm or less are used. They are usually self-supporting and uncooled. At hard X-ray energies ($>$ 3 keV) Be windows are generally used, generally 0.25–0.5 mm in thickness. Water cooling is essential but at these thicknesses, foils measuring approximately 1 cm in the smallest dimension will support several atmospheres of pressure, which provides a convenient way of completely isolating the machine vacuum from the beam line.

1.3.4 *Multilayers*

Although still in the development stages, *multilayers* offer considerable potential in overcoming optical and thermal problems associated with pre-focussing optics, particularly where next generation machines are concerned. Multilayers are 'artificial crystals' fabricated on mirror surfaces by alternately depositing heavy and light atom coatings, e.g. W and C. With repeat lengths typically in the range 30–50 Å several hundred layers behave as a 2D crystal, diffracting at soft X-ray wavelengths with Bragg angles of 10° or more. Many of the optical aberrations encountered with mirrors should be considerably reduced with multilayers where the reflecting angles are much larger. Also the refractory nature of the coatings will definitely be of help in coping with heat load problems when the radiated power P, might be kilowatts rather than watts. In the same sense, multilayers provide coarsely monochromatic radiation and if used in conjunction with the perfect crystal monochromators, to be discussed next, will help eliminate the heat load. Finally some experiments, such as X-ray microscopy and lithography, require very wide band-pass radiation. In this context multilayers can replace the filter-mirror combination often used at present [7].

1.4 Monochromators

A wide variety of monochromators is employed at synchrotron radiation facilities. Many are used in conjunction with mirror systems to produce focussed single wavelength beams. In this section, we will describe the most common designs for monochromatising X-rays and UV radiation.

1.4.1 *Crystal monochromators*

The tunable wavelength, λ, of an X-ray crystal monochromator is simply given by Bragg's Law

$$n\lambda = 2d \sin \theta \tag{1.11}$$

where d is the crystal spacing and n the order of the reflection. Note that the photon energy, E, corresponding to wavelength, λ, is given by

$$E = \frac{hc}{\lambda} \approx \frac{12.4}{\lambda(\text{Å})} \quad \text{keV} \tag{1.12}$$

In Eqn. 1.11, θ is the angle between the incident X-ray beam from the storage ring and the reflecting crystal planes. Because synchrotron ration is white, the wavelength or energy of the monochromatised beam is continuously tunable by rotating the crystal to alter θ. The crystals most commonly used are silicon and germanium for which the most versatile reflecting planes are (220) and (111). For Si and 2d spacings are 3.84 Å and 6.27 Å, respectively, 4.00 and 6.53 are the corresponding values for Ge. Provided dislocation-free material is used, the intensity and band pass of the monochromatised X-ray beam is principally governed by the *Darwin width* of the crystal. This is the intrinsic width of the Bragg reflection and is the minimum angular acceptance of the monochromator. It is given by [8]

$$\omega_0 = 2.12 r_0 \lambda^2 \frac{CNF(\lambda)}{\pi \sin 2\theta} e^{-M} \tag{1.13}$$

where r_0 is the classical electron radius, N the atomic density and $F(\lambda)$ the scattering factor. The remaining parameters are a polarisation factor C and a temperature factor e^{-M}. For perfect crystals Eqn 1.13 can be written

$$\omega_0 = \frac{d\lambda}{\lambda} \tan \theta \tag{1.14}$$

where the band pass $d\lambda/\lambda$ ($= dE/E$) is a constant ranging from 5.6×10^{-5} for Si(220) to 1.4×10^{-4} for Si(111) and 3.3×10^{-4} for Ge(111). Equation 1.14 is

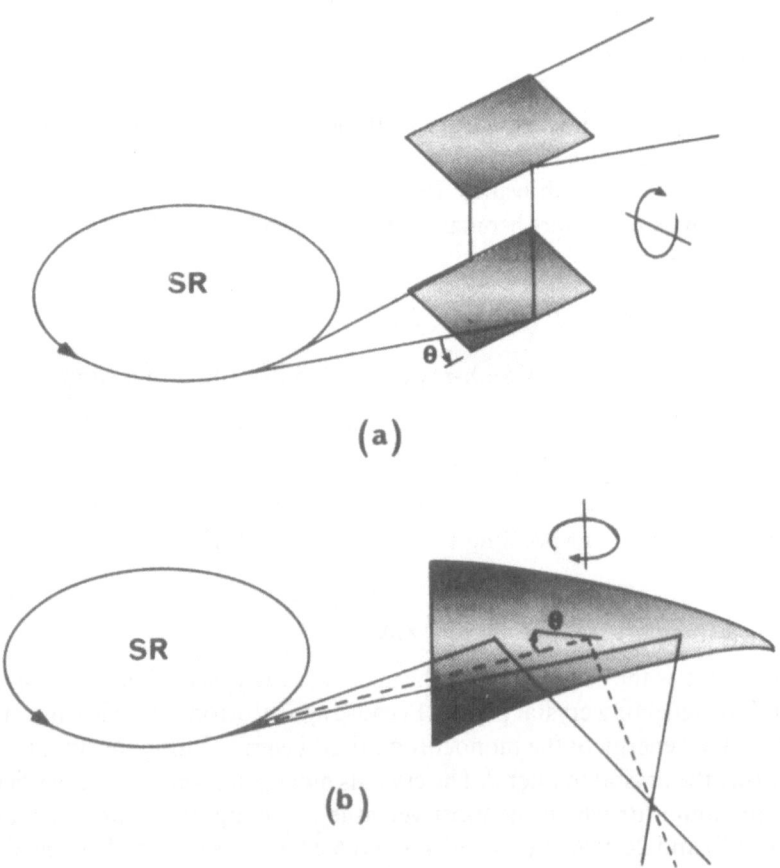

Figure 1.9. The two principal configurations for X-ray crystal monochromators. (a) The crystal arrangement diffracting vertically suitable for spectroscopy; (b) single bent crystal diffracting horizontally, suitable for diffraction.

simply obtained from Bragg's Law (Eqn. 1.11) by differentiation. The total intensity reflected by such a crystal is obtained by integrating over the Darwin width. Clearly for maximum intensity, ω_0 should be as large as possible, in which case Ge(111) is preferred to Si(111). For the highest spectral resolution, ω_0 should be as small as possible, so Si(220) is better than Si(111), although the reflected intensity is less.

There are principally two types of hard X-ray monochromator: the vertically dispersing two crystal configuration and the single crystal horizontally dispersing design. These are illustrated in Figure 1.9. In the two crystal arrangement (a) both crystals are aligned to within the Darwin width and inclined at an angle, θ, to the incident white radiation producing a parallel monochromatic beam. By rotating the crystal pair about a horizontal axis, the monochromator is rapidly and continuously turnable. If D is the gap between the crystal pair, the vertical displacement of the diffracted beam is $2D\cos\theta$. For $D \approx 10\,\text{mm}$, this displacement can be significant for large θ values, changing as the monochromator is scanned. This displacement can be compensated for by vertically translating the experiment. Alternatively, by translating the second crystal with respect to the first, a constant exit beam can be obtained. Both solutions are commonly employed. The double crystal monochromator is ideal for spectroscopy experiments such as EXAFS [9] (Chapters 6–8). It is also extremely convenient for high resolution powder diffraction (Chapter 2) and anomalous dispersion (Chapter 10).

If flat crystals are used, the monochromatised beam from a two crystal monochromator is simply a divergent extension of the incident fan of radiation from the tangent point in the storage ring. By bending one of the crystals sagittally, however, horizontal focussing can be realised. The radius, R_S, required is given by the mirror formula (Eqn. 1.10). Sagittally focussing crystal monochromators are still at the development stage. A practical intermediate solution is to combine a flat crystal monochromator with a toroidal mirror. Ideally the mirror should be placed after the monochromator to avoid the almost inevitable degradation in energy resolution (dE/E) brought about by optical aberrations in the reflected image. Focussed spot sizes of around 1 mm are readily achievable.

The energy resolution, dE/E, of a two crystal monochromator is determined not just by the Darwin width, ω_0 (Eqn. 1.14), but also by the storage ring source size, σ_V, and the acceptance angle of the crystals, $\Delta\theta$. For flat crystals $\Delta\theta$ is governed by the size of the monochromator entrance slits, s_V. If L is the source to monochromator distance the convoluted energy resolution is given by

$$\frac{dE}{E}\left(= \frac{d\lambda}{\lambda}\right) = \sqrt{\omega_0^{2+} + \left(\frac{\sigma_V}{L}\right)^2 + \left(\frac{s_V}{L}\right)^2}\cot\theta$$

As we have already pointed out, for storage rings, like the SRS, σ_V/L is comparable with ω_0. Clearly as $\sigma_{ph} \gg \omega_0$, in matching $\Delta\theta$ to ω_0 considerable intensity must be forfeited: sub-millimetre slits are required where the full beam height is typically several millimetres. A novel two crystal slitless monochromator where the crystals are bent about a horizontal axis with the curvature matched to the source divergence σ_{ph} achieves high energy resolution without limiting the vertical acceptance angle [10]. A toroidal mirror placed after the monochromator focusses the X-ray beam onto the sample. Focussed monochromatic beams approaching $10^{12}\,ps^{-1}$ are obtainable.

For a single crystal X-ray monochromator the diffracted beam is deflected through an angle 2θ. The basic arrangement is a horizontal two circle diffractometer and is the best configuration for diffraction experiments. By

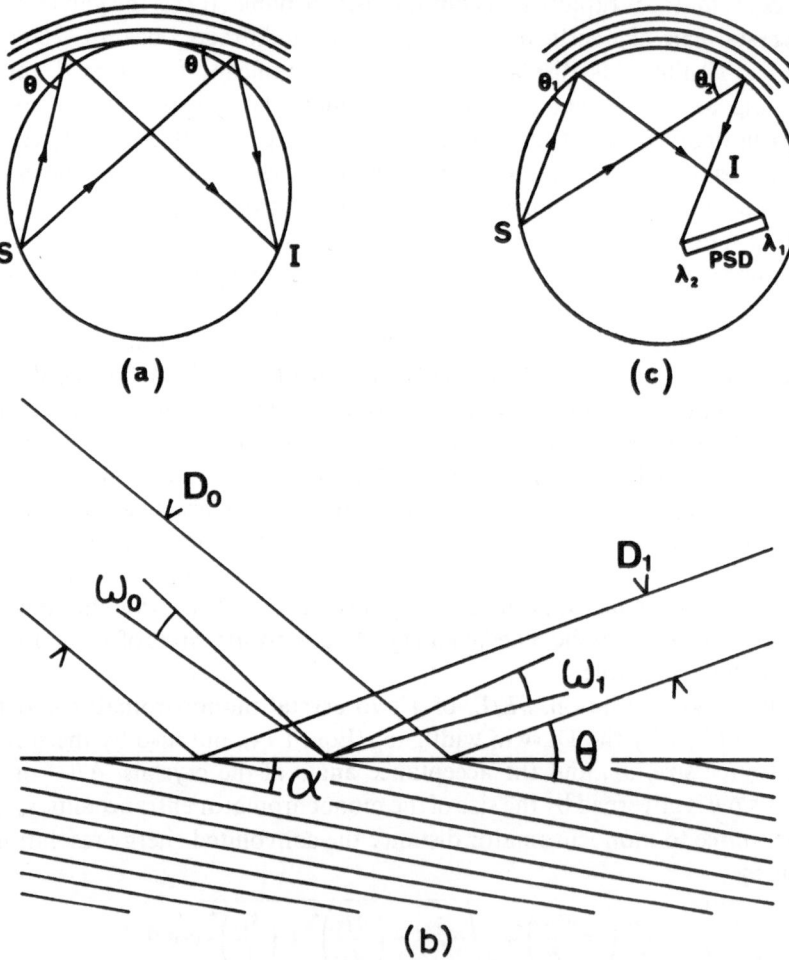

Figure 1.10. Focussing X-rays with crystals. (a) Rowland circle 'Johann' geometry; (b) asymmetric cut, α, beam is compressed left to right; (c) dispersive EXAFS geometry, PSD-position sensitive detector.

bending the crystal as shown in Figure 1.8(c), the monochromatic beam can be brought to a focus. The appropriate radius, R_M, can be obtained from the mirror formula given earlier (Eqn. 1.10) by replacing the grazing angle by the Bragg angle. Just as with a grating monochromator the choice of crystal curvature affects the band pass of the device. This is minimised in the *Rowland circle* geometry which is shown in Figure 1.10(a). For a symmetrically cut crystal, i.e. crystal planes are parallel to the crystal surface, image and source distances are the same ($u = v$). The radius of the circle is $R_M/2$. Provided the length of the crystal is not too great, most of the divergent rays from the source, S, lying on the Rowland circle all strike the crystal with the same Bragg angle, θ, minimising the band pass, $d\lambda/\lambda$, to form a focussed image I symmetrically on the opposite side of the circle [11]. This is the so-called Guinier setting. However, in the context of a synchrotron radiation source, where the tangent point may be up to 20 m away from the monochromator, symmetrical focussing would extend the experiment a further 20 m. The Guinier setting can be established for a shorter image distance by employing an asymmetrically cut crystal, for which the optical surfaces are cut at an angle α to the reflecting crystal planes [12]. This is illustrated in Figure 1.10(b). The angles of incidence, i, and reflection, r, are now different with respect to the crystal face and the formula for meridional focus (Eqn. 1.10) becomes

$$\frac{1}{R_M} = \frac{1}{2}\left[\frac{\sin i}{u} + \frac{\sin r}{v}\right] \tag{1.15}$$

where $i = \theta + \alpha$, and $r = \theta - \alpha$. The so-called Guinier condition is now given by

$$\frac{u}{v} = \frac{\sin i}{\sin r}$$

For $\theta = 14°$ and $\alpha = 12°$ this is close to 10:1. Accordingly the focus can be placed 2 or 3 m from the crystal [13] enabling the monochromator and experiment to be contained in the same work station hutch, which is always a desirable aim.

There are several other consequences of using an asymmetrically cut crystal. First the width of the beam is compressed from D_0 to D_1 as shown in Figure 1.10(b), namely

$$D_1 = D_0 \frac{\sin(\theta - \alpha)}{\sin(\theta + \alpha)}$$

This increase in image brilliance is, however, illusionary as the angular dispersion of the incident and diffracted beam are conversely affected. In particular the dispersion of the reflected X-rays, ω_1, which govern the band pass of the monochromator, is larger than ω_0

$$\omega_1 = \omega_0 \frac{\sin(\theta + \alpha)}{\sin(\theta - \alpha)}$$

multiplying this with D_1, we find

$$D_1 \, \omega_1 = D_0 \, \omega_0$$

which is just another manifestation of Liouville's theorem.

To maximise the diffracted brilliance of a horizontally focussing single crystal monochromator, a vertically focussing cylindrical mirror is usually placed upstream from the work station defocussing 1:1 or 2:1, both monochromator and mirror imaging the source at the same point [13]. With present synchrotron radiation sources, stigmatic images of around 0.3 mm in diameter can usually be achieved. This crystal/mirror combination is the preferred configuration for small angle scattering experiments (Chapter 4) and protein crystallography measurements (Chapter 10).

Although bent single crystal monochromators are often engineered with a moveable 2θ arm, this is a clumsy arrangement for spectroscopy. However, if the crystal is overbent bringing the focus inside the Rowland circle, the broadened band pass $\lambda_1 - \lambda_2$ is geometrically dispersed as shown in Figure 1.10(c). This offers a most interesting geometry for EXAFS experiments [15] and also anomalous dispersion. If l is the illuminated length of the bent crystal, the wavelength (energy) dispersion is simply given by

$$\frac{\lambda_1 - \lambda_2}{\lambda} = \frac{(E_1 - E_2)}{E} = \frac{l}{R_M \tan \theta} \approx \frac{l}{2v} \cos \theta \qquad (1.16)$$

where $v \ll u$ (Eqn. 1.10). By arranging λ_1 and λ_2 to span the wavelength of the absorption edge of interest, the transmission spectrum of a specimen placed at the focus can be recorded by a position sensitive detector (PSD) placed behind [18]. Alternatively, the spots of a single crystal diffraction pattern will be smeared and will record the change in intensity corresponding to the energy dependent scattering factor [19].

Given a bremsstrahlung X-ray source, *harmonics* ($n > 1$) will be present in the same diffracted beam as the fundamental (see Eqn. 1.11). The simplest technique for removing higher order X-rays in a crystal monochromator system is to employ a mirror reflection with a suitable cut-off, given by Eqn. 1.9

$$E_c = \frac{20\sqrt{\rho}}{\theta_c} \text{ keV}$$

In the single crystal monochromator configuration described in Section 1.4.1, harmonic rejection is a spin-off from the vertically focussing mirror and is generally adequate for minimising high order patterns in a diffraction experiment. However, Figure 1.7 demonstrates that reflectivity profiles are not step-functions but tail off considerably either side of E_c. A more efficient technique for harmonic rejection is required for X-ray spectroscopy experiments and for anomalous dispersion studies and is available for a two crystal monochromator if the sets of crystal planes are offset. This principal is

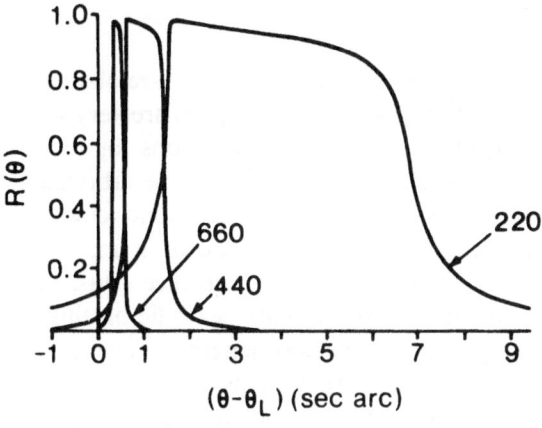

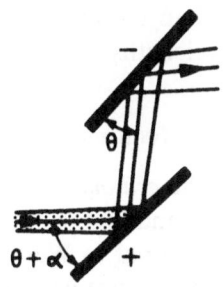

Figure 1.11. Single crystal reflectivities for Si(220), Si(440) and Si(660). Principle of harmonic rejection for a two crystal monochromator achieved by deflecting one member of the pair [16].

illustrated in Figure 1.11. The single crystal reflectivities for Si(220) and its harmonics (440) and (660) are plotted. The Darwin widths are given by

$$\omega_n = \frac{\omega_0}{(n+1)^2}$$

where ω_0 is given by Eqn. 1.13. ω_n not only decreases with the order of the reflection but is also angularly displaced. For the two crystal $(+1-1)$ arrangement shown the integrated reflectivity $R(\alpha)$ is the correlation of the individual crystal reflectivities R_1 and R_2

$$R(\alpha) = \int_0^{} R_1(\theta) \, R_2(\theta + \alpha) \, d\theta$$

where α is the angular displacement. Accordingly for identical crystals, by rocking one with respect to the other the overlap between the harmonics can be minimised for a small reduction in the fundamental [16]. This is conveniently done with a servo mechanism linking α to the output of the monochromator [17].

1.4.2 *Grating monochromators*

Whilst the physical optics of diffraction gratings are obviously different from those of crystal diffraction, the geometric optics are very similar: monochromatic light is reflected and the mirror equations for focussing (Eqn. 1.10) apply. As well as plane gratings, gratings can be fabricated in toroidal or spherical shapes and incorporated into Rowland circle geometries. The tangent point source may lie on this circle or it may be imaged by the monochromator pre-optics onto the entrance slits, as illustrated in Figure 1.17. The exit slits, however, will always lie on this circle as they are essential in selecting the photon wavelength or energy and in removing unwanted harmonics or higher orders. This can be seen in Figure 1.12 where the dispersion from a plane grating is shown schematically. In contrast to the convention for Bragg reflection, the angles for a grating are measured with respect to the normal rather than the diffracting surface. For an angle of incidence α the diffracted wavelengths, λ, and orders m, are geometrically dispersed through β where

$$m\lambda = d(\sin\alpha + \sin\beta) \qquad (1.17)$$

and d is the line pitch. This is the *grating equation*.

   It will be clear from Figure 1.12 and Eqn. 1.17 that α and β can be paired in a variety of ways. The reader is referred to West and Padmore [5] for a full discussion of the different practical grating configurations. The one most commonly used with synchrotron radiation sources is that in which the *deviation angle*, 2θ, is kept fixed

$$\alpha - \beta = 2\theta$$

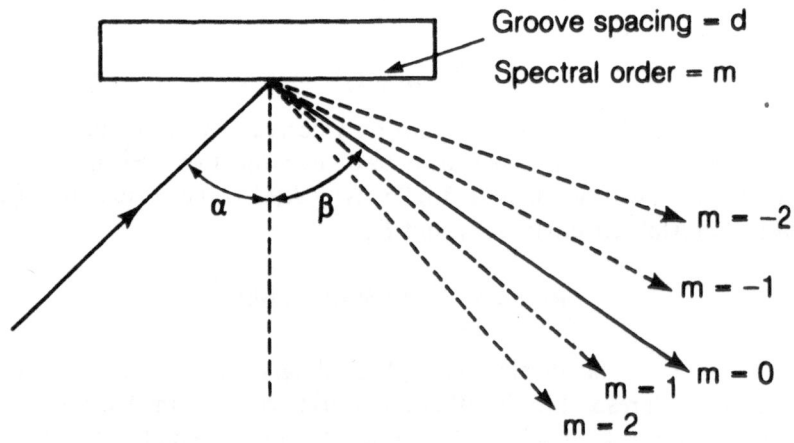

Figure 1.12. Notation for diffraction by a grating; α and β are the angles of incidence and diffraction, respectively. The deviation angle, α-β, is 2θ. The spectral orders are m.

in which case Eqn. 1.17 simplifies to

$$\lambda = \frac{2d}{m}\cos\theta\sin(\theta+\beta)$$

As $\theta+\beta$ is the angle turned from zero order ($\alpha=\beta$, $m=0$), the monochromator configuration is fixed apart from rotating the grating about an axis parallel to the grating rulings. Because α and β cannot exceed 90°, the limiting wavelength in this geometry, the *horizontal wavelength*, is given by

$$\lambda = \frac{2d}{m}\cos^2\theta$$

Taking typical line densities of 700–2000 lines mm^{-1}, 2θ is about 100° for VUV wavelengths in the range 100–1000 Å (120–12 eV).

Two fixed deviation angle geometries are shown in Figure 1.13. In the first (a), the white source is focussed onto the entrance slits. The grating is profiled

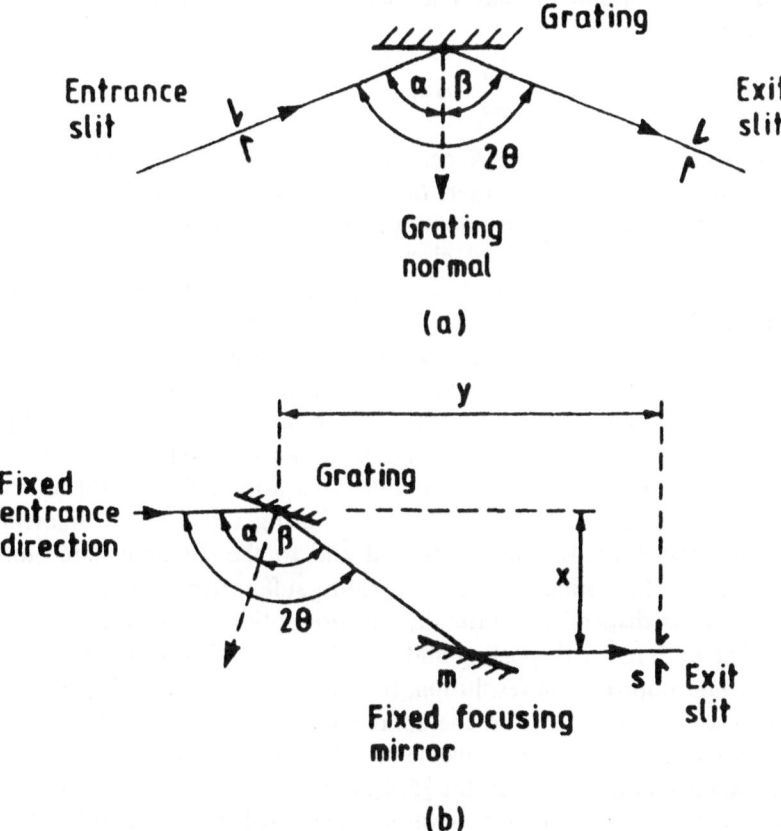

Figure 1.13. Two geometries for grating monochromators, both with fixed grating deviation angle, 2θ. (a) Single element, toroidal grating; (b) two elements, plane grating and mirror.

in the shape of a toroid so that a monochromatic stigmatic image is formed at the exit slits. This *toroidal grating monochromator* (TGM) is commonly used for photoemission experiments. A plane grating monochromator (PGM) is shown below (b). Here, the unfocussed monochromatic light is focussed by a fixed ellipsoidal mirror onto the exit slits. In this arrangement the entrance and exit beams are parallel which, as we have seen for a two crystal monochromator, can often be advantageous for spectroscopy.

The angular dispersion of a single element grating monochromator, the TGM shown in Figure 1.13(a), can be calculated by differentiating Eqn. 1.17 and maintaining α or β constant. Accordingly

$$\left(\frac{d\lambda}{d\beta}\right)_\alpha = \frac{d}{m}\cos\beta \quad \text{and} \quad \left(\frac{d\lambda}{d\alpha}\right)_\beta = \frac{d}{m}\cos\alpha$$

If s_1 and s_2 are the entrance and exit slit widths, respectively, in the dispersing direction, $d\alpha = s_1/u$ and $d\beta = s_2/v$, where u and v are the corresponding entrance and exit slit distances from the grating. The band pass of an otherwise optically perfect monochromator is therefore limited by source and image sizes, namely

$$\Delta\lambda_1 = \frac{ds_1}{mu}\cos\alpha, \quad \Delta\lambda_2 = \frac{ds_2}{mv}\cos\beta$$

For a Rowland circle of radius R, $\cos\alpha/u = \cos\beta/v = 1/R$ and the band passes are matched. Clearly the larger R is, the better the resolution of the monochromator. For an instrument operating in the VUV with a 1000 lines mm^{-1} grating and a Rowland circle radius of approx. 15 m, then, with entrance and exit slits matched to a 0.5 mm source, the best wavelength resolution, $\Delta\lambda/\lambda$, is about 10^{-3} i.e. 50 meV at 50 eV.

Of course grating monochromators attract optical aberrations associated with the grating shape and its pre-optics. As we have seen, in relation to crystal X-ray monochromators, the aim is not just to obtain a minimum band-pass instrument but also to maximise the angular acceptance to the white source. A curved grating will typically have a ruled area approaching 5000 mm^2 and subtend a solid angle of 100–200 mrad^2. Fitting this aperture inevitably generates optical aberrations which will add to the slit limited resolution outlined above. In particular a toroidal grating suffers from astigmatic coma if the aperture is dispersive. Optimally the pre-optics should focus approximately 1:1 both in the dispersing and non-dispersing directions. Magnifying optics would improve the resolution, but at the expense of the intensity.

TGMs will operate close to the source size limited resolution only for a limited wavelength range. Improvements can be made if the sagittal focussing is removed from the grating profile [5, 6]. Because the sagittal curvature is so much greater than the meridian curvature a spherical grating can be used. This is easier to fabricate compared to a cylinder or toroid even and enjoys better slope errors (> 0.2 arc s). The pre-optics of a spherical grating monochromator

(SGM) need only focus in the horizontal or non-dispersing direction on the exit slits, resulting in the entrance slits being effectively the tangent point source itself. Such 'slitless' SGM designs are beginning to be used on national facilities (e.g. station 1.1 at the SRS) and can produce monochromatic stigmatic images of the synchrotron source at the specimen with intensities in excess of $10^{12}\,\mathrm{ps}^{-1}$.

1.5 Detectors

Synchrotron radiation imposes severe demands on photon detecting devices owing to its intensity, wavelength range and collimation. Thus to exploit the potential of SR it is necessary to have detectors which will operate with intense photon beams from the IR to the hard X-ray regions of the spectrum. Experiments will sometimes use a white X-ray beam, but require energy analysis of the scattered radiation. Moreover, in many contemporary experiments rapid data collection is achieved by detection over a wide angle range using a *multidetector* device. In this section we will review the basic types of photon detectors, before considering multidetectors, and the special problems posed by the high intensity imaging.

In characterising detectors there are several useful parameters to be borne in mind. Suppose the number of incoming photons is N, the inherent photon limited noise is $\sqrt{N}$. The *quantum efficiency* (QE) of the detector per absorbed photon should be close to unity for the energy range of interest, unless the detection mechanism has intrinsic gain. The QE will clearly be degraded if the device noise is greater than $\sqrt{N}$. This is minimised in photon counting systems but detector *saturation* is often a problem. It should also be remembered that synchrotron radiation is pulsed, so detectors must possess count rates which are at least a decade greater than the value expected from the mean circulating beam current. Integrating devices allow for much higher maximum count rates than photon counting devices but offer little or no energy resolution, $\Delta E/E$. Both photon counting and integrating systems can be configured as PSDs. The optical resolution, Δl, or pixel size will vary from one family of detectors to another. As there is no 'universal' detector for synchrotron radiation, all these parameters need to be considered for optimum trade-off in matching a particular detector to a particular experiment. Table 1.2 lists the experiments most closely associated with the detectors described below.

1.5.1 *Low energy resolution detectors*

Ionisation chambers are the simplest of X-ray detectors. Their operation is based on the photoionisation of a rare gas. The detector which is illustrated

Table 1.2 Characteristics and uses of photon detectors for synchrotron radiation experiments.

Detector	Dimension	$\Delta E/E$	Spatial Resolution (μm)	Sensitivity	Count rate limit (ps^{-1})	Experiment[a]
Ionisation chambers	0	None	None	X-ray	10^{12}	XAS
Photoemission diodes	0	None	None	X-ray UV	10^{12} 10^{14}	XAS UVS
Scintillation counters	0	0.3	None None	X-ray UV	5×10^5 5×10^7	XAS, XRD UVS
Solid state detectors (SSDs)	0	0.03	None	X-ray	10^4	XAS, EDD, TEA
Cylindrical mirror analysers (CMAs)	0	0.003	None	Electrons	10^6	UVS, UPS, XPS
Proportional counters	0,1,2	0.5	100	X-ray	10^5 per wire	XAS, SAXS, XRD
Photodiode arrays (PDAs)	1	None	25	X-ray	10^9	EDE,SAXS
Charged couple devices (CCDs)	2	None	25	X-ray	10^9	PX, LD
Photographic film	2	None	1	X-ray	10^{12}	PX,SAXS, XRT
TV detector	2	None	5	X-ray	10^8	PX
Photoresist	2	None	0.005	Soft X-ray	10^{12}	SXRM
Image plates	2	None	50	X-ray	10^{12}	PX, LD, SAXS

[a] XAS, X-ray absorption spectroscopy; UVS, ultra violet spectroscopy; XRD, X-ray diffraction; EDD, energy dispersive diffraction; TEA, trace element analysis; SAXS, small angle X-ray scattering; EDE, energy dispersive EXAFS; PX, protein crystallography; LD, Laue diffraction; XRT, X-ray topography; SXRM, soft-X-ray microscopy.

schematically in Figure 1.14(a) comprises a gas filled chamber (typical gases are He and Ar), across which a moderate electric field is applied (a few $100 \, V \, cm^{-1}$). The incoming X-ray photons are photoelectrically absorbed; electrons may also be emitted by Auger processes. The fast electrons in turn cause electron-ion pair formation by inelastic collisions. The electric field applied across the device results in rapid ion and electron migration with minimal recombination.

The main advantages of the detector are its linearity and stability. Approximately 30 eV is needed to produce a single electron ion pair and the resultant ion current is directly proportional to the X-ray intensity. The ionisation chamber, however, is unsuitable for detecting low X-ray intensities, as the detector shows no intrinsic amplification of 'gain', the ion current being produced per photon, is not amplified by further ionisation processes. The maximum count rate of an ionisation chamber is around $10^{12}\,ps^{-1}$. In addition the detector has no energy discrimination; it can only detect higher energy photons, typically $> 2\,keV$.

For lower energies, photoemission diodes should be used. Like ionisation chambers these devices do not enjoy intrinsic gain. The emitter is usually a

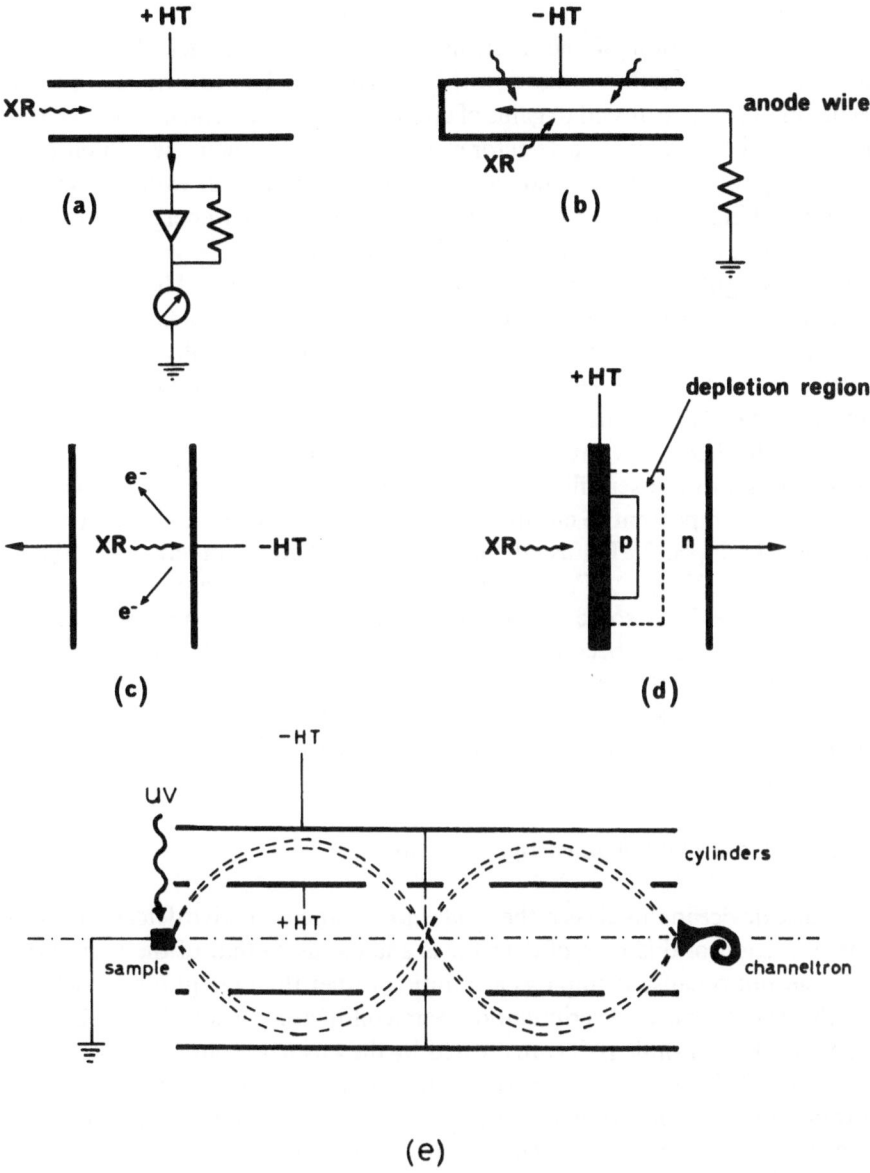

Figure 1.14. Schematics for various X-ray detectors. (a) Ionisation chamber; (b) proportional counter; (c) photoemission diode; (d) solid state detector; (e) cylindrical mirror analyser.

metal and the photocurrent is collected at a suitably placed positively biased collector as shown in Figure 1.14(c). Wavelength response is determined by the photon absorption characteristics of the metal and the corresponding electron escape depth. Cu is useful at soft X-ray energies whilst Al is best in the 10–250 eV UV regime.

Unlike the ion chamber or the photoemission diode, the *proportional counter* does exhibit gain. It exploits photomultiplication by operating at a far higher electric field. For a wire anode this is typically around $20 \, \mathrm{kV \, cm^{-1}}$. The resulting acceleration of the electrons produced by the inelastic collision processes causes further ionisation (Figure 1.14(b)). The recurrence of such secondary process can lead to gains of up to 5×10^3 above which, however, the detector will discharge. The detector clearly will detect much lower intensities but has poorer linearity and stability when compared with ionisation chambers. Unlike the ion chamber, though, it exhibits some energy resolution. The device is shown schematically in Figure 1.12(c).

For low light levels *scintillation counters* should be used. These detectors contain a fluorescent screen which absorbs X-rays or UV radiation and re-emits photons at lower energies. The fluorescence will normally be in the UV range and can be detected with high efficiency by a photomultiplier tube. Such detectors have high sensitivities and are used in experiments where it is necessary to detect low intensities with good signal to noise ratio.

The design of the scintillator screen is crucial. Grain size as well as film thickness are important to ensure a high quantum efficiency and a flat energy response. Tl doped NaI is usually employed for X-ray counters and sodium salicylate for UV and VUV detectors. Operating in photon counting mode, these devices exhibit some energy discrimination which is typically around 50% at X-ray energies becoming almost negligible in the UV. The count rate limit for 10 keV photons is a few $10^5 \, \mathrm{ps^{-1}}$ rising to approximately $10^8 \, \mathrm{ps^{-1}}$ for 100 eV photons. The detector QE is typically about 50% compared to the photomultiplier tube which might be as low as 10%.

1.5.2   *Detectors with high energy resolution*

The basic device in this class is the *solid state* photoconductive detector (SSD). The operation of this type of detector is analogous to that of the ionisation chamber but because it employs a counting rather than integrating mechanism this affords energy discrimination. Semiconductors, usually Si and Ge, are used and electron-hole pairs are created by the absorbed photons. Each pair requires $\sim 3 \, \mathrm{eV}$ of absorbed energy so the number of carriers and hence the current pulse is proportional to the photon energy. A multichannel analyser gives an energy spectrum of the absorbed radiation.

The most common type of SSD uses reversed bias p-n junctions as the detecting element, also shown in Figure 1.14(d). By applying the reverse bias to the junction a carrier depleted region is created thus reducing the background 'leakage' current of the detector and enhancing the signal-to-noise ratio of the device. The p-n junction is usually fabricated by diffusing Li into p type material. The detector is operated at low (liquid nitrogen) temperatures to prevent Li^+ diffusion, as well as to minimise thermal noise in the pre-amplifier.

SSDs may be used for detecting radiation from the visible to the X-ray regions, with Si devices being more appropriate for the softer and Ge for the harder radiation. Energy resolution of 0.03 ($\Delta E/E$) is possible in the X-ray regions making SSDs suitable for energy dispersive diffraction. The X-ray count rate limit of present SSDs is governed by the front-end electronics and is currently a few $10^4 \, ps^{-1}$. Multi-element SSDs have recently been introduced offering improvements in count rate of about a decade ($2 \times 10^5 \, ps^{-1}$).

1.5.3 *Photoelectron analysers*

The front end of an electron analyser system comprises an interaction region containing the target. The specimen may be a stream of gas or a single crystal mounted on a goniometer which intersects the incoming photon beam. In order to take advantage of the polarisation of synchrotron radiation, the target or the analyser or both will necessarily have some provision for orientation. Photoelectrons emitted in the interaction region are accelerated or retarded by the adjacent analyser and detected by a channeltron. (A channeltron is a miniature device, similar in operation to a photomultiplier, but monolithic in construction.) A variety of different electron analysers are available. The *cylindrical mirror analyser* (CMA) is illustrated in Figure 1.14(e). This has 'the best' collecting efficiency, intercepting a cone approaching 1 sr of the available photoemission. It comprises two concentric cylinders oppositely charged which accept and focus on to the channeltron monoenergetic electrons with a band pass that can be as small as 5 meV. The energy resolution is strongly dependent on the position of the specimen, which can impose considerable problems if reorientation is needed to explore polarisation phenomena. However, these effects are considerably reduced with the double pass CMA arrangement shown in Figure 1.14(e).

1.5.4 *Multidetectors*

The increasing demands for rapid data collection have led to the need to collect angularly dispersed data simultaneously. We review first one-dimensional devices and then two-dimensional or area detectors.

1.5.4.1 *One-dimensional, position sensitive detectors* One of the most successful of such devices has been the single wire anode proportional counter. The anode wire which may be straight or (more commonly) curved is enclosed

in a gas filled chamber as shown in Figure 1.14(b). The essential point is that the ionisation avalanche produced by each photon is localised and its position along the anode can be measured by timing the delay of the resulting pulse with respect to a start pulse. Wire chambers usually employ a reactive rare gas mixture (e.g. Ar/CH_4) and operate under pressure. The count rate which is space charge limited is approximately $10^5 \, ps^{-1}$. Such detectors are increasingly used in diffraction experiments.

An alternative to the proportional counter mode is the self-quenching streamer mode which can operate using a blade rather than a wire. The count rate is lower than 10^5 counts/s, but the device more rugged. An angular range of 120° in 2θ is possible with an angular precision of around 0.04°.

SSD arrays (e.g. photodiode arrays (PDA)) offer an alternative to PSDs, usually on a more miniature scale. Reticon S-series PDAs, for instance, are used for energy dispersive EXAFS and small angle scattering. They measure 25 mm in length, each pixel comprising a 2.5 mm × 0.025 mm photodiode. Hybrid devices can be much longer (100 mm) but with a coarser pixel pitch (0.25 mm) and are used for X-ray imaging. Operating in integrating mode these SSD arrays offer high count rates in excess of $10^9 \, ps^{-1}$ and an excellent dynamic range of 10^3–10^4 [20].

1.5.4.2 *Two-dimensional detectors* The principles described above for one-dimensional devices can be extended into two dimensions. By constructing a proportional counter comprising three parallel planes of closely spaced wires, two dimensionality can be achieved with total count rates in excess of 10^8 counts/s. Returning to Figure 1.14(b) cathode wires are placed above and below and orthogonal to the anode wires. The detector determines the *x, y* coordinates of the photon induced ionisation avalanches by locating their positions on the two cathode grids. The ultimate spatial resolution is governed by interpolation accuracy and parallax, but is typically 0.1 mm. *Multiwire proportional counters* are applicable to a single crystal diffraction where spatial resolution may be relaxed in favour of the inherent accuracy of counting statistics.

The most common two-dimensional solid state array is the *charge coupled device* (CCD). Pixel sizes (25 × 25 μm) are smaller than for PDAs, the devices are quieter and can run faster. However, they suffer badly from radiation damage. They are best used with a phosphor screen and fibre optic coupling but with some loss in the overall detector QE. Compared to multiwire proportional counters, CCDs offer much higher spatial resolution but at the expense of a much smaller array area. Trials are taking place to develop a system for protein crystallography data acquisition [21].

Finally there are a variety of detectors of sufficient size to be optimised for two-dimensional imaging of which the simplest and oldest is *photographic film*. The spatial resolution is high, being ultimately limited by the grain size in the emulsion. This is typically ≤ 1 μm. However there are a number of major disadvantages including the lack of temporal resolution and the need for subsequent development of the film. The dynamic range is poor because of the high intrinsic background due to chemical fog. Nevertheless photographic film is extremely easy to handle and is extensively used for macromolecular crystallography and also for X-ray topography.

Photoresists offer much higher spatial resolution than photographic film and are very effective for soft X-ray microscopy. The spatial resolution of resists is governed by the mean free path of photoelectrons and can be as small as 50 Å. Photoresists are developed by etching away the least polymerised material. This may be the exposed parts in the case of positive resists, or the unexposed parts in the case of negative resists. Images are usually viewed in a scanning electron microscope after the surface has been metallised. Compared to photographic film photoresists are exceptionally slow requiring exposure times of hours rather than minutes.

TV detectors involving the use of a phosphorescent screen which effects X-ray to visible light conversion and subsequent amplification and processing are becoming increasingly popular especially for protein crystallography. The phosphor of several square centimetres in area has good spatial resolution ($\sim 5\,\mu$m), but only a modest dynamic range of several 100:1 and a moderate count rate limit of around 10^8 cs. Nevertheless TV detector systems have been shown to give diffraction data of higher accuracy than film and in addition have the advantage of operating on-line.

Image plates are being developed to replace photographic film. Incident photons ionise Eu/Br/F centres in the plate forming stable colour centres. Under laser stimulation, these emit phosphorescence which can be recorded whilst scanning the plate, in a manner analogous to photographic film microdensitometry. Subjecting the plates to a flash of light bleaches the image for re-use. Image plates are attractive for use in a wide range of diffraction and imaging experiments. They have a large dynamic range of 10^5:1 and good linearity. Compared to TV detectors, the total usable area is larger.

1.6 Experimental layouts

Basically there are two geometries used for synchrotron radiation experiments. In the first the monochromator diffracts vertically and the monochromatic beam emerges parallel to the incoming white beam from the storage ring (Figures 1.9(a) and 1.13(b)). This configuration results in the best energy resolution because the vertical emittance of the storage ring is usually superior to the horizontal emittance. Monochromators involving a single deflection (Figures 1.9(b) and 1.13(a)), however, usually benefit from greater light gathering power and result in the most brilliant monochromatic beams. Examples of experimental layouts using these two geometries are shown in Figures 1.15–1.17. These cover three of the principal synchrotron radiation measurements: *X-ray absorption spectroscopy, X-ray diffraction* and *UV photoemission*. Additional details of variations for particular experiments will be covered in later chapters.

1.6.1 X-ray absorption spectroscopy

X-ray absorption spectroscopy experiments (Chapters 6, 7, 8 and 11) require a combination of high intensity and excellent energy resolution. Accordingly, a two crystal, vertically dispersing wide aperture monochromator, C, is used (Figure 1.15). It will generally operate behind a Be window unless soft X-rays are required. This should include harmonic rejection. The out-going monochromatic beam may be focussed with a toroidal mirror, M, at the specimen, S. Prior to this the incident beam intensity is registered by a semi-transparent reference ion chamber, I_0. This is necessary so that absorption spectra can be corrected for the decaying storage ring beam current and the instrument function of the monochromator/mirror combination.

If the element whose absorption is being measured is sufficiently concentrated, an ion chamber, I_t, placed behind the sample is sufficient to measure the transmitted beam. The X-ray absorbance of dilute systems is recorded from the characteristic X-ray emission using a scintillator or SSD I_f, placed at right angles to the monochromatic beam, the sample facing the incident radiation at an angle of 45°. This arrangement minimises the Compton scatter from the specimen. Alternatively, polished specimens can be inclined at grazing incidence to the in-coming X-rays, in which case the rear ion chamber measures the reflected beam or the X-ray fluorescence is collected above the surface. At the SRS stations 3.4 and 8.1 adopt the layout described here and illustrated in Figure 1.15.

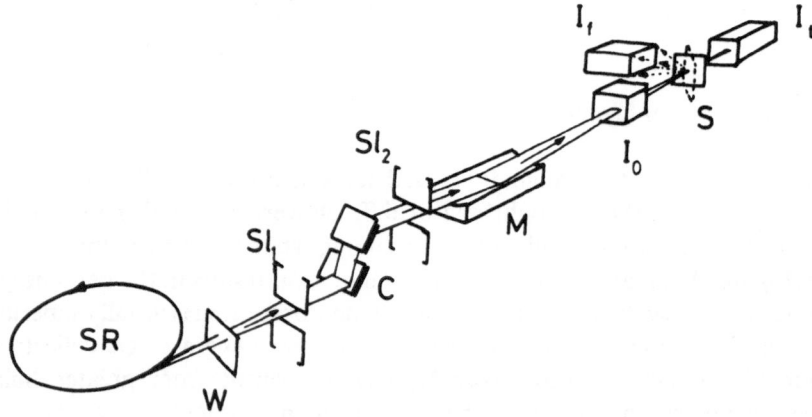

Figure 1.15. Layout for X-ray absorption spectroscopy experiments. W, Be window; $Sl_{1,2}$, slits; C, 2-crystal monochromator; M, toroidal mirror; I_0, I_t, I_f, ion chambers; S, specimen.

1.6.2 *X-ray diffraction*

X-ray diffraction from powders takes advantage of the excellent collimation of the synchrotron beam and is discussed in detail in Chapter 2. Where biological systems are concerned, however, brilliance is paramount and can be traded for some loss in energy wavevector resolution using the layout illustrated in Figure 1.16 (e.g. stations 2.1 and 9.6 at the SRS). With some small modifications, this geometry is suitable both for small angle X-ray scattering (SAXS) (Chapters 4 and 5) and protein crystallography (Chapter 10). Once again the experiment is separated from the storage ring ultra high vacuum (UHV) by a Be window, W. Vertical focussing is achieved with a cylindrical mirror, M, which also removes some of the unwanted high energy X-rays. Horizontal focussing and monochromatisation is obtained with a bent triangular-shaped

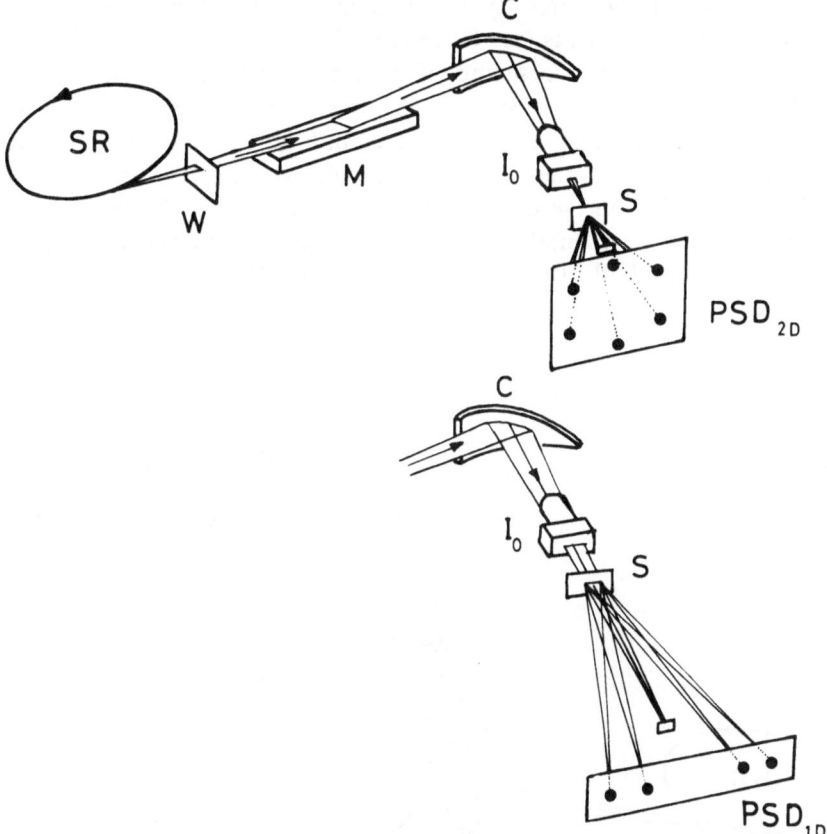

Figure 1.16. Layouts for small angle X-ray scattering and protein crystallography experiments. W, Be window; M cylindrical mirror; C, bent single crystal monochromator; I_0, reference ion chamber plus collimator; S, specimen; PSD, position sensitive detector.

crystal, C. The ordering of the two optical elements is sometimes changed, with the mirror placed second. In either event, both elements focus at the same position. For SAXS this is at the PSD detector and for macrocrystallography the focus is at the specimen, S. The incident beam intensity is monitored with an ion chamber, I_0, placed behind a fine collimator. As shown in Table 1.2, a variety of one- and two-dimensional PSDs are used to measure the scattered or diffracted X-rays. The undiffracted X-rays are removed with a beam stop. Compared to XAS (Figure 1.15), there will often be long optical paths separating the specimen from the detector; this is particularly true for SAXS. Vacuum spacers are used to remove the air attenuation and scatter. Data collection is generally sophisticated as experiments often involve in situ measurements where the structure may change rapidly.

1.6.3 *UV photoemission*

A combination of high intensity and high energy resolution is required for photoemission experiments (see Chapter 9). The relevant cross-sections are usually small and any worsening of photon energy resolution from the monochromator to gain intensity will degrade the overall resolving power of the electron analyser. The layout illustrated in Figure 1.17 is often used (e.g. stations 1.1. and 6.2 at the SRS). The toroidal grating, TG, disperses the synchrotron radiation vertically and the axis of the Rowland Circle (Figure 1.10(a)) is horizontal. Two cylindrical mirrors, one vertical, M_1, and the other horizontal, M_2, focus the white beam from the storage ring onto the

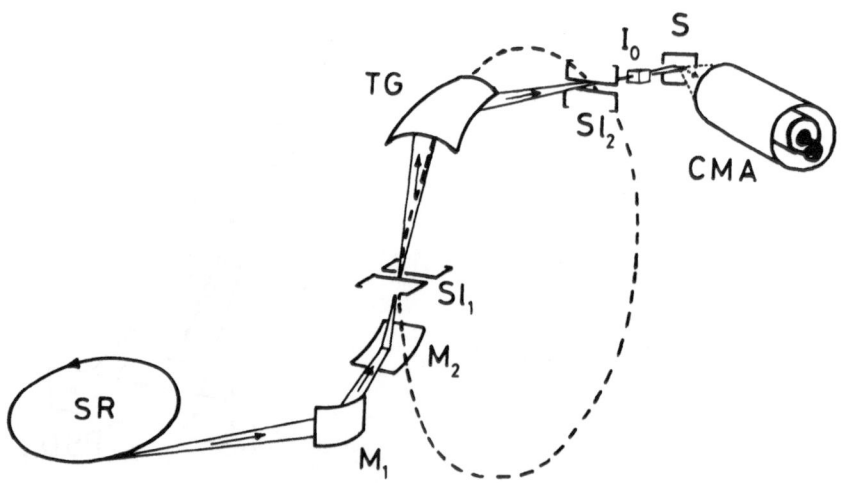

Figure 1.17. Layout for UV photoemission experiments. $M_{1,2}$, cylindrical mirrors; $Sl_{1,2}$, slits; TG, toroidal grating; I_0, reference monitor; S, specimen; CMA, cylindrical mirror analyser.

monochromator entrance slits, Sl_1. The doubly curved grating both disperses and focusses monochromatic UV onto the exit slits, Sl_2, close to the sample, S. A photoemission diode, I_0, monitors the intensity of the incident radiation whilst the photoemission from the specimen (usually mounted facing downwards) is collected by the electron analyser, CMA (often a cylindrical mirror analyser) at right angles to the in-coming photon beam. By scanning the field between the photoelectron detector plates, the photoemission can be energy analysed. Needless to say, experiments in this photon energy range must operate windowless to the tangent point, and necessitate a UHV environment which, in any case, is a prerequisite for surface science experiments.

1.7 Conclusion

This chapter is no more than an introduction to the characteristics of synchrotron radiation and beam line optics. Further details can be found in the reviews which have been cited. Nevertheless it is worthwhile bearing in mind the constraints and opportunities presented in optimising the instrument to the source to appreciate the excellent science which is now emerging from synchrotron radiation facilities world-wide and to which the remainder of this book is devoted.

References

1. Hofmann, A., in *Synchrotron Radiation Sources and Applications*, eds. G.N. Greaves and I.H. Munro, Scottish Universities Summer School in Physics, Edinburgh (1988) p. 1.
2. Suller, V., in *Synchrotron Radiation Sources and Applications*, eds. G.N. Greaves and I. H. Munro, Scottish Universities Summer School in Physics, Edinburgh (1988) p. 39.
3. Parratt, L.G. *Phys. Rev.* **95** (1954) 359.
4. MacDowell, A.A., West, J.B., Greaves, G.N. and Van der Laan, G., *Rev. Sci. Instrum.* **59** (1988) 843.
5. West, J.B. and Padmore, H.A., in *Handbook on Synchrotron Radiation*, Vol 2, North-Holland, Amsterdam (1987).
6. Howells, M., in *Synchrotron Radiation Sources and Applications*, eds. G.N. Greaves and I.H. Munro, Scottish Universities Summer School in Physics, Edinburgh (1988) p. 82.
7. Underwood, J.H., Thompson, A.C., Wu, Y., Giauque, R.D., *Nucl. Instrum. Methods A* **266** (1988) 296.
8. Warren, B.E., in *X-ray Diffraction*, Addison-Wesley, New York, (1969) Chapter 14.
9. Hastings, J.B., Kincaid, B.M. and Eisenberger, P., *Nucl. Instrum. Methods* **152** (1978) 167.
10. van der Hoek, M.J., Werner, P., van Zylen, P., Dobson, B.R., Hasnain, S.S., Worgan, J.S. and Luickx, G., *Nucl. Instrum. Methods A* **246** (1986) 385.
11. Du Mond, J.W.M., *Phys. Rev.* **52** (1937) 872.
12. Kohra, K., *J. Phys. Soc. Jpn.* **17** (1962) 589.
13. Lemonnier, M., Fourme, R., Rousseaux, F. and Kahn, R., *Nucl. Instrum. Methods* **152** (1978) 173.
14. Helliwell, J.R., Papiz, M.Z., Glover, I.D., Habash, J., Thompson, A. W., Moore, P.R., Harris, N., Croft, D. and Pantos, E., *Nucl. Instrum. Methods A* **246** (1986) 617.
15. Matsushita, T. and Phizackerley, R.P., *Jpn. J. Appl. Phys.* **20** (1981) 2223.
16. Hart, M. and Rodrigues, A.R.D., *J. Appl. Crystallogr* **11** (1978). 248

17. Greaves, G.N., Diakun, G.P., Quinn, P.D., Hart, M. and Siddons, D.P., *Nucl. Instrum. Methods* **208** (1983) 335.
18. Allinson, N.M., Baker, G., Greaves, G.N. and Nicoll, J.K., *Nucl. Instrum. Methods A* **266** (1988) 592.
19. Arndt, V., Greenough, T., Helliwell, J.R., Rule, S. and Thompson, A.W., *Nature* **298** (1982) 835.
20. Allinson, N.M. and Greaves, G.N., *Nucl. Instrum. Methods A* **273** (1988) 620.
21. Naday, I., Strauss, M.G., Sherman, I.S., Kraimer, M.R. and Westbrook, E.M., *Opt. Eng.* **26** (1987) 288.

2 X-ray diffraction from powders and crystallites

C.R.A. CATLOW

2.1 Introduction

X-ray diffraction is one of the most standard laboratory techniques. Powder diffraction has been used in a routine manner for phase identification, lattice parameter determination and for solving simple crystal structures, while single crystal techniques have been employed for many decades to provide accurate crystal structures. The scope of the field is, however, being enormously expanded by the availability of synchrotron radiation. Entirely new types of experiment in, e.g. high pressure studies and kinetic crystallography, become possible. The range and power of 'classical' powder diffraction methods has been greatly extended by the very high resolution data that can be obtained using synchrotron radiation, which is leading to the possibility of solving large complex crystal structures using powder data. Furthermore, in the area of single crystal diffraction, synchrotron radiation is opening up the possibility of the study of microcrystals whose small size would prohibit their investigation by conventional laboratory sources.

This chapter first summarises the fundamentals of X-ray diffraction and the features of synchrotron radiation that are of importance for crystallography, particularly in the area of powder diffraction, where to date there has been the greater use of synchrotron radiation. We then describe the basic instrumentation for different modes of instrumental operation. Applications of the techniques are then reviewed, with emphasis on high resolution powder diffraction (HRPD), energy dispersive (ED) studies, kinetic (or time-resolved) crystallography and anomalous scattering studies. We conclude with a brief account of the opportunities in the field of single crystal studies, which are subsequently discussed in greater detail in Chapter 10.

2.2 X-ray diffraction: basic features

Diffraction from crystalline materials is most simply described in terms of the Laue equations (of which a good discussion is given for example by Kittel [1], which summarise the conditions for diffraction as follows:

$$\mathbf{a} \cdot \Delta\mathbf{K} = 2\pi h$$
$$\mathbf{b} \cdot \Delta\mathbf{K} = 2\pi k \qquad (2.1)$$
$$\mathbf{c} \cdot \Delta\mathbf{K} = 2\pi l$$

where $\mathbf{a}$, $\mathbf{b}$ and $\mathbf{c}$ are the lattice vectors of the crystal. $\Delta\mathbf{K}$ is the scattering vector defined by

$$\Delta\mathbf{K} = \mathbf{k}' - \mathbf{k} \qquad (2.2)$$

where $\mathbf{k}$ and $\mathbf{k}'$ are the wave vectors of the incident and scattered radiation (note $|K| = 2\pi/\lambda$ and $|\Delta K| = 4\pi \sin\theta/\lambda$ where 2θ is the total angle through which the radiation of wavelength λ is scattered); h, k and l are integers. The Laue equations may be solved by defining three reciprocal lattice vectors, $\mathbf{a}^*$, $\mathbf{b}^*$ and $\mathbf{c}^*$, such that

$$\mathbf{a}^* = \frac{2\pi(\mathbf{b} \times \mathbf{c})}{\mathbf{a} \cdot \mathbf{b} \times \mathbf{c}}, \quad \mathbf{b}^* = \frac{2\pi(\mathbf{c} \times \mathbf{a})}{\mathbf{a} \cdot \mathbf{b} \times \mathbf{c}}, \quad \mathbf{c}^* = \frac{2\pi(\mathbf{a} \times \mathbf{b})}{\mathbf{a} \cdot \mathbf{b} \times \mathbf{c}} \qquad (2.3)$$

A reciprocal lattice may then be envisaged in which the reciprocal lattice points are defined by the vectors

$$\mathbf{G} = h\mathbf{a}^* + k\mathbf{b}^* + l\mathbf{c}^* \qquad (2.4)$$

where h, k and l are integers as in the Laue equations (2.1), the solutions to which are given by

$$\Delta\mathbf{K} = \mathbf{G} \qquad (2.5)$$

Each diffracted beam therefore corresponds, in reciprocal space, to the scattering vector touching a reciprocal lattice point—the basis of the well-known 'Ewald sphere' construction. It can be shown straightforwardly (see e.g. [1]) that Eqn. 2.5 is equivalent to the well-known Bragg equation

$$n\lambda = 2d\sin(\theta) \qquad (2.6)$$

where d the is spacing between layers of atoms in the crystal.

The next key concept in diffraction theory is the *structure factor* which may be defined as

$$S(G) = \sum f_i \exp(-i\mathbf{G} \cdot \mathbf{r}_i) \qquad (2.7a)$$

where the sum is overall i scattering centres each at point $\mathbf{r}$, in the unit cell. For X-rays, f_i has a strong dependence on θ; X-ray scattering is effected by the electrons and there is therefore interference between X-rays scattered from different points within the atoms. (In contrast f_i has no θ dependence for neutron scattering as neutrons are scattered by the nuclei which behave as point scattering centres.) We should also note that for X-ray diffraction it may, in certain contexts, be more useful to write $S(G)$ in an integral form, i.e.

$$S(G) = \int \rho(\mathbf{r}) \exp(-i\mathbf{G} \cdot \mathbf{r}) dV \qquad (2.7b)$$

where $\rho(\mathbf{r})$ is the electron density at point $\mathbf{r}$ in the unit cell and the integral is taken over the entire unit-cell volume.

The importance of the structure factors is that they control the intensity of the diffracted beams, the intensity I being given by

$$I = S(G) \, S^*(G) \qquad (2.8)$$

where the $S^*(G)$ is the complex conjugate of $S(G)$. Hence analysis of intensity data yields information on the structure of the unit cell. Indeed, Eqn. 2.7b may be Fourier transformed to yield the scattering density

$$\rho(r) = \frac{1}{V} \sum_G S(G) \exp(i\mathbf{G} \cdot \mathbf{r}) \qquad (2.9)$$

The $S(G)$ are then determined from the intensities via Eqn 2.8, which, however, only yields the *modulus* of the structure factor since I is a scalar quantity; the phase is indeterminate—the basis of the well-known 'phase problem' in crystallography.

The next important general observation is that the intensities of diffracted beams decrease with temperature. This can be understood simply in terms of the decrease in the effective scattering power of the atoms due to their increased thermal motions, and the effect can be described quantitatively by the relationship

$$I(T) = I(0) \exp(-8\pi^2 \, \bar{U}^2 \sin^2 \theta / \lambda^2) \qquad (2.10)$$

where $\bar{U}^2$ is the mean square of the amplitude of the atomic vibrations. The quantity $8\pi^2 \bar{U}^2$ is commonly referred to as the temperature factor, B, while the whole exponential factor in Eqn 2.10 is known as the Debye–Waller factor. Note that it is strongly dependent on θ and λ.

Analysis of Bragg intensities to yield structural information via Eqns. 2.6–2.9 can proceed in two ways. First, measured intensities may be used to fit via a least-squares routine, structural parameters (atomic coordinates and temperature factors). A starting model is of course required, and the procedure continues until the best possible agreement is obtained between measured and calculated intensities. The quality of this agreement is normally measured by the R factor defined as $R = \sum (I_{obs} - I_{calc}) / I_{obs}$, where the sum is over all observed intensities. Alternatively, given knowledge of (or assumptions as to) the phases of the structure factors, Eqn. 2.9 may be employed to obtain a Fourier map of the scattering density (i.e. electron density in X-ray studies). In practice, both techniques will commonly be used in the course of a structure refinement.

A final aspect of basic diffraction theory concerns *diffuse scattering*, which may be defined as scattering between (and around) Bragg peaks with an intensity greater than that of the background. In an infinite, perfectly ordered crystal in which the atoms are static, such scattering would be zero, that is, all the scattering intensity would be in the Bragg peaks. In practice, this condition is never achieved. Thermal motions give rise to instantaneous deviations from perfect periodicity, hence the phenomenon of thermal diffuse scattering (TDS). Static disorder, due to defects and impurities also destroys the condition of ideal periodicity and results in diffuse scattering. Such scattering is commonly weak, and synchroton radiation data with its high intensity and resolution offers the opportunity of collecting accurate diffuse scattering data and hence gaining information on disorder in crystals.

There are two broad classes of experimental diffraction technique, single crystal and powder, although each class has several sub-divisions. Single crystal studies explore the full three-dimensional geometry of reciprocal space, and are the preferred technique when adequate single crystals are available. In contrast, powder diffraction simply measures intensity as a function of (2θ); and all reciprocal lattice vectors with the same value of $|G|$ (i.e. of $h^2 + k^2 + l^2$) are superimposed. Moreover, at higher scattering angles, the spacing in 2θ between Bragg peaks decreases and for larger, lower symmetry unit cells, they will overlap in the powder pattern, leading to severe loss of structural information. However, we shall see later in this chapter that these latter problems with powder methods are being overcome with the availability of synchrotron radiation, a development which is of considerable significance as many important materials can only be prepared as powders.

Many of the points made in this brief introductory survey are amplified later in this chapter and in Chapters 3 and 10. We now continue with an account of the role of synchrotron radiation in the field of powder diffraction studies.

2.3   Synchrotron radiation and powder diffraction

The unique features of synchrotron radiation discussed in Chapter 1 provide the following major advantages in powder studies:

(i) The *high intensity* allows data to be collected rapidly (in some cases in times of ~ 1 s) allowing time dependent processes (e.g. phase transitions and solid state reactions) to be studied. In addition the intensity of the synchrotron beam allows small samples (of a few milligrams) to be studied.

(ii) The *tunability* of the source confers an advantageous flexibility in the choice of the optimum wavelength for an experiment. Moreover, the short wavelengths available from wiggler radiation allows 'high Q' data to be collected; i.e. data from a greater volume of reciprocal space are obtained, which yield more accurate and highly resolved structures. In addition, *white beam* (rather than monochromatic) experiments using *energy dispersive* techniques may be performed. As discussed further below, the sample, in such experiments, is exposed to the full synchrotron beam, and the scattered radiation is collected at a fixed detector angle and energy analysed.

The tunability of the source also allows us to exploit the phenomenon of *anomalous dispersion* which is based on the variation of the real and imaginary parts of the X-ray scattering power at wavelengths close to the absorption edge. As discussed in Section 2.4.4, the effect allows us to determine the contribution of a particular atom to the overall structure factor. Its study clearly requires a source with a wide range of wavelengths.

(iii) The *collimation* of the beam results in very narrow Bragg peaks with well-defined peak shapes. Thus, with synchrotron sources, it is possible to obtain Bragg peaks with full width at half maximum (FWHM) of $< 0.03°$ in 2θ. Data on Si are shown in Figure 2.1; in addition to its narrowness the peak shape is accurately described as a convolution of a Gaussian with a Lorentzian, the 'pseudo Voigt' function. In contrast the more divergent laboratory X-ray

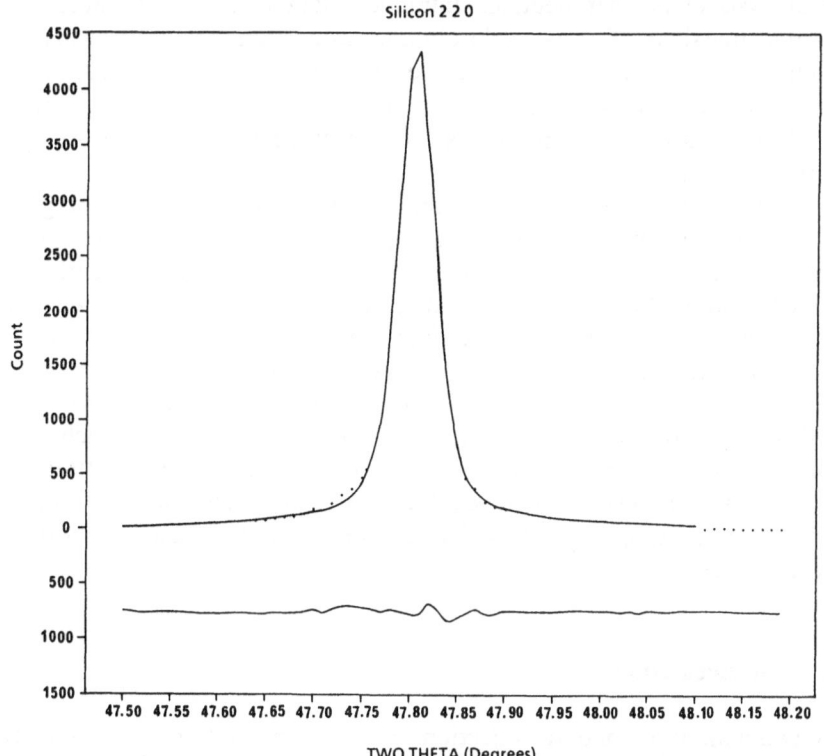

Figure 2.1. High resolution data for Si collected on Station 9.1 at the SRS Daresbury Laboratory. The figure shows the (220) peak with a full width at half maximum of $\sim 0.04°$ in 2θ. The deviations from linearity of the lower line show the small extent of the discrepancies of the fit of the peak shape to the pseudo-Voigt function.

sources give broader peaks (with FWHMs of typically 0.15° in 2θ), and the peak shapes are described by more complex functions with larger numbers of parameters.

With narrow, well-defined Bragg peaks, the field of *high resolution powder diffraction* (HRPD) is opened up. In traditional work with laboratory sources, the structural information, other than cell dimensions, available from X-ray powder diffraction is very limited. For structures with large unit cells and/or low symmetry, Bragg peaks will overlap at higher scattering angles. Accurate intensities, and hence structure factors, for these reflections cannot be refined. If, however, peak overlap is removed, owing to the reduction in peak widths, Bragg intensities may be measured and structures may be solved. In practice, it is commonly not possible to remove peak overlap for large structures with large low symmetry cells, in which case an alternative strategy is available provided the peak shape is well-defined. Each point in the scattering profile may be written as a sum of contributions from several reflections, and the structure is refined by fitting to the whole profile. This technique, known as *profile analysis* was developed originally by Rietveld [2] and was applied to the analysis of powder neutron diffraction data owing to the accurately Gaussian line shape of Bragg peaks obtained using monochromatised neutron beams from reactor sources. Indeed the development of profile refinement of powder neutron data was a milestone in the development of crystallography in the 1970s. With synchrotron sources, X-ray profile refinement is beginning to enjoy similar success.

(iv) The *polarisation* of the beam may be exploited in studies of magnetic structure as in the work of Cooper *et al.* [3]. Effects of polarisation may also need to be taken into account in data analysis of non-magnetic structures.

In concluding this section we should note that the advantages of intensity and tunability apply equally well to single crystal as to powder work. To date there have been relatively few single crystal studies of inorganic materials using synchrotron radiation. Studies of crystallites (i.e. small single crystals of $\sim 10\,\mu m$ size) are however possible given the high intensity of synchrotron radiation. Some examples are given later in this chapter. Such studies, in addition to single crystal work exploiting the phenomenon of anomalous dispersion (discussed later in this chapter) are expected to grow in importance in the next few years.

2.4 Instrumentation

The two main types of data collection mode are *angle* and *energy* dispersive. The former is similar to conventional X-ray powder diffraction and uses a monochromatised X-ray source, similar to the EXAFS layout described in Chapter 1. The latter, as noted above, employs white beam and fixed detector angle with energy analysis of the scattered radiation.

2.4.1 *Angle dispersive scans*

Figure 2.2(a,b) illustrates schematically this mode of operation. White beam is monochromatised, typically by a channel cut Si(111) monochromator and the monochromatic beam is incident on the sample which may be mounted in flat plate mode or as a capillary. In the former case the sample will normally rotate (through θ) as the detector scans (through 2θ), i.e. the normal $\theta/2\theta$ mode is used as in conventional powder diffraction. However, as we shall see later, there are experiments in which it is desirable to have a fixed sample angle. It is also normally desirable to spin the sample (either flat plate or capillary) about the axis. The small synchrotron beam will be scattered by relatively few crystallites; spinning helps to ensure the averaging of crystal orientation which is necessary for high quality reproducible data.

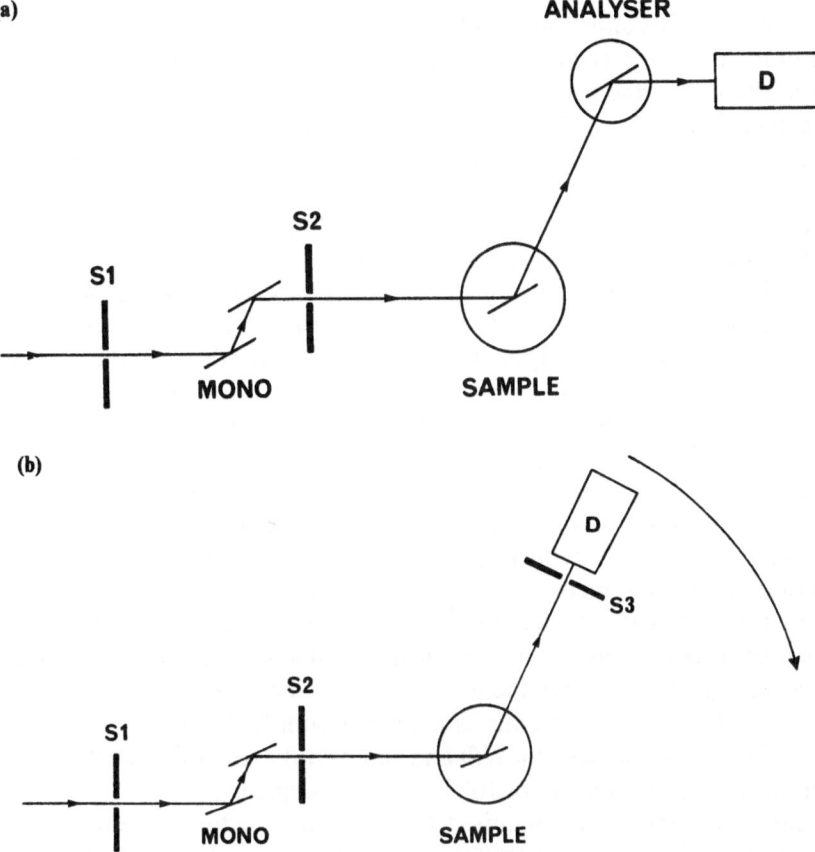

Figure 2.2.   (a) Schematic illustration of the essential features of an angle dispersive powder diffraction instrument with crystal analyser. S1 and S2 represent slits and D the detector. (b) Schematic illustration of a high resolution powder diffractometer involving the use of fine pre-detector slits.

Proportional, scintillation and solid state detectors (SSDs) may all be used to measure the scattered radiation. These are described in Section 1.5. SSDs have definite advantages if the monochromatic beam is significantly contaminated by harmonics, as the energy discrimination of this detector may be used to eliminate from detection any harmonic radiation. The geometry of the detector system is of considerable importance in achieving high resolution data. It is necessary to reduce as far as possible the angular range of the scattered radiation entering the detector at a given position. There are two strategies for achieving this. In the first, a *crystal analyser* is used as illustrated in Figure 2.2(a). The detector receives the radiation satisfying the Bragg condition for the crystal, which thereby fixes very precisely the angle of the scattered radiation from the sample (crystal rocking curve widths are discussed in Section 1.4). Detection via an analyser crystal has been very successfully used in the high resolution powder diffraction instrument at the Brookhaven NLS; data of very high quality (with FWHMs of $\sim 0.02°$) have been obtained. The disadvantage of this method is that it entails a considerable loss of intensity and correspondingly long data collection times.

The alternative and simpler strategy is to use slit systems. Good results can be obtained by using fine post-monochromator and pre-detector slits (see Figure 2.2(b)), optimisation of which and of the sample-detector separation can yield narrow pseudo-Voigt peaks with widths of $\sim 0.03°$. An ingenious system has been designed by Hart and Parrish [4] (and implemented on both the Stanford and Daresbury synchrotron sources). As shown in Figure 2.3 a relatively large beam is incident on the sample; but between the sample and the detector there are a set of long, fine diffracted beam collimators. The angle of the diffracted beam is strictly controlled, as to enter the detector the scattered photons must fulfil the conditions of scattering direction imposed by the collimators. The intensity remains good, however, as the data are collected from a large area of sample, and peak widths of 0.04 and 0.05° with good intensity can be obtained. The optics of an instrument with this design results in the data being free from many of the geometric errors that occur with conventional X-ray diffractometers. Consequently diffractometers with this design are ideal for precision lattice parameter measurements; highly accurate location of peak positions is also needed if ab initio structure determination is to be attempted, as will be discussed further below. Such instruments require a high level of precision engineering. Thus the diffractometer (shown in Figure 2.3(b)) that was built at Daresbury recently is equipped with high resolution encoders linked directly to the detector (2θ) and sample (θ) axes of the diffractometer with precisions respectively of 10^{-4} and 10^{-3} degrees.

If rapid data collection is required, the scanning detector can be replaced by a position sensitive detector (PSD) of the type discussed in Section 1.5.4. As illustrated in Figure 2.4, data are collected over a wide angle range simultaneously. Loss of resolution is inevitable with such detectors, although data with peak widths of 0.1° can be collected. Data collection time may, however,

(a)

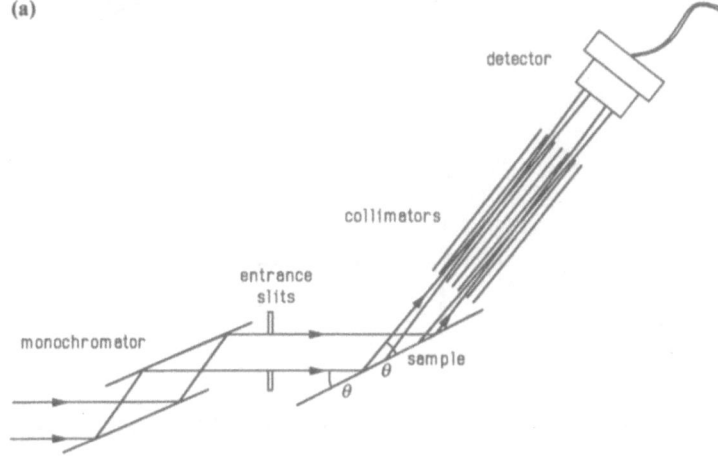

(b)

Figure 2.3. (a) Schematic illustration of a high resolution powder instrument based on the Hart–Parrish design. (b) Station (8.3), based on the Hart–Parrish design implemented at the SRS at the Daresbury Laboratory.

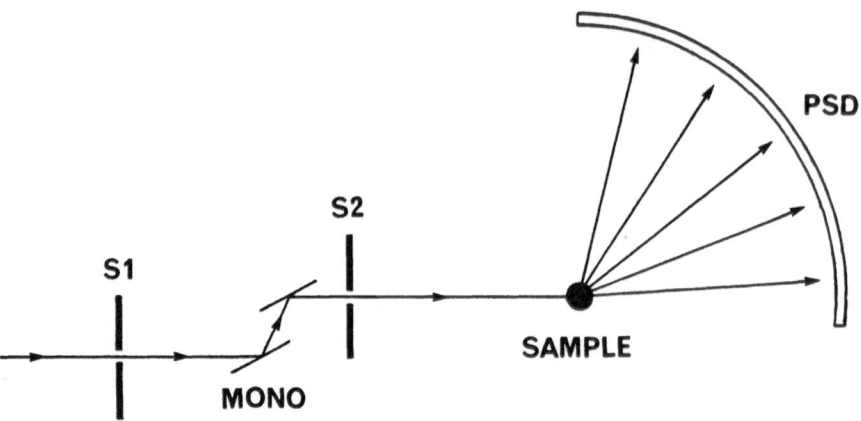

Figure 2.4. Schematic illustration of the operation of a powder diffractometer employing a PSD.

be dramatically short, i.e. of the order of a few seconds, thus allowing kinetic phenomena to be studied.

2.4.2 *Energy dispersive studies*

The configuration for this type of experiment is particularly simple. The white beam after passing through a slit system (and in some experiments a 'pin-hole' to achieve very small beam sizes) is incident on the sample. Data are collected by a *fixed* detector usually sited at a low scattering angle. A solid state detector is used; this is coupled to a multichannel analyser (MCA), enabling the scattered radiation to be energy analysed. The resulting diffraction pattern thus comprises a plot of scattered intensity versus photon energy rather than scattering angle as in a conventional angle dispersive experiment. To understand this further we refer to the Bragg equation: $\lambda = 2d \sin \theta$. In an angle dispersive experiment θ is varied with fixed λ in order to collect data corresponding to different d spacings. However, we can recast the Bragg equation by making the substitution $\lambda = ch/E$ (where E is the photon energy), thus giving $E = ch/(2d \sin \theta)$. Thus in energy dispersive measurements, θ is constant, and the Bragg condition for different values of d is satisfied by radiation with different energies. It is worth noting that related procedures are used in neutron studies using pulsed sources, where a white beam is scattered from a sample, and the scattered beam is energy analysed using time-of-flight techniques.

The choice of the detector angle θ is governed by the trade-off between resolution and intensity. Low values of $\sim 10°$ are typical as in the instrument at Daresbury which is shown in Figure 2.5.

The advantages of energy dispersive experiments are two-fold; first by using the full white beam and collecting the whole pattern simultaneously,

Figure 2.5.   The energy dispersive, white beam Station 9.7 at the SRS, Daresbury Laboratory.

data collection times are short and time dependent processes may be studied. Secondly, the use of a fixed detector angle greatly facilitates the design of environmental stages as only one exit port for the scattered beam is needed. This is a major advantage in high pressure studies where ED methods are being widely used in conjunction with diamond anvil cells.

The method does, however, have serious drawbacks. At X-ray energies, the best resolution available with current SSDs is $\sim 180 \, \text{eV}$ yielding a resolution for the resulting diffraction pattern that is an order of magnitude less than obtained using angle dispersive techniques. Intensity data are, moreover, rarely sufficiently reliable to be used quantitatively and ED methods have largely been used to study lattice parameter variations. Finally, the high flux of the scattered radiation may cause problems by saturating the detector (for which present count rate limits are typically 5×10^4 counts/s).

An alternative approach to white beam energy dispersive instrumentation is to use monochromatic radiation and to scan the monochromator angle and hence the wavelength. This retains the advantages of the fixed detector arm but improves the resolution and removes the problem of detector saturation. A number of successful experiments of this type have been reported and this promising method will almost certainly be used increasingly in the future.

2.5   Applications

2.5.1   *High resolution powder diffraction*

Several recent studies have shown how good high quality data with narrow well-defined peaks can be collected using synchrotron radiation sources. Typical examples are given in Figures 2.6 and 2.7. To illustrate the power of the method we take a recent example from the work of McCusker [5], where synchrotron powder X-ray methods were used to solve, using ab initio techniques, the structure of a novel zeolite clathrasil phase 'sigma-2'. This compound is essentially a polymorph of SiO_2 with a cage-like structure in which a large organic molecule (1-amino adamantane, $C_{10}H_{17}N$) is incorporated. As with many zeolitic materials, single crystals are unavailable. Nevertheless an accurate structure for a complex unit cell containing over two hundred atoms, has been obtained using a powder sample with data collected on the diffractometer designed by Cox *et al.* [6] on beam line X13A at the NLS Brookhaven. The procedure used in solving the structure can be considered as an archetypal example of an approach which will become increasingly common with the expansion in synchrotron powder diffraction. It comprised the following steps.

   (i) Data, which are shown in Figure 2.8, were subjected to routine processing involving background subtraction, normalisation and precise

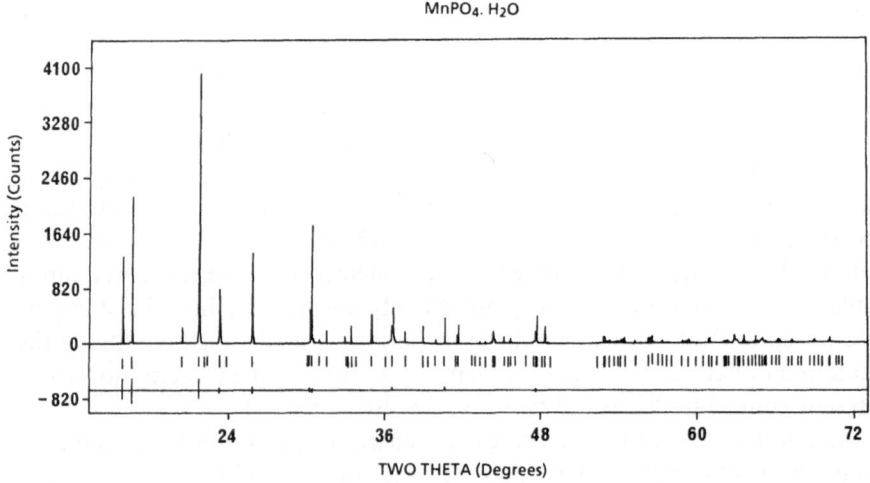

Figure 2.6.   Recent data collected by Cheetham and coworkers on $MnPO_4H_2O$, using the powder diffractometer on the NSLS at the Brookhaven National Laboratory.

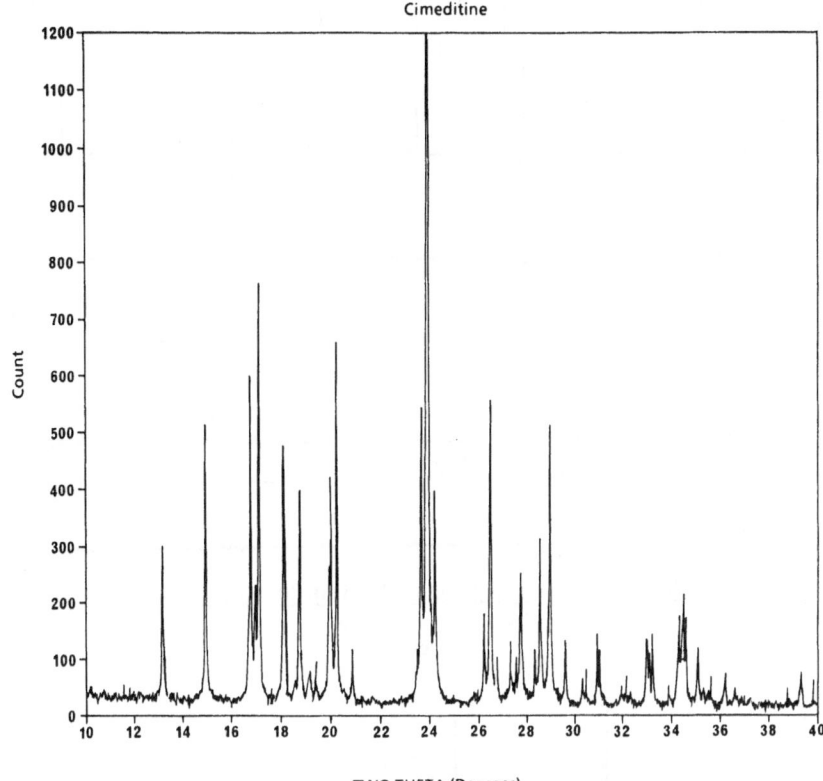

Figure 2.7.   Recent data collected by Cernik and coworkers on the drug cimeditine using Station 8.3 on the SRS at the Daresbury Laboratory.

location of the peak maxima. Despite the high resolution there is significant peak overlap at higher angles.

(ii) The peak positions were input into an *autoindexing* program. Several such programs are now available; McCusker's study used the TREOR code of Werner *et al.* [7]. The code examines the data in a systematic manner in order to identify systematic absences which will help the space group to be determined. The results of this analysis led to a choice of three possibilities; subsequent detailed examination of the intensity data (see below) showed that the structure was centro-symmetric which fixed the space group of $I4_1/amd$. The indexing exercise of course yields lattice parameters for the unit cell.

(iii) The next step is to extract peak intensities from the data; and it is here that the high resolution and well-defined peak shapes of synchrotron powder patterns are vital. For these allow the extraction of the intensities of individual Bragg reflections except for the limited number of reflections where there is still severe peak overlap. Thus by fitting the data using a standard pseudo-Voigt peak shape, single crystal-like intensities are recovered from the powder pattern; the program ALLHKL of Pawley [8] was used in the present study.

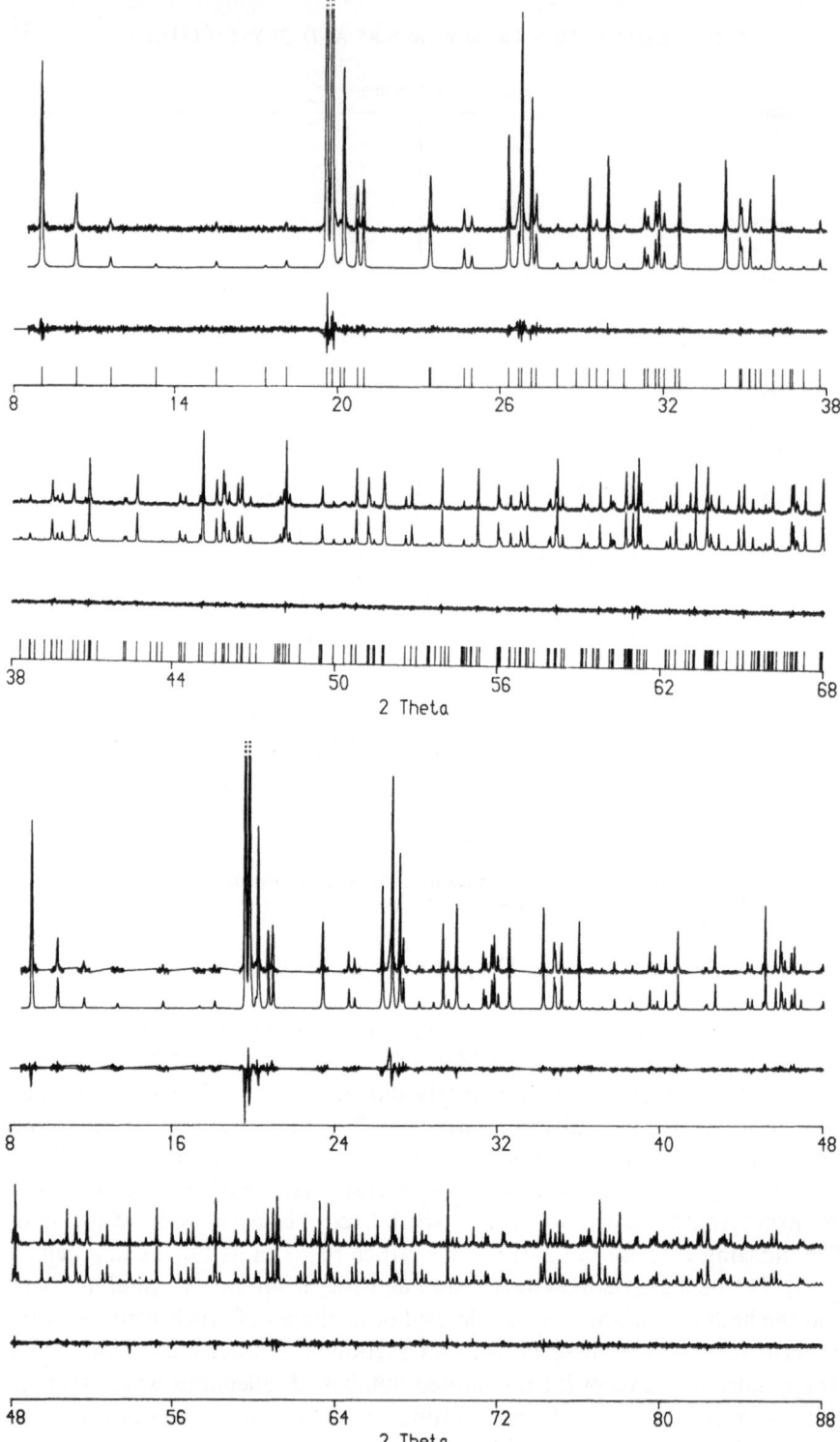

Figure 2.8. Data collected by McCusker [5] on sigma-2 using the powder diffractometer on the NSLS at the Brookhaven Laboratory. The top pattern shows an observed (upper) calculated (middle) and difference (lower) pattern for the ALLHKL refinement of the data. The lower pattern shows a similar comparison for the Rietveld refinement.

Of course the quality of the intensity data will be inferior to that of a good single crystal experiment; but it is sufficient for the generation of an approximate structure.

(iv) The intensity data are input into a direct methods, single crystal structure refinement package (in this case the XTAL program of Hall and Stewart [9]). Direct methods (the basis of which is beyond the scope of this chapter, but which are discussed by Karle [10]) allow structures to be solved without prior approximate knowledge of the structure. McCusker was able to use such techniques successfully with the sigma-2 data, yielding a structure in which the complex SiO_2 framework was reasonably well-defined.

(v) The final stage is to refine this approximate structure using the *Rietveld technique* that was discussed earlier in this chapter. The Rietveld method does not attempt to extract component Bragg intensities but fits the whole profile. The refinement improved the accuracy of the framework structure and yielded information on the location of the encapsulated 1-amino adamantane. This information cannot be too precise as it is clear that the molecule has a disordered distribution.

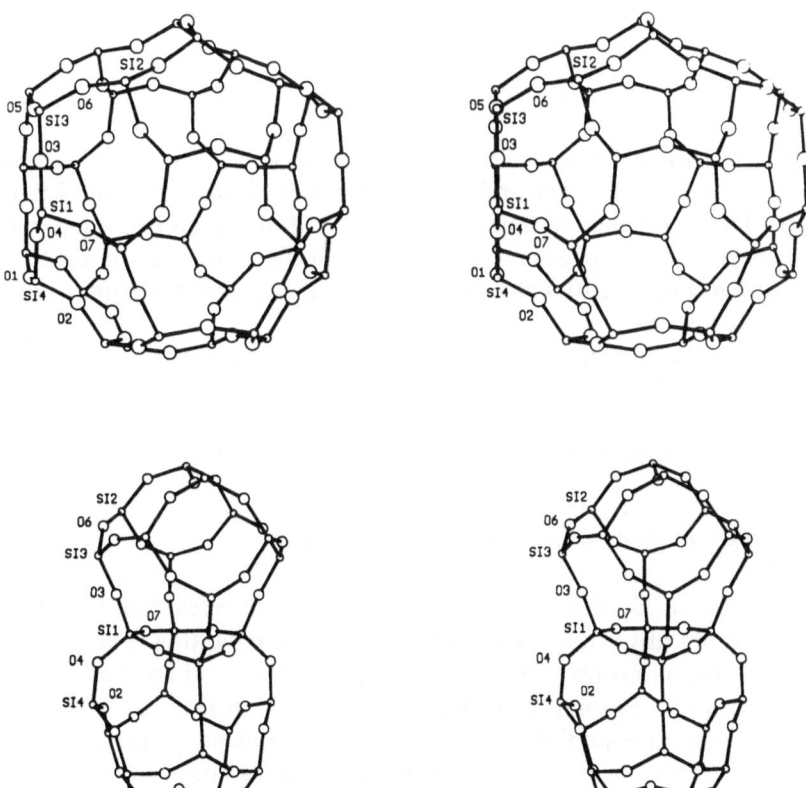

Figure 2.9. Cage structures of sigma-2 (after Ref. 5).

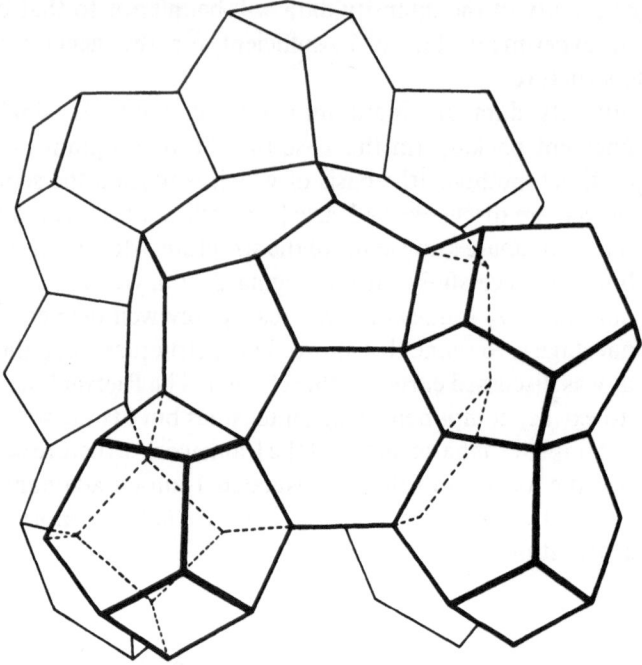

Figure 2.10.   Cage linkages in sigma-2 (after Ref. 5).

The cage structure revealed by the refinement is particularly fascinating. Two types of cage, a larger and a smaller one as shown in Figure 2.9, are present. The larger cage comprises rings of 5 and of 6 Si atoms, while in the smaller cage 4 or 5 atoms are present. The way in which the cages fit together is illustrated in Figure 2.10 which shows the smaller cages clustering round a central larger cage.

However, as commented by McCusker, the most significant feature of this work is the fact that the structure was solved directly from powder diffraction data. Ab initio structure deformation using synchrotron data promises to become an increasingly important crystallographic technique.

2.5.2   *Time-resolved (or kinetic) crystallography*

As discussed in Section 2.3.3, time dependent processes can be studied crystallographically using synchrotron radiation by collecting data in the ED mode or in the monochromatic mode using a PSD. A good illustration of the latter mode of operation is provided by recent work at Daresbury of Barnes *et al.* [11], who have studied a number of topical problems concerning kinetic processes in materials. The first example of their work concerns the dynamic crystallisation of glassy metals comprising Fe, Ni, Mo, B and Si. These are important materials as they are both amorphous and magnetic; the experiment concerned their recrystallisation at high temperatures. The material was heated up from room temperature to 550°C, with data collected regularly during its rise in temperature. Each data set took 3 min to collect, and some of the most interesting patterns are shown in Figure 2.11. The pattern is initially typical of an amorphous material; we note that the peaks at 17–20 keV are not due to diffraction but result from fluorescence of the element present in the multicomponent glass. Crystallisation, with the separation out of various iron related phases is evident from the various peaks appearing at 550°C.

Another problem recently studied by the same group concerned the onset of crystallisation of zirconia from zirconium hydroxide on heating. Data for this system are shown in Figure 2.12. Analysis of the results shows that zirconia first crystallises at around 500°C into the tetragonal form, as shown by the splitting of the 202 and 220 peaks. This is interesting as we know that the final product required for a high performance ceramic is in the partially stabilised monoclinic form. Subsequent work by the same group has observed the tetragonal → monoclinic transition at ~ 1000°C.

These results illustrate the very considerable potential for kinetic crystallography with powder samples using ED techniques. The field is expected to expand greatly in the near future, with the use of monochromatic radiation and PSDs in addition to the energy dispersive methods.

2.5.3   High pressure studies

We have noted that the energy dispersive mode of data collection is particularly suitable for work with diamond anvil cells, and several examples of the use of ED techniques in high pressure studies have been reported in recent years [12]. The applications include studies of minerals, where knowledge of high pressure phase transitions is of obvious geophysical relevance, KNO_3 [13], where a range of intriguing phase transitions have been identified, and most recently on superconducting oxides. The latter is a good illustration of what can be achieved in relatively simple experiments, as will now be described in further detail.

La_2CuO_4 when doped with divalent ions such as Sr^{2+} and Ba^{2+} shows superconductivity in the range 20–40 K. The discovery of this effect by Bednorz and Muller [14] has led to an explosion of work on this and related materials. High pressure studies of the material are of particular interest as they allow us to investigate theories of bonding in these solids and to test interatomic potential models.

Akhtar et al. [15] therefore performed a series of high pressure experiments on the parent compound La_2CuO_4 using diamond anvil cells and the energy dispersive facilities at the SRS, Daresbury. ED diffraction patterns were collected for pressures varying from ambient to 150 kbar; a typical spectrum is shown in Figure 2.13. NaCl was used as the pressure calibrant as the variation

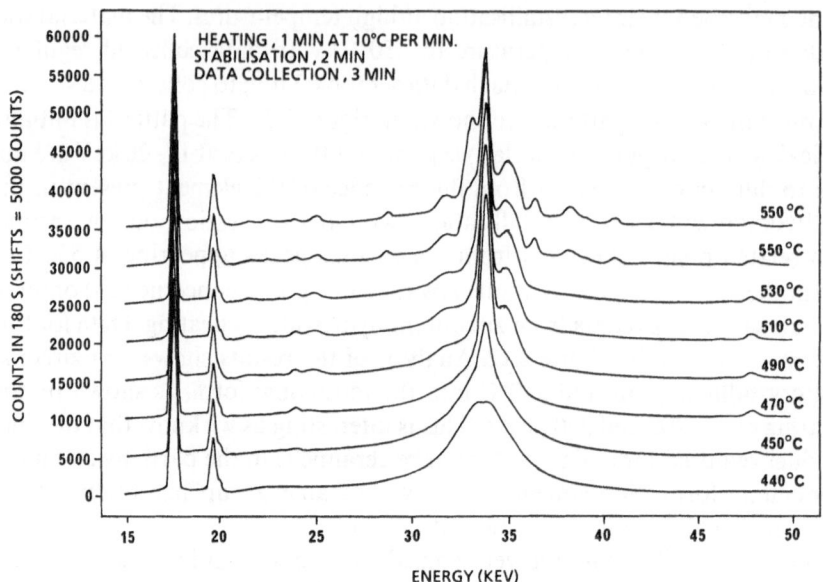

Figure 2.11.   Data collected in ED mode on glassy metals at several temperatures by Barnes *et al.* [11] using Station 9.7 at the SRS, Daresbury Laboratory.

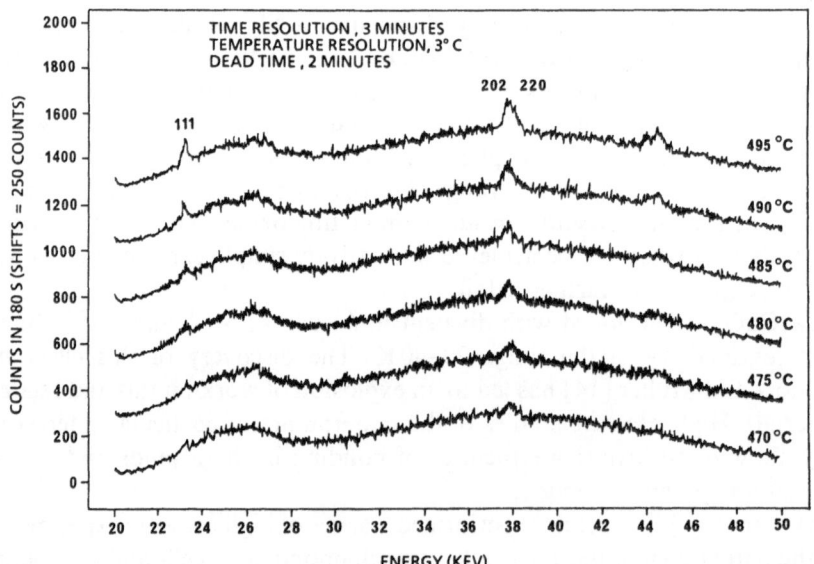

Figure 2.12.   Data collected by Barnes *et al.* [11] showing the transformation from zirconium hydroxide to zirconia on heating. The experiment again used Station 9.7 at the SRS, Daresbury Laboratory.

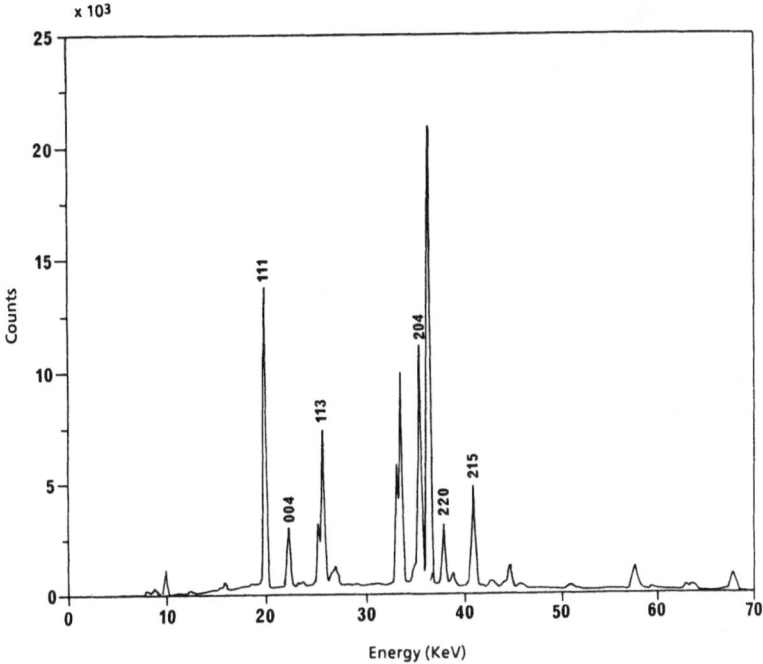

Figure 2.13. Energy dispersive spectrum for La_2CuO_4 at ambient pressure (after Ref. 15).

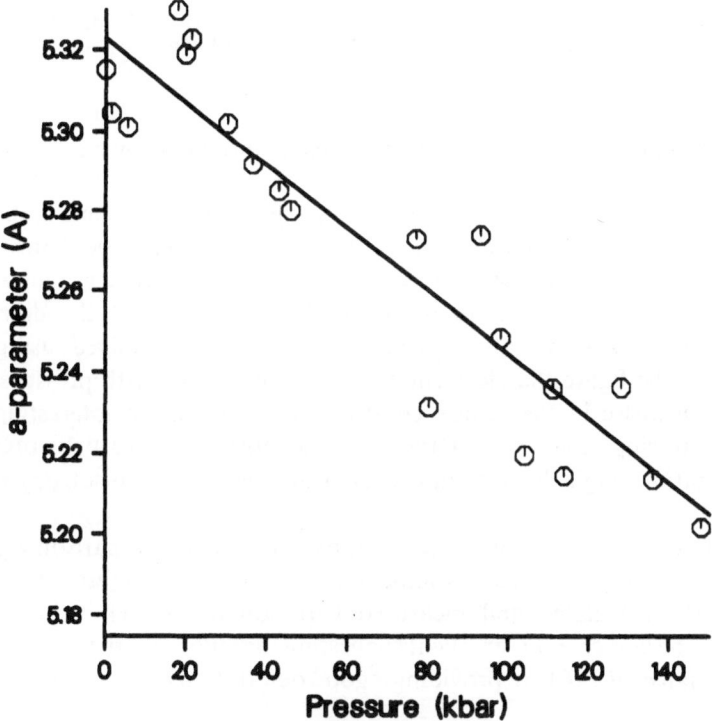

Figure 2.14. Variation of the 'a' lattice parameters with pressure for La_2CuO_4. Key: ⊙ experimental; —fitted line.

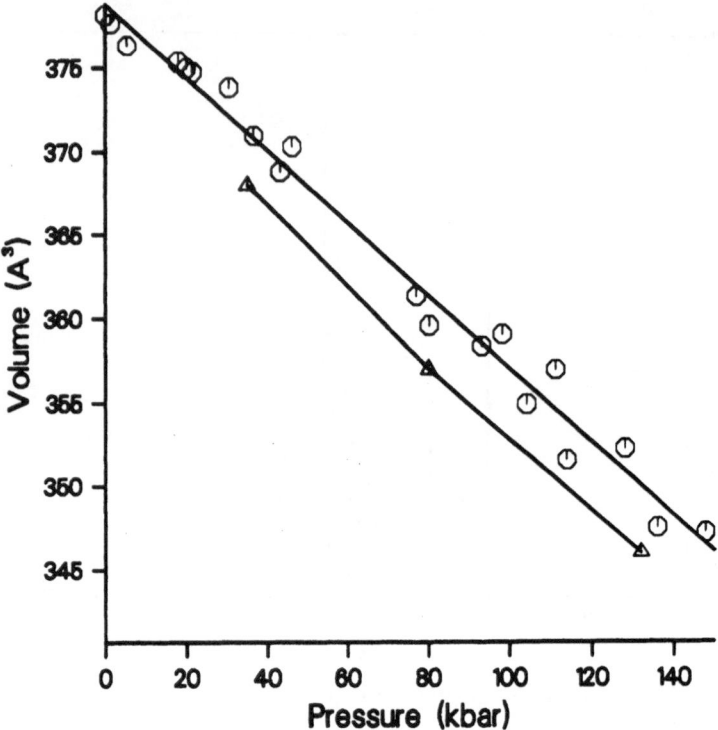

Figure 2.15. Calculated and experimental variation in cell volume for La_2CuO_4. Key: ⊙ experimental; △ theoretical; — fitted line.

with pressure of the lattice parameter of this material is known accurately to high pressures.

After indexing the diffraction pattern it was possible by measuring the shift with the pressure in the energy of the Bragg peaks to calculate the changes in 'd spacing' an hence in lattice parameter. Results are displayed graphically in Figure 2.14. La_2CuO_4 has an orthorhombic unit cell, but the difference between the a and b lattice parameters cannot be resolved using ED techniques; the figure therefore shows the variation of a with pressure. The expected decrease in the lattice constant is observed. An interesting and rather surprising result is that the a/c ratio does not vary with pressure. Such variation might have been expected in view of the anisotropy of the structure.

One of the most important aspects of the work was the comparison with the results of band theory calculations. This is shown in Figure 2.15 which displays the calculated and measured variation of the cell volume with pressure. The agreement is excellent suggesting that the band theory studies of Temmerman *et al.* [16] are providing a good description of the cohesion of the material.

2.5.4 *Anomalous dispersion studies*

The phenomenon of anomalous dispersion refers to the variation with wavelength of the effective scattering power of an atom for X-rays at wavelengths close to an absorption edge. The X-ray scattering power, f, of an atom is a complex quantity; i.e. $f = f' + if''$, the imaginary component arising because of the possibility of absorption of X-rays by the atom. Both f' and f'' vary close to an edge as shown in Figure 2.16. Clearly in order to exploit the effect, it is necessary to have a tunable source as provided by synchrotron radiation.

The value of anomalous dispersion studies is that by varying the effective scattering power of an atom, we can deconvolute the contribution of that particular atom to the overall structure factor. The ability to do this is especially valuable in disordered materials. For example, a conventional diffraction study on an amorphous solid yields the total radial distribution function (rdf). Thus in, say, a glassy alloy containing two types of atom A and B, the total rdf that is measured is a superposition of $A \cdots A$, $A \cdots B$ and $B \cdots B$ correlations. However, if the effective scattering power of both atom types is varied by collecting data both close to and away from the absorption edges, then the partial rdfs due to these three types of correlation can be extracted. The practical problems associated with performing such experiments are non-trivial. It is necessary to get close to the absorption edge; thus a very narrow band pass monochromator is needed; 4-bounce designs are commonly used. Nevertheless a number of groups at synchrotron radiation sources have managed to make progress in this difficult field.

A simple and elegant application of the use of anomalous dispersion was reported recently by Moroney *et al.* [17]. They studied the important ionically conducting ceramic Y-substituted ZrO_2. The material contains Y atoms substituted for Zr in a fluorite structure host, with the lower valence of the Y leading to vacancy compensation. Conventional diffraction experiments yield information only on the average scattering from this cation site. Moroney *et al.*, by collecting data close to both the Zr and Y edges were able to obtain information on the individual contributions of the two types of atom. In particular they were able to show that the temperature factor, B, of the Zr atom is considerably higher than that of Y. This difference, which is almost certainly due to the greater static disorder of the Zr ions, supports models in which the vacancies in the lattice preferentially occupy sites close to the Zr rather than Y. Such models had been suggested by earlier EXAFS studies [18] and by calculations. They have important consequences for our understanding of the defect structure of the material. Anomalous dispersion can, of course, be exploited with single crystals as well as powders; and the reader should refer to Chapter 10 for a discussion of the exploitation of the effect in single crystal studies of proteins.

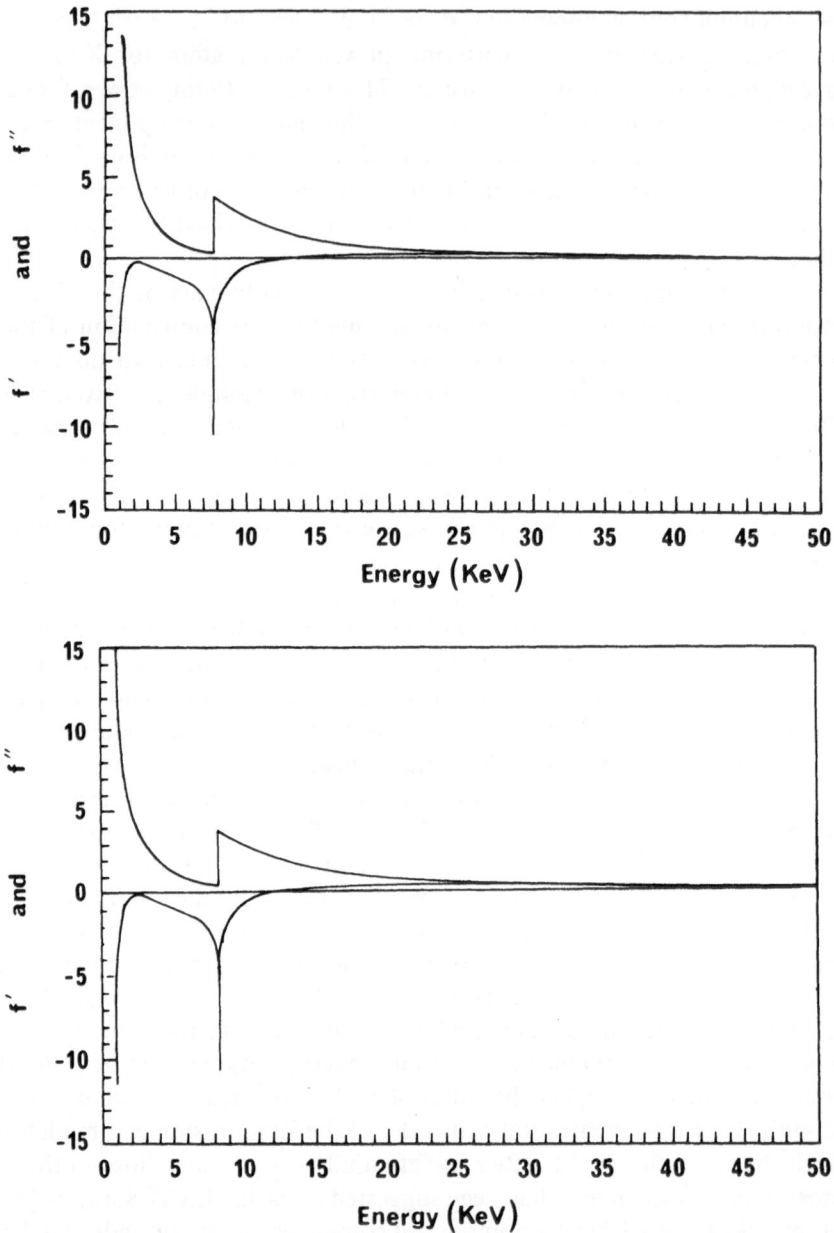

Figure 2.16.　Schematic variation in f' and f'' for atoms or wavelengths close to the absorption edge. Top, Co 27; bottom, Ni 28.

2.5.5   Studies of surface films

We conclude our survey of applications of synchrotron radiation powder techniques with an example of an experiment that profits directly from the precise beam optics, which is possible owing to the high collimation of the synchrotron beam. The application concerns the determination of stresses generated in oxide films grown on metal substrates, and as such is of obvious relevance to the study of corrosion. The strains in the surface oxide film can be measured by the ingenious technique shown in Figure 2.17. First a reflection from the film is measured using the standard $\theta/2\theta$ geometry as shown in Figure 2.17(a). The resulting Bragg scattered radiation will be due to those crystalline planes which are parallel to the plane of the surface layer. The sample 'axis' is then rotated by a small angle ψ. Bragg reflection is therefore now due to planes which are inclined at an angle to the surface plane, as shown in Figure 2.17(b). If the oxide layer is subject to a lateral stress, the corresponding strain will mean that the d spacing of such layers is slightly shifted; the 2θ value of the corresponding Bragg reflection will therefore also move. This strain in the oxide layer will be manifested by a shift in 2θ as ψ is varied. Given a detailed measurement of the strain, and knowledge of the elastic properties of the oxide, the stress may be evaluated, as discussed by Noyan [19], who also presents greater details of the technique.

Although laboratory sources have been used with this technique, synchrotron radiation allows far greater precision. The quality and utility of the data can be seriously reduced by the geometric aberrations which are present in laboratory sources; and indeed the experiment is ideal for the very precise beam optics provided by the Hart–Parrish design discussed in Section 2.3.1. We should also note that the technique requires decoupling of the θ and 2θ axes, as discussed earlier. The ability to tune the synchrotron source also allows different depths of the film to be probed, as the harder, shorter

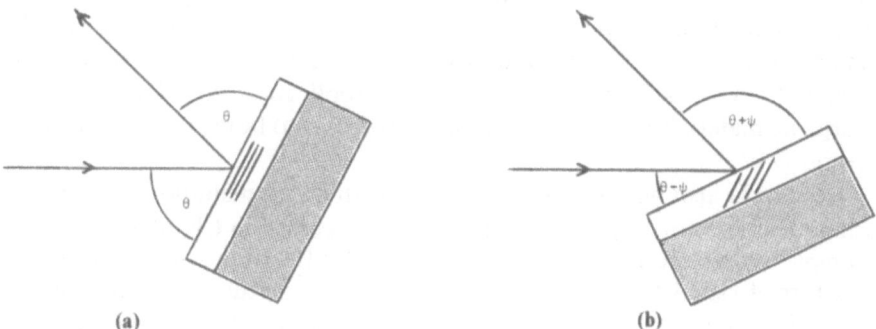

(a)   (b)

Figure 2.17.   The $\sin^2 \psi$ technique. (a) Bragg reflection from surface film in $\theta/2\theta$ geometry. (b) Bragg reflection from surface film rotated by ψ away from $\theta/2\theta$ geometry.

wavelength radiation will penetrate more strongly than the softer X-rays. Synchrotron applications of the technique are currently in their infancy; but useful data have been collected on NiO films by Fitch [20] at the SRS Daresbury.

2.6   Single crystal studies

Single crystal techniques using synchrotron radiation have been used most extensively in the field of protein crystallography, detailed discussion of which is presented in Chapter 10. Inorganic applications are, however, growing, with exciting recent developments in *microcrystalline* diffraction (referred to in Chapter 1) and in the application of *Laue* methods. Again, both these topics are discussed in Chapter 10, in the context of applications to protein structure determination; a brief account is, however, presented here with emphasis on the role of the techniques in inorganic structural studies.

2.6.1   *Microcrystalline diffraction*

There is considerable incentive for developing techniques to allow single crystal experiments to be performed on very small crystals of size 10 μm or less. For many materials larger single crystals are unavailable. In addition, small crystals may facilitate special sample environments, e.g. high pressure cells. Moreover, microcrystalline samples have distinct advantages in that corrections due to absorption, thermal diffuse scattering and most importantly, extinction are minimal. Such experiments are feasible with synchrotron radiation owing to its very high intensity, although there are considerable difficulties in mounting such small crystals, and the experiments place high demands on the positional stability of the beam.

An elegant recent demonstration of the power of such methods was provided by the work of Bachmann *et al.* [21] who collected and refined data on a 6 μm microcrystallite of CaF_2. The experiment used conventional single crystal methodologies with monochromatic radiation. It is, of course, essentially a 'demonstration' experiment; application to more complex inorganic materials such as zeolites and ceramics is to be expected in the near future.

We should also note that microcrystalline diffraction studies are possible using area detectors, e.g. the FAST detector discussed in Chapter 1. A good example is the recent work of Andrews *et al.* [22] who determined the structure of a small ($8 \times 18 \times 175 \, \mu m^3$) crystal of a molecular sieve compound, piperazine silicate using the FAST area detector diffractometer at the Daresbury Laboratory. Further discussion of this type of experiment is given in Chapter 10.

2.6.2 *Laue methodologies*

Laue crystallography, directly exploits the white beam of the synchrotron; and as discussed in greater detail in Section 10.5.1 allows integration of the reflected intensities over wavelength. The advantage of the technique is that again it may be used to study small crystals, but in addition, data collection times are short, with millisecond exposures being possible with strongly scattering crystals. Kinetics experiments are therefore possible and notable success has been achieved in the study of biological materials as discussed in section 10.5.2.

In the inorganic field, Laue techniques are beginning to be applied to, for example, small crystallites of organometallic compounds as illustrated by the recent work of Harding *et al.* [23] on the iron/rhodium compound $(FeRhCl(CO)_5$ dppee) (where dppee is $Ph_2PCH(=CH_2)PPh_2$). Studies of kinetic processes, e.g. sorption in zeolites, are to be expected in the near future.

2.7 Summary and conclusion

In this chapter we have shown that in the area of powder diffraction, synchrotron radiation techniques both enhance existing powder diffraction techniques, and lead to entirely new types of experiment. The greatest success to date has concerned the development of high resolution powder techniques, as in the work of McCusker (see also the reviews in [24]). But kinetic crystallographic studies, high pressure work, anomalous dispersion experiments and investigations of surface films are all expected to develop rapidly in the future. Equally exciting is the progress in microcrystalline diffraction, where both monochromatic and white beam techniques promise to make a substantial contribution to structural inorganic chemistry.

Acknowledgements

I am grateful to Drs. A.N. Fitch, P. Pattison, R. Cernik, S.M. Clark, E. Dooryhee, P. Barnes and A.K. Cheetham for several useful discussions and for permission to refer to unpublished work.

References

1. Kittel, C., *Solid State Physics*, 5th edn., John Wiley Interscience, New York (1976).
2. Rietveld, H.M., *J. Appl. Crystallogr.* **2**, (1969) 65.
3. Cooper, M., *Physica B* **159** (1989) 137.
4. Parrish, W. and Hart, M., *Z. Kristallogr.* **179** (1987) 161.
5. McCusker, L., *J. Appl. Crystallogr.* **21** (1988) 305.
6. Cox, D.E., Hastings, J.B., Cardoso, L.P. and Finger, L.W., *Mater. Sci. Forum* **9** (1986) 1.

7. Werner, P.E., Erickson, L. and Westdahl, M., *J. Appl. Crystallogr.* **18** (1985) 367.
8. Pawley, G.S., *J. Appl. Crystallogr.* **14** (1981) 357.
9. Hall, S.R. and Stewart, J.M., *XTAL2.1, Users Manual*, University of Western Australia and Maryland, USA (1986).
10. Karle, J., *Angew. Chem. Int. Ed. Engl.* **25** (1986) 614.
11. Barnes, P., Hausermann, D., Mamott, G.T., Tarling, S.E. and Vrcelj, R. to be published.
12. Hatton, P.D., *Mater. Sci. Forum* **81** (1986) 21.
13. Adams, D.M., Hatton, P.D., Heath, A.E. and Russell, D.R., *J. Phys. C* **21** (1988) 505.
14. Bednorz, J.G. and Muller, K.A., *Z. Phys. B* **64** (1986) 189.
15. Akhtar, M.J., Catlow, C.R.A., Clark, S.M. and Temmerman, W.M., *J. Phys. C.* **21** (1988) L917.
16. Temmerman, W.M., Stocks, G.M., Durham, P.J. and Sterne, P.A., *J. Phys. F: Met. Phys.* **17** (1987) L135.
17. Moroney, L.M., Thompson P. and Cox, D.E., *J. Appl. Crystallogr.* **21** (1988) 206.
18. Catlow, C.R.A., Chadwick, A.V., Greaves, G.N. and Moroney L.M., *J. Am. Ceram. Soc.* **69** (1986) 272.
19. Noyan, I.C., *Metall. Trans.* **14A** (1983) 249.
20. Fitch, A.N., to be published.
21. Bachmann, R., Kohler, H., Schulz, H. and Weber, H.-P., *Acta Crystallogr. A* **41** (1985) 35.
22. Andrews, S.J., Papiz, M.Z., McMeeking, E., Blake, A.J., Lowe, B. M., Franklin, K.R., Helliwell, J.R. and Harding, M.M., *Acta Crystallogr. B* **44** (1988) 73.
23. Harding, M.M., Maginn, S.J., Campbell, J.W., Clifton, I. and Machin, P., *Acta Crystallogr B* **44** (1988) 142.
24. Catlow, C.R.A., ed., *Mater. Sci. Forum* **9** (1986).

3 X-ray topography

J. MILTAT and M. DUDLEY

3.1 Introduction

X-ray topography is the generic name for diffraction based X-ray imaging techniques. Its main aim is the visualisation and characterisation of localised strains, such as those arising from defects such as dislocations and precipitates contained in an otherwise perfect crystalline medium, or phase boundaries (in the wave-optical sense) such as stacking faults, rotational or misfit boundaries.

A typical X-ray topography experiment first involves obtaining a nearly parallel beam of X-rays, to be characterised by its angular divergence D_A and wavelength spread D_λ. If a crystal is immersed in the beam so that it satisfies Bragg's law

$$2\,d_H \sin \theta_B = n\lambda \qquad (3.1)$$

the incident beam is reflected (diffracted). An X-ray topograph is simply a record on a suitable detector (photographic plates, video detector) of the two-dimensional intensity distribution in the diffracted beam. In Eqn. 3.1, d_H is the lattice plane spacing for the operating reflection $H = (h,k,l)$, θ_B is Bragg's angle, λ is the X-ray wavelength and n is the order of the reflection.

Figure 3.1 displays the two basic geometries of X-ray topography, namely the transmission (or Laue) geometry (Figure 3.1(a)) and the reflection (or Bragg) geometry (Figure 3.1(b)). Due to various limitations, amongst which source size, detector performances and counting statistics are the most significant, the spatial resolution of X-ray topography is, at best, of the order of 1 μm. Therefore, when compared to dark field transmission electron microscopy, X-ray topography suffers from a drastically inferior spatial resolution. However, due to the low absorption of X-rays, fairly massive single crystals may be imaged and X-ray imaging techniques are characterised by a unique strain sensitivity, as shown below.

Figure 3.2 shows four different topographs of the same area of a silicon wafer containing piled-up dislocations (D_1) as well as precipitates (P). They differ by the choice of the active reflection, the operating wavelength or the imaging technique. For instance, strong contrast variations are observed between topographs differing only by the wavelength choice (Figure 3.2(b)

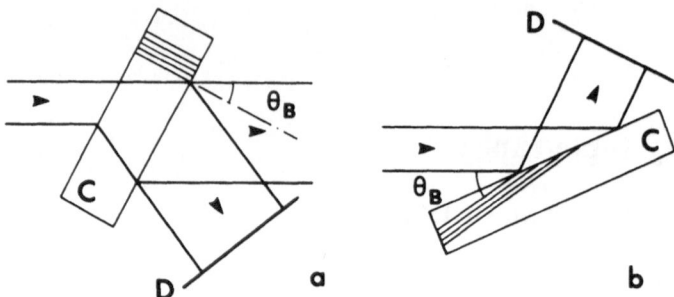

Figure 3.1. Basic geometries of X-ray imaging. (a) Transmission or Laue geometry; (b) reflection or Bragg geometry. C, crystal; D, detector.

vs. (c)). On the other hand, only small differences are seen between topographs in Figure 3.2(a) and (b), except for the fact that some dislocations are nearly invisible in Figure 3.2(b) (e.g. D_2). Finally, apart from the usual contrast complementarity between the H and the $\bar{H}$ reflections under intermediate absorption conditions, only little difference may be detected between topographs in Figure 3.2(c) and (d) although the imaging techniques were different, namely the laboratory based Lang and the synchrotron based Laue techniques, respectively.

In a conventional laboratory environment, the Laue technique is routinely used for crystal orientation purposes. It requires a white beam, i.e. in analogy with visible light, a beam with a large wavelength spread. Each time Eqn. 3.1 is satisfied through a proper combination of lattice plane spacing and operating wavelength, the incident beam is diffracted as shown schematically in Figure 3.3. As recognised long ago [1], the Laue technique becomes an imaging technique as soon as the beam divergence becomes low. Small divergence and large wavelength spread characterise the X-ray beams emitted by storage rings (see Chapter 1). The Laue technique has thus become the simplest of all synchrotron based imaging techniques. Further, only about 1 min was necessary to record the white beam topograph in Figure 3.2(d) whereas several hours proved necessary for its Lang technique counterpart: the high intensity, another characteristic of synchrotron sources, has rendered feasible experiments monitoring slowly evolving phenomena such as crystal growth (see Section 3.4).

The large intensity of synchrotron sources also allows for the insertion in the experimental set-up of beam conditioners, such as pseudo plane wave monochromators, without extending exposure times beyond unacceptable limits. Section 3.3 briefly describes X-ray imaging techniques, separating integrated intensity from plane wave imaging methods. Illustrations of their use can be found in Sections 3.5 and 3.6 devoted to dislocation analysis and strain mapping. In Section 3.8, the applications of topographic techniques to the study of solid state reactions are reviewed.

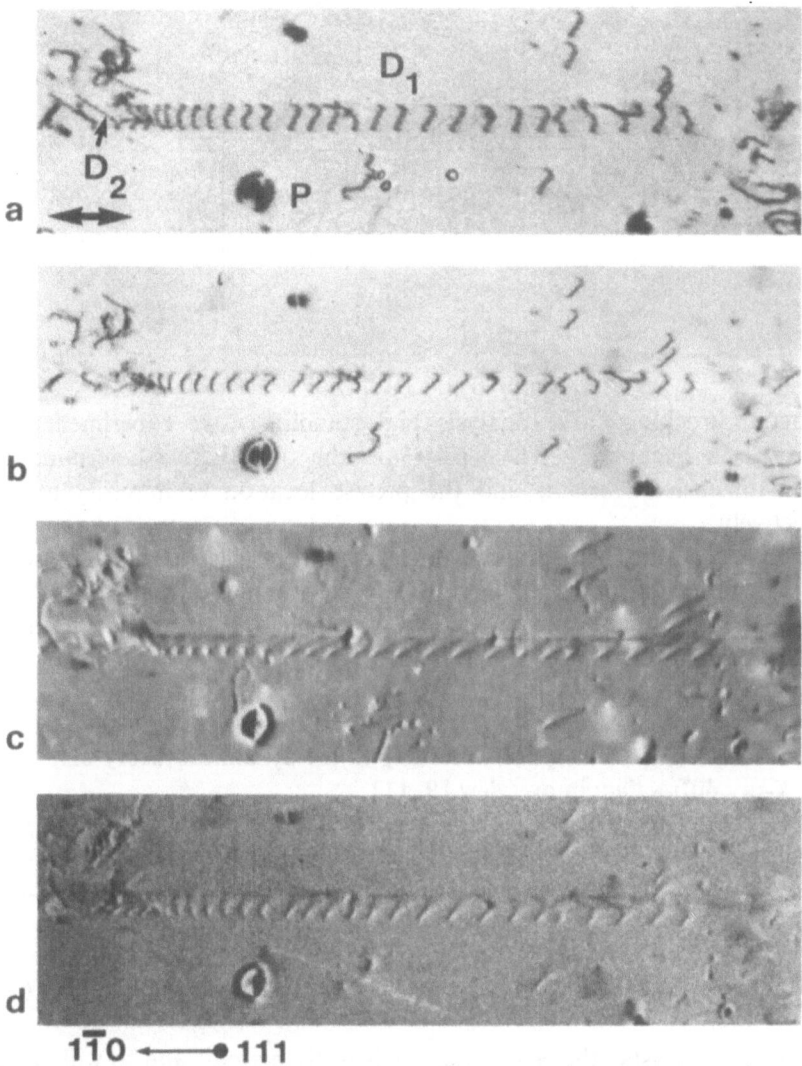

Figure 3.2. Dislocation pile-up in silicon. (a) Lang Topograph, MoKα radiation (low absorption), $1\bar{1}1$ reflection; (b) $2\bar{2}0$ reflection; (c) Lang topograph, CuKα radiation (intermediate absorption), $2\bar{2}0$ reflection; (d) white beam topograph, $\lambda \sim 1.5\,\text{Å}$, $\bar{2}20$ reflection. Scale mark, $200\,\mu m$.

Because topographic methods may not be understood without reference to the dynamical theory of X-ray diffraction, a brief account of this essential theory is first presented (Section 3.2). Finally, it is necessary to include in this chapter an introduction to techniques which are the parent of topography in the sense that their background is also the dynamical theory of X-ray

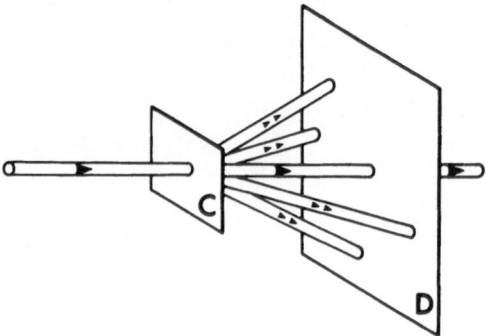

Figure 3.3. The Laue method.

diffraction; rocking curve analysis and standing wave experiments are described in Section 3.7. Their aims are the quantitative determination of one-dimensional strains and the precise location of impurity atoms, respectively.

Several excellent treatises or reviews have in the past been dedicated to the applications of topographic and related techniques to materials science [2–8]. They are highly recommended as complementary to the summary presented in this chapter.

3.2 Dispersion and absorption according to the dynamical theory of X-ray diffraction:an overview [9–11]

Classical dispersion theory indicates that the dielectric susceptibility χ is proportional to the charge density, and related to the refraction index, n, through

$$\chi = \frac{n^2 - 1}{n^2} \tag{3.2}$$

In a perfect crystal, the charge density is triply periodic and so will be the dielectric susceptibility. χ may therefore be written as a Fourier expansion over all vectors $\mathbf{H}$ of the reciprocal lattice

$$\chi(\mathbf{r}) = \sum_H \chi_H \exp\left[-2\pi i (\mathbf{H} \cdot \mathbf{r})\right] \tag{3.3a}$$

where

$$\chi_H = -\frac{R\lambda^2}{\pi V} F_H \tag{3.3b}$$

and R is the classical radius of the electron (2.818×10^{-5} Å), V is the volume of the unit cell, F_H the usual structure factor and $\mathbf{r}$ the position vector.

As with F_H, χ_H is a complex quantity. The zero order term in the expansion, χ_0, is the average value of the dielectric susceptibility. Its real part is negative whereas its imaginary part, also negative, is related to the usual linear absorption coefficient, μ_0, through

$$\mu_0 = -\frac{2\pi}{\lambda}\chi_{io} \tag{3.4}$$

A simple inspection of Eqn. 3.3b indicates that χ_H coefficients are extremely small quantities. It follows that the refractive index for X-rays is only slightly different from unity

$$n \approx 1 + \frac{\chi_0}{2} \tag{3.5}$$

This is the basic reason why lenses in the classical sense are not available in the hard X-ray case. X-ray topographs will therefore be unmagnified images of the crystalline medium under investigation.

Let us now consider a plane wave (wave vector $\mathbf{k}$:$1/\lambda$ in vacuo) impinging on a material satisfying Eqn. 3.3. It excites in the crystal waves which are solutions of Maxwell's equations, where it is customary to neglect magnetic interactions as well as possible electrical conductivity. The general solution for the electric displacement $\mathbf{D}(\mathbf{r})$ is a Bloch wave, or wavefield

$$\mathbf{D}(\mathbf{r}) = \sum_j \mathbf{D}_j(\mathbf{r}) = \sum_j \sum_H \mathbf{D}_{Hj} \exp[-2\pi i(\mathbf{K}_{Hj}\mathbf{r})] \tag{3.6}$$

where the wave vectors $\mathbf{K}_{Hj}$ are all deduced from one another through the Laue relation

$$\mathbf{K}_{Hj} = \mathbf{K}_{Oj} + \mathbf{H} \tag{3.7}$$

A simple choice for $\mathbf{K}_{Oj}$ consists of relating $\mathbf{K}_{Oj}$ to the wave vector in vacuo, $\mathbf{k}$, by means of the usual laws of refraction.

It is fortunate that, in the X-ray case, a two beam approximation ($j = 1,2$) is usually sufficient to properly describe diffraction phenomena. Re-injecting the general solution (3.6) into Maxwell's equations provides two independent relations linking the wave vector and electrical displacement amplitudes; their secular equation reads

$$[K_{Oj}{}^2 - k^2(1 + \chi_0)][K_{Hj}{}^2 - k^2(1 + \chi_0)] = k^4 C^2 \chi_H \chi_{\bar{H}} \tag{3.8}$$

where all quantities have been previously defined, except the variable C which describes the polarisation of the incident beam ($C = 1$, $\cos 2\theta_B$ for σ and π polarisations, respectively).

Equation 3.8 is the basic dispersion relation of the dynamical theory of X-ray diffraction. Let us consider its possible solutions.

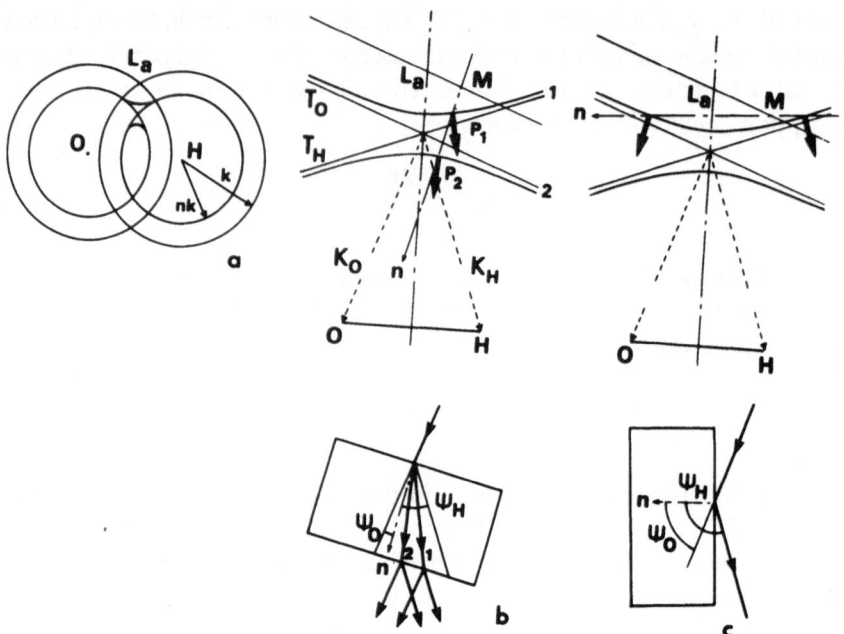

Figure 3.4.   (a) Dispersion surface construction; (b) blow-up, Laue geometry; (c) Blow-up, Bragg geometry. Top, reciprocal space; bottom, direct space.

(i) Dividing both members of Eqn. 3.8 by k^4, it appears that the right-hand side is an infinitesimal quantity of the second order in χ. Equation 3.8 will therefore be satisfied if (either) K_{Oj} (or K_{Hj}) is equal to

$$K_{Oj} \simeq k\left(1 + \frac{\chi_0}{2}\right) = nk \qquad (3.9)$$

Under such conditions, only one wave travels in the crystal: it is obviously the refracted wave corresponding to the plane wave in vacuo.

(ii) If both K_{Oj} and K_{Hj} are close to $k[1 + (\chi_0/2)]$, the right-hand side of the dispersion relation may no longer be neglected. Equation 3.8 then indicates that the solution becomes degenerate with two possible solutions for the coupled variables K_{Oj}, K_{Hj}. In the reciprocal space (k space), Eqn. 3.8 represents a hyperboloid of revolution, the section of which in the $\mathbf{K}_O$, $\mathbf{K}_H$ plane (incidence plane) is shown in Figure 3.4.

The dispersion surface construction involves a modification of the more familiar Ewald sphere construction in that the excited reciprocal lattice points are located at the origins of the spheres rather than at their surfaces. Initially, spheres of radius equal to the magnitude of the free space wave vector, $\mathbf{k}$, are constructed. Refraction is taken into account in constructing a set of concentric spheres with a slightly shorter radius, nk. In the vicinity of the intersections of those last two spheres, solution degeneracy implies that two wavefields propagate in the crystal, represented by their tie-points P_1 and P_2. The location of P_1 and P_2 is determined by the deviation $\Delta\theta$ from exact Bragg angle (L_a M in the figure) and the condition of continuity of the tangential components of the wave vectors, i.e. by the direction **n** of the normal to the crystal entrance surface for X-rays. It should be noted that there exists one dispersion surface per polarisation state (σ or π).

The relation between wave and ray theories is obtained through the computation of the time averaged Poynting vector. It turns out that the Poynting vector is normal to the dispersion surface. A full relation between the reciprocal and direct spaces is now established as summarised in Figure 3.4.

Several additional results of the dynamical theory of X-ray diffraction need to be mentioned.

(1) Each wavefield is composed of two coherent plane waves which interfere with a periodicity equal to the lattice plane spacing. It may be shown that, in most materials, the nodes (antinodes) of the interference pattern associated with wavefield P_1 (P_2) coincide with the maxima of the electronic density. Wavefields of types 1 and 2 will therefore suffer different absorptions: wavefields 1 are usually less absorbed than they would be under the influence of normal photoelectric absorption, resulting in an anomalous transparency to X-rays, first discovered by Borrmann [12,13]. The contrary holds for wavefields 2. This result is extremely important in practice, since it proves most useful in the determination of lattice curvature as shown in Section 3.3. Wavefields 1 and 2 may also interfere, with a periodicity (the Pendellösung periodicity) equal to

$$\Lambda = \frac{\Lambda_0}{(1 + \eta^2)^{1/2}}, \quad \Lambda_0 = \frac{\cos\theta_B}{C|\chi_H\chi_{\bar{H}}|^{1/2}}\lambda \qquad (3.10)$$

where η is related to the departure from Bragg angle $\Delta\theta$ through

$$\eta = \frac{\sin 2\theta_B}{C|\chi_H\chi_{\bar{H}}|^{1/2}}\Delta\theta \qquad (3.11)$$

Λ_0 values range typically from a few micrometres to several tens of micrometres, although in some large unit cell organic material, values in excess of 100 micrometres are possible. Such interferences give rise, for instance, to the contrast oscillations observed along certain dislocation line images in Figure 3.2(a) and also to the fringe pattern associated with twin lamellae observed in Figure 3.18.

(2) Bragg reflection occurs in a finite range of $\Delta\theta$ values, defined for thick crystals ($t \gg \Lambda_0$) by

$$\delta = \delta_0 \left|\frac{\cos\psi_h}{\cos\psi_0}\right|, \quad \delta_0 = \frac{2C|\chi_H\chi_{\bar{H}}|^{1/2}}{\sin 2\theta_B} \qquad (3.12)$$

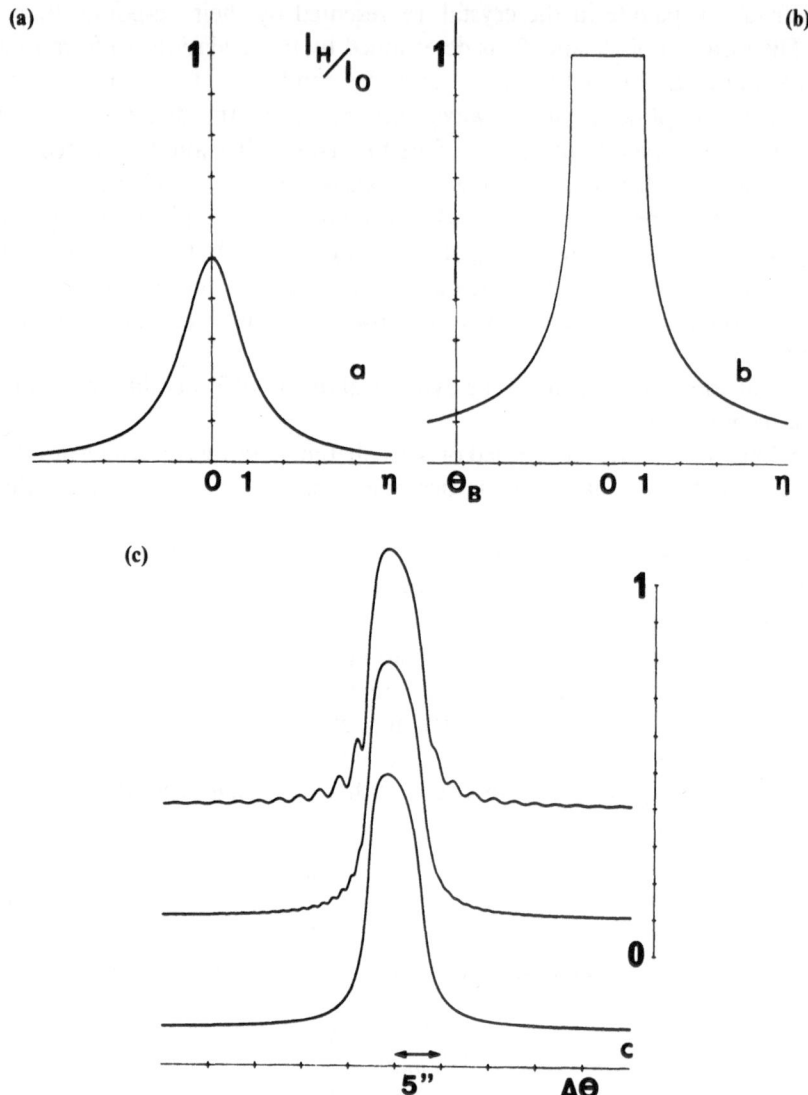

Figure 3.5. Reflecting power: (a) Laue geometry, thick crystal and zero absorption; (b) Bragg geometry; (c) from the thick to the thin absorbing crystal, Bragg geometry. In ascending order $t/\Lambda_0 = 10, 1, 0.5$ (gadolinium gallium garnet, 444 reflection at 1.2378 Å).

where ψ_H and ψ_0 are defined in Figure 3.4. δ_0 is typically equal to a few seconds of arc. The distribution versus $\Delta\theta$(or η) of the reflecting power P, i.e. the ratio of the diffracted to the incident intensities, is known as the reflection profile or rocking curve. As shown in Figure 3.5, rocking curves in the Laue and Bragg geometries are markedly different, even in the case of zero absorption. In the Bragg case and for thick crystals, the range of total reflection extends from $\eta = -1$ to $\eta = +1$, but Bragg reflection occurs at an angle slightly different from θ_B due to refraction. The flanks of the rocking curve are extremely steep, this remark being the basis of strain mapping techniques to be described in Section 3.6.

In the Laue case, the reflecting power is at most equal to 0.5 as the result of destructive interferences between the coherent waves propagating in the crystal. Finally, as shown also in Figure 3.5, the rocking curves of thin crystals ($t \ll \Lambda_0$) exhibit oscillations, another direct result of the dynamical theory.

3.3 Topographic techniques and contrast formation mechanisms

The diffraction of an incident plane wave has, for the sake of clarity, been solely considered in the preceding section. However, in the absence of routinely available X-ray lasers, X-ray beams delivered by classical fine focus or synchrotron sources possess both angular divergence D_A and wavelength spread D_λ. The result $D_{A\lambda}$ of the convolution of D_A and D_λ always greatly exceeds the natural acceptance angle δ_0 of single crystals. Only through the use of beam conditioners may the natural X-ray beams be transformed into nearly plane waves. X-ray topographic techniques may therefore be separated into integrated intensity techniques ($D_{A\lambda} \gg \delta_0$) and pseudo plane wave techniques ($D_{A\lambda} \ll \delta_0$).

3.3.1 *Integrated intensity techniques*

The most widespread laboratory X-ray imaging technique is the Lang method [14] which uses a fine focus source ($\sim 100\,\mu\text{m}$) and a beam collimated by $\sim 100\,\mu\text{m}$ slits located at $\sim 50\,\text{cm}$ from the source. Such a collimation is necessary for the separation of the $\alpha_1\,\alpha_2$ doublet of K_α characteristic lines. A simple calculation shows that the angular divergence amounts to $\sim 150''$ whereas the wavelength spread of characteristic line $\Delta\lambda/\lambda$ is of the order of 10^{-3}.

A complete picture of the crystal under study may be obtained by a translation of the crystal and detector in front of the slit. Topographs in Figure 3.2(a–c) were obtained by means of the Lang technique for which the incident wave is well represented by a spherical wave [15]. As a consequence, the full dispersion surface is excited and the intensity at any point along the exit surface for X-rays is the result of the interferences of conjugate wavefields propagating in the same direction (Figure 3.7(a)).

The Laue technique also belongs to the class of integrated intensity techniques because the divergence of existing synchrotron beams is at best comparable to δ_0 and the wavelength spread is extremely large ($\Delta\lambda/\lambda \approx 1$–2).

Bragg's law (Eqn. 3.1) readily indicates that various combinations of the d_H, $n\lambda$ lead to the same θ_B value. Since, for a given reflection, d_H belong to a discrete set of values, it appears that several harmonics may superimpose in a given

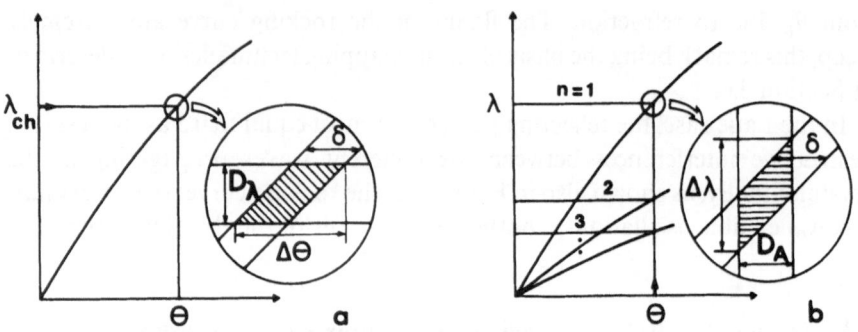

Figure 3.6. Du Mond diagrams pertaining to: (a) the Lang method; (b) the white beam technique.

diffraction spot (or topograph) [16, 17]. DuMond diagrams [18] provide a clear picture of the experimental conditions. In a characteristic line experiment (e.g. the Lang method), the spectral range D_λ together with δ determines the angular acceptance $\Delta\theta$ of the crystal (Figure 3.6(a)). In a white beam experiment, the beam divergence D_A together with δ defines the spectral acceptance $\Delta\lambda$ of the crystal (Figure 3.6(b)). In both cases, the hatched area is a measure of the integrated intensity diffracted by the crystal for the operative wavelength.

Let us now assume that non-uniform elastic displacements $\mathbf{u}(\mathbf{r})$ do exist in the crystalline material under investigation. There exists one basic assumption behind all contrast theories, namely, the possibility of defining locally a triply

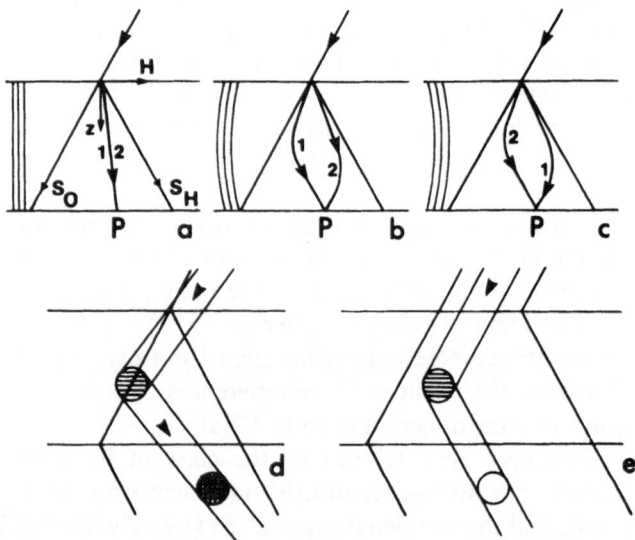

Figure 3.7. Contrast formation mechanisms. Ray paths for integrated intensity techniques: (a) perfect crystal; (b), (c) positive (negative) strain gradients. Orientation contrast in (d) integrated intensity techniques, (e) plane wave imaging.

periodic crystal, characterised, however, by a modified reciprocal lattice vector set

$$\mathbf{H}'(\mathbf{r}) = \mathbf{H}(\mathbf{r}) - \nabla_r[\mathbf{H} \cdot \mathbf{u}(\mathbf{r})] \tag{3.13}$$

Keeping the origin O constant in Figure 3.4, Eqn. 3.13 implies that the whole dispersion surface glides along its T_0 asymptote. For slowly varying strains, it may equivalently be considered that the tie-point of a given wavefield migrates along the dispersion surface.

For a symmetrical Laue reflection, it may be shown [19] that migration is governed by the differential equation (see Figure 3.7 for the choice of the coordinates system)

$$\pm\frac{d\eta}{dz} = \frac{1}{\cos^2\theta_B}\frac{\partial^2}{\partial s_0 \partial s_H}[\mathbf{H}\cdot\mathbf{u}(\mathbf{r})] \tag{3.14}$$

where the $+$ and $-$ signs apply to wavefields of types 1 and 2, respectively.

The meaning of Eqn 3.14 is that the direction of propagation of wavefields is modified by the presence of a non-uniform displacement field $\mathbf{u}(\mathbf{r})$. In other words, rays are bent similarly to what happens in ordinary geometrical optics in the presence of a varying index of refraction [20].

Defining the dimensionless variable $Z = z/\Lambda_0$, Eqn. 3.14 can be rewritten as

$$\pm\frac{d\eta}{dZ} = \frac{2}{\cos\theta_B}\frac{\Lambda_0}{\delta_0}G \tag{3.15}$$

where

$$G = \frac{\partial}{\partial s_0}\delta(\Delta\theta)$$

$\delta(\Delta\theta)$ is the additional departure from exact Bragg angle associated with the existence of $\mathbf{u}(\mathbf{r})$ [21]

$$\delta(\Delta\theta) = \frac{1}{k\sin 2\theta_B}\frac{\partial}{\partial s_H}[\mathbf{H}\cdot\mathbf{u}(\mathbf{r})]$$

The quantities $\delta(\Delta\theta)$ and G are equal to

$$\delta(\Delta\theta) = \left[\frac{\partial}{\partial z} + \tan\theta_B\frac{\partial}{\partial y}\right](u_\perp)$$

$$G = \frac{1}{\cos^2\theta_B}\left[\cos^2\theta_B\frac{\partial^2}{\partial z^2} - \sin^2\theta_B\frac{\partial^2}{\partial y^2}\right](u_\perp) \tag{3.16}$$

where $u_\perp$ is the displacement normal to the reflecting lattice planes.

$\delta(\Delta\theta)$ appears as a linear combination of the strains σ_{ij} and rotations ω_{ij} deduced from the displacement field $\mathbf{u}(\mathbf{r})$. G will be referred to as the (effective) strain gradient.

Equation 3.14 or 3.15 implies that wavefields belonging to branch 1 and 2 of the dispersion surface have opposite curvatures, hence the drawing of the ray paths in Figure 3.7 is valid for an integrated intensity imaging technique and specialised to the case of a constant strain gradient (the ray trajectories are then hyperbolae): wavefields of type 1 bend as the bending of the net planes whereas wavefields of type 2 bend in the opposite direction. Remembering that wavefields 1 and 2 suffer different Borrmann absorptions, it may be inferred that the intensity distribution at the exit surface will be different from the perfect crystal case (Figure 3.7(a). It may be shown that for small G values, a positive (negative) strain gradient yields, in most materials (see Section 3.2), a black (white) contrast in Lang or white beam topographs provided the product of crystal thickness and linear absorption coefficient is of the order 2–3 (black means excess X-ray intensity). Such conditions hold for topographs in Figure 3.2(c,d) and indeed both white and black contrast defect images are seen. Contrast reversal is expected between the H and $\bar{H}$ reflections, as shown in Figure 3.2(c,d).

This short discussion shows that ray bending provides a very powerful tool for strain analysis in integrated intensity experiments [22–26].

In classical optics, it is well known that the limit of validity of geometrical optics may be expressed as

$$\lambda/\rho \ll 1 \qquad (3.17)$$

where ρ is the ray radius of curvature. In the X-ray case, it may be shown that the equivalent relation is [27]

$$\Lambda_0/\rho \ll 1 \qquad (3.18)$$

i.e. a very similar relation, where, however, the characteristic length is the Pendellösung distance instead of the radiation wavelength. Since the ray radius of curvature is simply $\rho^{-1} = d\eta/dz = (1/\Lambda_0)\,(d\eta/dZ)$, Eqn. 3.15 readily indicates that the inequality (Eqn. 3.18) may be written as

$$\frac{\Lambda_0}{\delta_0}G \ll 1 \qquad (3.19)$$

For a dislocation, for instance, the geometrical optics approximation would cease to be valid at a distance typically equal to $10\,\mu m$ from the dislocation core. Therefore, ray bending is insufficient to explain all contrast phenomena.

The contrast of dislocations in X-ray topography has received lasting attention [e.g. 28–30], mainly in conjunction with the Lang technique. In the Laue geometry, the creation of new wavefields in the highly distorted regions around the dislocation core has been demonstrated to be the source of the fringe pattern sometimes associated with dislocation images (so-called intermediary image). It has recently been demonstrated that geometrical results can also be used in the presence of strong strain gradients if taking into account the creation of those new wavefields [31].

Let us finally consider regions of the crystal where the strains and rotations are sufficiently high so that the local departure from Bragg angle $\delta(\Delta\theta)$ exceeds δ. When the beam is divergent and/or not monochromatic, such regions will be able to pump out of the incident beam rays or wavelengths not participating in the diffraction of the surrounding perfect crystal.

If such regions diffract kinematically, their image will show an excess of intensity, hence will be black (Figure 3.7(d)). This is the mechanism responsible for the relatively narrow dislocation images observed under low absorption conditions (Figure 3.2(a,b)) [21,32]. It has been shown that, to a fair accuracy, dislocation image width is determined by the projected width of the volume around the dislocation core satisfying $\delta(\Delta\theta) \approx \delta$ [33].

It is clear that direct and intermediary images overlap and are therefore often impossible to separate: for instance, the contrast oscillations observed along certain dislocation line images in Figure 3.2(a) contain both the intermediary and the direct images.

The simple mechanism of orientation contrast gives an indication of strain sensitivity in X-ray imaging experiments: since δ is typically of the order of a few seconds of arc, X-ray imaging methods easily reveal lattice rotations of the same order, or, according to Bragg's law, lattice strains $\Delta d/d = \delta/\tan\theta_B \approx 10^{-5}$.

3.3.2 *Pseudo plane wave techniques*

A precise description of the methods utilized for the production of specific X-ray waves is beyond the scope of this review. However, an inspection of DuMond diagrams in the two reflection case proves instructive. The so-called $(+, +)$ and $(+, -)$ settings are represented in Figure 3.8. Clearly, the whole angular acceptance of crystal C_2 is always covered in the $(+, +)$ setting, whatever θ_{B1} and θ_{B2} (Figure 3.8(a)). In the $(+, -)$ setting, this remark also proves true (Figure 3.8(b)) unless $\theta_{B1} = \theta_{B2}$ (Figure 3.8(c)). One then refers to the parallel setting. A pseudo plane wave is, however, only obtained if the angular spread after the first reflection is much smaller than the width of the rocking curve of crystal C_2. This requirement may be fulfilled by the use of a strongly asymmetric first crystal reflection [34].

More generally, mixtures of asymmetric and parallel reflections allow both quasi plane waves and harmonics rejection to be attained, the latter being an important requirement when dealing with synchrotron sources [35,36]. For instance, the triple reflection monochromator $(+ 333, + 131, - 131)$ available at LURE-DCI delivers a harmonic free beam with an angular divergence $\sim 2 \times 10^{-6}$ rad ($\sim 0.3''$) and a wavelength spread $\Delta\lambda/\lambda \approx 10^{-6}$ at $\lambda = 1.238$ Å [36, 5].

In a plane wave experiment, orientation contrast is an important and simple contrast mechanism: contrary to the previous case, however, a diffracted

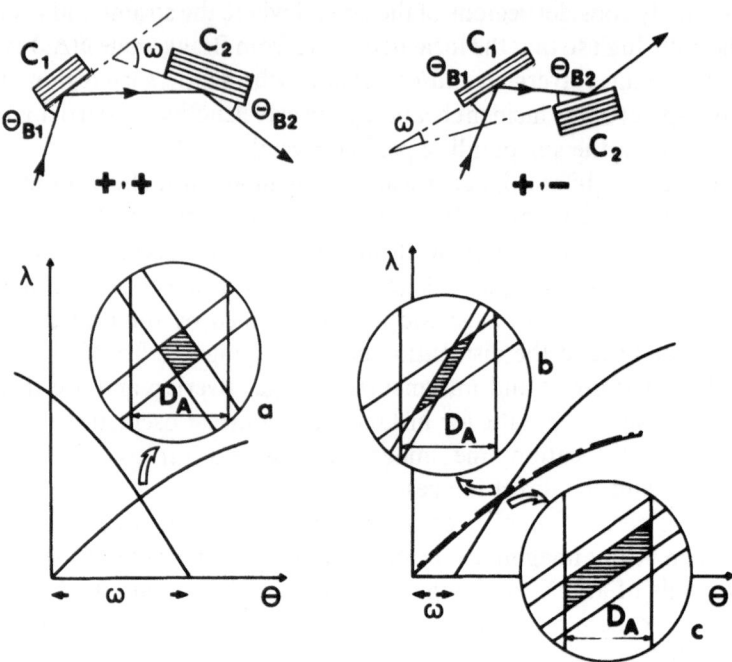

Figure 3.8. Du Mond diagrams in the two reflection case: (a) (+ , +) setting; (b) (+ , −) setting; (c) parallel setting.

intensity deficiency indicates the location where $\delta(\Delta\theta)$ exceeds δ_0 (white contrast) (Figure 3.7(e)). The study of dislocation contrast in the Bragg geometry [37–39] has shown that the now familiar concept of wavefields creation is also relevant [40b]. Finally, ray bending may also take place in the Bragg case, outside, however, the range of total reflection [40a].

3.4 Crystal growth defects

Figure 3.9 shows two topographs of nearly perfect single crystals: the first (Figure 3.9(a)) exhibits the classical defect structure found in solution, gel or even hydrothermally grown crystals [e.g. 41–45], namely, dislocations, growth sector boundaries, growth bands and, eventually, inclusions. The second (Figure 3.9(b)), shows the typical defects observed in metallic crystals grown from the melt, including sub-grain boundaries, dislocations and, also, inclusions or precipitates [46]. Growth bands are also often characteristic of non-metallic melt grown crystals [47].

Although laboratory based topographic techniques are still well suited to the analysis of factors affecting solution or gel growth, such as the influence of screw or edge dislocations on growth rate [42, 48–50], or the influence of dopants on crystal quality [e.g. 44], the advent of synchrotron sources has

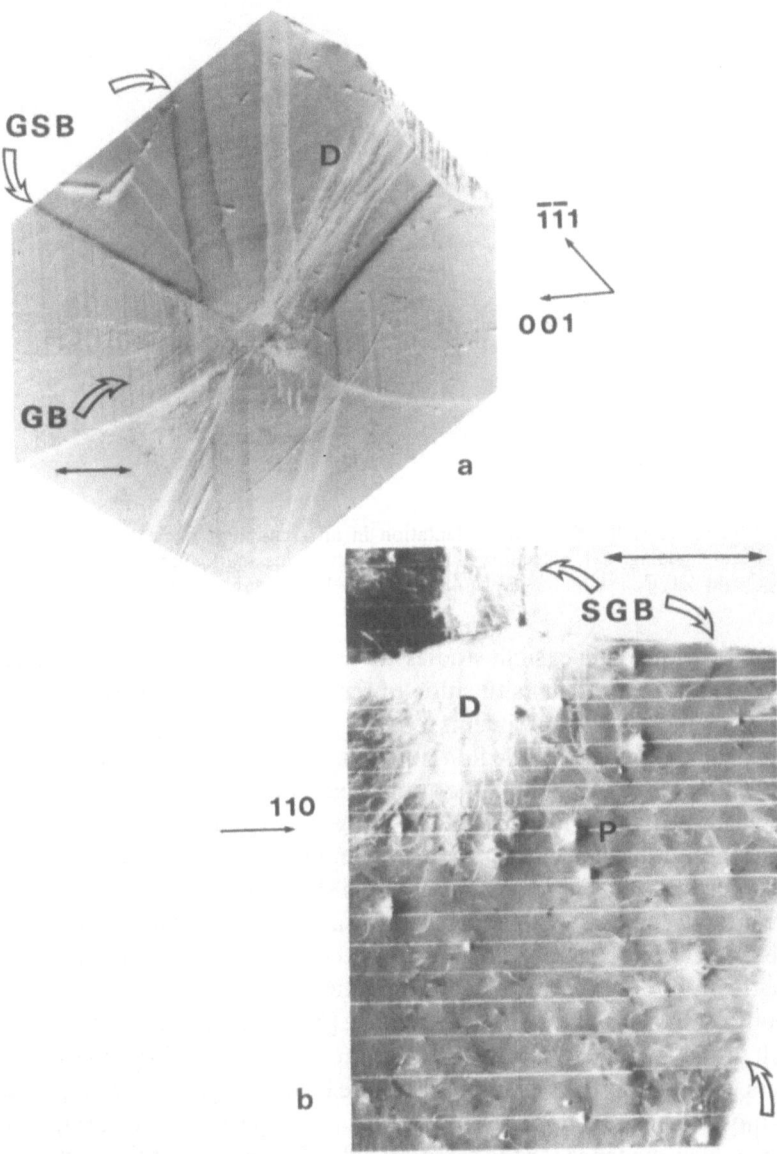

Figure 3.9. Lang topographs: (a) (1$\bar{1}$0) slice of a Sr(NO$_3$)$_2$ crystal. AgKα radiation, $\bar{2}\bar{2}$2 reflection (after [70]). Scale mark, 3 mm; (b) (001) slice of a Fe 3.5% Si crystal. MoKα radiation, 110 reflection. The horizontal white lines are 90° magnetic domain walls. D, dislocations; P, precipitates; SGB, subgrain boundaries; GSB, growth sector boundaries; GB, growth bands. Scale mark, 1 mm.

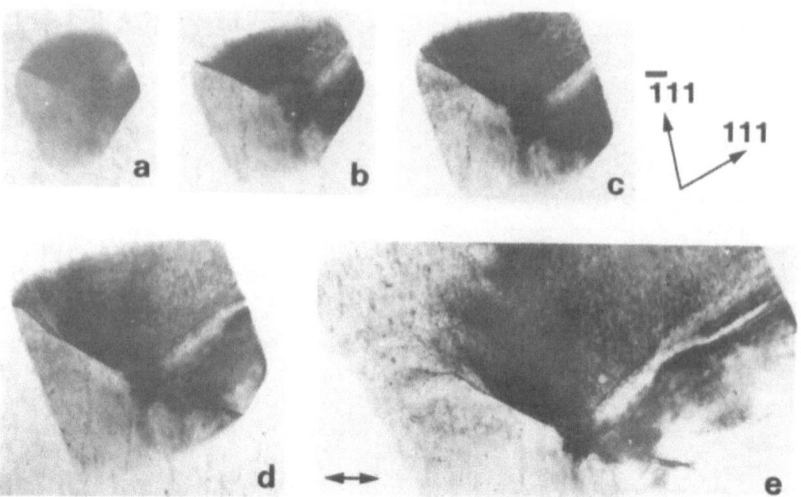

Figure 3.10. Recrystallisation after indentation in an aluminium single crystal (after [52]). White beam imaging, $\bar{1}11$ reflection, Laue geometry. Scale mark, 1 mm. (a) 404°C, 13 min; (b) 420°C, 25 min; (c) 450°C, 43 min; (d) 450°C, 47 min; (e) 450°C, 56 min.

allowed previously unfeasible studies to be undertaken [see 51]. This is the case, for example, in the in situ study of recrystallisation or growth of metals from the melt. In such experiments, contrary to solution or gel growth from a nucleus, the orientation of the first nuclei is, a priori, unknown. The ability of white beam imaging to pick up the appropriate wavelength from the incoming beam ensures that the nucleus will be imaged as soon as its size exceeds the resolution limit of the detector: white beam imaging appears essential under such circumstances. Figure 3.10 shows one striking example of the similarity between solution growth and recrystallisation in an indented aluminium single crystal [52]; faceting and growth sectors are clearly revealed. It has also been shown in other recrystallisation studies that moving grain boundaries may possess a long range periodic structure [53]. In addition, the morphology of melt grown metallic crystals, as well as the role of dislocations in the growth process have been analysed [54–56]. White beam imaging has also proved to be a most useful tool for the analysis of polytypism, a fascinating subject [e.g. 57]. New evidence for the essential role of screw dislocations clusters (or giant dislocations) [58] in the growth of zinc sulphide crystals has recently been obtained [59 and loc. cit.].

3.5   Dislocation analysis: integrated intensity techniques

As outlined above, dislocations play an important role in crystal growth. Dislocation generation and propagation are also the basic mechanisms of plasticity in crystalline materials. It should therefore not be surprising that dislocation analysis is a common task for researchers utilising X-ray imaging.

Topologically, a dislocation is uniquely characterised by its line direction **L** and its Burgers vector **b**, including direction and sign. Line directions can usually be easily determined through the use of several reflections. Noteworthy is the capacity of white beam X-ray imaging to produce several reflections simultaneously: line direction determination may then simply be obtained by means of elementary geometry [60].

Equations (3.13) or (3.16) indicate that the strainfield of any kind of defect should prove invisible when

$$\mathbf{H} \cdot \mathbf{u(r)} = 0 \tag{3.20}$$

Equation 3.20 simply means that the displacement normal to the reflecting planes is null. This condition is fulfilled for screw dislocations if

$$\mathbf{H} \cdot \mathbf{b} = 0 \tag{3.21}$$

An additional condition is required for edge dislocations, namely

$$\mathbf{H} \cdot (\mathbf{b} \wedge \mathbf{L}) = 0 \tag{3.22}$$

Equations 3.21 and 3.22 are the classical relations enabling the determination of the direction of Burgers vectors. It needs to be mentioned that, in many instances, complete contrast extinction proves impossible to achieve; in the case of dislocations, this impossibility may be ascribed to decoration by impurities and to the influence of surface stress relaxation which may break the symmetry of the displacement field.

The determination of the sign of Burgers vectors is often a more delicate task, as illustrated in the following example [61]. Although it belongs to the early days of synchrotron based imaging techniques, this example has been chosen for two reasons. First, it is the only one known to the authors where the same dislocations have been imaged by means of the Laue technique and plane wave reflection topography. Second, the Laue topographs have benefited from the wavelength tunability of the synchrotron beam: by choosing a main wavelength at 1.2 Å, i.e. just above the K absorption edge of gallium (1.19 Å), absorption has been minimised whereas the not too short wavelength favours image quality.

Figure 3.11 shows various types of epitaxial dislocations, i.e. dislocations generated in order to relax the lattice mismatch between the substrate and the epitaxial layer grown on top of it. In the present example, the substrate orientation is [001] and dislocations with line direction [110] may possess any of the four $\frac{1}{2} \langle \bar{1}01 \rangle$ Burgers vectors in the $(\bar{1}1\bar{1})$ and $(\bar{1}11)$ glide planes. Such dislocations are therefore 60° dislocations.

Sample curvature after the growth of the epitaxial layer is schematically drawn in Figure 3.11(a). Such a curvature implies that the lattice parameter of the layer is larger than that of the substrate. It follows that, in order to

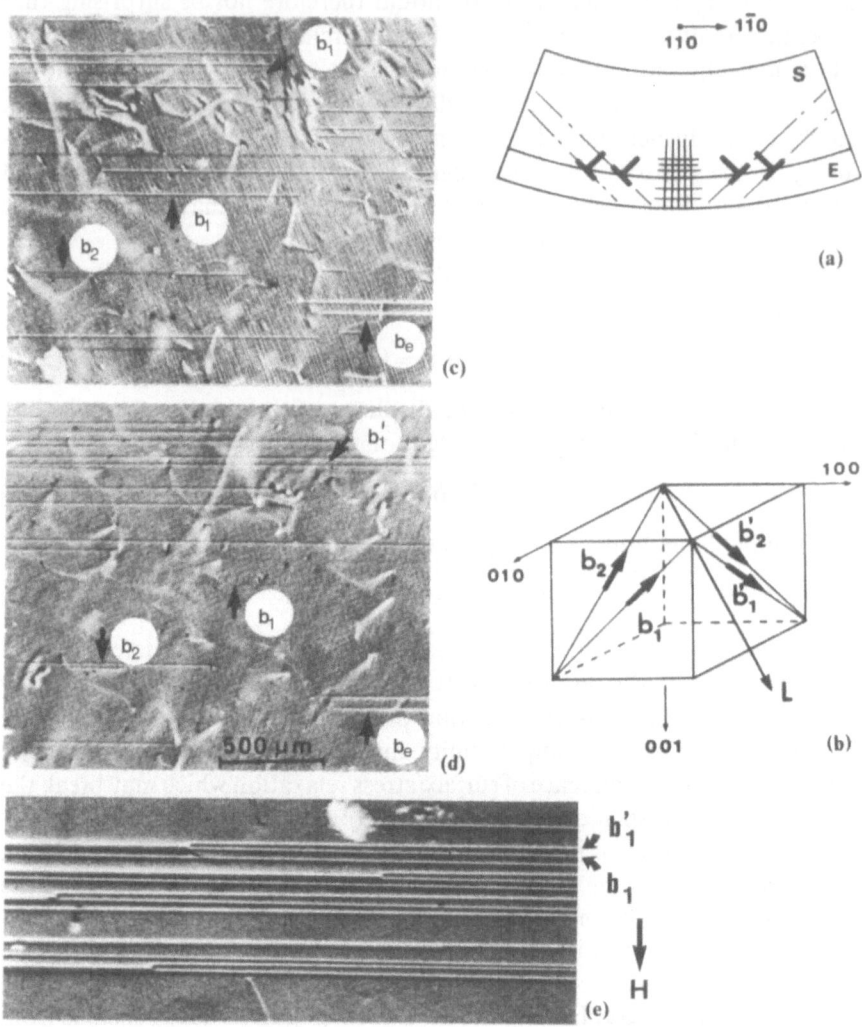

Figure 3.11. Epitaxial dislocations in $Ga_{1-x}Al_xAs_{1-y}P_y/GaAs$ heterojunctions (after [61]). White beam imaging at $\lambda \sim 1.2$ Å. Laue geometry. (a) Bending due to lattice mismatch; (b) Burgers vectors distribution; (c) $\overline{4}2\overline{2}$; (d) 242; (e) $\overline{2}20$ reflections. Dislocations labelled $\mathbf{b}_e$ are pure edge dislocations.

accommodate lattice mismatch, dislocations must have their extra half-planes directed towards the interior of the substrate (Figure 3.11(a)). Therefore, in this example, the sign of the Burgers vectors of dislocations is known from arguments independent of contrast analysis, hence the drawing of Burgers vectors in Figure 3.11(b).

After noticing, as an example of extinction rules, that dislocations with Burgers vectors $\mathbf{b}_1$ and $\mathbf{b}_1'$ are invisible in the 242 and $\overline{4}2\overline{2}$ reflections

(Figure 3.11 (d,c)), respectively, let us consider the contrast of dislocations with Burgers vectors $\mathbf{b}_1$ and $\mathbf{b}_2$. According to Figure 3.11(b), they should have the same edge ($\mathbf{b}_e$) and opposite screw ($\mathbf{b}_s$) Burgers vector components. Although their respective contrasts in the $\overline{4}\overline{2}\overline{2}$ reflection (Figure 3.11(c)) are markedly different, it is anticipated that, in a reflection satisfying $\mathbf{H} \cdot \mathbf{b}_s = 0$, both types of dislocations should exhibit similar contrasts. A similar argument applies to dislocations with Burgers vectors $\mathbf{b}_1'$ and $\mathbf{b}_2'$. Further, Figure 3.11(e) indicates that the contrast of dislocation with Burgers vectors $\mathbf{b}_1$ and $\mathbf{b}_1'$ (or $\mathbf{b}_2$, $\mathbf{b}_2'$) are essentially identical.

Let us now attempt to apply the results of Section 3 to the case of a simple edge dislocation. If reflection occurs off the slip plane of edge dislocations, a mostly white or black contrast is expected as shown in Figure 3.12(a, b) [62]. In this configuration, $\mathbf{H} \cdot \mathbf{b}_e = 0$ and contrast formation is due to the buckling of lattice planes parallel to the slip plane ($\mathbf{H} \cdot (\mathbf{b} \wedge \mathbf{L}) \neq 0$). If reflection occurs off planes normal to the Burgers vectors, the situation appears complex because any reflecting plane in the vicinity of the dislocation core is submitted to opposite curvatures of comparable magnitude (Figure 3.12(c, d)). Although the contrast is expected to reverse with the sign of the Burgers vector, it may not easily be predicted.

In the $\overline{2}20$ reflection, reflection occurs off planes which are neither the dislocations slip planes nor planes perpendicular to the edge components of the Burgers vectors. However, a simple inspection of Figure 3.11(a) indicates that lattice plane buckling is certainly not the most efficient lattice

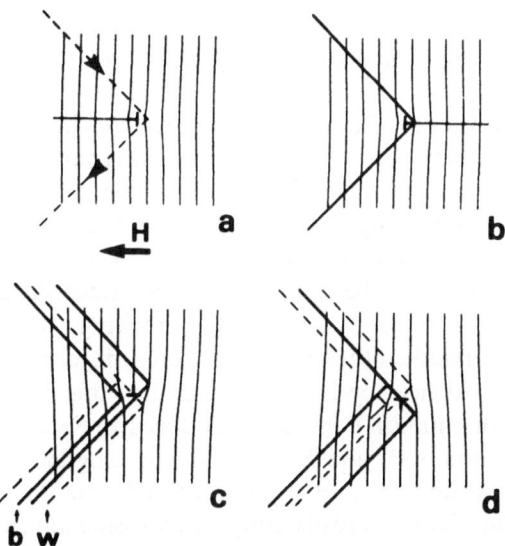

Figure 3.12.   Contrast formation around an edge dislocation (a,b) reflection from planes parallel to the slip plane; (c,d) reflection from planes perpendicular to the Burgers vector. Solid lines, positive curvature (black contrast); dotted lines, negative curvature (white contrast).

deformation for contrast formation since dislocations with Burgers vectors $\mathbf{b}_1$, $\mathbf{b}_2$ and $\mathbf{b}'_1$, $\mathbf{b}'_2$ should then exhibit opposite contrasts, in contradiction with experimental evidence.

The contrast of these epitaxial dislocations in the $\bar{2}20$ reflection may only be simply understood if it is assumed that the observed contrast is controlled by wavefield curvature in the immediate vicinity of the exit surface for X-rays. Surface stress relaxation should then be taken into account. Under these assumptions, the observed contrast is compatible with the Burgers vectors drawn in Figure 3.11(b) (rays w, b in Figure 3.12(c)).

This long discussion pinpoints the difficulties associated with the determination of the sign of Burgers vectors from the sole inspection of dislocation image contrast in integrated intensity imaging techniques. Moreover, the discussion has failed to take into account possible harmonic superposition. However, to a fair accuracy, defect contrast in a Laue experiment may be reconstructed by a properly weighted addition of the intensities specific to each operating wavelength as demonstrated in [45]. It is generally accepted that the presence of (short wavelengths) harmonics increases the strain sensitivity [17].

Finally, arguments based on lattice plane curvature are not applicable to the analysis of screw dislocations, unless surface stress relaxation is accounted for [63]. Let us also mention that, in specific geometries and under low absorption conditions, the sign of the Burgers vectors of edge [64, 65] and even screw [66] dislocations may be deduced from the analysis of direct or intermediary images.

3.6   Dislocation analysis and strain mapping: plane wave imaging

Due to the usual existence of lattice mismatch, the rocking curve of an epilayer is, in reflection geometry, separated from the rocking curve of its substrate (see Figure 3.11(a)). An example of this common behaviour will be shown in the next section. In the following, dealing again with the images of the epitaxial dislocations described in the previous section [67,68], dislocation contrast in the substrate reflection will be solely considered. Since the crystal is bent, a single topograph incorporates the whole rocking curve as shown in Figure 3.13. In the middle of the reflection range, dislocation images appear white, indicating a deficiency in X-ray intensity: this contrast is a simple manifestation of orientation contrast, introduced in Section 3. For η values corresponding to points P_1 and P_2 along the rocking curve, the dislocation contrast is black-white and reverses between $\eta(P_1)$ and $\eta(P_2)$.

Let us again consider the case of a simple edge dislocation and consider first the case where reflection occurs off its slip plane (Figure 3.13 (b, c)). Depending on the sign of the Burgers vector, the local Bragg angle is larger than $\theta_B(\delta(\Delta\theta) > 0)$ to the left or right of the dislocation core. If the extra half-plane points

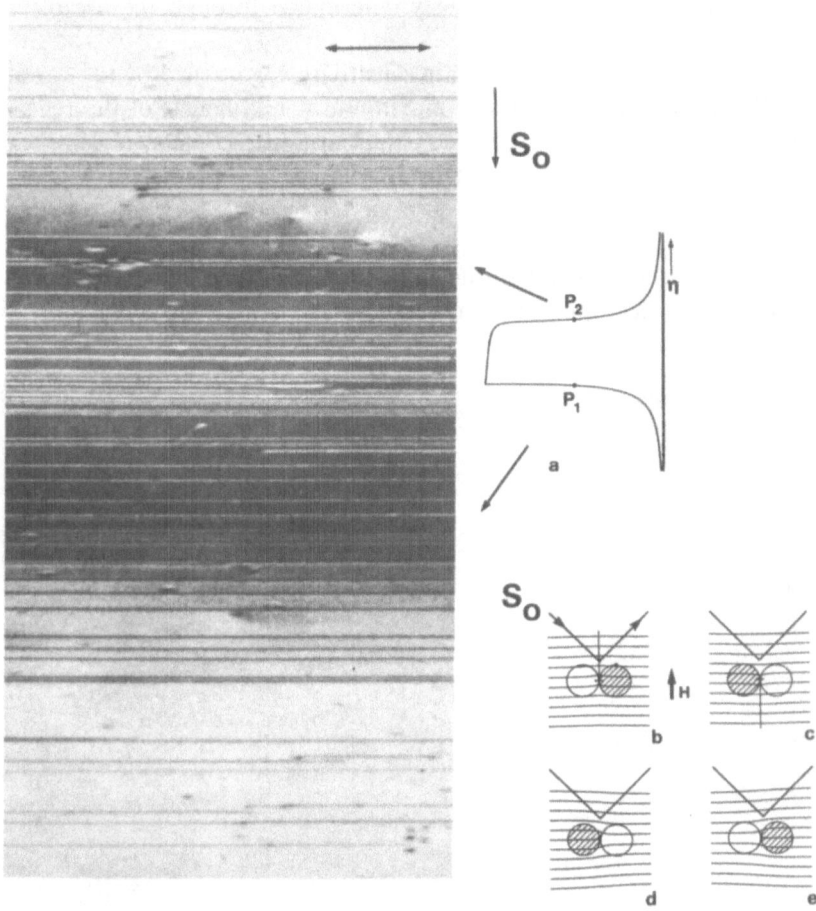

Figure 3.13. Epitaxial dislocations in $Ga_{1-x} Al_x As_{1-y} P_y$/Ga As heterojunctions (after [67]). Plane wave imaging at $\lambda = 1.2378$ Å. **Bragg geometry.** (a) 004 substrate reflection and schematic reflection profile. Scale mark, 200 μm. Orientation contrast for an edge dislocation: (b,c) reflection from planes parallel to the slip plane; (d,e) reflection from planes perpendicular to the Burgers vector $\delta(\Delta \theta) > 0$ (< 0) in the hatched (empty) circles.

towards the interior of the substrate, regions satisfying $\delta(\Delta\theta) > 0$ should appear black (white) when the incidence angle corresponds to $\eta(P_1)$ ($\eta(P_2)$). The contrary applies to regions satisfying $\delta(\Delta\theta) < 0$. Therefore, along the projection of the incident beam direction, S_0, dislocation images should appear black-white for $\eta(P_1)$ and white-black for $\eta(P_2)$, in agreement with the experimental observations. If reflection occurs off planes perpendicular to the Burgers vector (Figure 3.13(d, e)), similar arguments apply. However, regions satisfying $\delta(\Delta\theta) > 0$ or $\delta(\Delta\theta) < 0$ are now defined as the regions of tensile or compressive strain rather than misorientation. These simple arguments demonstrate that the sign of Burgers vectors of edge (and also screw dislocations) may be unambiguously determined through the use of plane wave imaging techniques.

In experiments of this kind, strain sensitivity is governed by the variation $dP/d\eta$ (or $dP/d\theta$). It is obviously maximum at mid-height of the reflection profile and higher in Bragg geometry than in Laue geometry (Figure 3.5). It has been shown [69, and loc. cit.] that the use of extremely asymmetrical reflections can increase the strain sensitivity up to $\sim 10^{-8}$, a really remarkable figure.

Figure 3.14 shows the various appearances of the same crystal as in Figure 3.9(a), when imaged at different incidence angles [70] (reflection

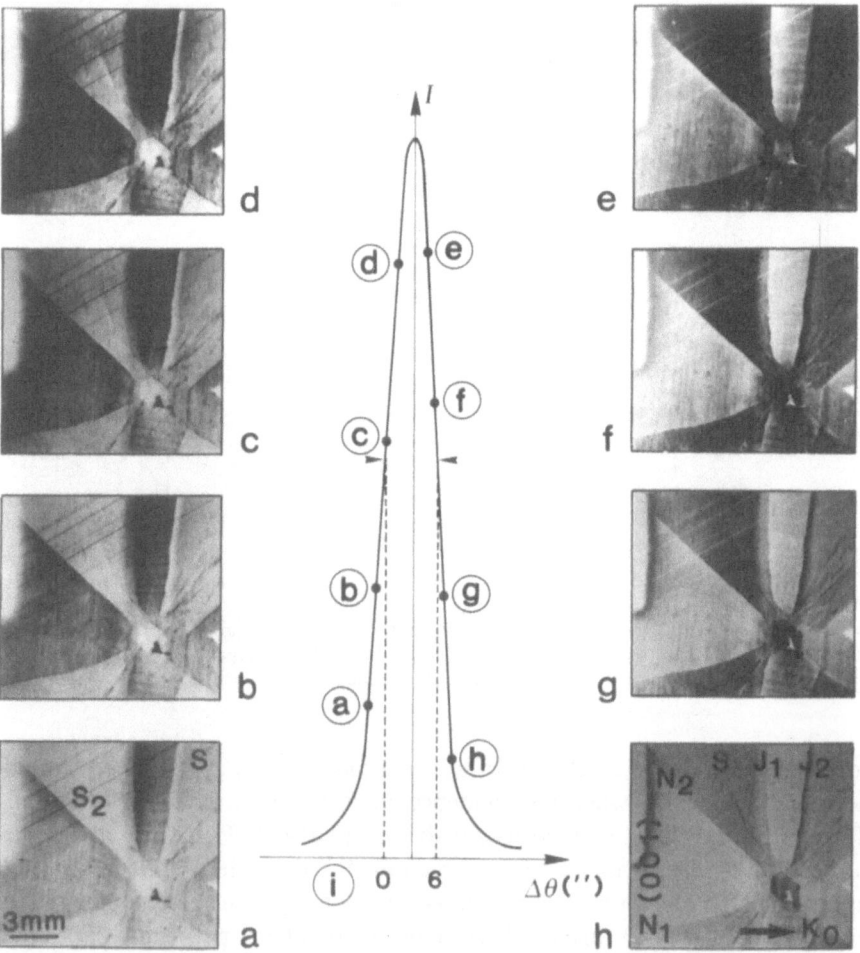

Figure 3.14. Plane wave topographs of the $Sr(NO_3)_2$ crystal shown in Figure 3.9, as a function of the angle of incidence (after [70]). Bragg geometry, $8\bar{8}0$ reflection at 1.2378 Å.

geometry, plane incident wave). Any growth sector, for instance, may appear black or white according to the value of η. It is straightforward to show that the use of two rotation axes in the plane of the sample allows for a full determination of the lattice strains and rotations of a crystal area with respect to a reference region of the same crystal. In the strontium nitrate crystal that was investigated, typical lattice parameter variation between growth sectors amount to $\sim 10^{-6}$ whereas rotations are of the order of 0.3″. Similarly, the relative lattice parameter variation in growth bands was estimated to reach $\sim 6 \times 10^{-6}$, a probably not uncommon figure. Finally, the inclined bands with black or white contrast in Figure 3.14 are locations in the crystal which suffered a prolonged exposure to a finely collimated X-ray beam: the lattice swelling associated with radiation induced damage proved to be about 10^{-5}.

It may be concluded from the previous example that plane wave imaging in reflection geometry allows for the quantitative determination of lattice strains and rotations at a spatial resolution only limited by detector performances. Although the first experiments of this type were performed with conventional X-ray sources [e.g. 71], synchrotron sources equipped with pseudo plane wave monochromators allow for such measurements to be performed on a fast and nearly routine basis.

3.7 From the analysis of one-dimensional strains to the precise location of impurity atoms

An important practical aspect of materials characterisation is the analysis of one-dimensional strains, as the result of e.g. epitaxial growth [72–75] or materials processing. As an example, ion implantation will be considered in this section. Similarly to radiation induced lattice damage, ion implantation damage at fairly low doses (10^{14}–10^{15} ions/cm^2 or less) may essentially be described as the formation of interstitials and vacancies in the host lattice resulting in lattice swelling.

Figure 3.15(a) shows the experimental and simulated rocking curves of a magnetic garnet epilayer deposited by liquid phase epitaxy on a non-magnetic garnet substrate (gallium gadolinium garnet, GGG) [76, and loc. cit.]. The epitaxial layer has been implanted with neon ions at energy 200 keV and dose 2×10^{14}/cm^2. Here, contrary to the case of the epitaxial system considered in Sections 3.5 and 3.6, the lattice parameter of the epitaxial layer is slightly smaller than that of the substrate; hence, the substrate reflection is recorded at a glancing angle smaller than that of the epitaxial layer.

At the scale of Figure 3.15(a), only minute differences separate the implanted from similar but not implanted epitaxial systems. On top of these rather conventional features, a close inspection of the rocking curves indicates, however, that some extra diffracted intensity may be recorded on the side of the low incidence angles. This extra intensity amounts to less than 1% of the

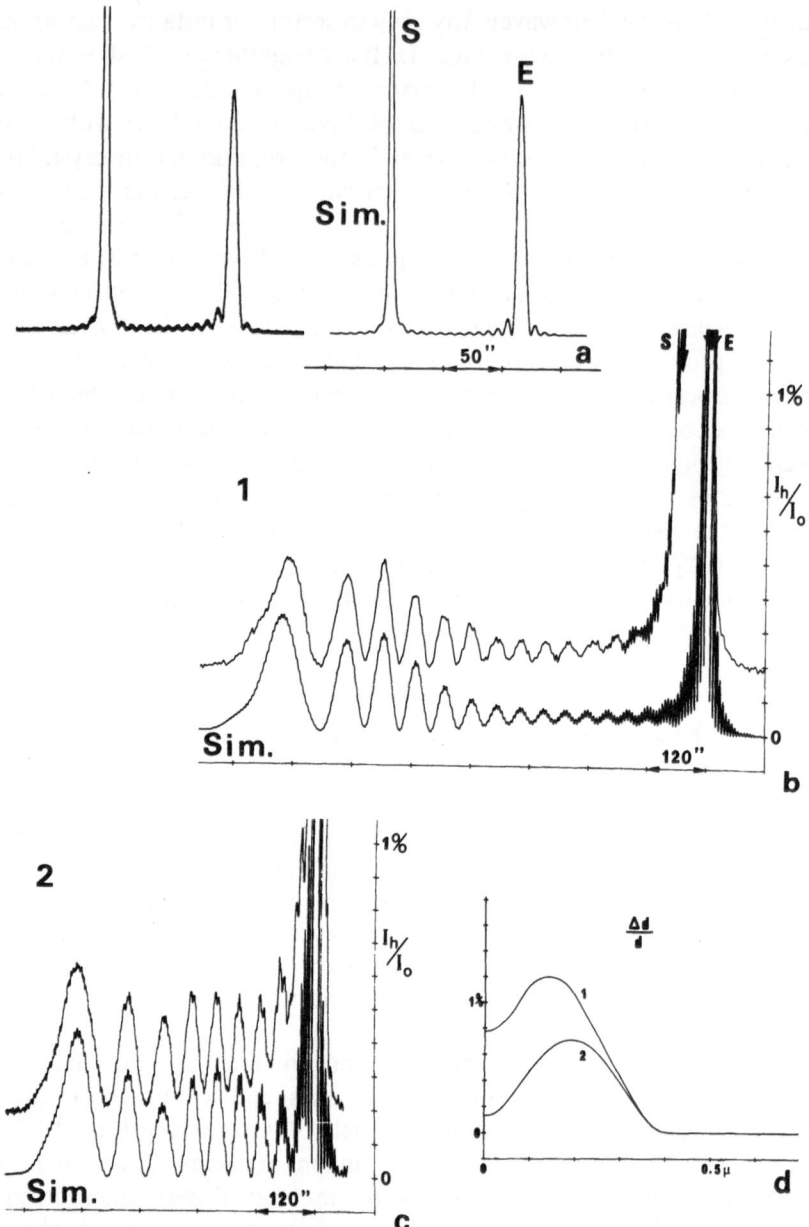

Figure 3.15. Experimental and simulated (sim.) rocking curves of a neon implanted garnet epilayer on GGG.111 orientation. Bragg geometry. $\lambda = 1.2378$ Å. (a) Large scale, E, S epitaxial and substrate peaks, respectively. 888 reflection; (b) small scale, 444 reflection; (c) after a 1 h anneal at 400°C; (d) lattice expansion normal to the surface.

main peaks intensity (Figure 3.15(b, c)). The simulation of rocking curves then allows for a precise determination of the lattice expansion in the direction normal to the epilayer surface (Figure 3.15(d)). It ought to be mentioned that the strain determination proves extremely precise. In the above example, the determination of lattice expansion was a prerequisite to the computation of the strain distribution around implantation boundaries. Plane wave imaging was subsequently used to monitor these strains and a good agreement between theory and experimental observations was achieved [76].

Rocking curve analysis has also allowed the determination of the temperature distribution as a function of time in ion-implanted silicon after the application of annealing laser pulses [77, 78]. These experiments provide the first example in this review of the use of the time structure of synchrotron X-ray beams. Stroboscopic imaging has also been developed [e.g. 79].

Finally, a few words must be said about standing wave experiments [80, 81]. Equations (3.6) and (3.7) readily indicate that a standing wave pattern

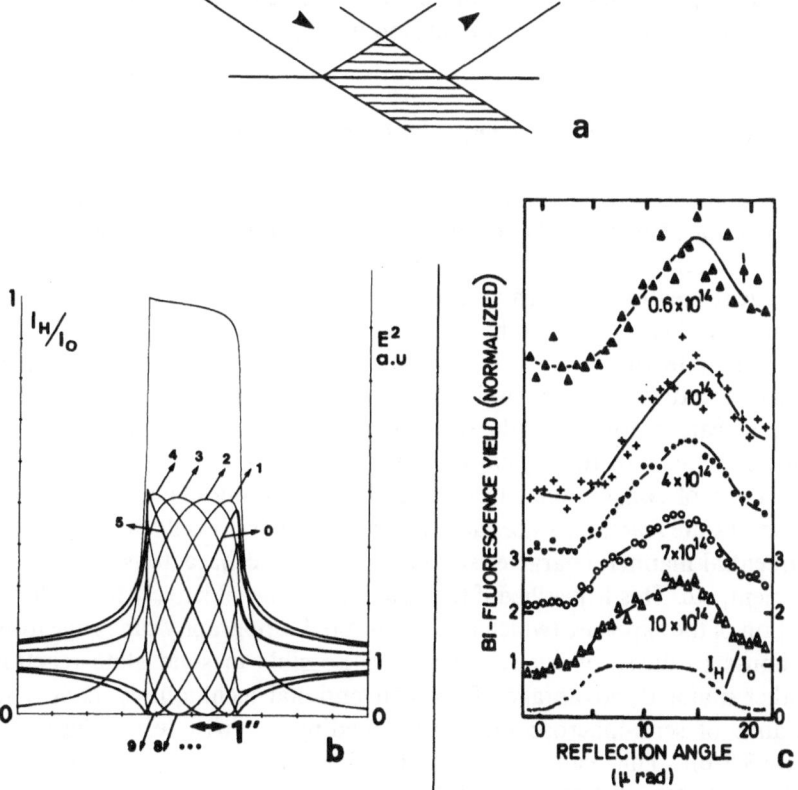

Figure 3.16. (a) Standing waves (Bragg geometry); (b) reflecting power and distribution of the electric vector intensity. Silicon, 220 reflection at $\lambda = 0.8$ Å. Curves $n = 0$–9 refer to distances $n/10$ from a reflecting lattice plane close to the free surface; (c) fluorescence yield of a Bi implanted silicon crystal as a function of dose (after [90]). 220 reflection at ~ 0.8 Å (1μrad $\simeq 0.2''$).

develops in a crystalline medium at Bragg reflection (see Section 3.2). In the Bragg case, the standing wave pattern exists both inside the crystal, down to the penetration depth, and outside the crystal as shown in Figure 3.16(a).

Figure 3.16(b) shows the distribution of the electric vector amplitude as a function of η and the distance between two reflecting planes (silicon, 220 reflection). If a foreign atom is located at a substitutional position in the lattice, its fluorescence signal as a function of η would follow the curve labelled 0. If it sits in between two reflecting planes, then the curve labelled 5 should be followed. Finally if the distribution is random, the fluorescence yield should be similar to the rocking curve profile. Since the standing wave pattern extends outside the crystal, adsorbates on surfaces and interfaces may also be studied [82–84]. For more details, the reader should refer to Refs. 85–89 (and loc. cit.). A sole example of the application of the standing wave technique is given below: it deals again with ion implantation in silicon at low energies [90]. After a suitable anneal, Figure 3.16(c) indicates that most of the bismuth ions are located in substitutional positions at low doses whereas the substitutional fraction drops to about 0.4 at the highest dose considered. This extremely elegant technique is expected to develop widely.

3.8 Applications of X-ray topography to the study of solid state reactions

Conventional Lang topography has been used by several groups to study solid state reactions [91–93]. The technique can be used to characterise the initial defect structure in the reactive crystals, and following this the influence of subsequent treatment (thermal, radiolytic, photolytic) on the solid can be examined in detail. This can be achieved via the monitoring of the development of the substructure at various stages in the reaction and the correlation of this substructure with the well-defined initial structure. In this way topography can be utilised for the study of the influence of defects on reactivity.

However, as with the crystal growth studies, limitations can arise in the application of conventional techniques in conducting experiments dealing with solid state reactions since such studies can generally only be conducted in a sequential manner; characterisation, treatment, recharacterisation, further treatment, etc. This is justified if the reaction can be effectively frozen for the duration of the exposure (which in conventional topography can range from a few hours to days), which is not always possible in solid state reactions. Another major disadvantage of the conventional technique is its very low tolerance of semi-macroscopic lattice bending or twist (e.g. subgrains in Figure 3.9(b)). This can become particularly problematic in solid state reactions, where reaction induced strains, or sometimes strains present in the crystals prior to reaction, can lead to the formation of Bragg contours, i.e. prevent imaging of sufficiently large volumes of crystal.

The drawbacks deriving from the poor tolerance of lattice distortion, as well as those potentially deriving from the relatively long exposure times can be effectively eliminated by utilising white beam synchrotron topography [4,6,17]. Let us first, however, describe the conventional topographic work of Begg *et al.* [93] on the thermal decomposition reaction of sodium chlorate single crystals. Although the strain induced during this reaction prevents monitoring of reaction beyond about 1% conversion, the information obtained in the early stages of reaction serves as an excellent example of the applicability of the general topographic technique in problems of this nature.

The thermal decomposition reaction of sodium chlorate can be considered to be single crystal to polycrystal plus gas, or single crystal to amorphous (crystallite size too small to yield diffraction peaks) plus gas, and can be modelled using a classical nucleation and growth mechanism. In their study, Begg *et al.* found no evidence for preferential reaction occurring at dislocations, rather evidence for preferential reaction occurring at 'strained volumes' in the crystal was obtained. This evidence initially consisted of the modification of the strain field associated with the original strained volumes, discernible as a change in contrast. These changes are illustrated in Figure 3.17, which shows a series of Lang topographs recorded before reaction (a), and at various stages after the onset of thermal decomposition (b)–(d). In Figure 3.17(a) growth dislocation images propagating from the seed crystal to the main growth faces can be seen. Also visible in Figure 3.17(a) are contrast features characteristic of inclusions, which are the 'strained volumes' referred to earlier. The nature of the original contrast from these strained volumes, indicated by the letter 'I' on Figure 3.17(a) (black/white lobe contrast with the the lighter lobe on the side of positive **H** vector), suggests that the 'inclusion' or 'precipitate' responsible for the strain, occupies a smaller volume than the space provided for it by the crystal [94]. In other words the surrounding lattice is under tension. It was speculated that the original strain centres could be impurity centres or vacancy clusters. Another possibility which was not considered, was that these strain centres could be germ nuclei which were present in the crystal initially [95], following localised decomposition which may have occurred at low temperature before the decomposition proper, or following decomposition during heating to the proposed decomposition temperature. This possibility would perhaps be just as feasible as the original speculations, since these too would have the correct kind of strain field associated with them.

Following the initial contrast change, observations indicated that as reaction progresses, dislocation loops are punched into the lattice surrounding the strained volumes (note the formation of cross-like patterns which progressively grow in size, e.g. X on Figures 3.17(b)–(d)), as the strain energy associated with the nucleation processes occurring in the vicinity of the interfaces of the germ nuclei is relaxed via plastic deformation, which becomes possible at the elevated reaction temperature (sodium chlorate is brittle at room temperature).

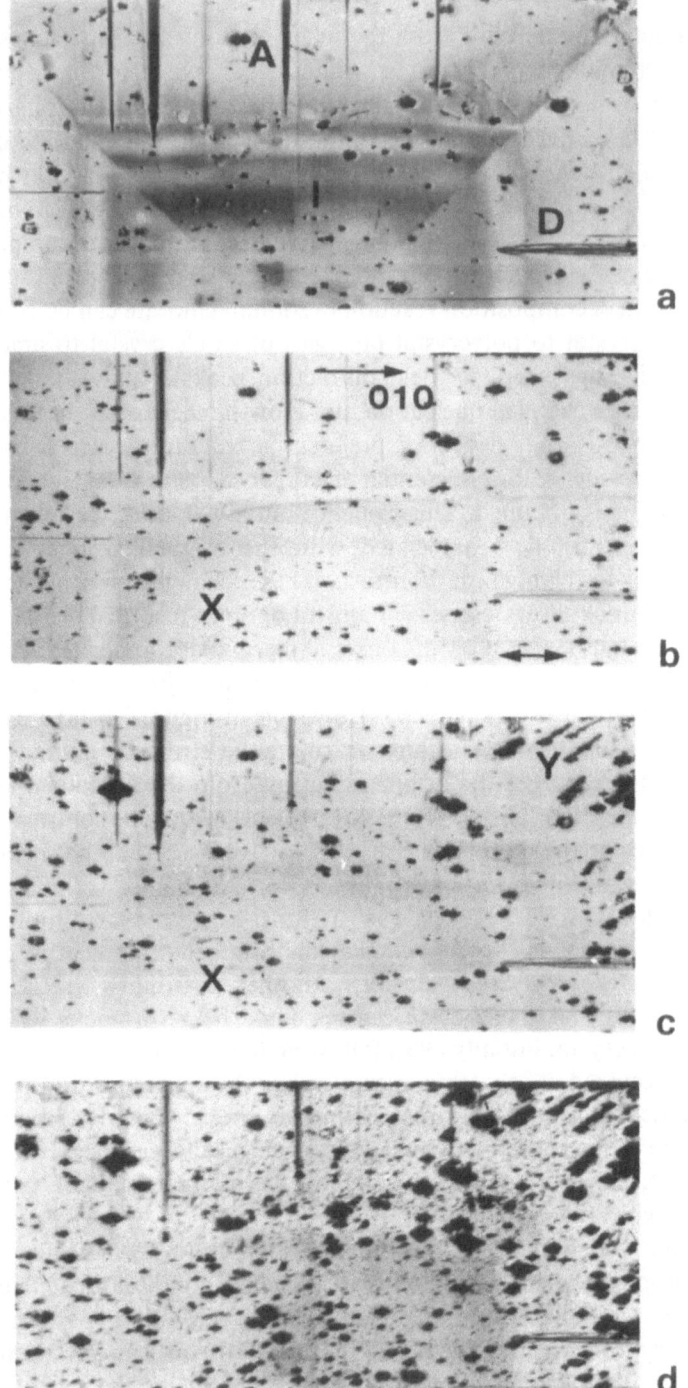

Figure 3.17. Lang topographs of a sodium chlorate single crystal. 020 reflection, Ag Kα radiation (after [93]). (a) Room temperature; (b) 1 h heating at 420 K; (c) 1 h at 500 K; (d) 1 h at 523 K. A, inactive strain centre; I, active strain centres; D, dislocations. Scale mark, 1 mm.

In this case it appears that the volumes surrounding the precipitates are preferred to those surrounding dislocations, as far as promoting preferential reaction is concerned. If, as suggested earlier, these precipitates are in fact germ nuclei, then interfacial energy considerations would predict that the reaction should occur preferentially at the interface between reacted and unreacted volumes, and this would explain the observed effects. On the other hand strain energy considerations would predict that reaction would be more difficult in the dilated regions surrounding the germ nucleus, since compressive rather than tensile stresses would be required to relax reaction induced stresses. In addition, the generation of gas during the reaction must also be considered. Overall, the combination of favourable interfacial energy considerations, the possibility of plastic deformation to relax the stresses generated during the solid to solid part of the reaction, and the ready accommodation of gas in the dilated regions surrounding the germ nuclei, lead to conditions conducive to nucleation in the vicinity of these sites in the crystal. This analysis, coupled with the observation that the growth dislocations do not appear to become involved in the reaction, may provide evidence that in this particular reaction, strain energy developed due to the volume changes incurred in the solid components is not the all-important factor in determining sites for nucleation of reaction, but that a combination of several factors comes into play.

To illustrate the applicability of white beam synchrotron topography to the study of solid state reactions let us now consider the solid state polymerisation of the diacetylene PTS [96–100]. This polymerisation reaction can be effected using UV radiation, X-radiation, gamma radiation, heat, and mechanical stress. In an earlier optical microscopic study [101], dislocations were found not to influence reactivity in the radiation induced reaction, although they were found to play a role in the thermally induced reaction. This can be explained in terms of the activation barriers to reaction. Generally speaking the solid state polymerisation reaction in PTS can be divided into two basic steps: chain initiation and chain propagation. The barrier to the former of these steps is approximately an order of magnitude larger than that to the latter. It has been shown that the measured activation energies to thermally induced polymeris- ation are primarily associated with chain initiation, whereas the measured barriers to the photo-induced reaction are primarily associated with chain propagation [102, 103]. This can be explained in terms of the photon energies, in the photo-induced case, being much larger than the barrier to chain initiation, so that in this case the effective barrier is that to chain propagation. Once reaction has been initiated at room temperature, chain propagation is then readily achieved, so that dislocations do not have the opportunity to become involved. However, in the thermal case, the relaxation of reaction induced stress at dislocations can effectively lower the barrier to chain initiation, since reaction induced stress is a necessary contributor to any activation barrier. In this way, local reaction kinetics can be influenced.

Since polymerisation can be effected by X-irradiation, it would appear at first sight that X-ray topography could not be used as a non-destructive technique to monitor strain fields generated during reaction. However, the potential problem of unwanted X-ray induced polymerisation can be effectively eliminated by filtering the incident synchrotron white beam to remove the more highly absorbed, and therefore more damaging (i.e. more effective in inducing reaction), longer wavelength components. The resultant incident beam spectrum is still sufficiently broad to be able to record a partial Laue pattern, although the removal of the longer wavelength components inhibits X-ray induced reaction sufficiently that the crystal can be exposed to the filtered X-ray beam for several hours without introducing any unwanted radiation induced effects.

Should preferential reaction occur around dislocations, a localised strain field is expected to be set up due to the volume change which occurs upon polymerisation. The long range effects of this strain field are expected to give rise to topographic contrast. No nucleation as such is expected, merely a volume in which reaction is more advanced than neighbouring regions, i.e. a more dense solid solution, since polymer forms a solid solution in monomer. Eventually the long range strain field is expected to anneal out as reaction in the bulk equilibrates with that in the vicinity of dislocations, or perhaps via other mechanisms such as crack propagation. On the other hand if no preferential reaction occurs then no such localised strain fields should be induced, and therefore no such contrast within Laue spots should be discernible. The Laue pattern of the monomer structure would then be expected to distort gradually into that corresponding to the polymer structure.

Results obtained comply with the above models. For example in the X-ray induced reaction, no contrast changes are discernible within the Laue spots during reaction. The Laue pattern of the monomer distorts in a continuous manner into that of the polymer, with the exact defect structure present in the monomer being frozen into the polymer crystal. An example of this defect

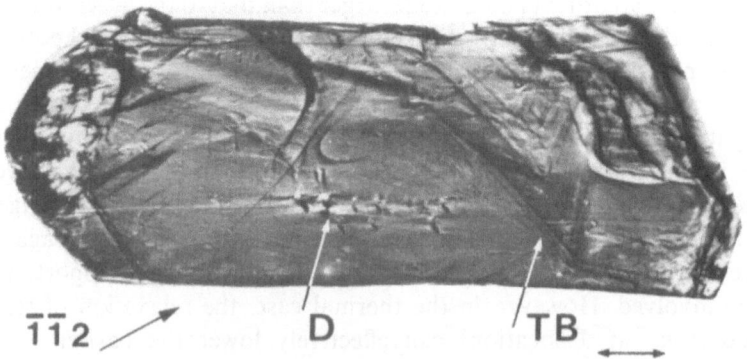

Figure 3.18. White beam topograph of a PTS crystal polymerised in the X-ray beam. $\bar{1}\bar{1}2$ reflection at $\lambda \sim 1$ Å. D, dislocations; TB, twin boundary. Scale mark, 1 mm.

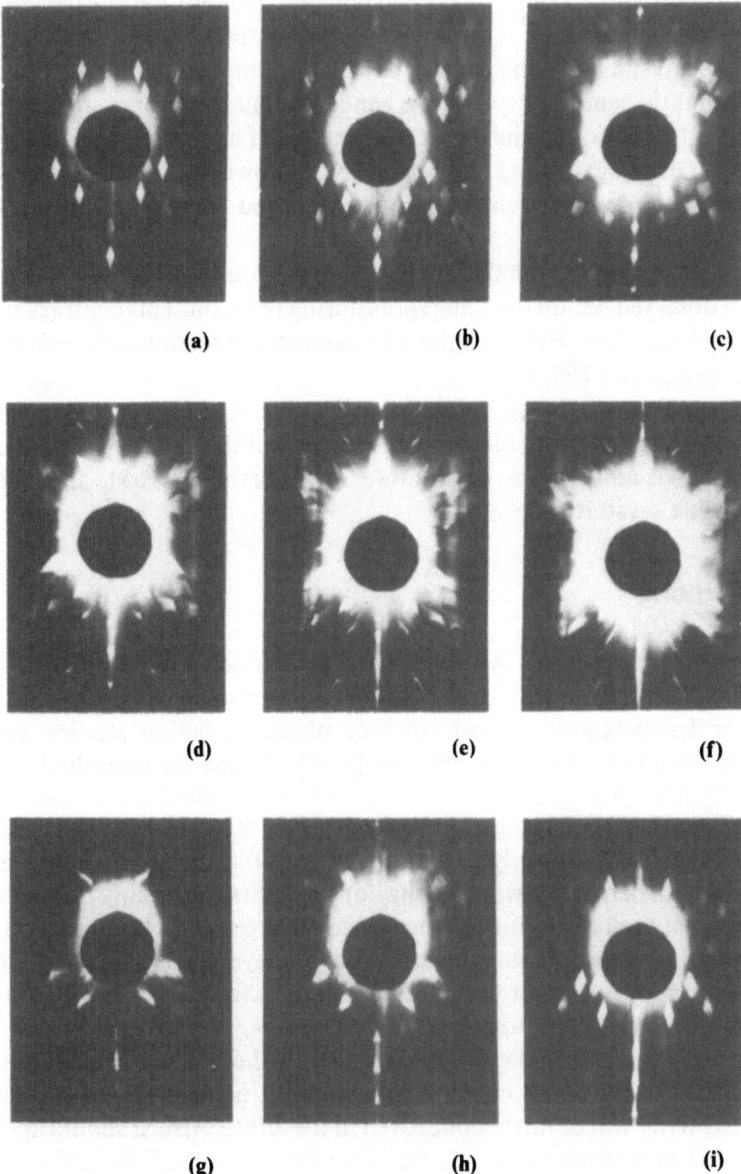

Figure 3.19. Laue patterns of a PTS monomer crystal undergoing X-ray induced polymerisation. Cumulative X-ray doses, (a)–(i) 0.46, 1.02, 2.50, 3.19, 3.49, 4.09, 4.39, 5.41, 11.63 M rad.

structure is shown in Figure 3.18, recorded from a polymer crystal, polymerised in the X-ray beam. This defect structure was identical to that observed
in the original monomer crystal [96]. An example of a series of Laue patterns
recorded during X-ray induced reaction is shown in Figure 3.19. The extra
asterism visible in the figure arises from the slight residual non-uniformity of

the absorption profile of the X-rays as a function of depth, despite the filtering, leading to a slightly faster reaction rate on the entrance surface than the exit surface. As a result, a strain field ensues which, in the case of filtered radiation, tends to slightly bend the crystal (this bending is much more severe in the case of unfiltered radiation). This strain manifests itself as lattice curvature which shows up as asterism on the Laue pattern. Note, however, that microstructural information is preserved on topographs recorded throughout most of the reaction.

On the other hand in the thermally induced reaction, well-defined contrast has been observed within the Laue spots during reaction. This contrast, due to preferential reaction in the vicinity of clustered dislocations, anneals out as reaction progresses [99, 100].

Other synchrotron based studies of solid state reactions include investigations of decomposition reactions in nickel sulphate hexahydrate single crystals [104], ammonium perchlorate single crystals [105], and sodium nitrate single crystals [106].

3.9   Conclusion

Numerous applications of X-ray imaging techniques have been necessarily omitted in this short review. Amongst these, plastic deformation and domain (ferro- or ferrimagnetic, ferroelectric) or phase transition studies deserve particular mention. Previous reviews [4–8] should be consulted for an introduction to these topics.

The authors have tried to avoid dissociating conventional topographic imaging techniques (mainly the Lang method) from synchrotron based experiments. There are two reasons for this: first, in many instances, a continuity between conventional and synchrotron based imaging experiments appears obvious. Second, many of the concepts developed in order to explain contrast mechanisms in for example the Lang technique also apply to white beam synchrotron topography.

The main aim of this paper was to demonstrate the extreme strain sensitivity of diffraction based X-ray imaging techniques. It is most probable that this strain sensitivity will be further improved in the future. Also, it should again be emphasised that plane wave imaging, rocking curve analysis and standing wave experiments are all quantitative techniques. In this area, and in the field of real time experiments, the advent of synchrotron sources has proved to be a major step forward.

Acknowledgement

The authors are deeply indebted to their colleagues for allowing them to include some of their results in this review.

References

1. A. Guinier and J. Tennevin, *Acta Crystallogr.* **2** (1949) 133.
2. *Modern Diffraction and Imaging Techniques in Material Science*, eds. S. Amelincks, R. Gevers, G. Remaut and J. Van Landuyt, North-Holland, Amsterdam (1970).
3. Lecture notes of the International Summer School on X-ray Dynamical Theory and Topography, Limoges, France (1975).
4. *Characterization of Crystal Growth Defects by X-ray Methods*, eds. B.K. Tanner and D.K. Bowen, Plenum Press, New York (1980).
5. M. Sauvage and J.F. Petroff, in *Synchrotron Radiation Research*, eds. H. Winick and S. Doniach, Plenum Press, New York (1980).
6. *Applications of X-ray Topographic methods to Materials Science*, eds. S. Weissmann, F. Balibar and J.F. Petroff, Plenum Press, New York (1984).
7. M. Sauvage-Simkin, in *Defects in Solids*, eds. A.V. Chadwick and M. Terenzi, Plenum Press, New York (1986).
8. C. Malgrange, Lecture notes, Aussois Winter School on Synchrotron Radiation, CNRS (1986).
9. W.H. Zachariasen, *Theory of X-ray Diffraction in Crystals*, John Wiley (1945) and Dover Publications (1967).
10. R.W. James, *The Optical Principles of the Diffraction of X-rays*, Bell and Sons (1948).
11. M. Hart, in *Characterization of Crystal Growth Defects by X-ray methods*, eds. B.K. Tanner and D.K. Bowen, Plenum Press, New York (1980) pp. 216–263.
12. G. Borrmann, *Z. Phys.* **42** (1941) 157.
13. G. Borrmann, *Z. Phys.* **127** (1950) 297.
14. A.R. Lang, *Acta Crystallogr.* **12** (1959) 249.
15. N. Kato, *Acta Crystallogr.* **14** (1961) 526, 627.
16. T. Tuomi, K. Naukkarinen and P. Rabe, *Phys. Status Solidi A* **25** (1974) 93.
17. M. Hart, *J. Appl. Crystallogr.* **8** (1975) 436.
18. J.W.M. DuMond, *Phys. Rev.* **35** (1937) 872.
19. N. Kato, *J. Phys. Soc. Jpn.* **18** (1963) 1785; **19** (1964) 67; **19** (1964) 971.
20a. H. Hashizume and K. Kohra, *J. Phys. Soc. Jpn.* **31** (1971) 204.
20b. H. Hashizume, *J. Phys. Soc. Jpn.* **31** (1971) 1124.
21. A. Authier, *J. Phys.* **27** (1966) 57.
22. Y. Ando and N. Kato, *J. Appl. Crystallogr.* **3** (1970) 74.
23. N. Kato and J.R. Patel, *J. Appl. Phys.* **44** (1973) 965.
24. J.R. Patel and N. Kato, *J. Appl. Phys.* **44** (1973) 971.
25. J. Miltat and M. Kleman, *Philos. Mag.* **28** (1973) 1015.
26. J. Miltat, *Philos. Mag.* **33** (1976) 225.
27. A. Authier and F. Balibar, *Acta Crystallogr.* **A26** (1970) 647.
28. F. Balibar and A. Authier, *Phys. Status Solidi* **21** (1967) 413.
29. A. Authier, F. Balibar and Y. Epelboin, *Phys. Status Solidi* **41** (1970) 22.
30. A. Authier and Y. Epelboin, in *Applications of X-Ray Topographic Methods to Materials Science*, eds. S. Weissmann, F. Balibar and J.F. Petroff, Plenum Press, New York (1984) pp. 111–23.
31. C. Malgrange and F. Balibar in *Applications of X-ray Topographic Methods to Materials Science*, eds. S. Weissmann, F. Balibar and J.F. Petroff, Plenum Press, New York (1984) pp. 111–123.
32. A. Authier, in *Modern Diffraction and Imaging Techniques in Material Science*, eds. S. Amelincks, R. Gevers, G. Remaut and J. van Landuyt, North-Holland, Amsterdam (1970).
33. J. Miltat and D.K. Bowen, *J. Appl. Crystallogr.* **8** (1975) 657.
34. K. Kohra, H. Hashizume and J.I. Yoshimura, *Jpn. J. Appl. Phys.* **9** (1970) 1029.
35. M. Hart, in *Characterization of Crystal Growth Defects by X-Ray Methods*, eds. B.K. Janner and D.K. Bowen, Plenum Press, New York (1980).
36. T. Matsushita and H. Hashizume, in *Handbook of Synchrotron Radiation*, Vol. 1, ed. E.E. Koch, North-Holland, Amsterdam (1983) pp. 261–314.
37. T. Bedynska, *Phys. Status Solidi A* **18** (1973) 147.
38. T. Bedynska, R. Bubakova and Z. Sourek, *Phys. Status Solidi A* **36** (1976) 509.

39.   J. Gronkowski, *Phys. Status Solidi A* **57** (1980) 105.
40a. J. Gronkowski and C. Malgrange, *Acta Crystallogr.* **A40** (1984) 507.
40b. J. Gronkowski and C. Malgrange, *Acta Crystallogr.* **A40** (1984) 515.
41.   H. Klapper, in *Characterization of Crystal Growth Defects by X -Ray Methods*, eds. B.K. Tanner and D.K. Bowen, Plenum Press, New York (1980).
42.   F. Lefaucheux, M.C. Robert and E. Manghi, *J. Cryst. Growth* **56** (1982) 141.
43.   H.L. Bhat, K.J. Roberts and J.N. Sherwood, *J. Appl. Crystallogr.* **16** (1983) 390.
44.   S. Gits, M.C. Robert and F. Lefaucheux, *J. Cryst. Growth* **71** (1985) 203.
45.   N. Herres and A.R. Lang, *J. Appl. Crystallogr.* **16** (1983) 47.
46.   G. Champier, in *Characterization of Crystal growth Defects by X -Ray Methods*, eds. B.K. Tanner and D.K. Bowen, Plenum Press, New York (1980) pp. 97–132.
47.   D.T.J. Hurle and B. Cockayne, in *Characterization of Crystal growth Defects by X -Ray Methods*, eds. B.K. Tanner and D.K. Bowen, Plenum Press, New York (1980) pp. 46–72.
48.   F. Lefaucheux, M.C. Robert and A. Authier, *J. Cryst Growth* **19** (1973) 329.
49.   S. Gits-Leon, F. Lefaucheux and M.C. Robert, *J. Cryst Growth* **44** (1978) 345.
50.   J.N. Sherwood and T. Shripathi, *J. Cryst. Growth* **88** (1988) 358.
51.   M.C. Robert and F. Lefaucheux, *J. Cryst. Growth* **65** (1983) 637.
52.   J. Gastaldi and C. Jourdan, *J. Cryst. Growth* **52** (1981) 949; **54** (1981) 361.
53.   J. Gastaldi and C. Jourdan, *Philos. Mag.* **A50** (1984) 309.
54.   O. Nittono, T. Ogawa, S.K. Gong and S. Nagakura, *Jpn. J. Appl. Phys.* **23** (1984) L 581.
55.   G. Grange, J. Gastaldi and C. Jourdan, *J. Appl. Phys.* **62** (1987) 1202.
56.   G. Grange, J. Gastaldi and C. Jourdan, *J. Cryst. Growth* **87** (1988) 325.
57.   G.R. Fisher and P. Barnes, *J. Appl. Crystallogr* **17** (1984) 231.
58.   F.C. Frank, *Philos. Mag.* **A56** (1987) 263.
59.   S. Mardix, A.R. Lang, G. Kowalski and A.P.W. Makepeace, *Philos. Mag.* **A56** (1987) 251.
60.   J. Miltat and M. Dudley, *J. Appl. Crystallogr* **13** (1980) 555.
61.   J.F. Petroff and M. Sauvage, *J. Cryst. Growth* **43** (1978) 628.
62.   A.R. Lang and M. Polcarová, *Proc. R. Soc.* **A285** (1965) 297.
63.   E. Dunia, C. Malgrange and J.F. Petroff, *Philos. Mag.* **A41** (1980) 291.
64.   J.I. Chikawa, *J. Appl. Phys.* **36** (1965) 3496.
65.   B.K. Tanner, D. Midgley and M. Safa, *J. Appl. Crystallogr.* **10** (1977) 281.
66.   A.R. Lang, *Z. Naturforsch, A* **20** (1965) 636.
67.   J.F. Petroff, M. Sauvage, P. Riglet and H. Hashizume, *Philos. Mag.* **A42** (1980) 319.
68.   P. Riglet, M. Sauvage, J.F. Petroff and Y. Epelboin, *Philos. Mag.* **A42** (1980) 339.
69.   U. Bonse and I. Hartmann, *Z. Kristallogr.* **156** (1981) 265.
70.   M.C. Robert, F. Lefaucheux, M. Sauvage and M. Ribet, *J. Cryst. Growth* **52** (1981) 976.
71.   J. Yoshimura, T. Miyazaki, T. Wada, K. Kohra, M. Hosaka, T. Ogawa and S. Taki, *J. Cryst. Growth* **46** (1979) 691.
72.   V.S. Speriosu and H.L. Glass, in *Applications of X-Ray Topographic Methods to Materials Science*, eds. S. Weissmann, F. Balibar and J.F. Petroff, Plenum Press, New York (1984) pp. 413–420.
73.   M. Sauvage-Simkin and J.F. Petroff, in *Applications of X -Ray Topographic Methods to Materials Science*, eds. S. Weissmann, F. Balibar and J.F. Petroff, Plenum Press, New York (1984) 421–434.
74.   S. Bensoussan, C. Malgrange and M. Sauvage-Simkin, *J. Appl. Crystallogr.* **20** (1987) 222.
75.   S. Bensoussan, C. Malgrange, M. Sauvage-Simkin, K. N'Guessan and P. Gibart, *J. Appl. Crystallogr.* **20** (1987) 230.
76.   J. Miltat, *Philos. Mag.* **A57** (1988) 685.
77.   B.C. Larson, C.W. White, T.S. Noggle and D.M. Mills, *Phys. Rev. Lett.* **48** (1982) 337.
78.   B.C. Larson, C.W. White, T.S. Noggle, J.F. Barhorst and D.M. Mills, *Appl. Phys. Lett.* **42** (1983) 282.
79.   H. Cerva and W. Graeff, *Phys. Status Solidi A* **82** (1984) 35; **87** (1985) 507; **93** (1986) K129.
80.   B.W. Battermann, *Phys. Rev.* **133** (1964) A759; *Phys. Rev. Lett.* **22**.(1969) 703.
81.   S. Kjaer Andersen, J.A. Golovchenko and G. Mair, *Phys. Rev. Lett.* **37** (1976) 1141.
82.   J.A. Golovchenko, J.R. Patel, D.R. Kaplan, P.L. Cowan and M.J. Bedzyk, *Phys. Rev. Lett.* **49** (1982) 560.
83.   P. Funke and G. Materlik, *Surf. Sci.* **188** (1987) 378.

84. A.E.M.J. Fisher, E. Vlieg, J.F. van der Veen, M. Clausnitzer and G. Materlik, *Phys. Rev. B* **36** (1987) 4769.
85. J.R. Patel and J.A. Golovchenko, *Phys. Rev. Lett.* **50** (1983) 1858.
86. G. Materlik, *Z. Phys. B* **61** (1985) 405.
87. M.J. Bedzyk, G. Materlik and M.V. Kovalchuk, *Phys. Rev. B* **30** (1984) 2453.
88. M.J. Bedzyk and G. Materlik, *Phys. Rev. B* **32** (1985) 6456.
89. A. Authier, *Acta Crystallogr.* **A42** (1986) 414.
90. N. Hertel, G. Materlik and J. Zegenhagen, *Z. Phys. B* **58** (1985) 199.
91. J.M. Schultz, *J. Mater. Sci.* **12** (1976) 2258.
92. R.G. Rosemeier, R.E. Green, Jr. and R.H. Baughman, *J. Appl. Phys.* **52** (1981) 7129.
93. I.D. Begg, P.J. Halfpenny, R.M. Hooper, R.S. Narang, K.J. Roberts and J.N. Sherwood, *Proc. R. Soc. London A* **386** (1983) 431.
94. E.S. Meieran and I.A. Blech, *J. Appl. Phys.* **36** (1965) 3162.
95. D.A. Young, in *Decomposition of Solids,* Pergamon, Oxford (1966) p. 11.
96. M. Dudley, J.N. Sherwood, D. Bloor and D. Ando, *J. Mater. Sci. Lett.* **1** (1982) 479.
97. M. Dudley, J.N. Sherwood, D. Bloor and D. Ando, *Mol. Cryst. Liq. Cryst.* **93** (1983) 223.
98. M. Dudley, J.N. Sherwood, D. Bloor and D. Ando, in *Polydiacetylenes,* eds. D. Bloor and R.R Chance, NATO ASI Series E (Applied Sciences), no. 102, Martinus Nijhoff, Dordrecht, (1985) p. 87.
99. M. Dudley, J.N. Sherwood and D. Bloor, *Proceedings of ACS, Division of Polymeric Materials: Science and Engineering,* **54** (1986) 426.
100. M. Dudley, in *Proceedings of the Materials Research Society, Symposium on Synchroton Radiation,* Fall 1988 MRS Meeting, Boston (1989) to appear.
101. W. Schermann, G. Wegner, J.O. Williams and J.M. Thomas, *J. Polym. Sci., Polym. Phys. Ed.* **13** (1975) 753.
102. H. Eckhardt, R.R. Chance and T. Prusik, in *Polydiacetylenes,* eds. D. Bloor and R.R. Chance, NATO ASI Series E (Applied Sciences), No. 102, Martinus Nijhoff, Dordrecht (1985).
103. H. Gross, W. Neumann and H. Sixl, *Chem. Phys. Lett.* **95** (1983) 584.
104. D.B. Sheen and J.N. Sherwood, *Chemistry in Britain* June (1986) 535.
105. H.L. Bhat, P.J. Herley, D.B. Sheen and J.N. Sherwood, in *Applications of X-Ray Topographic Methods to Materials Science,* eds. S. Weissmann, F. Balibar and J.F. Petroff, Plenum Press, New York (1984).
106. J.R. Laia, Ph.D Thesis, Stony Brook (1986).

4 Small angle X-ray scattering and the study of microemulsions

H.C. GERRITSEN and C. ROBERTUS

4.1 Introduction

Mixtures of oil and water are unstable. Due to their different nature phase separation occurs instantaneously. Addition of surface active agents, surfactants, can stabilise a water and oil mixture, resulting in the formation of a so-called emulsion. These systems are, on a macroscopic scale, homogeneous dispersions of two otherwise immiscible liquids [1]. Emulsions have been a focus of attention in colloid physics and chemistry, due to the great diversity in properties of emulsions and their wide application in the pharmaceutical, oil recovery and food industries [2]. A large range of different surfactants exists, leading to the existence of many different types of emulsions. The most well known are milk and mayonnaise, mixtures of water and oil or fatty acids stabilised by proteins.

Some water/oil/surfactant systems form a special type of emulsions, so-called microemulsions. A microemulsion system is a transparent, isotropic, thermodynamically stable fluid. On a microscopic scale, at low water-oil ratios it has been shown that the water is dispersed in small spherical droplets ($R = 5–100\,\text{Å}$) in the oil phase [3]. The droplets are small compared to the wavelength of visible light, so that the system appears transparent. These systems are especially suitable for study with small angle neutron or X-ray scattering techniques. These techniques provide information concerning shape, size, polydispersity and particle interactions in inhomogeneous systems.

In imaging techniques such as microscopy discussed in Chapter 12, real space information is realised by focussing optics to reconstruct directly the image of the sample structure. In scattering techniques the diffraction pattern is measured directly without focussing optics. A recorded diffraction pattern consists of intensities of scattered radiation, the square of the amplitude, as a function of position or scattering angle. A scattering experiment therefore does not yield any information on the phase of the scattered radiation. Due to this loss of phase information, only average correlations, or fluctuations, in the sample are obtainable.

The intensities in the interference pattern and the correlations in the sample are related by Fourier transformation resulting in a reciprocity relation between scattering angle and correlation lengths in the sample. The reciprocity can be readily identified in Bragg's Law where the first order reflection ($n = 1$) of radiation of wavelength λ, from atomic planes separated by a distance d, is observed at angles θ given by

$$n\lambda = 2d \sin(\theta) \tag{4.1}$$

Large d spacings induce reflections at small θ. At the same time if d is fixed a decrease of the wavelength induces the scattered radiation to be observed at smaller angles. To observe reflections at finite angles the wavelength used should be of the same order of magnitude as the dimensions of the correlations in the sample. To study structures at atomic level, X-rays are used. Measurement of larger structures in principle needs radiation of longer wavelength. However, a large part of the electromagnetic spectrum is unsuitable for scattering techniques because of strong absorption. This is predominant between 10 and 1000 Å. Structures of these dimensions can therefore only be studied using X-rays scattered at small angles.

During the past two decades good progress has been made in the field of both theory and experiment of small angle X-ray scattering (SAXS). Excellent books on the subject are by Guinier and Fournet [4], Kratky and Glatter [5], Hosemann and Bagchi [6], Vainstein [7] and Kerker [8]. These books contain numerous results of a wide variety of experiments on biological and colloidal systems as well as the basic theory.

One of the experimental problems in SAXS is the large dynamic range of the scattering pattern. While the scattered intensity close to the primary beam can be very high, the intensity at high angles rapidly decays to very low values. This means that to obtain good statistics in the high angle region very powerful X-ray sources are required. Until the end of the 1970s only X-ray tubes, with their intrinsic moderate intensity, were available. This meant that unpractical long measuring times were required for moderately scattering samples and that experiments on very weak scatterers were virtually impossible.

When synchrotron radiation sources became available to experimentalists a great leap forward was made in small angle scattering. The high intensity of these sources allowed the investigation of a much broader range of samples. As was noted in Chapter 1, many dedicated synchrotron radiation facilities are available to scientists of many disciplines. For the SAXS community this means that even structures of the weakest scatterers can be examined routinely by means of SAXS using synchrotron radiation. Nowadays even time resolved studies with a time resolution in the millisecond region are possible, thus opening the way to dynamical studies.

In this chapter we do not give a complete review of the theory, applications, experimental practice and data interpretation of small angle X-ray scattering. This would be an impossible task to do in a single chapter of a book. For this we refer to the books mentioned above. We hope to illustrate in the next section some of the hurdles encountered in small angle scattering experiments and the way they are tackled using synchrotron radiation. The chapter is therefore divided into two sections. The first section gives a description of the hardware requirements when a synchrotron radiation source is used for small angle scattering experiments. In the latter section we present results of experiments, performed using the experimental facilities described in the first section, on Aerosol-OT (AOT) microemulsions. It has turned out that the AOT microemulsion is a useful example to illustrate the use of small angle X-ray scattering techniques to obtain structural information in liquid-like systems at physiological temperatures.

4.2 SAXS hardware requirements

In this section we describe the hardware requirements for a typical SAXS experiment using a synchrotron radiation source. Besides the large difference in intensity between conventional X-ray sources and synchrotron radiation sources there are other differences which dictate a different layout for the experiment. In practice, the synchrotron radiation source (tangent point) and the experiment are separated by a considerable distance of 10 m or more, this in contrast to conventional SAXS cameras such as the Kratky camera, where distances are less than 1 m. Other differences include the high total beamload on optical surfaces in case of synchrotron radiation sources and the differences in emission characteristics. Typically the vertical divergence of 1–2 Å synchrotron radiation is about 0.3 mrad. The horizontal divergence available to the experiments is in practice up to 40 mrad. Furthermore the source size is of the order of less than 1 mm vertically and several millimetres horizontally.

If we look at Bragg's law again we see that the scattering angle actually reflects the ratio λ/d. In principle this gives us the opportunity to match the wavelength to the dimensions we want to observe. However, in practice there are some severe limitations to the usable wavelength range and in the soft X-ray region $(\lambda > 10\,\text{Å})$ for instance, because of absorption one would be confined to samples with a thickness of the order of 1 μm, which is extremely impractical. Thus only shorter wavelength radiation, X-rays, are generally used for SAXS. On the other hand the scattering power of samples decreases for very short wavelengths. Other aspects that need to be considered are the source emission characteristics and the spectral response of the detector system. The optimum wavelength is usually 1–2 Å and often a wavelength of about 1.5 Å is the best compromise. This corresponds to the wavelength of the Cu-Kα emission line of conventional X-ray tubes.

In the ideal case the beam of photons would be perfectly collimated and infinitely small. Real sources, however, have finite dimensions and divergence. Consequently a collimation system is required. In its simplest form this can be a set of pinholes or slits but the price of using a simple collimation system like this is a great loss of intensity. This can be avoided by using focussing optics in combination with slits.

Focussing can only be obtained by reflection and/or diffraction optics (see Sections 1.3 and 1.4). Often a combination of two cylindrical elements is used, one being a mirror for the vertical focussing and the other a crystal used both for monochromatisation and horizontal focussing. This configuration confers the advantage of independent horizontal and vertical focussing. In practice, the cylindrical shape is obtained by bending flat elements. Although cylindrical optics will produce spherical aberrations this is hardly a problem in practice since the size of the image of the source is usually larger than the aberration level. An advantage of this configuration is that the focal length of the system can easily be adapted to the experimental requirements.

The bent mirror/monochromator arrangement has already been described in Section 1.4.1. In the context of SAXS experiments we will make some additional remarks specific to the technique.

4.2.1 *Monochromator*

The wavelength resolution requirements in SAXS are not very high. Usually a resolution $d\lambda/\lambda < 1\%$ is sufficient. A resolution of this order assures that no significant broadening of scattering peaks and overlap of scattering orders occurs. Since the intrinsic resolution of crystal monochromators is much better than 1%, one usually concentrates on maximising the total intensity rather than minimising the band pass when designing a monochromator system. Often crystals like Ge(111) or Si(111) are used. Some of the properties of these materials are summarised in Table 4.1. Obviously Ge(111) is better in terms of integrated reflecting power, however, it is much more difficult to produce high quality Ge single crystals of large enough dimensions. Due to crystal faults and impurities the efficiency of Ge monochromators is less than expected on theoretical grounds. Therefore the difference in integrated reflecting power between Ge and Si is less than expected from Table 4.1. Si has the additional advantage that it is produced in large quantities of high quality for the semiconductor industry. Consequently Si monochromators are much cheaper than Ge monochromators. Furthermore Ge produces fluorescence when radiation shorter than about 1.2 Å is used.

Table 4.1 The 2d spacing of Ge and Si, and the intrinsic Bragg reflection width, ω_o (see Section 1.4.1 and Eqn. 4.12) wavelength resolution, $d\lambda/\lambda$, and the integral reflecting power, I, of these materials at 1.5 Å.

Material	2d Å	W_s (arc s)	$d\lambda/\lambda$	I
Si(111)	6.271	7.395	1.41×10^{-4}	3.99×10^{-5}
Ge(111)	6.533	16.338	3.26×10^{-4}	8.59×10^{-5}

Often crystals are used which are asymmetrically cut (Chapter 1). Commonly, asymmetry angles α of up to 11° are used in order to compress the beam. This has the advantage that the beam width at the sample position is significantly reduced (Section 1.4.1), so that smaller samples can be used. The drawback of compression is that the divergence of the diffracted beam due to the intrinsic crystal divergence ω_1 is increased such that the beam size at the detector is larger than without compression (see Eqns. 1.12, 1.17).

Crystal monochromators, not only diffract the first order of the selected wavelength, but also higher harmonics. This may, depending on the intensity of the harmonics, be a significant problem. However the vertically focussing mirror can act as a low pass filter. This is accomplished by choosing a glancing angle such that the harmonics of the radiation to be used have a shorter wavelength than the critical wavelength ϕ_c (Eqn. 1.9).

4.2.2 *Collimation*

As we have pointed out above, a SAXS set-up requires collimation in order to produce a well-defined beam and a low background. One must bear in mind that the intensities that are measured in SAXS are typically 10^6 times lower than that of the primary beam. Therefore great care has to be taken in the design of the slit system to keep the background as low as possible. In Figure 4.1 an example of a SAXS beamline, based on beamline 8.2 at the Daresbury SRS, is given.

In front of the first optical element, a Ge(111) monochromator, a water cooled aperture is present. The purpose of this aperture is to reduce the beamload on the monochromator. Furthermore it reduces the amount of unused radiation. Any unused radiation will produce more scattering background and fluorescence than is strictly necessary, and should therefore be avoided.

The monochromator is asymmetrically cut with an asymmetry angle of 8°, triangularly shaped and 30 cm long. It is aligned such that radiation of 1.5 Å is selected (Bragg angle = 13.3°). The asymmetry angle of 8° compresses the

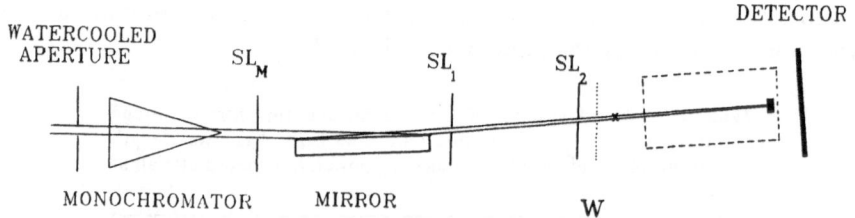

Figure 4.1.   An example of a small angle scattering beamline, including a watercooled aperture, a horizontally focussing Ge(111) monochromator, a quartz mirror for vertical focussing and a position sensitive detector. Sets of slits are marked SL, W is an X-ray window and the sample position is identified by a cross.

diffracted beam by a factor of 4 (Eqn. 1.16). Furthermore the monochromator is positioned 20 m from the synchrotron radiation source, which means that at as much as 5.4 mrad of radiation is accepted horizontally. The bending mechanism allows focussing over a wide range of image distances.

Behind the monochromator, slits are located to cut off scattering and fluorescence from the monochromator SL_M. Furthermore these are used to make sure that the X-ray beam does not hit the edges of the mirror, which would produce serious scattering. The uncoated silica mirror is located at 22.4 m from the source and consists of a piece of quartz 700 mm long, 90 mm wide and 40 mm thick. It operates at a glancing angle of about 4 mrad, which gives a vertical acceptance angle of 0.14 mrad. The mirror is focussed by bending it to the required radius of curvature to focus at the detector.

Close behind the mirror another set of slits SL_1 is located to cut down scattering from the previous elements. After SL_1 at least one more set, SL_2, is required in order to suppress the scattering from the collimation upstream. The pair of slits, SL_1 and SL_2, determine the opening angle for scattered radiation. In this example, the slit sets are about 5 m apart.

Up to this point the whole system is contained inside a vacuum. Next a window, W, is required as interface between the optics and slit section and the sample area, S. Since window materials also contribute to the background of the system great care has to be taken in the choice of window material. Mica is in practice a very suitable window material. Because of its crystalline structure it mainly produces scattering in well-defined directions at high angles. Thickness of down to $15 \mu m$ can be used such that absorption is low.

The necessary air path at the sample position should be kept as short as possible since air scatter also contributes to the background. For the most sensitive experiments the sample area can be flushed with helium to reduce this.

The next element is an evacuated pipe between sample and detector, marked by the dotted box in Figure 4.1. It has a variable length of 1–4 m and is evacuated in order to reduce air scatter. The window material at the entrance of the pipe can be mica again. Inside the pipe, close to the exit, a beamstop is located to prevent the primary beam from hitting the detector and also to avoid scattering from the exit window. Due to the large diameter of the pipe, several tens of centimeters, a different exit window material has to be chosen. In this case mylar or kapton of 0.1–0.2 mm thickness are good candidates.

If designed and aligned properly, the background of the system described above can be very low and is mainly determined by the opening angle for stray radiation as determined by the last two slit sets. The geometry of the last two slits, sample and detector are illustrated in more detail in Figure 4.2. The usable range of the small angle scattering setup is determined by the angle β, as seen from the sample, over which scattered light can be detected. In Figure 4.2, L_1, L_2 and L_3 are the distances between SL_1 and SL_2, SL_2 and S, and the

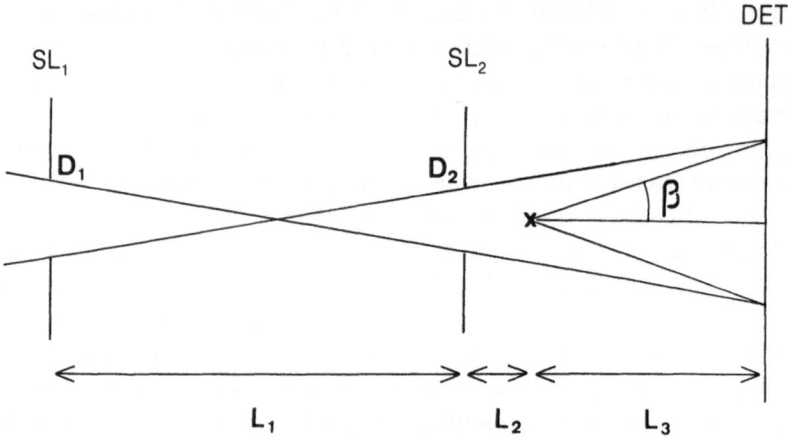

Figure 4.2. The opening angle β over which background scattering can be observed is determined by: the slit widths D_1 and D_2 of SL_1 and SL_2 respectively, the distance L_1 between the two slit sets, the distances L_2 between slit 2 and the sampled marked X and the sample to detector distance L_3.

sample S to detector (DET), respectively. The slit widths are denoted by D_1 and D_2. Using this notation the following expression can be derived for the angular region, β, over which scattering will be observed:

$$\beta = \text{atan}\left[\frac{1}{2L_3}\left(\frac{(D_1 + D_2)(L_1 + L_2 + L_3)}{L_1} - D_1 \right) \right] \qquad (4.2)$$

For a given fixed first slit to detector distance, $L_1 + L_2 + L_3$, optimum position for the second slit, SL_2, is at about two thirds of the total distance, L_1, measured from the first slit. When carefully aligned the above collimation setup can be used to record Bragg spacings of up to 2000 Å.

In this chapter, we do not go into detail about detector systems. In common with macromolecular diffraction, usually photographic film or linear wire-detectors are used in SAXS set-ups. These detectors are summarised in Section 1.5.4 with further details in Section 11.1.2. This chapter continues with an account of the application of SAXS to the study of microemulsions.

4.3 Experiments on AOT microemulsions

A sodium-bis(2-ethylhexyl)sulphosuccinate molecule, commonly known as Aerosol-OT or AOT, is made up of a sulphate headgroup and two hydrocarbon tails. The molecule belongs to the class of ionic surfactants. When dissolved in an apolar solvent the polar headgroups aggregate, with the hydrocarbon tails extended into the apolar phase, forming micelles. The micelle core forms a hydrophylic environment in which a polar solvent can be solubilised, forming a swollen micelle. In this way large amounts of two otherwise immiscible liquids, for example water and oil, can form a thermodynamic stable dispersion known as a microemulsion.

The average droplet size in a microemulsion is mainly determined by the area per surfactant molecule and the molar ratio of water to surfactant, W_0. The molar ratio of oil to surfactant, S_0, determines the droplet or microemulsion concentration. Variation of the water-oil ratio can induce dramatic changes in the macroscopic properties of the microemulsion, similar to critical phase transitions in one-component fluid systems. For example, a conductivity threshold and a maximum in the permittivity as a function of water concentration and/or temperature is observed in the AOT/H_2O/iso-octane microemulsion. The behaviour of the dielectric properties of this AOT/H_2O/iso-octane microemulsion can be well understood within the framework of percolation theory [10, 11].

A concentration increase can induce different types of microscopic structural transitions. The microemulsion can change from a droplet structure at low concentration to a bicontinuous structure with irregular regions of water interspersed in oil regions of similar shape [12, 13]. On the other hand the microemulsion may retain its droplet nature at higher concentrations where large aggregates of droplets exists [14–16]. The type of transition will most probably be governed by the geometry of the surfactant molecule and the way it is incorporated in the water-oil interface.

In this section we discuss two aspects of the small angle X-ray scattering from the AOT-microemulsion. The first part concerns the high angle region of the scattered intensity curve, classically known as the Porod region. This region contains information about variations in electron density over small distances. Importantly, fluctuations across the water-oil interface contribute to the scattered intensity. In contrast to the latter part of this section, no assumptions concerning the structure of the microemulsion on long-range length scales need to be made. Measurement of the high angle part of the scattering curve requires good signal-to-background ratio in the experiment as solvent scattering is of the same order of magnitude as the desired scattered intensities from the solutions studied. This demonstrates the need for a stable, high intensity source for small angle X-ray scattering made available by synchrotron radiation. The second part of this section discusses a model from liquid state theory enabling the description of the scattering from the microemulsion over a wide range of concentrations. Here the assumption is made that the microemulsion consists of water droplets surrounded by a monomolecular layer of surfactant molecules dispersed in the oil phase. The droplet model can give a quantitative description of the effects of polydispersity in size on the experimentally observed scattered intensity curves, enabling extraction of structural information from experimental data.

Throughout we will keep in mind the following: small angle scattering techniques have until now been the only techniques available to obtain directly structural information about the microemulsion system on the scale of 1–100 Å. Secondly, we note that the microemulsion is a comparatively simple system so that it can be expected that simple models will describe the observed scattering. In turn these models are instructive to outline the power and limitations of small angle scattering as a tool to extract structural parameters of more complicated systems.

4.3.1 High angle limit

X-rays are scattered at small angles by fluctuations in electron density of the medium. The fluctuations can be described using the distance correlation function [5, 17]:

$$\gamma(\mathbf{r}) = \langle \eta(\mathbf{x} + \mathbf{r})\eta(\mathbf{x}) \rangle / \overline{\eta^2} \qquad (4.3)$$

with $\eta(\mathbf{x}) = \rho(\mathbf{x}) - \bar{\rho}$, where $\rho(\mathbf{x})$ is the electron density at point $\mathbf{x}$, and $\overline{\eta^2}$ is the average of the squared electron density about some average value $\bar{\rho}$. For solutions and other centro-symmetric systems $\gamma(\mathbf{r})$ is a function of r only. If no long range order exists $\gamma(r)$ will decay to zero for large r. The scattered intensity, can then be given in terms of $\gamma(r)$ as

$$I(q) = I_e V \overline{\eta^2} \int 4\pi r^2 \gamma(r) \frac{\sin(qr)}{qr} \, dr \qquad (4.4)$$

with $q = (4\pi/\lambda)\sin(\theta/2)$, V the volume irradiated and

$$I_e = 7.9 \times 10^{-26} I_0 \frac{(\cos^2\theta + 1)}{2R_s^2}$$

where I_0 is the incident radiation intensity of wavelength λ, R_s the sample to detector distance and θ the scattering angle. $I_e V$ can be taken arbitrarily equal to unity. We note that $qI(q)$ and $r\gamma(r)$ is a sine Fourier pair.

Porod [18–20] has derived an expression for $I(q)$ in the limit of large q, known as 'Porod's Law', for a system composed of two phases with a difference in electron density

$$I(q) = 4\pi \overline{\eta^2} \frac{S}{q^4} \qquad (4.5)$$

with S the total surface between the phases. Equation 4.5 is, however, valid only for two phase systems with an ideal interface, at which the electron density changes discontinuously.

In the case of AOT microemulsions it is doubtful whether this condition is satisfied. The AOT-surfactant is made up of a SO_4 headgroup attached to two hydrocarbon chains. These chains will have approximately the same electron density as iso-octane and therefore not contribute to an electron density fluctuation. The sulphate polar headgroup, however, has a relatively high electron density compared to either water or iso-octane. There will thus be a strong variation in electron density across the oil-water interface over a distance of approximately 1–5 Å. Consequently we do not expect to find the q^4 decay of the scattering curve at high angles as described by Eqn. 4.5.

An expression for the scattered intensity at high q can nevertheless be obtained for a general pseudo two phase system. In this system arbitrary shaped regions of electron densities ρ_o and ρ_w are separated by a small layer of thickness ∂ and electron density ρ_s. See, for example Figure 4.7. $\gamma(r)$ can now be calculated from Eqn. 4.3 in an analogous way to that of Weigel [21]. From this it can be shown that the scattered intensity is given as

$$q^4 I(q) = b' \quad \text{for } 1/q < \partial \tag{4.6a}$$

$$q^4 I(q) = dq^2 + 2b \quad \text{for } \partial < 1/q < 2R \tag{4.6b}$$

with b', $2b$ and d directly proportional to S and functions of only the differences

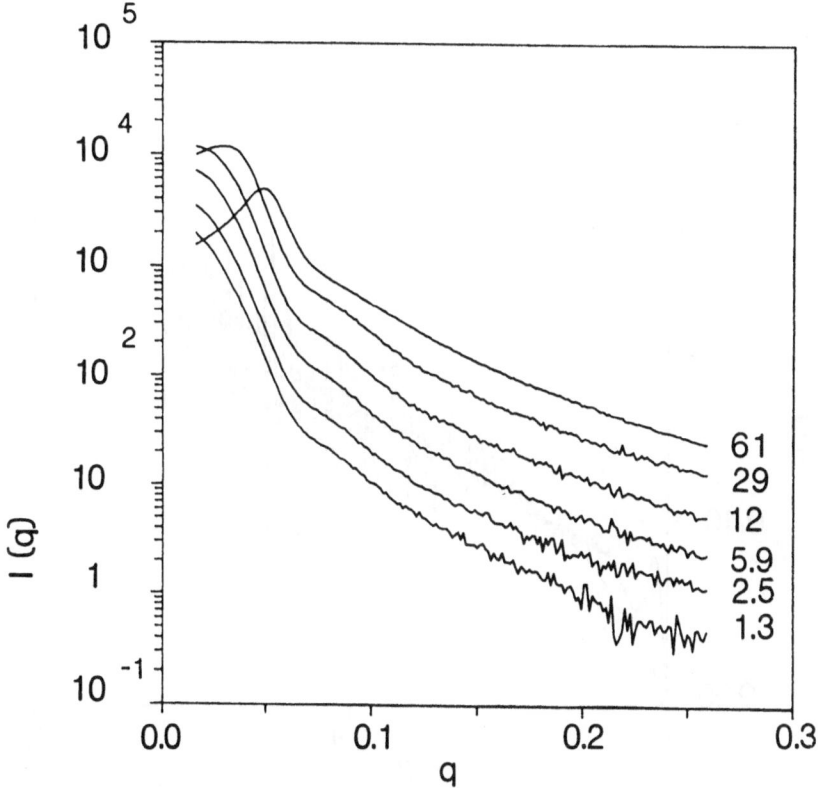

Figure 4.3. SAXS from an AOT/H_2O/iso-octane microemulsion (I(q) in arbitrary units, 9 in Å^{-1}), at $T = 22°C$, with constant water-surfactant ratio $W_0 = 35$ and vol.% AOT + H_2O increasing from 1.3 (bottom trace) to 61% (top trace).

in electron densities, the average radius of curvature, R, of the interface and the transition layer ∂. It is also possible to show that Eqns. 4.6 provide a useful approximation for the scattering from a general pseudo two phase system not containing finite closed volumes.

The expression for $\gamma(r)$, obtained prior to the calculation of Eqn. 4.6 differs from the expression for an ideal two phase system only in appearance of a $1/r$ term. It is this term that introduces the $1/q^2$ dependence in the high angle scattering limit, Eqn. 4.6b. It can be seen from Eqns. 4.6 that for the pseudo two phase system considered here, the high angle limit of the scattered intensity can be written in terms of the total area of interface between components, the relative electron densities and the radius of curvature of the interface. These quantities can be extracted from experimental data on plotting q^4I versus q^2, a so-called modified Porod plot. The slope, d, and intercept, $2b$, in Eqn. 4.6b are both linear functions of the area of interface, S, and their ratio is characteristic of the diffuse interface. For example, for $\partial/R \ll 1$, i.e. a lamellar-like system, this ratio is given by [22, 23]

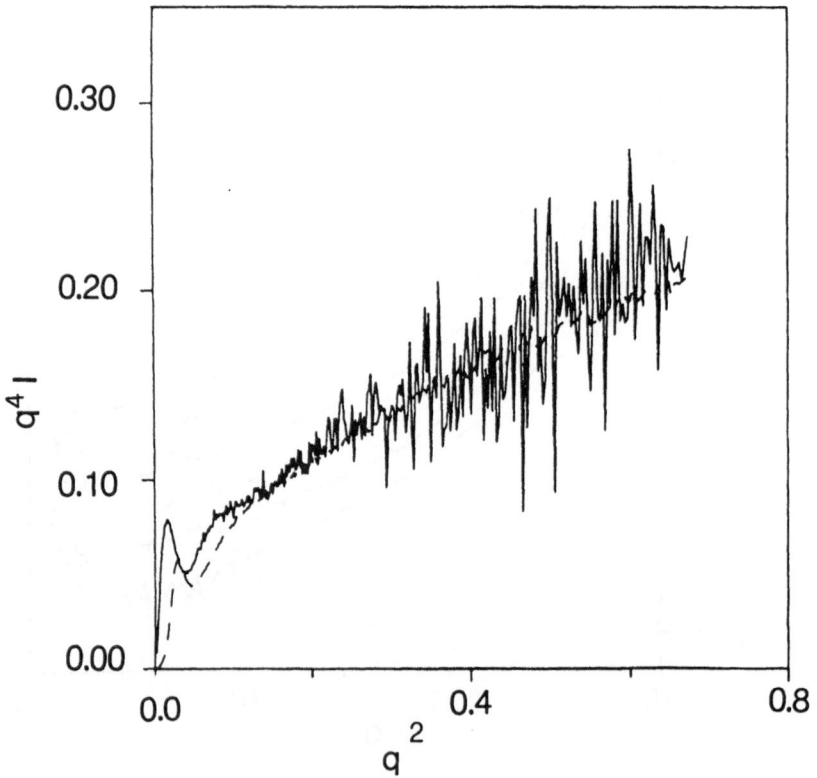

Figure 4.4. Modified Porod plot, Eqn. 4.6b, of the 2.5 (——) and 61 (– – –) vol.% AOT + H₂O microemulsion of Figure 4.1, $W_0 = 35$ (q^2 in Å^{-2}).

$$\frac{d}{2b} = \frac{(\rho_s - \rho_w)(\rho_s - \rho_o)}{(\rho_o - \rho_w)^2} \partial^2, \quad \partial/R \ll 1 \tag{4.7}$$

Unfortunately the total area of interface can only be obtained from the measurements on calibration of the intensity. When no calibration of the instrument is performed an absolute determination of the area per surfactant molecule is not possible, and only relative values can be determined. This is not the case for the ratio $d/2b$ which is independent of the experimental factors characteristic of the apparatus.

Figure 4.3 gives a typical example of SAXS data for a microemulsion concentration series with constant W_0. In the low angle region ($q < 0.1 \text{ Å}^{-1}$) interference effects appear on increasing the water content giving rise to a broad peak at approximately $5.3 \times 10^{-2} \text{ Å}^{-1}$, which is typical of microemulsion scattering. In contrast the shape of the scattered intensity for $q > 0.1 \text{ Å}^{-1}$

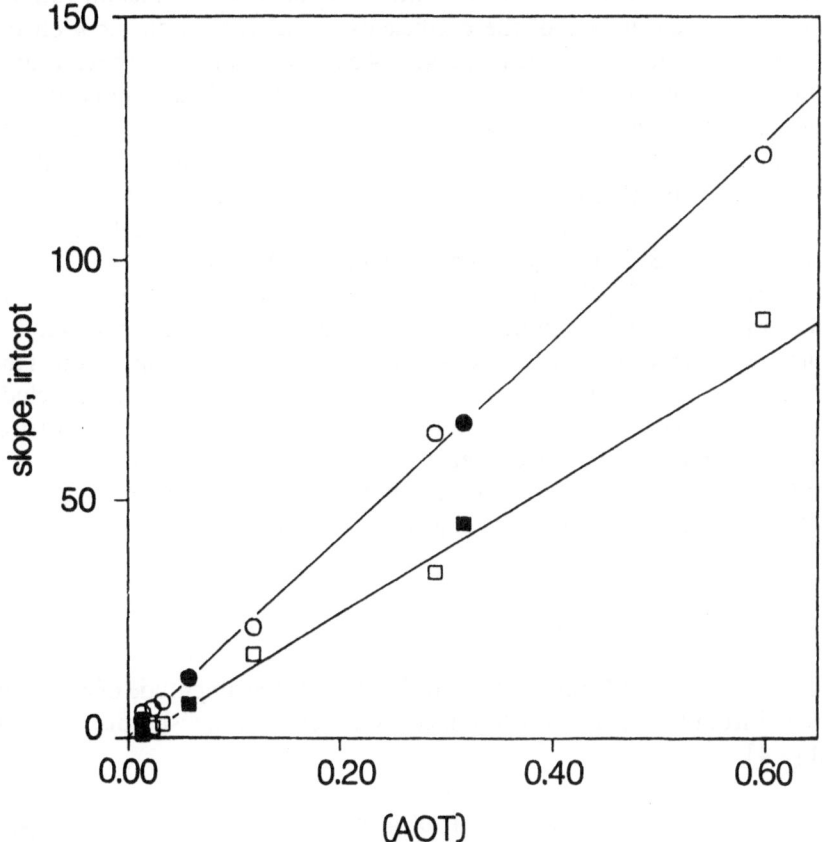

Figure 4.5.   Coefficients d and $2b$ of Eqn. 4.6b at different AOT concentrations (in mmol/l) and constant water-surfactant ratio $W_0 = 26$ (closed symbols) and 35 (open symbols). d and $2b$ in arbitrary units are a measure for the total area of interface.

appears to be the same over the concentration range used. The scattered intensity here differs only by a multiplication factor for different concentrations. This is manifested as a vertical shift on the logarithmic plot of Figure 4.3. The modified Porod plots for a high (61 vol.% AOT + water) and a low (2.5 vol.% AOT + water) concentration of microemulsion are shown in Figure 4.4. Note that the intensity is scaled by the concentration of AOT. It can be clearly seen that the plot of $q^4 I$ vs. q^2 follows closely the modified Porod Law given by Eqn. 4.6b. This behaviour extends over the entire high angle region observed in our experiments. The simple $1/q^4$ dependence does not describe the experimental data. The slope and intercepts, d and $2b$, Eqn. 4.6b, of the modified Porod plots are shown in Figure 4.5 as a function of AOT concentration, for the two different concentration series. Both the slope and the intercept are found to be linearly dependent on the surfactant concentration. No absolute determination of intensity was undertaken in order to calibrate these quantities. The calibration independent variable, $d/2b$, was found to be independent of the concentration and water/surfactant ratio.

The linear dependence of the coefficients d and $2b$ of Eqn. 4.6b, on the surfactant concentration, shown in Figure 4.5 demonstrates that the total area of interface is proportional to the total number of surfactant molecules in solution. If all the surfactants are localised in the water-oil interface then the average area per surfactant, Σ_s, is independent of the microemulsion concentration. Furthermore we find that Σ_s is independent of W_o, as the increase of the total interfacial area for every additional AOT molecule is the same for a $W_0 = 35$, and $W_0 = 26$ microemulsion (Figure 4.5).

The invariance of $d/2b$, Eqn. 4.6b, of the modified Porod plot to changes in concentration or in W_0, indicates that the manner in which the AOT is incorporated in the oil-water interface is independent of the composition of the microemulsion. The observation of a constant Σ_s and $d/2b$ does not preclude structural changes in the microemulsion when the concentration is varied. However the structural changes that can take place are limited. For example, a description of the microemulsion using a polydisperse spherical droplet model will have the restriction imposed that $\bar{R}^2$ is constant.

4.3.2   Scattering from particle systems

4.3.2.1   *General*   As X-rays interact with electrons the amplitude of radiation scattered from a medium is determined by the electron density distribution, $\rho(\mathbf{x})$ [3,4]

$$A(\mathbf{q}) = \int_V \rho(\mathbf{r}) e^{-i\mathbf{q}\cdot\mathbf{r}} d^3\mathbf{r} \qquad (4.8)$$

with $\mathbf{q}$ the scattering vector, equal to the difference between the vector of incident and scattered radiation, with magnitude $4\pi/\lambda \sin(\theta/2)$. V is the total scattering volume. Evidently only fluctuations in $\rho(\mathbf{r})$ produce scattering at non-zero angles. A constant $\rho(\mathbf{r})$ will only contribute to the scattering in the direction of incident radiation, $\mathbf{q} = \mathbf{0}$, as then all waves are in phase. Therefore an arbitrary constant can be subtracted from $\rho(\mathbf{r})$, for example the medium density. In a system consisting of particles dispersed in a medium of homogeneous density, ρ_0, this will only leave the integral over the particle volumes, i.e.

$$A(\mathbf{q}) = \sum_{\substack{i,\,\text{all} \\ \text{particles}}} \int_{V_i} (\rho(\mathbf{r}) - \rho_0)e^{-i\mathbf{q}\cdot\mathbf{r}} d^3\mathbf{r} \qquad (4.9)$$

Assuming that the particles contain a centre of symmetry located at $\mathbf{R}_i$ relative to an arbitrary origin the integration over V_i gives the total scattered amplitude

$$A(\mathbf{q}) = \sum_{\text{all particles}} A_n(\mathbf{q})e^{-i\mathbf{q}\cdot\mathbf{R}_n} \qquad (4.10)$$

where $A_n(\mathbf{q})$ is the amplitude scattered by the nth particle when located at the origin. Unfortunately radiation detectors are not phase sensitive so that they do not measure amplitudes, but rather the square of the amplitude intensities:

$$I(\mathbf{q}) = A(\mathbf{q})\,A^*(\mathbf{q}) = \sum_n \sum_m A_n(\mathbf{q})A^*_m(\mathbf{q})e^{-i\mathbf{q}\cdot(\mathbf{R}_n - \mathbf{R}_m)} \qquad (4.11)$$
$$\text{all particles}$$

Equation 4.11 gives the intensity of scattered radiation at one instant of time for a particular configuration of particles. During the time it takes to make a X-ray measurement the particle will move due to for example Brownian motion. Therefore Eqn. 4.11 must be averaged over all possible particle orientations and configurations to obtain the expression for the actual observed intensity. To simplify this we will assume that the orientation of a single particle does not depend on the position of the particle relative to other particles. This constraint is, however, unnecessary for particles with centro-symmetric electron density distributions. For the sake of simplicity we will assume, for the moment, that the particles are identical with number density x. The averaged intensity then takes the following form

$$\langle I(q) \rangle = xP(q)S(q) \qquad (4.12)$$

where $P(q) = \langle A_1(\mathbf{q})A^*_1(\mathbf{q}) \rangle$ and

$$S(q) = 1 + \int_V 4\pi r^2 g(r) \frac{\sin qr}{qr} dr$$

Averaging over all orientations causes the intensity to be dependent on the magnitude of $\mathbf{q}$ only, rendering a circular symmetric scattering pattern. The function $g(r)$ is known as the radial pair correlation function. It describes the probability of finding a particle at distance r from a chosen particle. This function plays a central role in liquid state theory. Knowledge of $g(r)$ enables calculation of the thermodynamic properties of the liquid state such as free energy or chemical potential, compressibility and phase behaviour. The fact that $g(r)$ for a colloidal system is accessible through small angle scattering experiments makes these techniques valuable.

In the form given by Eqn. 4.12 it is clear that the intensity is made up of two contributions. The first part, $P(q)$, is determined by the internal structure of the particles. The latter part, $S(q)$, the static structure factor, is determined by the configuration of the particles and therefore depends on the inter-particle interactions. Note that $S(q)$ becomes unity for dilute systems as the probability of finding particles separated by finite distances becomes zero. Thus when one is interested only in the internal particle structure, $P(q)$, the intensity must be measured at 'infinite' dilution. This experimental condition cannot always be realised, especially if the particles, for example, are known to undergo concentration dependent size and/or shape changes. Liquid state theory has progressed enormously in the past 10 years. For various interaction potentials closed expressions have been found for $g(r)$ and therefore for $S(q)$, enabling a separation of inter- and intra-particle interference effects [24–27]. The availability of analytical expressions for intensities enables comparison of measured and calculated intensities to extract structural parameters in fitting procedures.

4.3.2.2 *Polydispersity* A common feature of colloidal particle systems is that they exhibit a continuous diversity in shape and size. This latter polydispersity is often a major obstacle in the analysis of the experimental scattering curve. Details of the particle shape or size and interactions will be averaged out over the particle size distribution. The resulting scattered intensity will therefore only be dependent on the main features of the particle structure or inter-particle interaction potential.

A region where polydispersity in particle size becomes dominant is the low angle region of the scattering curve of dilute particle systems. In this region Guinier's Law is often applied. It states that at low q the scattered intensity decays according to [4]

$$I(q) \approx \exp\left[-(R_g q)^2/3\right] \tag{4.13}$$

where R_g is the radius of gyration. For a homogeneous spherical particle R_g is equal to $\sqrt{3/5}$ of the particle radius. Thus the slope of a plot of $\ln(I(q))$ versus q^2, a Guinier plot, gives the particle diameter. In a polydisperse system an averaged intensity over the size distribution is measured. It can be shown that R_g^2 obtained from a Guinier plot for polydisperse homogeneous spherical particles is proportional to the quotient of $\langle R^8 \rangle / \langle R^6 \rangle$ where $\langle \rangle$ is the average over the particle size distribution [8]. Therefore not the average particle size is obtained but the quotient of the eighth and sixth moment of the size distribution. These can differ significantly from the average value of R. This is demonstrated in Figure 4.6 where we have plotted the apparent

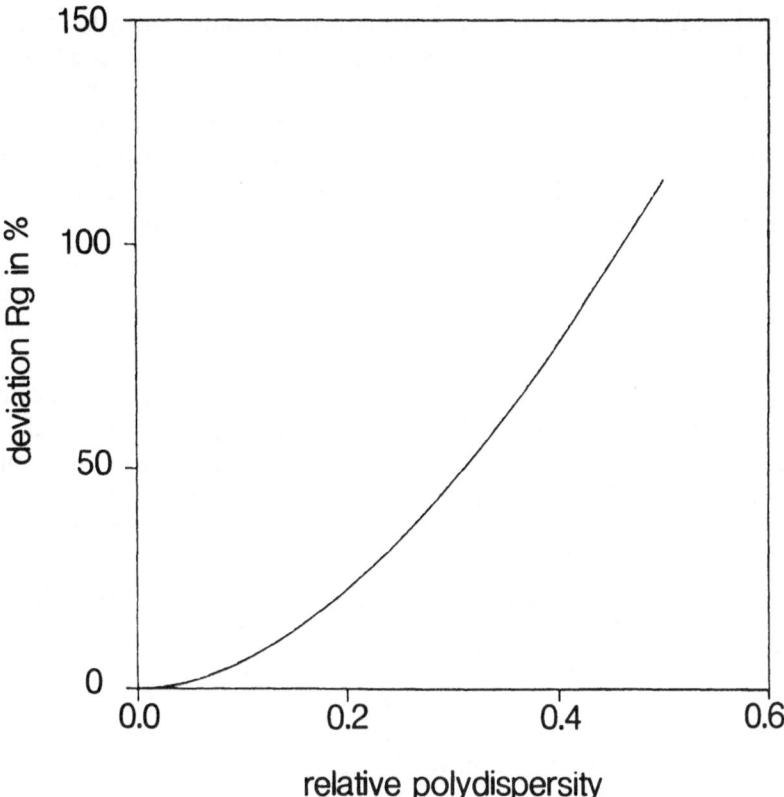

Figure 4.6.   Effect of a gaussian size distribution on the apparent radius of gyration of a system of spherical homogeneous particles.

R_g obtained from a Guinier plot of spherical homogeneous particles with a gaussian size distribution in particle radius as a function of the relative width of the distribution. We have given this example to demonstrate the effect of polydispersity which is so often underestimated or neglected.

The phenomenon of polydispersity can be considered as a curse and blessing at the same time. On the one hand specific details in the scattered intensity from the system will be averaged out resulting in a loss of information. On the other hand polydispersity reduces the large number of parameters necessary to describe the many particle problem; only statistical averages over the size distribution remain. In the theory summarised above polydispersity can be incorporated by using discrete size distributions. This reduces the summation over all particles in Eqn. 4.11 to a summation over a finite number of particle classes. The number density in each class is determined by the given size distribution. In the limit of one class the equations reduce to Eqn. 4.12.

To describe the small angle X-ray scattering from the AOT microemulsion we have to assume a model for the interaction between water droplets. A potential available from liquid state theory is the so-called sticky hard sphere potential [28, 29]. This potential has a hard repulsive core in combination with a short range attractive part. The range of the attraction is in fact infinitely small so that one speaks of surface adhesion. This means that at distances smaller than the radius of interaction, two droplets repel each other. Only when in contact with each other do the droplets attract each other. At distances larger than the droplet radius no interaction takes place. The measure of attraction at contact is described by a single parameter τ. For large τ there is negligible adhesion between droplets. As τ becomes smaller the adhesion becomes larger. For this reason τ is often referred to as a quasi-temperature of the system.

The sticky hard sphere potential is of course quite pathological. We cannot expect the microemulsion droplets to interact through this potential. The reason for choosing this potential, however, is twofold: first of all we must keep in mind that because of polydispersity we will not be able to extract from the experimental data more than the general features of the interaction in the system. The sticky hard sphere potential contains the features of a liquid-like system in terms of the two parameters, hard sphere radius and stickiness parameter τ. Second to this is the fact that a lot of theoretical work has been done with this potential and it gives the opportunity to calculate pair correlation functions analytically for some special forms of polydispersity, and numerically for more general forms of polydispersity [30–32]. The distribution we have chosen to describe the statistical fluctuations in size of the water droplets is the Schulz distribution [33]. It is described by two parameters, mean and width of the distribution. This distribution is similar to gaussian size distribution but exhibits a slight skewness which is a function of the width σ of the distribution.

4.3.2.3 *Results*

Section 4.3.2.2 we have shown that the area per surfactant molecule is independent of the composition of the microemulsion. Consequently it can be shown that in the polydisperse droplet model at fixed molar ratio of water and surfactant, W_0, the quotient

$$\langle R^3 \rangle / \langle R^2 \rangle = 3v W_0 / \Sigma_s \tag{4.14}$$

remains constant at all concentrations. Here R is the droplet radius, Σ_s the area per surfactant molecule, v the volume of a single water molecule, 30 Å^3. The electron density profile taken for a single droplet based on the chemical structure of the AOT molecules is shown in Figure 4.7. This electron density profile provides the simplest model accounting for the behaviour of the scattering curve at high angles. The surfactant tails have approximately the same electron density as the oil phase and therefore do not contribute to the scattering of a single droplet. However, it can be expected that they will have to be accounted for in the interaction radius, R_{hs} (see Figure 4.7). The difference $d_s = R_{hs} - R_w - \partial$ will be an estimate of the contribution of the surfactant tails

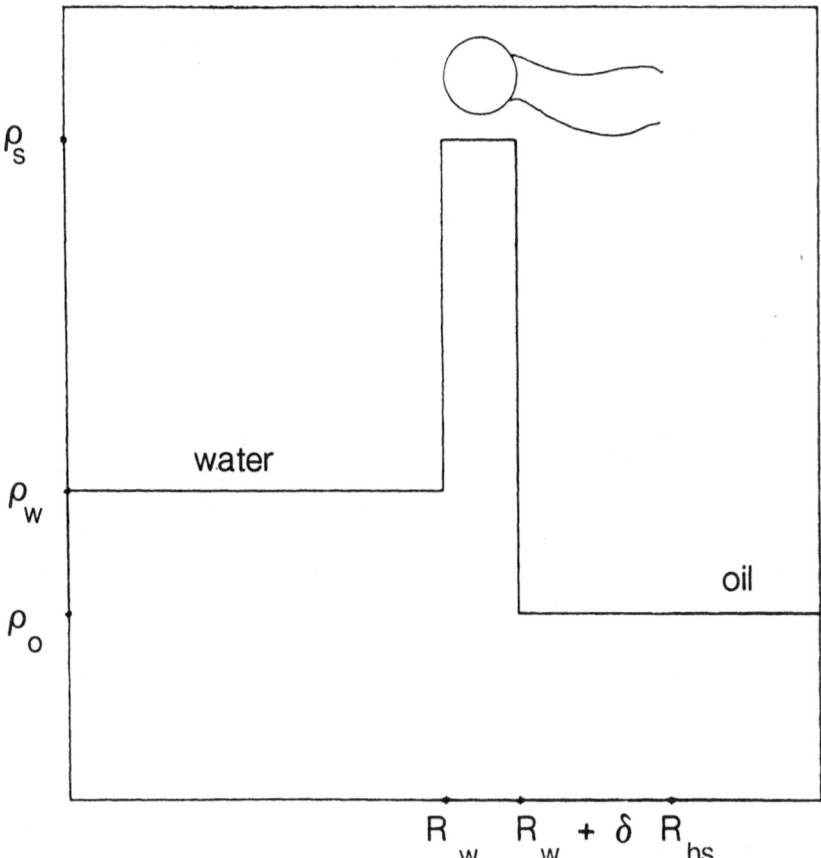

Figure 4.7.   Model for the electron density profile, $\rho_e(r)$, of an AOT-microemulsion droplet.

to the interaction radius. We now have the following adjustable parameters describing the scattered intensity: (1) R, σ, Σ_s coupled by Eqn. 4.14; (2) ρ_s (the electron density of the headgroup), ∂; (3) τ; (4) d_s.

It is important to realise that the parameters ρ_s and ∂ cannot be determined uniquely from the data. Firstly, ρ_s can only be determined relative to the difference in electron density between the water and the oil phase, which were taken to be those of the pure bulk phases. Secondly, the values of ρ_s and ∂ are mainly determined by the high angle portion of the scattering curves. As pointed out previously, to a first approximation, only their product enters the calculations so that the two parameters are strongly correlated. We have subsequently kept ρ_s and ∂ constant in our analysis.

The small angle X-ray scattering curves from the AOT/H$_2$O/iso-octane microemulsion at volume fractions of dispersed phase, ϕ, between 2 and 50% with constant water surfactant ratio, $W_0 = 30$ are shown in Figure 4.8. The data were obtained at 33°C where the microemulsion is known to exhibit a

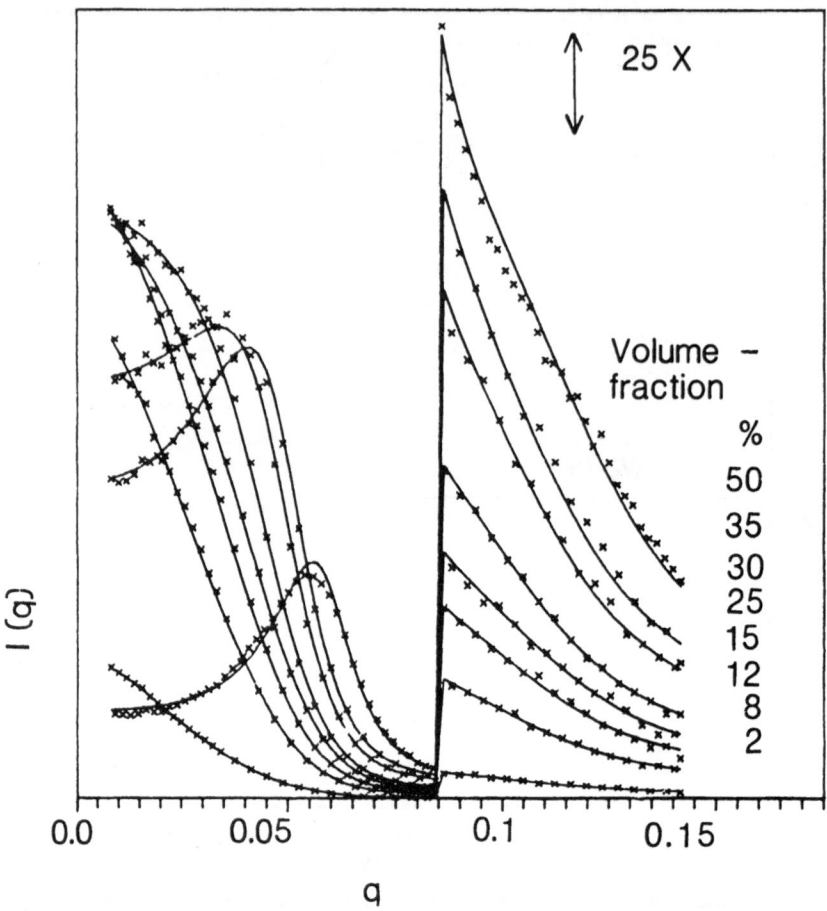

Figure 4.8. Observed (×) and calculated (———) small angle X-ray scattering intensities from an AOT-microemulsion with $W_0 = 30$, $T = 33°C$ at volume fractions, ϕ, of dispersed phase (water + AOT) between 2 (bottom trace) and 50% (top trace) (I(q) in arbitrary units, q in Å^{-1}).

Table 4.2 Parameters calculated for the AOT/H$_2$O/iso-octane micro-emulsion, with $W_0 = 30$ at $T = 33°C$.

$\phi(\%)$	$R_{hs}(\text{Å})$	σ/R_{hs}	$\Sigma_s(\text{Å}^2)$	$d_s(\text{Å})$	τ
2	45.2	0.230	56.8	0.0	0.16
8	46.8	0.232	54.4	0.0	0.65
12	46.9	0.234	54.1	0.0	1.8
15	45.9	0.232	55.5	0.0	5.4
20	48.2	0.233	52.7	0.0	5.0
30	45.2	0.219	54.5	5.7	95
35	43.0	0.215	55.7	6.6	85
50	37.1	0.203	60.8	7.6	99

critical behaviour in its dielectric properties. The theoretical intensities calculated using the polydisperse droplet model are also shown in Figure 4.8. The values extracted for the parameters are given in Table 4.2. From Figure 4.8 it is clear that the scattered intensities calculated using the droplet model give a good description of the observed scattered intensities. Only at the highest particle density do some deviations occur.

Two striking features appear in the values of the extracted parameters in Table 4.2. Firstly the invariance of both the area per surfactant, Σ_s, and the relative polydispersity σ/R. This implies that the average droplet size remains constant over the range of concentrations studied. The value of Σ_s found here, $53-57\,\text{\AA}^2$, is in good agreement with that reported previously from other techniques. The second striking feature of Table 4.2 is the dependence of both τ and d_s on the volume fraction ϕ. At volume fractions, below 0.3, it seems there is no contribution of the surfactant tails to the hard sphere radius, i.e. $d_s \approx 0$, as opposed to at higher volume fractions where d_s is found to be between 5 and 8 Å. This fact can be explained in the following way: the value of the hard sphere radius is mainly determined by the position of the first maximum in the static structure factor. This maximum is pronounced only at high particle densities. The value obtained for the hard sphere radius at low particle densities will therefore not be very sensitive to small deviations due to the surfactant tails. At higher densities the first maximum in the structure factor and therefore in the scattered intensity will be much more pronounced so that an accurate hard sphere interaction radius will be obtained. We have fitted the low density scattering curves with fixed values of d_s ranging between 6 and 9 Å. These values for d_s are in good agreement with the size of a AOT-molecule. It was found that the extracted parameters, τ, σ, R and Σ_s, are not sensitive to these small changes in d_s.

We can see from the analysis that the microemulsion droplets do not interact solely through a hard sphere potential. An attractive interaction between the droplets appears to exist as values of τ are found ranging between 0.15 and 99. From this it can be concluded that the attraction between the particles decreases with increasing volume fraction. It indicates the presence of a negative cooperative effect whose origin is not yet clear.

Acknowledgements

We are grateful to the Netherlands Organisation for Scientific Research (NWO) for supporting this work.

References

1. T.P. Hoar and J.H. Schulman, *Nature* **152** (1943) 102.
2. K.L. Mittal and B. Lindman, eds., in *Surfactants in Solution*, Plenum Press, London and New York.

3. V. Degiorgio and M. Corti, eds., in *Physics of Amphiphiles: Micelles, Vesicles and Microemulsions*, North-Holland, Amsterdam (1985).
4. A. Guinier and G. Fournet, in *Small Angle Scattering of X-rays*, John Wiley, New York and Chapman and Hall, London (1955).
5. O. Kratky and O. Glatter, in *Small Angle X-Ray Scattering*, Academic Press, London (1982).
6. R. Hosemann and S.N. Bagchi, in *Direct Analysis of Diffraction by Matter*, North-Holland, Amsterdam (1962).
7. B.K. Vainstein, in *Diffraction of X-Rays by Chain Molecules*, Elsevier, Amsterdam (1966).
8. M. Kerker, in *The Scattering of Light and Other Electromagnetic Radiation*, Academic Press, New York and London (1969).
9. T. Matsushita and H. Hashizume, in *Handbook on Synchrotron Radiation*, Vol 1A, ed.E.E. Koch, North-Holland, Amsterdam (1983) Chap. 4.
10. M.A. van Dijk, G. Casteleijn, J.G.H. Joosten and Y.K. Levine, *J. Chem. Phys.* **85** (1986) 85.
11. M.A. van Dijk, E. Broekman, J.G.H. Joosten and D. Bedeaux, *J. Phys.* **47** (1986) 727.
12. E.W. Kaler, H.T. Davis and L.E. Scriven, *J. Chem. Phys.* **79** (1983) 5685.
13. E.W. Kaler, K.E. Bennet, H.T. Davis and L.E. Scriven, *J. Chem. Phys.* **79** (1983) 5673.
14. M. Kotlarchyk and S.W. Chen, *J. Chem. Phys.* **79** (1983) 2461.
15. M. Kotlarchyk and S.W. Chen, *J. Phys. Chem.* **86** (1982) 3273.
16. M. Kotlarchyk and S.W. Chen, J.S. Huang and M.W. Kim, *Phys. Rev. A* **29** (1984) 2054.
17. P. Debye and A.M. Bueche, *J. Appl. Phys.* **20** (1949) 51.
18. G. Porod, *Kolloid Z.* **124** (1951) 89.
19. G. Porod, *Kolloid Z.* **125** (1952) 51.
20. G. Porod, *Kolloid Z.* **125** (1952) 109.
21. D. Weigel, A. Renouprez and B. Imelik, *J. Chim. Phys.* **62** (1965) 125.
22. C. Robertus, J.G.H. Joosten and Y.K. Levine, in *Progress in Colloid and Polymer Science.*
23. L. Auvray, J.P. Cotton, R. Ober and C. Taupin, *J. Phys.* **45** (1984) 913.
24. L. Blum and G. Stell, *J. Chem. Phys.* **71** (1979) 42.
25. P. van Beurten and A. Vrij, *J. Chem. Phys.* **74** (1981) 2744.
26. A. Vrij, *J. Chem. Phys.* **69** (1978) 1742.
27. A. Vrij, *J. Chem. Phys.* **71** (1979) 3267.
28. R.J. Baxter, *J. Chem. Phys.* **49** (1968) 2770.
29. R.J. Baxter, *J. Chem. Phys.* **52** (1969) 4559.
30. C. Robertus, W.H. Philipse, J.G.H. Joosten and Y.K. Levine, *J. Chem. Phys.* (in press).
31. J.W. Perram and E.R. Smith, *Chem. Phys. Lett.* **35** (1975) 138.
32. B. Barboy, *Chem. Phys.* **11** (1975) 357.
33. G.V. Schulz, *Z. Phys. Chem.* **43** (1935).

5 Time-resolved small angle X-ray scattering on polymers

G. UNGAR

5.1 Introduction

The combination of a bright source of X-rays and an efficient detector has been exploited in a variety of dynamic experiments, e.g. to study the contraction of muscle, liquid crystal transitions, etc. Here we describe the usefulness of small angle X-ray scattering (SAXS) using synchrotron radiation to study synthetic polymers and model chain compounds. Time-resolved SAXS now provides unique information about the pathways of various structural transformations of polymers. On-line combination with other dynamic techniques, such as differential scanning calorimetry (DSC), is an area of growing interest.

The spatial extent of the scattering pattern is inversely related to the size of the scattering object. X-rays scattered or diffracted at high angles give information about the detailed structure of the molecules or crystals at the atomic resolution. SAXS, however, provides information on the overall shape and size of the scattering object, its density, orientation, packing with other objects, etc. Since X-ray scattering arises from electron density fluctuations in the specimen, a scattering object can be any inhomogeneity—a molecule in solution, a small crystal, a part of a large molecule, a solid or a liquid particle, a rubber particle in a glassy polymer matrix, etc.

Synthetic polymers have been one of the principal areas of application of small angle X-ray scattering. This is because the periodicities encountered in these systems are often in the SAXS range, i.e. several tens to about a thousand angstroms. In the present article we are not concerned with scattering on solutions of macromolecules. Parameters such as radius of gyration can be extracted from the SAXS curve of a dilute polymer solution. In such cases the approach is similar to that described for microemulsions in Chapter 4. The emphasis of this article is on polymeric solids.

In order to follow rapid structural changes in a scattering pattern, the data need to be collected quickly. This requires a high photon flux on the sample. To follow such changes in the small angle region, an X-ray beam of small area and small angular divergence is required. Taken together, these requirements imply the use of a bright X-ray source, where brightness is defined as the number of photons per unit area, per unit solid angle, per second. Thus the synchrotron radiation source is extremely useful for these types of experiments.

Experimental hardware, including monochromatisation and collimation of the beam, as well as detection of scattered photons, is dealt with in Chapters 1 and 4. The reader is also referred to the reviews by Elsner, et al. [1] and Russell [2]. To take full advantage of synchrotron radiation, ancillary equipment is required in addition to the basic hardware. Ancillary equipment usually controls sample environment and is custom designed for specific experiments. Thus, for example, time resolution of temperature induced processes is at present limited not by the limited photon counting period, but rather by the limited rate of change in specimen temperature. Hot and cold gas blowers [3, 4] and piston driven T-jump assemblies [5] are used to attain the desired temperature rapidly. Ever more elaborate ancillary equipment is being used, such as stretching frames [6], combined X-ray diffraction and DSC cells [7, 8], shearing cells [9], etc.

Some basic principles of SAXS theory are given in Chapter 4. Several classical monographs deal with SAXS in general terms [10–12], and the more specific subject of SAXS on polymers is treated in books on X-ray diffraction in polymer science by Vainstein [13], Alexander [14], Kakudo and Kasai [15] and recently by Balta-Calleja and Vonk [16]. The application of synchrotron radiation to polymers has been reviewed fairly extensively in the aforementioned articles by Riekel [1] and Russel [2]. The present article does not attempt to give a comprehensive review of the field, but rather deals with certain illustrative examples and new development.

5.2 SAXS and polymers

SAXS is extensively used to study the size and shape of crystallites in semicrystalline polymers. The two most common morphological forms in semicrystalline polymers are lamellar and fibrillar, the first occurring in solution grown single crystals and melt-grown spherulites (spherical aggregates of radiating lamellae), the second in oriented systems such as spun or drawn fibres. The lamellae, in conventional synthetic polymers, consist of a crystal core through which straight molecular stems cross perpendicularly (or nearly perpendicularly), and of a less ordered layer containing chain folds, entanglements and chain ends [17]. Regularly alternating regions of higher (crystal core) and lower (surface) electron density give rise to small angle diffraction. The periodicity, called the long period, L, is in the range 50–1000 Å.

Before describing the interesting results of time resolved experiments, we shall briefly mention the different kinds of information about the lamellar morphology that SAXS can provide. The simplest and most frequent usage of SAXS is undoubtedly the determination of the L value from the diffraction peak position. However, an analysis of the scattering peak profiles also gives information about the size of the coherently scattering unit (size of lamellar stacks) and about the distribution of lamellar thicknesses, i.e. how well the periodicity is maintained within a stack. It must be realised that the lamellar stacks in semi-crystalline polymers are metastable products of random events during crystallisation. Although the distribution of lamellar thicknesses can be narrow, there is no long range order. A stack of lamellar polymer crystals is sometimes referred to as a one-dimensional paracrystal [11]. There is no driving force for a stack to regularise, as there is for a true three-dimensional crystal to increase its perfection. This poor order is manifested in the general lack of higher small angle diffraction peaks.

Further, the analysis of the integral intensity of the diffraction peaks, either relative or absolute, can provide information about the electron density profiles. One important property of the scattering function is that, irrespective of how the scattered intensity is distributed over the reciprocal space, i.e. irrespective of the shape of the scattering curve, the total scattered intensity per unit volume of the sample is determined entirely by the mean square electron density fluctuation in the specimen. The so-called SAXS invariant Q [18], is defined as

$$Q = \int_0^\infty q^2 I(q) dq = \overline{\Delta\eta^2}$$

Here q is the scattering vector, $q = (4\pi/\lambda)\sin(\theta/2)$, θ the scattering angle, I the intensity, and $\overline{\Delta\eta^2}$ is the mean square electron density fluctuation. Applied to the simple two-phase (crystalline – amorphous) lamellar polymer model, this means, for example, that the total scattered intensity in the low angle region will be highest if the crystalline and amorphous layers are of equal thickness. Thin amorphous layers separated by thick crystalline lamellae will produce weak scattering, as will the reverse case of thin dense layers separated by thick layers of lower density (Babinet principle). Thus, by assuming certain models, SAXS can reveal not only the overall value of the long period L, but also its partition between the crystalline and the amorphous components.

Measurement of absolute SAXS intensity is not trivial but can be very useful (see example of n-alkanes in Table 5.1). The recommended method is the use of the standard sample of low density polyethylene, obtainable from Antun Paar A.G., Graz, Austria. The scattered intensity at a particular angle has been calibrated versus the primary beam intensity by the method of Kratky [19]. It is worth noting that the parameters of this reference sample have been defined for both slit collimation (Kratky-type cameras) [19] and for pinhole collimation [20], the latter being effectively the SAXS geometry used with synchrotron radiation. One definite advantage of the highly monochromatic synchrotron radiation beam for absolute SAXS intensity measurement is that the irradiated mass of the specimen per unit beam area, the necessary normalisation parameter, can be determined accurately by measuring the transmittance of the primary beam and applying the simple Lambert-Beer's

Table 5.1 Intercrystalline layer parameters for integer forms of long n-alkanes from absolute SAXS intensities [30].

	E melt crystd. (C_{198}, C_{246})	F_2 quenched from melt (C_{198}, C_{246})	F_2 solution crystd. (C_{198})	Linear polyethylene, melt crystd.
Electron deficiency per unit area of interlayer, κ (e/A^2)	0.66	0.73	0.40	2.4
Average thickness of non-crystalline layer from κ, t_1 (Å) (assuming amorphous density $= 0.85\,\mathrm{g/cm^3}$)	13	14.5	8	48
Non-crystalline fraction, t_1/L (%)	5.5	10.5	6.3	21
2nd moment of density deficiency profile across non-crystalline layer, σ^2 (Å^2)		25	11	
Thickness of non-crystalline layer, from σ^2, t_2 (Å) (assuming two-phase model)		17	11	

law. Applying the same procedure to a non-monochromatic beam can lead to serious errors.

Even without measuring the absolute values, the appropriately corrected relative intensity distribution $I(q)$ can give considerable insight into the nature of density fluctuations within the polymer. Here we mention two methods applicable to lamellar systems such as crystalline polymers: the one-dimensional correlation function coupled with simple or more complex two phase models due to Vonk [16] and the method of Strobl [21] which determines the second moment of the electron density profile across a layer, as well as the total electron deficiency of the layer (see also below).

5.3 Time-resolved SAXS in polymer crystallisation and annealing

The origin of lamellar crystal morphology of synthetic polymers is in crystallisation kinetics: thin chain-folded crystals grow faster, even if they are thermodynamically less stable, than the thick extended-chain crystals. However, the as-formed metastable crystals often thicken subsequently, either while still at the crystallisation temperature, or during heat annealing at some later time. SAXS studies of both of these subsequent processes have been restricted by the long exposures required when using conventional X-ray generators. The two examples below illustrate how synchrotron radiation can help in overcoming these difficulties.

In order to help understand the complex process of polymer crystallisation it is important to be able to measure the thickness of the primary as-grown crystals as a function of crystallisation temperature T_c. This provides one of the main tests for any crystallisation theory of polymers. However, in the case of the 'prototype' polymer, polyethylene, the initial L could until recently be measured only for solution crystallisation at large supercoolings, where crystal thickening does not occur. Recent time-resolved SAXS experiments by Barham [22], using synchrotron radiation, show that the initial lamellae that form on crystallisation from the melt are only about half as thick as had been assumed previously on the basis of conventional time-averaged SAXS measurements. Figure 5.1 shows the long period in a sample of linear polyethylene as a function of crystallisation time at $T_c = 123.4°C$. The first crystals that form have a periodicity of only 160 Å, but a second diffraction peak at nearly double that periodicity appears within seconds. After about 1 min only the thickened crystals remain and their slow thickening continues. If the sample is kept at T_c for longer, the thickened lamellae become more regularly packed and a second order reflection appears (see Figure 5.1). The previous measurements of L vs. T_c have invariably overlooked the initial 160 Å periodicity and hence much controversy arose in connection with the

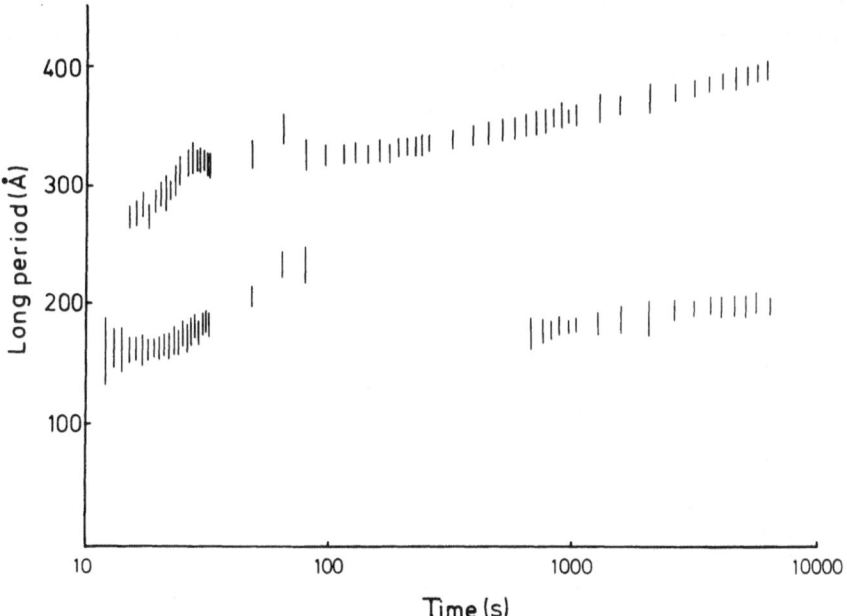

Figure 5.1. Long period L in a sample of linear polyethylene as a function of crystallisation time at $T_c = 123.4°C$ [22]. The values between 100 and 200 Å appearing around 1000 s correspond to the second order diffraction peak.

inexplicably large L in melt grown crystals as compared to that in solution grown crystals.

The second example deals with the actual process of lamellar thickening on heat annealing of polyethylene, where there is ambiguity as to whether the thickening process involves solid state diffusion of molecular chains or melting and subsequent recrystallisation. It should be remarked that annealing greatly affects mechanical properties of polymeric products. Grubb *et al.* [23] have performed real time SAXS experiments on single crystal polyethylene mats using the Cornell high energy synchrotron (CHESS) and a two-dimensional detector based on a Vidicon camera. Isothermal sequences recorded at the annealing temperature T_A show that the sharp arc in the pattern of the crystal mat (which corresponds to the orientation of lamellae in the polymer) spreads towards lower angles producing a continuous streak approx. 10 s after reaching T_A. This eventually transforms into an intense spot at a lower angle, the spacing of which (L) continues to increase with time. Spells [3] has performed similar experiments using a one-dimensional detector. A hot air blower system, remotely controlled through a solenoid valve was used to heat the sample rapidly to T_A. The measured L is plotted against annealing time in Figure 5.2. There is an initial jump in layer spacing, after which L continues to increase logarithmically with time.

Before synchrotron radiation was available, such rapid changes could not be observed and only the logarithmic thickening was known. Nevertheless, the relative roles of the two mechanisms in lamellar thickening are not fully understood although it appears that, for a given T_A, fast heating favours melting/recrystallisation, and slow heating, the solid diffusion mechanism.

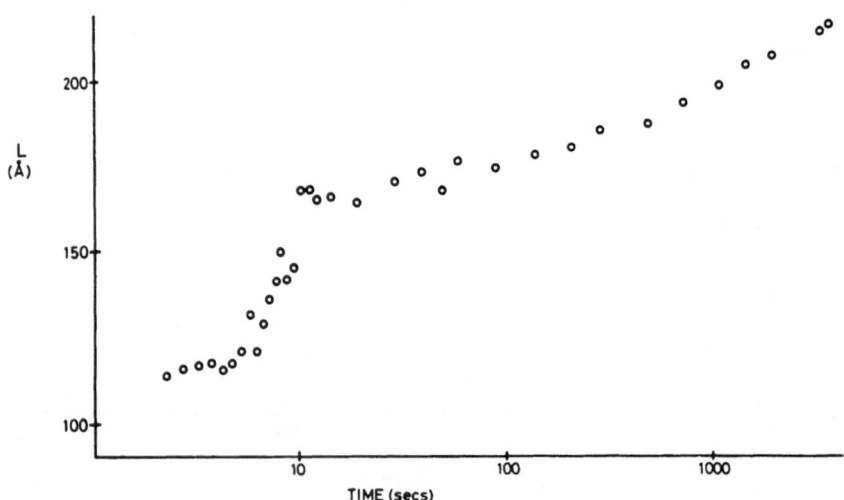

Figure 5.2. Long period L of polyethylene single crystals shown as a function of the annealing time at $T_A = 123°C$ [3].

An interesting new development concerning polymer crystal thickening is the observation that in certain disordered and highly mobile hexagonal phases (now these are recognised as columnar liquid crystal phases [24]), lamellar thickening proceeds with great speed. A synchrotron SAXS and WAXS study of a substituted polyphosphazene by Magill and Riekel [25] shows very convincingly the difference in annealing behaviour of the ordered crystalline phase and of the columnar mesophase: while there is virtually no change in L below the transition temperature, as this was exceeded, L increased to almost a thousand angstroms within seconds.

In addition to crystallisation and annealing studies, other time-resolved SAXS experiments have been performed on crystalline polymers using synchrotron radiation. Thus transformations in lamellar morphology accompanying rapid stretching have been monitored in situ [6]. Polymorphic phase transition was one of the first phenomena in synthetic polymers to be studied by synchrotron radiation, using the synchrotron at Novosibirsk [26], although in this case wide- rather than small-angle diffraction was required. Some examples of polymorphic transitions are mentioned further below.

Amorphous polymers are also amenable to time-resolved SAXS studies. One example is monitoring the progress of phase separation in a system of immiscible polymers, a subject briefly dealt with further below. Another interesting application is the study of time development of the so-called crazes, or microcracks in glassy polymers subjected to stress. Two-dimensional recording of the progress of formation of microfibrils and microvoids during crazing, using synchrotron radiation, yielded some very interesting new information on the mechanism of this important process in commercial polymeric products [27].

5.4 Model polymers: ultra-long *n*-alkanes

An important recent development in polymer science is the synthesis of *n*-alkanes of uniform chain length in the range of several hundred carbon atoms [28]. In many respects these compounds bridge the gap between low molecular weight substances and macromolecules. Until recently the longest *n*-alkanes available contained about 100 carbon atoms and they invariably crystallised in the extended chain (E) form. The new alkanes, the first in the series being $C_{150}H_{302}$, display chain folding, which is the characteristic mode of crystallisation of polymers. Using conventional SAXS and Raman spectroscopy [29] the fold length L in the *n*-alkanes was found to vary with crystallisation temperature T_c as it does in polyethylene, but in contrast to polyethylene it is an integral reciprocal of the full chain length. This means that the chains are either fully extended (E), or else folded exactly in two (F_2), three (F_3), etc (see Figure 5.3). It also shows that the methyl end groups are at the

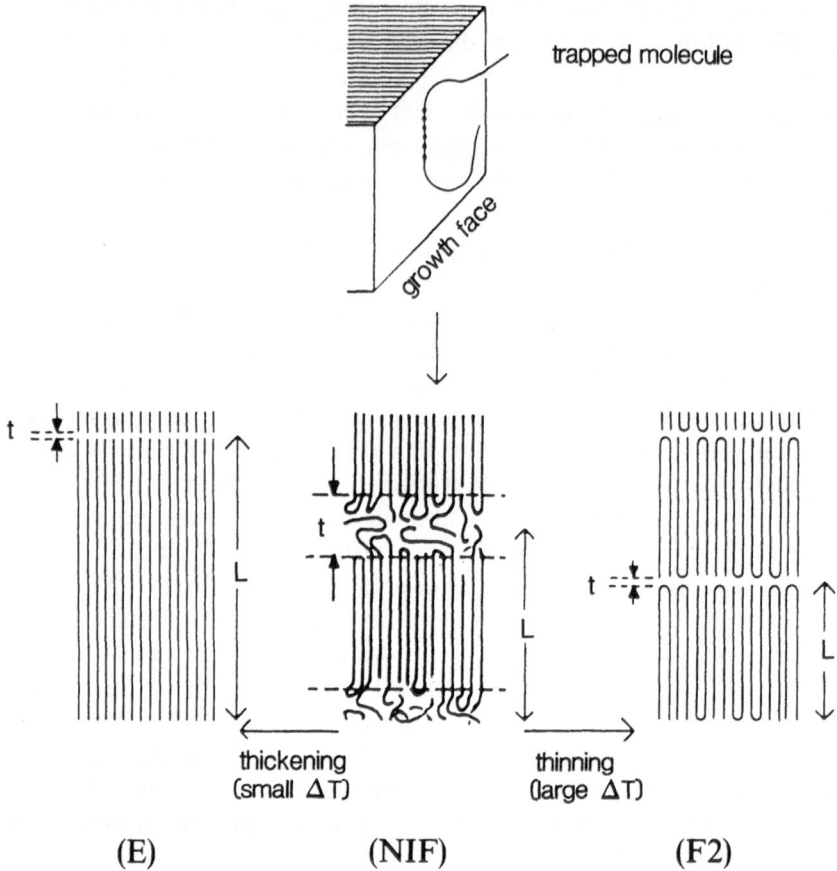

Figure 5.3. Schematic representation of the primary crystallisation and subsequent transformations in very long *n*-alkanes as revealed by time-resolved SAXS using synchrotron radiation. L is the long period and t the thickness of the non-crystalline layer. Fully extended form (E), noninteger form (NIF), form folded in two (F2).

surface of the molecular layers, and that the folds are sharp and contain only a few methylene groups.

Although the L-value for alkane crystals can be measured accurately due to the existence of several diffraction orders, a still better approach to investigating the thin intercrystalline layers is measurement of absolute diffracted intensities. For this purpose the method of Strobl [21] has been applied, whereby the extrapolated limiting structure factor of the zeroth diffraction order, B_0, is equal to the total electron deficiency per unit area of the non-crystalline layer, κ. The extrapolation is performed by fitting the measured structure factors of several diffraction orders ($B_1, B_2, \ldots$) to an even order (usually second order) polynomial. The curvature gives the parameter σ^2, which is the second moment of the electron density profile, and can be regarded as the one-dimensional analogue of the radius of gyration. Note that this method is only applicable if $t \ll L$, where t is the thickness of the non-crystalline layer, and if t is fairly uniform throughout the sample. Note that the absolute SAXS intensity is a very sensitive probe for fold layer thickness t; in the first approximation the extrapolated zero order intensity increases with t^2.

Table 5.1 (p. 124) lists the fold layer parameters of two integer forms, E and F_2, for n-alkanes $C_{198}H_{398}$ and $C_{246}H_{494}$, as obtained from the absolute intensity measurements [30]. While the parameters κ and σ^2 are obtained directly, t_1, t_2 and the non-crystalline fraction, t_1/L, are based on the assumption of a discrete two phase model, i.e. of a rectangular density profile. Data for a melt-crystallised sample of high-density polyethylene are also included for comparison. The thinness of the fold layer in n-alkanes, particularly in the solution crystallised sample, indicates the high crystallinity and, accordingly, "tightness" of chain folds.

While the final states in mature alkane crystals tend to be those with integer folding (IF), described above, the initial crystals, as grown from the melt, can in fact have non-integer layer periodicities, as revealed by the time-resolved in situ crystallisation studies [31]. Figure 5.4 shows a set of SAXS diagrams recorded at 30 s (0.5°C) intervals during the cooling of the alkane n-$C_{246}H_{494}$ from the melt at a constant rate of $-1°C/min$. The intense diffraction peak that appears rapidly at 120°C and then disappears almost completely corresponds to a non-integer form (NIF) with L in between those of the chain extended (E) and the folded-in-two (F_2) forms.

NIF gives way both to E and F_2 forms by subsequent lamellar thickening or thinning, respectively. As room temperature is reached, neither SAXS nor Raman spectroscopy reveal any significant amount of NIF and without synchrotron experiments it might have remained undetected.

A number of isothermal crystallisation experiments have subsequently been performed on the long alkanes. It turns out that NIF crystallisation takes place only below a certain temperature T^*, which is close to the melting point of NIF. Above T^* only direct crystallisation in the E form can occur. If the crystallisation temperature is only slightly below T^*, NIF–E transformation takes place involving isothermal lamellar thickening. If, however, T_c is well below T^* only thinning of the original NIF lamellae occurs. Both processes can occur simultaneously at intermediate temperatures.

Such behaviour suggests that the integer forms E, F_2 etc. are the stable or metastable ones in pure monodisperse alkanes, but that the non-integer form is kinetically favoured, i.e. it crystallises more rapidly. This result highlights the importance of the mode of initial deposition of individual molecular stems onto the growing crystal face in polymer crystallisation. Figure 5.3 illustrates the primary crystallisation step and subsequent changes in long alkanes. Sections of the crystallising molecule attach and detach at random on the crystal growth face. If an integer form crystallises directly, growth proceeds only when the attaching molecule matches the substrate molecules longitudinally.

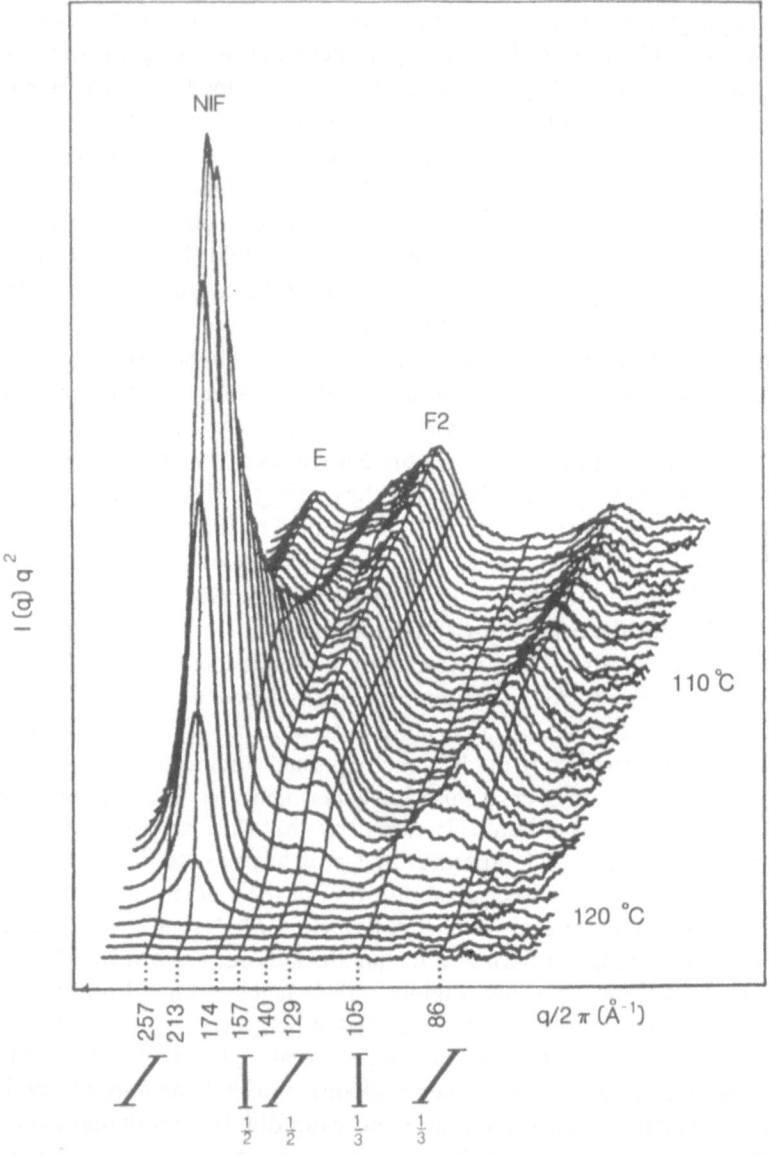

Figure 5.4. Time-resolved SAXS curves (Lorentz corrected vs. wavevector q) of the n-alkane $C_{246}H_{494}$ during cooling from the melt at $1°C/min$. Successive time frames are recorded at 30 s ($0.5°C$) intervals [31]. The spacings (in Å) marked on the abscissa correspond to expected L for different integer forms with or without chain tilt.

NIF, however, allows a larger number of successful placings since chain ends need not lie exactly on a smooth layer surface. Also, if a molecule becomes attached with its central portion, leaving two long dangling ends, or cilia, uncrystallised, new molecules can deposit on top and around it before there is a chance for a more favourable repositioning. The cilia are then left to form a non-crystalline layer of lower density and of considerable thickness, as sketched schematically in Figure 5.3. The presence of the thick non-crystalline layer will be responsible for the high diffraction intensity observed (Figure 5.4), as has been explained further above. The process of lamellar

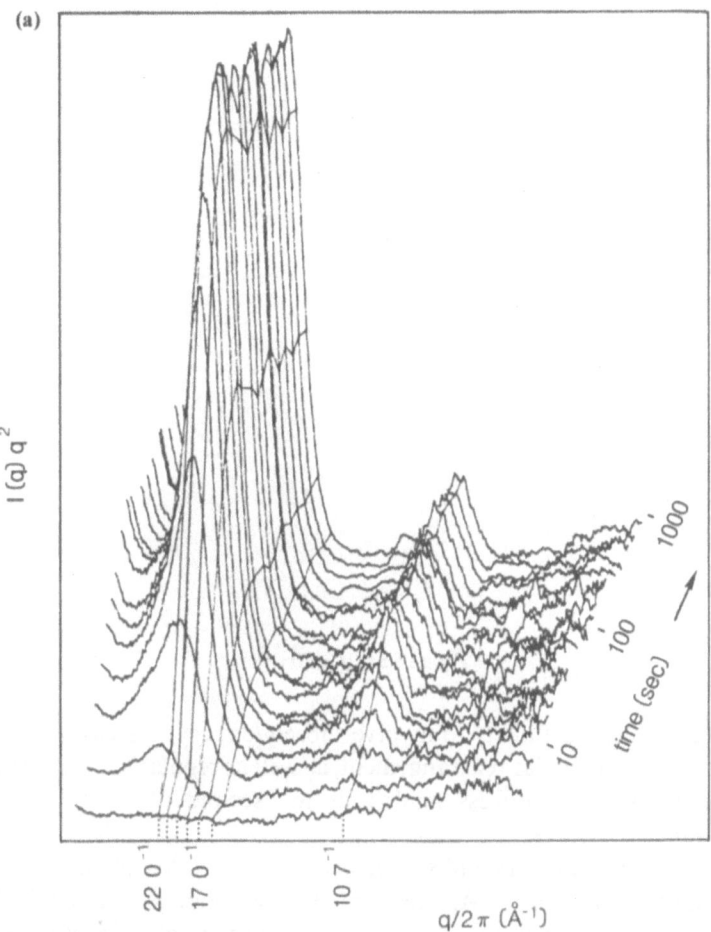

Figure 5.5. SAXS curve evolution during isothermal crystallisation (a) of the chain-extended from (n-C$_{198}$H$_{398}$, 115°C) and (b) chain-folded form (n-C$_{246}$H$_{494}$, 104°C). In (b) both the diffraction intensity and the layer spacing decrease after primary crystallisation due to ordering of fold interlayers (secondary crystallisation). L decreases from 190 to 157 Å (equal to half chain length). Note log time scale [32].

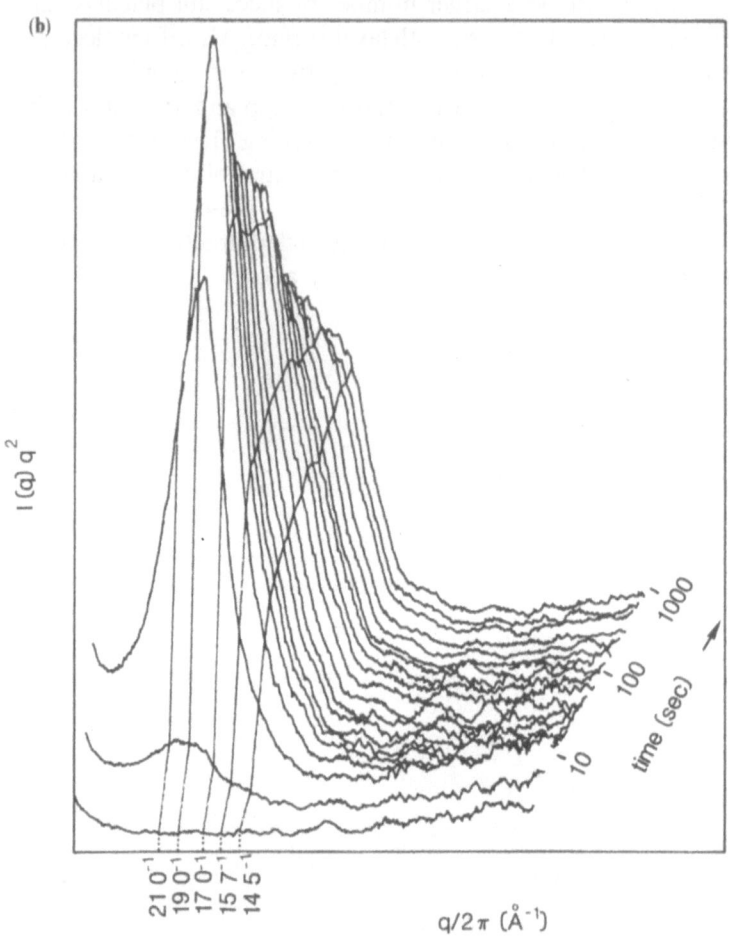

Figure 5.5.   (*Continued*)

'thinning' is thus most likely to involve subsequent crystallisation of the cilia following slow molecular rearrangement, as well as regularisation (tightening) of chain folds, in a process analogous to "secondary crystallisation" of conventional polymers.

No such secondary crystallisation occurs if extended chain crystals are formed directly above T^*. At these small supercoolings only chains which are positioned 'properly' and crystallised along their entire length are stable; lamellae thinner than the length of a full chain will not grow. This difference between extended and folded chain crystallisation is clearly illustrated in Figure 5.5 [32]. While the initial NIF peak in Figure 5.5(b) decreases in intensity and moves with time to lower spacings, the original E peak in Figure 5.5(a), once fully developed, maintains constant position and intensity.

5.5   Phase separation in liquid polymer mixtures

Due to their low entropy of mixing most polymer pairs are immiscible. However, due to specific interactions, such as those between polar groups, miscibility can be found in certain systems, and in such cases there is usually an upper critical dissolution temperature. Below this temperature, interesting critical phenomena may be found around the so-called spinodal line in the phase diagram: spinodal decomposition into separate phases takes place through coherent concentration fluctuations whose amplitude increases with time. The scale of these fluctuations is usually in the region of the wavelength of light and critical opalescence can be seen. However, in certain systems the scale is much smaller. It was found to be within the range of SAXS periodicities in the system poly(methyl methacrylate)/solution chlorinated polyethylene.

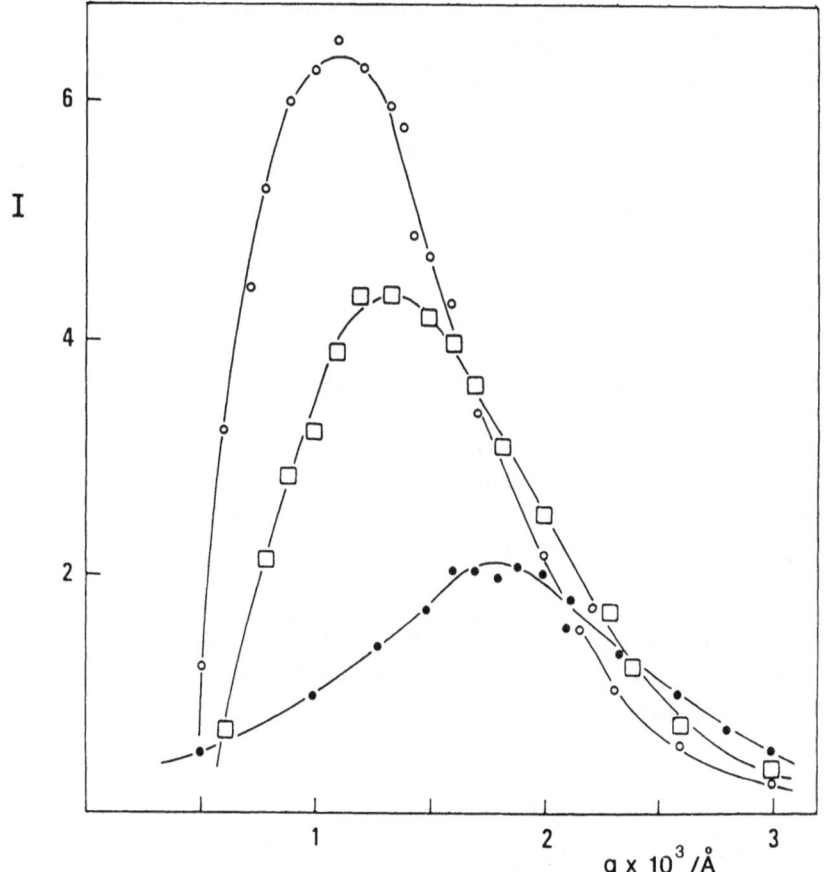

Figure 5.6.   SAXS curves recorded during phase separation in the binary system: high molecular weight poly(methyl methacrylate)/solution chlorinated polyethylene at 117°C. Data are collected after 50 s (○), 100 s (□) and 170 s (●). The increasing scattered intensity and spacing indicate, respectively, the increasing amplitude and wavelength of density fluctuations [33].

Time-resolved SAXS experiments, as well as neutron scattering, have been performed by Hill *et al.* [33]. Figure 5.6 shows the development of a broad scattering peak due to the increasing amplitude of electron density fluctuations during the phase separation. Fundamental parameters such as the correlation length and critical exponent can be obtained from such measurements. Experiments similar to that described above were also performed by Russel [34] on different binary systems.

5.6   Simultaneous diffraction and calorimetry experiments

While time-resolved scattering proves to be extremely useful in studying rapid structural changes, it is often even more desirable to perform another dynamic experiment simultaneously with the recording of time-resolved diffractograms. Russel and Koberstein [7] have combined the rapid collection of SAXS curves with DSC. While the specimen is heated or cooled, and the diffractograms collected, the heat flow into or out of the specimen is being simultaneously recorded. Thus a uniquely defined one-to-one correspondence is established between the structural changes as revealed by diffraction, and thermal events such as endotherms, exotherms or changes in heat capacity. The technique has been applied, e.g. to monitor the complex melting process in solution crystallised polyethylene [7] and to studies of segmented polyurethanes [35].

We have improved the simultaneous X-ray diffraction/DSC technique (XDDSC) by selecting a more X-ray transparent material for specimen cells and have applied the method successfully to studies of phase behaviour in a number of molecular and liquid crystal systems, both polymeric and non-polymeric [8]. Small as well as wide angle diffraction ranges were covered. The material we use for sample pans and cuvettes is either boron nitride or graphite. Both materials are highly transparent to X-rays, have good thermal conductivity and can be easily machined. The use of sample pans made of these materials enables the achievement of a tenfold increase in transmitted X-ray intensity over the previously used aluminium pans. The resulting signal-to-noise ratio and the scan speed can thus be greatly increased.

Two examples of application of the XDDSC technique are given. The first example is the heating scan of an oligomer of the commercially important poly(aryl ether ketone) (PEK). From the scientific point of view it shows

PEK

certain interesting properties, such as multiple melting, which are not yet fully understood [36, 37]. The oligomer with six phenylene rings shows a particularly complex melting behaviour, with four closely spaced endotherms for a sample crystallised from the melt at a moderate rate. The lower melting phases are metastable, since annealing can reduce the number of endotherms to only two or possibly even one.

The melting thermogram of PEK-6 oligomer recorded during an XDDSC heating scan is shown in Figure 5.7(a), and the corresponding diffractograms in Figure 5.7(b). Only the section of the scattering curve showing the second order layer reflection is shown, although the angular range actually recorded extended from low to wide angles. Note that the DSC output was fed into the computer and handled by the same time frame generator as the signal from the linear position sensitive detector; thus there can be no confusion as to the correspondence of the calorimetric and diffraction data. The bold diffraction curves correspond to the arrowed positions in the thermogram, numbered accordingly by roman numerals.

There are no major changes in the wide angle diffraction region associated with the individual endotherms, apart from an overall stepwise reduction in diffracted intensity. Thus the crystalline subcell, i.e. local arrangement of phenylene groups, remains unchanged. However, associated with each endotherm is a small discontinuous change in position of the layer reflections (Figure 5.7(b)). Simultaneously with each of the endothermic transitions seen in Figure. 5.7(a) there is a reduction in intensity and a shift to lower spacings (marked in Figure 5.7(b) in angstroms). There are other subtle but distinct changes in the diffraction pattern, and their analysis is still in progress. It should be noted that the layer thickness in the highest melting form, which is found in the narrow region between 221 and 222°C, closely corresponds to that calculated for orthogonal layers of extended molecules.

The second example of the use of the XDDSC technique is the heating scan of a liquid crystal forming polyether, PHMS-7. This polymer, synthesised by V. Percec, contains rigid methylstilbene mesogen units separated by flexible $-O(CH_2)_7O-$ spacers. In addition to their nematogenic properties, this and the other members of the same series display interesting polymorphism in the crystalline state [38].

A sample of PHMS-7 was quenched from the nematic melt to room temperature and a XDDSC heating scan was performed at 5°C/min (see Figure 5.8(a) and (b)) [39]. The thermogram (Figure 5.8(a)) suggests a complex phase behaviour and, indeed, the series of diffractograms (Figure 5.8(b)) reveals three crystal forms, denoted I, II and III, appearing in succession with increasing temperature. Form I is definitely metastable and its structure is not well understood. Its transformation to Form II is marked by the broad exotherm (b in Figure 5.8(a)). This in turn melts and again recrystallises (sharp exotherm at 135°C) in Form III which finally melts into the nematic state (sharp endotherm d). Attention is drawn to the low angle diffraction region.

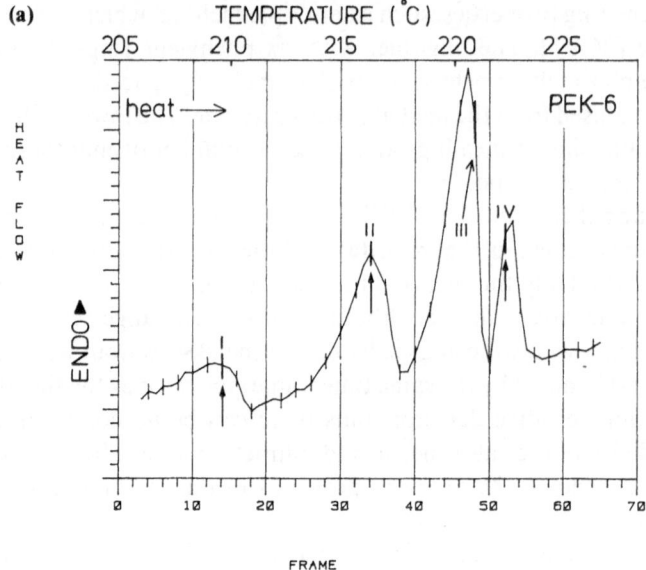

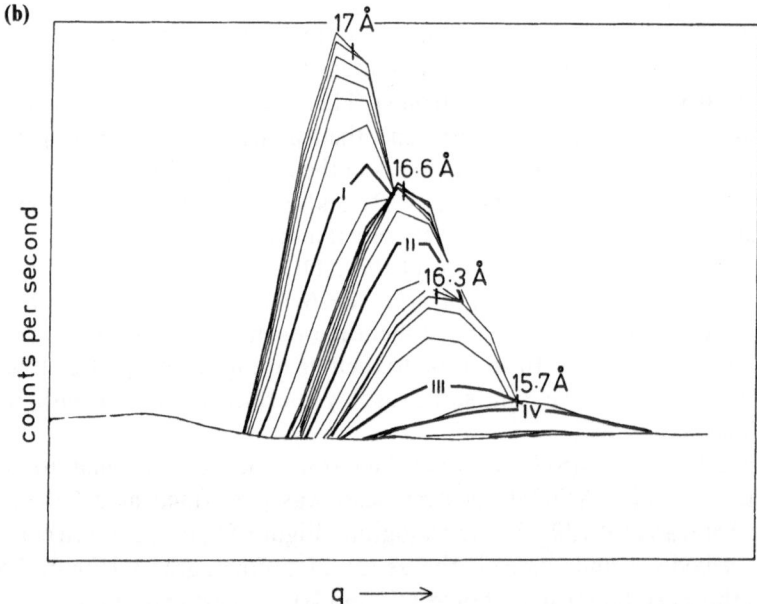

Figure 5.7. Heating XDDSC scan of PEK-6 oligomer in the melting range. Heating rate 0.5 deg/min. (a) Thermogram, (b) simultaneously recorded diffractograms showing the second order layer reflection (002). Two adjacent frames, each recorded for 40 s, were co-added, hence each curve in Figure 5.7(b) covers a temperature interval of 2/3°C. Bold curves, designated by roman numerals, correspond to arrowed positions in the thermogram. A stepwise decrease in intensity and shift to smaller d-spacings occurs at each endotherm [8].

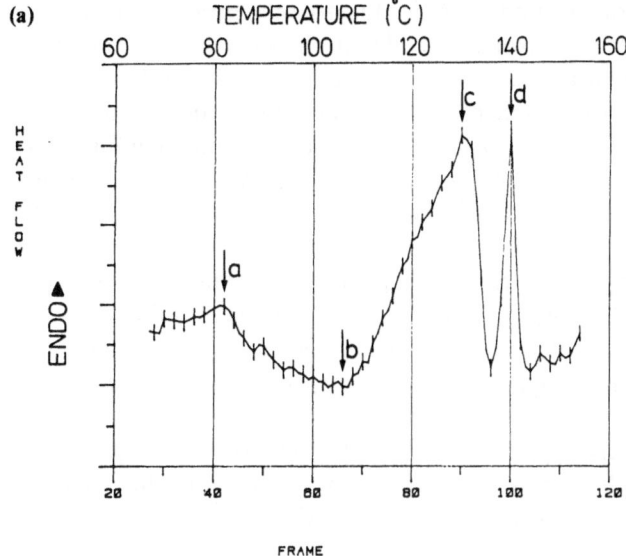

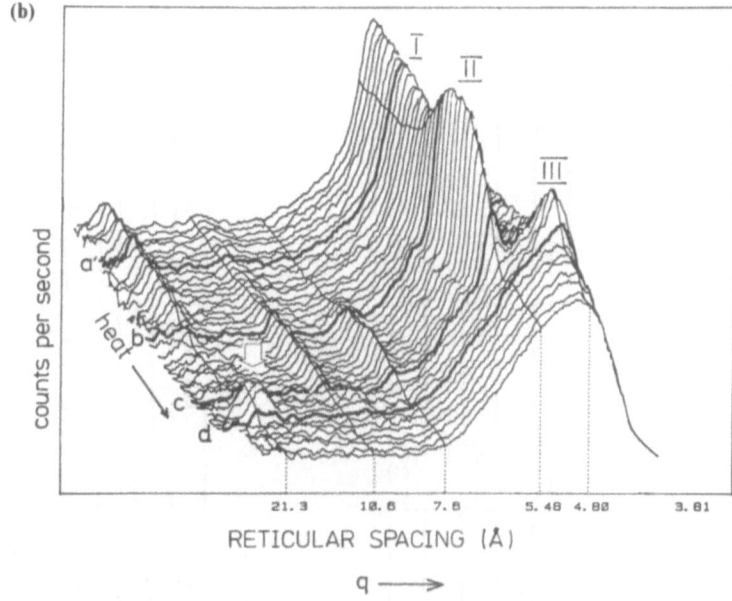

Figure 5.8. Heating XDDSC scan (5°C/min) of the main chain polyether PHMS-7 quenched from the nematic melt. (a) Thermogram, (b) diffractograms covering the low to wide angle region. Two adjacent frames, each recorded for 12 s, were co-added, hence each curve in Figure 5.8(b) covers a temperature interval of 2°C. Bold curves, designated by letters a–d correspond to arrowed positions in the thermogram. Three successive crystal forms are observed (I, II and III). Endotherm a is not associated with any visible structural transformation. Note the reappearance of the ∼ 20 Å reflection in Form III (wide arrow) [39].

The reflection around 20 Å is due to the intramolecular periodicity, the spacing agreeing reasonably well with the length of the monomer unit. Note, however, the disappearance of this reflection in Form II and its reappearance in Form III. Complementary fibre diffraction studies have shown [38] that Forms II and III have structures schematically drawn in Figure 5.9, where the rectangles represent the rigid mesogen units. As seen, the existence of the glide plane in the 'intermeshed' form II extinguishes the reflection corresponding to the monomer repeat. In the 'layer' form III, on the other hand, a very intense reflection appears, corresponding to the monomer repeat (note that only a fraction of the polymer had a chance to crystallise in Form III during the heating run). This high intensity is due to the large difference in density between the mesogen and spacer layers, and to the fact that the thicknesses of the two layers are comparable (see above).

We have also applied the XDDSC method to considerable advantage in studies of other polymeric liquid crystals [40], metallo liquid crystals [41], cyanoterphenyls [42] and hexagonal columnar polymeric phases [43].

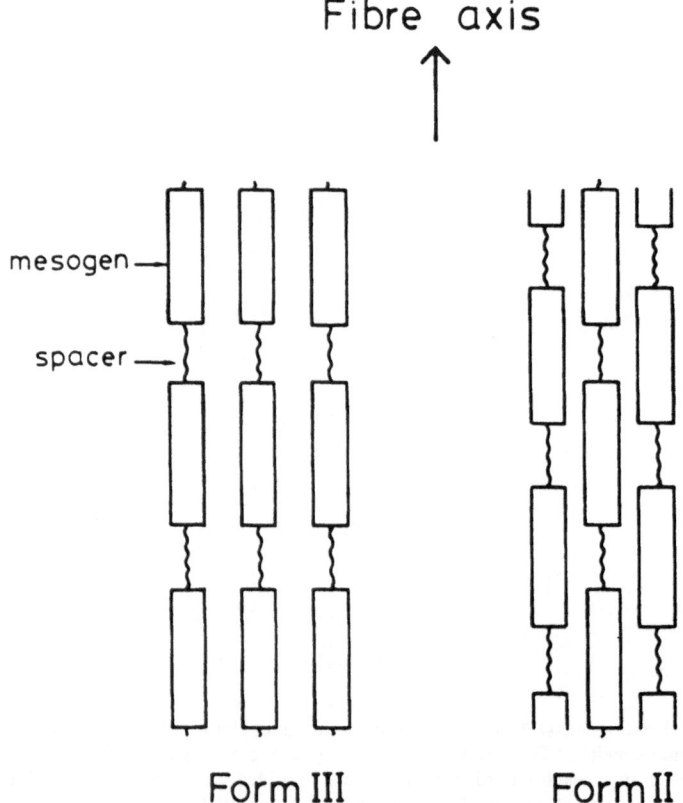

Figure 5.9.   Schematic representation of the mesogen-spacer packing in Forms II (intermeshed) and III (layer) in polymers of the PHMS series [38].

References

1. G. Elsner, C. Riekel and H.G. Zachmann, *Adv. Polym. Sci.* **67** (1985) 1.
2. T.P. Russell, *Adv. Polym. Sci.*, in press.
3. S.J. Spells and M.J. Hill, to be published.
4. G. Ungar, unpublished.
5. R. Zeitz, J. Heuer and C. Riekel, unpublished, quoted in Ref. 1.
6. Wen-li Wu, C. Riekel and H.G. Zachmann, *Polym. Commun.* **25** (1984) 76.
7. T.P. Russell and J.T. Koberstein, *J. Polym. Sci., Polym. Phys. Ed.* **23** (1985) 1109.
8. G. Ungar and J.L. Feijoo, *Molec. Cryst. Liq. Cryst.*, in press.
9. J.A. Odell, A. Keller, E.D.T. Atkins, M.J. Nagy, J.L. Feijoo and G. Ungar, *Mat. Res. Soc. Symp. Proc.* **134** (1989) 223.
10. A. Guinier and G. Fournet, in *Small Angle Scattering of X -Rays*, John Wiley, New York (1955).
11. R. Hosemann and S.N. Bagchi, in *Direct Analysis of Diffraction by Matter*, North-Holland, Amsterdam (1962).
12. O. Kratky and O. Glatter, in *Small Angle X -Ray Scattering*, Academic Press, London (1982).
13. B.K. Vainshtein, in *Diffraction of X -Rays by Chain Molecules*, Elsevier, Amsterdam (1966).
14. L.E. Alexander, in *X-Ray Diffraction Methods in Polymer Science*, Wiley-Interscience, New York (1969).
15. M. Kakudo and N. Kasai, in *X -Ray Diffraction by Polymers*, Kodansha, Tokyo and Elsevier, Amsterdam (1972).
16. F.J. Balta-Calleja and C.G. Vonk, in *X -Ray Scattering of Synthetic Polymers*, Elsevier, Amsterdam (1989).
17. B. Wunderlich, in *Macromolecular Physics*, Academic Press, New York (1976).
18. E.W. Fischer, H. Goddar and G.F. Schmidt, *J. Polym. Sci. B* **5** (1967) 619.
19. O. Kratky, I. Pilz and P.J. Schmitz, *J. Colloid Interface Sci.* **21** (1966) 24.
20. I. Pilz and O. Kratky, *J. Colloid Interface Sci.* **24** (1967) 211.
21. G.R. Strobl, *Kolloid. Z.-Z. Polym.* **250** (1972) 1039.
22. P.J. Barham and A. Keller, *J. Polym. Sci., Polym. Phys. Ed.* **27** (1989) 1029.
23. D.T. Grubb et al., *J. Polym. Sci., Polym. Phys. Ed.* **22** (1984) 367.
24. G. Ungar, *Polymer*, submitted.
25. J.H. Magill and C. Riekel, *Makromol. Chem., Rapid Commun.* **7** (1986) 287.
26. P. Forgacs et al., *J. Polym. Sci., Polym. Phys. Ed.* **18** (1980) 2155; P. Forgacs, B.P. Tolochko and M.A. Sheromov, *Polym. Bull.* **6** (1981) 127.
27. H.R. Brown and N.G. Njoku, *J. Polym. Sci., Polym. Phys. Ed.* **24** (1986) 11.
28. I. Bidd and M.C. Whiting, *Chem. Commun.* (1985) 543; K.S. Lee and G. Wegner, *Matromol. Chem., Rapid Commun.*, **6** (1985) 203.
29. G. Ungar, J. Stejny, A. Keller, I. Bidd and M.C. Whiting, *Science* **229** (1985) 386.
30. G. Ungar and S. Rastogi, unpublished.
31. G. Ungar and A. Keller, *Polymer* **27** (1986) 1835.
32. G. Ungar, in *Integration of Fundamental Polymer Science and Technology*, Vol. 2, eds. P.J. Lemstra and L.A. Kleintjens, Elsevier Applied Science, London (1988) pp. 346–362.
33. R.G. Hill, P.E. Tomlins and J.S. Higgins, *Macromolecules* **18** (1985) 2555.
34. T.P. Russell, G. Hadziioannou and W. Warburton, *Macromolecules* **18** (1985) 78.
35. J.T. Koberstein and T.P. Russell, *Macromolecules* **19** (1986) 714.
36. D.J. Blundell, *Polymer* **28** (1987) 2248.
37. D.C. Bassett, R.H. Olley and I.A.M. Al Raheil, *Polymer* **29** (1988) 1745.
38. G. Ungar and A. Keller, *Mol. Cryst. Liq. Cryst.* **155** (1988) 313.
39. G. Ungar, J.L. Feijoo, A. Keller and V. Percec, to be published.
40. G. Ungar, J.L. Feijoo, A. Keller, R. Yourd and V. Percec, *Macromolecules*, submitted.
41. A.S. Cherodian, J.L. Feijoo, R.M. Richardson, G. Ungar and P.M. Unitt, presented at the Annual Conference of the British Liquid Crystal Society, Sheffield, 10–12 April 1989.
42. A.S. Cherodian, R.M. Richardson, G. Ungar, J.L. Feijoo and G.W. Gray, presented at the Annual Conference of the British Liquid Crystal Society, Bristol, 9–11 April 1990.
43. J.L. Feijoo and G. Ungar, *Amer. Chem. Soc., Polym. Prepts.* in press.

6 EXAFS and structural studies of glasses

S.J. GURMAN

6.1 Introduction

The availability of intense beams of continuous X-radiation from synchrotron sources such as electron storage rings has stimulated the development of a technique of structure determination which is particularly suited to studies of short range interatomic correlations. It has long been observed that in condensed matter, the X-ray absorption cross-section for the photoexcitation of an electron from a deep core state exhibits oscillations as a function of photon energy. This structure is known as the extended X-ray absorption fine structure, or EXAFS. The oscillations are a final state electron effect, arising from the interference between the wavefunction of the outgoing photoelectron and that small part of itself which is scattered back from neighbouring atoms. Thus, conceptually, EXAFS may be considered as a type of electron diffraction where the source of electrons lies within the atom which participates in the electron absorption event. Since the use of this technique for structure determination is so recent, and the theory less commonly described than that of X-ray or neutron diffraction, we will describe it in some detail.

The primary objective of EXAFS studies is to determine the local atomic environment of the excited atom by analysing the measured oscillatory structure. The interference which gives rise to the EXAFS reflects directly the total phase of the backscattered wave, which is largely made up of the product of the photoelectron wavevector and the distance travelled but which also includes contributions from the scattering process and from the passage of the photoelectron out and back through the potential of the excited atom. (The major contribution to the absorption matrix element comes from regions of space very close to the nucleus of the excited atom since the core state is highly localised.) The amplitude of the oscillations depends on the number and electron scattering strength of the scattering atoms. Consequently, analysis of the EXAFS yields information not only on the distance but also on the number and chemical type of the near neighbours of the excited, or central, atom. Also, since the EXAFS spectrum is measured on a known absorption edge, due to an atom of a known chemical type, the technique is chemically specific, giving the coordination of a known type of atom. The EXAFS depends only on the local atomic environment by reason of the fact that only elastically scattered electrons can contribute to the interference and the elastic mean free path of the electrons is short. The technique is therefore particularly useful for the study of amorphous materials where it is the local atomic environment, involving such parameters as chemical order, which interests us. EXAFS spectra typically contain information on atoms up to about 5 Å from the central atom.

In this the first of a series of chapters on the applications of EXAFS we shall begin by considering the fundamentals of the EXAFS process, the theoretical description of which shows what information is contained within the experimental spectrum. We shall then briefly consider data analysis techniques to see how this information may be extracted. The remainder of this chapter will review the application of EXAFS to the determination of glass structure. Subsequent chapters will deal with EXAFS of ionically conducting solids (Chapter 7), catalysts (Chapter 8), surfaces (Chapter 9) and biological molecules (Chapter 11). More detailed information on the EXAFS technique may be obtained from any one of several specialised texts and review articles [1–4].

6.2 Basic principles of EXAFS

The attenuation of X-rays passing through a medium occurs via three principal processes: scattering, pair production and photoelectric absorption. In the energy range of interest for EXAFS studies (1–40 keV) photoelectric absorption dominates the attenuation process. In photoelectric absorption a single X-ray photon is absorbed by an atom, giving up its energy to a single electron which is thereby excited into a higher energy level. The energy balance is expressed by

$$E_f = \hbar\omega - E_b \tag{6.1}$$

where $\hbar\omega$ is the energy of the X-ray photon, E_b the (positive) binding energy of the electron in its initial state and E_f its energy after emission from the atom. We assume that the final state electron is unbound, i.e. that it has a continuous distribution of allowed energies: Rydberg transitions to weakly bound states are not involved in EXAFS.

In order to understand the mechanism which gives rise to the EXAFS spectrum, let us consider the K absorption edge, corresponding to excitation of an electron from the 1s level: this is the edge most often used in EXAFS spectroscopy. In the dipole approximation the probability of X-ray absorption is given by [5]

$$P = \frac{2\pi^2 e^2}{mc^2\omega} |\langle f|\boldsymbol{\varepsilon}\cdot\mathbf{r}|i\rangle|^2 \rho(E_f) \tag{6.2}$$

where $|i\rangle$ is the wavefunction of the initial 1 s state, $|f\rangle$ that of the final state and $\rho(E_f)$ is the density of allowed states at the final state energy E_f. ε is the electric field polarisation vector of the X-ray beam. The selection rules in the dipole approximation force $|f\rangle$ to be a p state if the initial state is s-like. For X-ray energies well above the absorption edge, the density of states $\rho(E_f)$ is a smooth function and may be approximated by that of a free electron of momentum k and energy $\hbar^2 k^2 / 2m$ where

$$\frac{\hbar^2 k^2}{2m} = E_f + \bar{E} = \hbar\omega - E_{\text{edge}} + E_o \qquad (6.3)$$

in which k is the photoelectron momentum and $\bar{E}$ is the energy of a free electron of zero momentum and corresponds to the effective mean potential experienced by an excited electron in the medium. The various energies are shown diagrammatically in Figure 6.1. Note that we measure the photo-electron energy and momentum not from the position of the edge, but from a lower energy. The absorption edge occurs at the Fermi energy (or at least at the energy of the lowest unoccupied state) and corresponds to a photon energy E_{edge}. The offset E_o is usually called the threshold energy: it is of order $10\,\text{eV}$. With the free electron assumption for $\rho(E_f)$ the only factor that can give rise to the EXAFS signal is the matrix element. Now the initial state wavefunction is fixed and does not vary with ω. The final state wavefunction $|f\rangle$ does vary with ω and it is this variation which gives rise to the fine structure.

The wavefunction $|f\rangle$ may be considered as a sum of two contributions. If the atom were isolated, then the photoelectron would be in a solely outgoing state relative to the central atom. In this case the matrix element, and hence the absorption coefficient, shows no fine structure and the X-ray absorption coefficient varies monotonically with ω. Such is the case for a monoatomic gas such as argon. If now the central atom is surrounded by other atoms, as in a

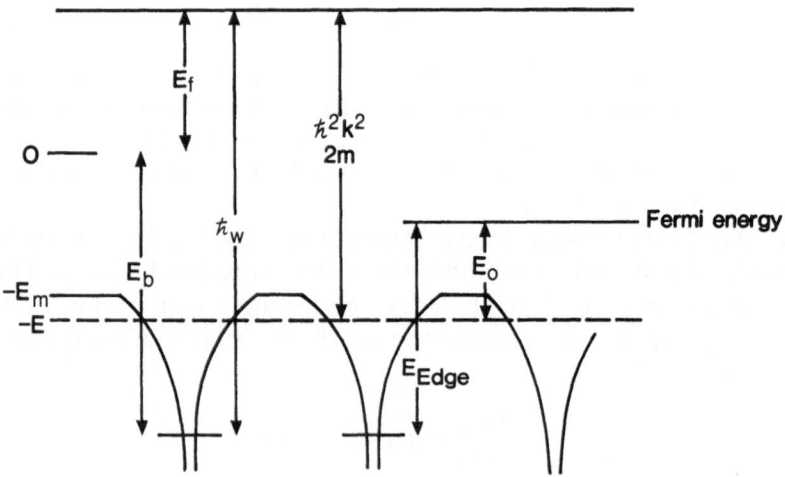

Figure 6.1.   Energies in the X-ray absorption process.

molecular gas or any condensed phase, then the outgoing electron will be scattered by the surrounding atoms, giving rise to incoming waves. These incoming, or backscattered, waves can interfere constructively or destructively with the outgoing wave. This interference gives rise to an oscillatory variation in the matrix element as a function of ω, since the wavelength of the photoelectron is a function of ω according to the Eqn. 6.3. The significant region of space for the interference is the region where $|i\rangle$ exists, close to the nucleus. Constructive interference there increases the matrix element, and hence the absorption coefficient, whilst destructive interference lowers the matrix element below the free atom value.

The absorption coefficient for a condensed sample can therefore be written as

$$\mu(k) = \mu_0(k)[1 + \chi(k)] \qquad (6.4)$$

where k, the photoelectron momentum, may be expressed in terms of the photoelectron energy, and hence of the photon energy, by use of Eqn. 6.3. $\mu_0(k)$ is the smoothly varying background which physically corresponds to the absorption coefficient of an isolated atom. The fine structure, or EXAFS function, arising as a consequence of the interference between outgoing and backscattered electron waves is therefore defined by

$$\chi(k) = [\mu(k) - \mu_0(k)]/\mu_0(k) \qquad (6.5)$$

We note that $\mu_0(k)$ is only that part of the atomic absorption coefficient which is due to transitions from the initial state of interest, i.e. the contribution of one particular edge. The first step in data reduction is to remove the contributions of the other edges from the measured absorption coefficient, which may be done by fitting a smooth function, such as the Victoreen expression [6], to the absorption below the edge of interest and extrapolating to higher energies. Similarly we obtain $\mu_0(k)$ by fitting such an expression to the edge contribution and so obtain the EXAFS function $\chi(k)$. The various stages of this process are illustrated in Figure 6.2.

6.2.1   The EXAFS function

The theory of the EXAFS function has a long history, going back to the original work of Kronig [7]: indeed EXAFS is still sometimes called Kronig structure. The first successful theory of EXAFS, based on the physical ideas discussed above, was developed in the early and middle 1970s and has subsequently been refined and generalised by several workers [8–12]. We shall limit ourselves here to a brief description of the single-scattering theory appropriate for amorphous materials with an occasional comment on multiple-scattering contributions and single-crystal studies, applications of which can be found in Chapters 8 and 9.

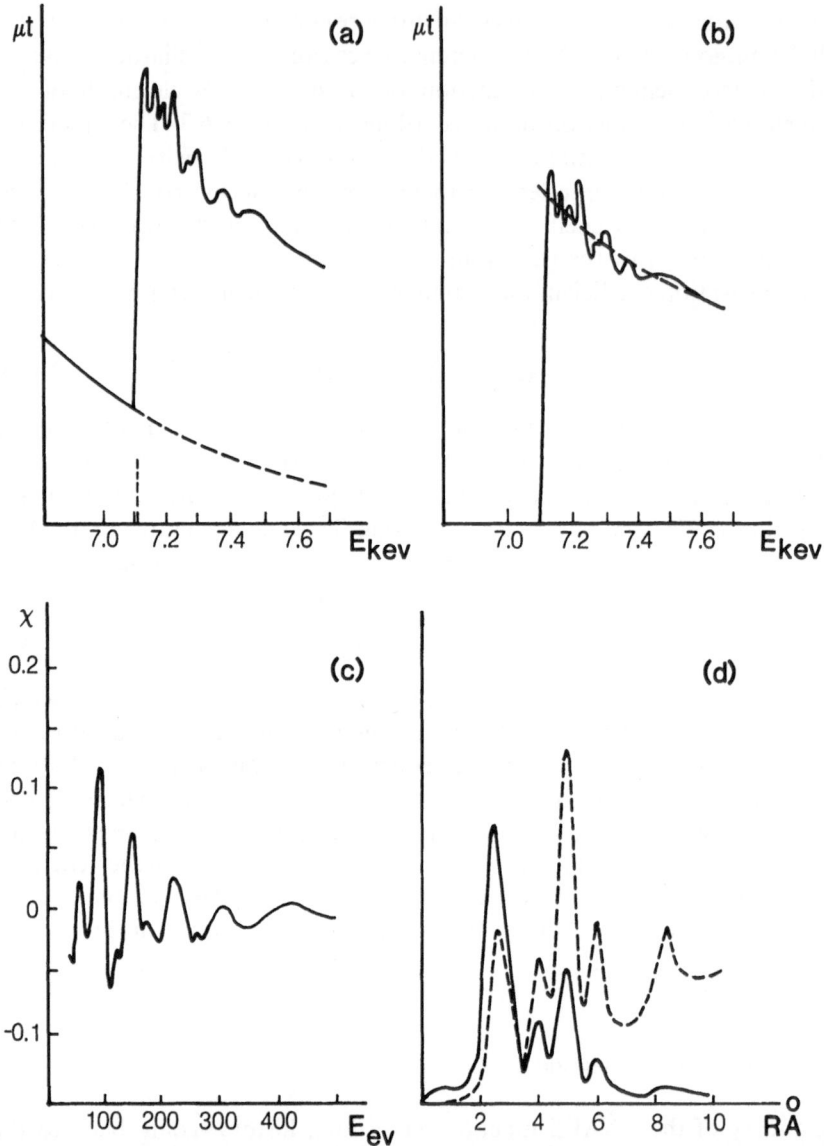

Figure 6.2. Extraction of EXAFS data. (a) Total absorption spectrum. The dashed line is the extrapolated contribution from lower energy edges which is subtracted from the total absorption to obtain the contribution from the edge of interest, which is shown in (b). The dashed line in (b) is the smooth atomic absorption factor μ_0: with this the EXAFS function shown in (c) may be obtained by use of Eqn. 6.5. The Fourier transform of the EXAFS function is shown in (d) where the solid line is the Fourier transform of $k \chi (k)$ and the dashed line R^2 times this, which corresponds to the partial RDF (Eqn. 6.13) when the phase factors are taken into account.

EXAFS is essentially an electron scattering problem and as is usual in such problems we expand our wavefunctions in angular momentum eigenfunctions: this is particularly useful here since the initial state has a well-defined angular momentum. Thus we write the matrix element for an isolated atom as

$$\mu_0(E) \approx \sum_{m_0} \sum_{lm} |\langle lm|\boldsymbol{\varepsilon}\cdot\mathbf{r}|l_0 m_0 \rangle e^{i\delta_l}|^2 \qquad (6.6)$$

where l_0 and m_0 are the angular momentum quantum numbers of the initial state, the m_0 sum being over the degenerate sub-levels of total angular momentum l_0. For a K edge we have $l_0 = 0$ and so $m_0 = 0$ only. l and m are the angular momentum quantum numbers of the final state wavefunction, whose possible range will be limited by selection rules: for a K edge $l = 1$ only, δ_l is the phaseshift introduced by the central atom potential: with this factored out the matrix element may be shown to be real [13].

In a condensed sample the outgoing wave part of the photoelectron wavefunction is scattered by the atoms surrounding the central atom, giving rise to EXAFS. The scattered wave may also be written as a sum over angular momenta. A single angular momentum component of the outgoing wave will give rise to many components in the scattered wave. In this situation the final state wavefunction may be written as

$$|f\rangle = \sum_{LM} (I + Z)^{L'M'}_{LM} |LM^4\rangle \qquad (6.7)$$

where the unit matrix represents the outgoing wave and the Z matrix the scattered wave. With this form of wavefunction the EXAFS function $\chi(k)$ defined by Eqn. (6.5) can be written down as a linear function of the matrix Z:

$$\chi(E) = \frac{1}{\mu_0} \sum_{m_0} \sum_{\substack{lm \\ l'm'}} \langle l_0 m_0|\boldsymbol{\varepsilon}\cdot\mathbf{r}|lm \rangle\, 2\,\mathrm{Re} Z^{lm}_{l'm'}\, c^{i(\delta_l + \delta_{l'})} \langle l'm'|\boldsymbol{\varepsilon}\cdot\mathbf{r}|l_0 m_0 \rangle \qquad (6.8)$$

The matrix elements now refer only to isolated-atom wavefunctions. All of the effects of the scattering atoms are contained in the matrix Z and the EXAFS function is known once Z is known. All of the different forms of EXAFS theory arise from the different forms and approximations used for Z.

The Z matrix can be expanded in a series over the different orders of scattering and usually only the single-scattering contribution need be considered. Also, for a given order of scattering, Z may be written as a sum over electron paths away from and back to the central atom. Since the electron mean free path is short, only a few paths contribute to Z. Further, since χ is a linear function of Z there is no interference between the different paths in χ. Thus χ is also a sum over electron paths where we may treat each path separately. Multiple scattering contributions may therefore be simply added on to the single-scattering results if this proves to be necessary or useful. Such multiple scattering contributions contain information on bond angles and analysis of them enables us to determine these. Such analyses are described in Chapter 8 e.g. $Rh_6(CO)_{16}$ in Section 8.4. The limited mean free path means that the path sum is calculable, but the usefulness of the technique is limited to local structure.

At very low photoelectron energies, i.e. very close to the edge, the electron mean free path is long and in this region performing the path sum becomes impractical. This is the region of XANES, X-ray absorption near edge structure, for the study of which we need a full multiple-scattering theory [14]. We shall not discuss this topic here.

In the case of a single crystal sample the direction of the X-ray beam, and hence of ε, is fixed relative to directions in the crystal. The polarised character of synchrotron radiation is discussed in Chapter 1. We take this into account in the analysis of Eqn. 6.8 by writing $\varepsilon \cdot r$ in terms of angles, the result being an expression for $\chi(k)$ which depends on the beam direction and polarisation. In surface studies we can use this dependence to identify the attitude of adsorbate molecules on crystalline surfaces: this point is considered in more detail in Chapter 9.

For amorphous samples the angle between the electric field vector of the beam ε and r is randomly distributed and averaging over this angle leaves only the diagonal elements of Z to contribute to χ [10]. Also, for a K edge, the limitation to $l = 1$ only means that the atomic matrix elements cancel out of Eqn. 6.8 and we have

$$\chi(E) = \frac{2}{3} \operatorname{Re} \sum_m \sum_{lm} Z_{lm} e^{2i\delta_l} \qquad (6.9)$$

The form of Z can be calculated exactly from theory. In a single-scattering process the outgoing spherical wave travels from the central to the scattering atom, is scattered there, and then travels back to the central atom. On working through the detailed algebra of this process we have, for a scattering atom at a distance R from the central atom

$$\sum_m Z_{lm} = \sum_{L_1 L_2} [h^{(1)}_{L_1}(kR)]^2 T_{L_2} \tfrac{1}{2}(2L_1 + 1)(2L_2 + 1)[C(L_2 L_1 l; \text{ooo})]^2 e^{2i\delta_l} \qquad (6.10)$$

The two Hankel functions come from the propagation of spherical waves along the two legs of the single-scattering path and the scattering t-matrix, T, gives the atomic scattering. The other factors come from the angular momentum expansions necessary in following the process. The origins of this form for $\chi(k)$, the curved wave theory, are explained in detail elsewhere [10].

The curved wave theory is exact, but does not give us much physical insight. To obtain this we consider the plane wave theory, the earliest form of EXAFS theory. In this simplified theory, the two Hankel functions of Eqn. 6.10 are replaced by their asymptotic forms, which are simple exponentials. The two L

sums may then be performed analytically to obtain

$$\chi(k) = \frac{1}{kR^2} |f(\pi)| \sin(2kR + \psi|2\delta)$$ (6.11)

$f(\pi)$, whose phase is ψ, is the atomic backscattering factor for electrons. Electron scattering by atoms is a complex process and $f(\pi)$ shows a strong dependence on electron energy and also on the chemical type (atomic number) of the scattering atom: we use this variation to identify atoms in EXAFS spectroscopy. Examples of $f(\pi)$ are shown in Figure 6.5.

The plane wave form of the EXAFS function may easily be extended to describe multiple-scattering by following the progress of the electron from atom to atom. For a double-scattering path involving atoms 1 and 2 at distances R_1 and R_2 from the central atom we have [15]

$$\chi_2(k) = \frac{-1}{k} \frac{1}{R_2} |f_2(\theta_2)| \frac{1}{|R_2 - R_1|} |f_1(\theta_1)|$$

$$\times \frac{1}{R_1} \sin[k(R_1 + |R_2 - R_1| + R_2 + \psi_1 + \psi_2 + 2\delta)]$$ (6.12)

The extension of the exact curved-wave theory to include multiple-scattering contributions has also been made [11]. Multiple-scattering contributions to EXAFS are considered in Chapter 8.

6.2.2 Structural information in EXAFS

The expression for the EXAFS function $\chi(k)$ given as Eqn. 6.11 relates to a single scattering atom at a distance R from the central atom. In an amorphous material or powder solid there will be many such atoms and we must sum over them to obtain the total EXAFS signal. The definition of the partial radial distribution functions (RDF) $P_{\alpha\beta}$ as the number of atoms of a given chemical type β at a given distance from an atom of a particular type α shows that these are the functions which we must use in such a sum. Since we know the chemical type of the central atom, because we are looking at the EXAFS on a particular edge, we have for the EXAFS on an absorption edge of an atom of the type α

$$\chi_\alpha(k) = -\sum_\beta |f_\beta(\pi)| \int_0^\infty \frac{dr}{kr^2} P_{\alpha\beta}(r) \sin(2kr + \psi_\beta + 2\delta) e^{-2r/\lambda}$$ (6.13)

where the sum is over the different types of atom present in the sample, including α. In this equation we have also introduced a term involving the elastic mean free path of the electron, λ, in order to allow for the effects of inelastic scattering.

A comparison of the form of Eqn. 6.13 with those obtained for the diffuse scattered radiation in diffraction experiments shows that EXAFS gives very

148 APPLICATIONS OF SYNCHROTRON RADIATION

similar structural information as do such techniques: this is not surprising, since Eqn. 6.13 is based on a single-scattering theory, this time of electrons. The great advantage which EXAFS has over diffraction techniques is that we may simply measure the structure on the absorption edges of all of the different types of atom in the sample, thus directly obtaining the local environment of each type of atom separately.

However, EXAFS experiments are not usually interpreted using Eqn. 6.13, essentially because the presence of the various phase factors ψ_β and 2δ, and the strong energy dependence of the backscattering factors $|f_\beta(\pi)|$ (see Figure 6.5) make taking a Fourier transform difficult. Also the plane wave approximation, on which Eqn. 6.13 is based, is not usually sufficiently accurate for use in structural studies. Such a 'real space' analysis is possible and is described below. The multiple scattering contributions further complicate matters in this technique. It is more usual to assume that the partial radial distribution functions $P_{\alpha\beta}(r)$ consist of a set of peaks of simple shape and then to perform the integration in Eqn. 6.13 analytically. In this case the area under each peak corresponds to a partial coordination number N and our analysis is then described in terms of 'shells of atoms'. A typical shell contains N_j atoms, all of the same chemical type, lying at an average distance R_j from the central atom. If we assume that the peak shape is gaussian with a mean square variation σ_i^2 in interatomic distance, we obtain

$$\chi_\alpha(k) = -\sum_j \frac{N_j}{kR_j^2}|f_j(\pi)|e^{-2\sigma_j^2 k^2}e^{-2R_j/\lambda}\sin[2kR_j + \psi_j + 2\delta] \quad (6.14)$$

for the single scattering contribution. The form of the multiple scattering contribution can be obtained equally simply.

The peak shape we have used gives rise to the usual Debye-Waller factor. In a crystal, the mean square variation in interatomic distance is due to thermal motion, but we note that amplitudes of the two atoms, because of these thermal motions, are to some extent correlated: this is due to the contribution from long wavelength phonons which have little effect on short interatomic distances [16]. In an amorphous system there will also be an additional contribution to σ_j^2 from the static variability in the distances due to structural disorder. Such a contribution may also appear in complex crystals due to a variety of slightly different sites (see Chapter 7 for some examples). The two contributions may be separated by making temperature-dependent measurements. The assumption of a gaussian peak shape is by far the commonest and is generally valid for thermal motion and for weak structural disorder. It may, however, be a poor approximation for highly disordered solids, since the difference between the rapidly varying, hard-sphere repulsive part of the interatomic potential and the flatter attractive part tends to lead to asymmetric peak shapes in these materials, especially for the nearest neighbours. A notable example of an anharmonic system (α-AgI) is discussed in Section 7.2.

6.3   EXAFS data analysis

EXAFS spectroscopy is exclusively used to determine the local structure in disordered solids and liquids and in complex molecules and thus our interest in data analysis is concentrated on how to best obtain the atomic coordinations from an experimental spectrum. The methods of data analysis in use at the present time may be grouped into two classes: *real space analysis*, where we work on the Fourier transform of the EXAFS spectrum $\chi(k)$; and *k-space analysis* where we work with the spectrum itself. In either case, the first step is to extract the EXAFS function from the measured data using the techniques of background subtraction and normalisation described in Section 6.1 and illustrated in Figure 6.2. This process may be carried out using standard fitting routines and is usually straightforward, although the small size of $\chi(k)$ compared to unity means that the fitting procedure used to give $\mu_0(k)$ must be rather accurate.

6.3.1   *Real space analysis*

The simplest method of analysis, and the one which was almost exclusively used in the early days of EXAFS work (1970–1975), is to Fourier transform $k\chi(k)$ with respect to $\sin(2kR)$ or $\exp(-2ikR)$: the latter is to be preferred since taking the modulus of a complex transform removes some of the problems associated with the finite data range. In particular, the modulus does not show the rapid oscillations of period $\pi/k_{\max}$ which appear in the sine transform (see Figure 6.3); the presence of these oscillations severely limits the usefulness of many of the early EXAFS analyses. The multiplication of $\chi(k)$ by k before transforming is simply to remove the factor of k^{-1} in Eqn. 6.13. In many cases the spectrum is weighted by k^n, with $n = 3$ being most common, in order to counteract the decrease in the amplitude of $\chi(k)$ with increasing k due to $f(\pi)$ and $\exp(-2\sigma^2 k^2)$: a constant amplitude function gives a sharper Fourier transform.

The result of the simple Fourier transform is $\sum_\beta P_{\alpha\beta}(r)$ with the r values of the peaks somewhat shifted due to the effects of the extra phase contribution $2\delta + \psi$, weighted by the various backscattering factors and including truncation effects from the finite data range. As is the case with diffraction data, the latter may be minimised at the cost of some peak broadening by the use of a window function. The effects of the extra phase usually appear as a simple shift in the peak position which is always negative and usually between 0 and 0.5 Å. This occurs because the total phase $2\delta + \psi$, which depends on both the central and scattering atoms, is usually quite accurately represented by a linear function

$$2\delta + \psi = ak + b \tag{6.15}$$

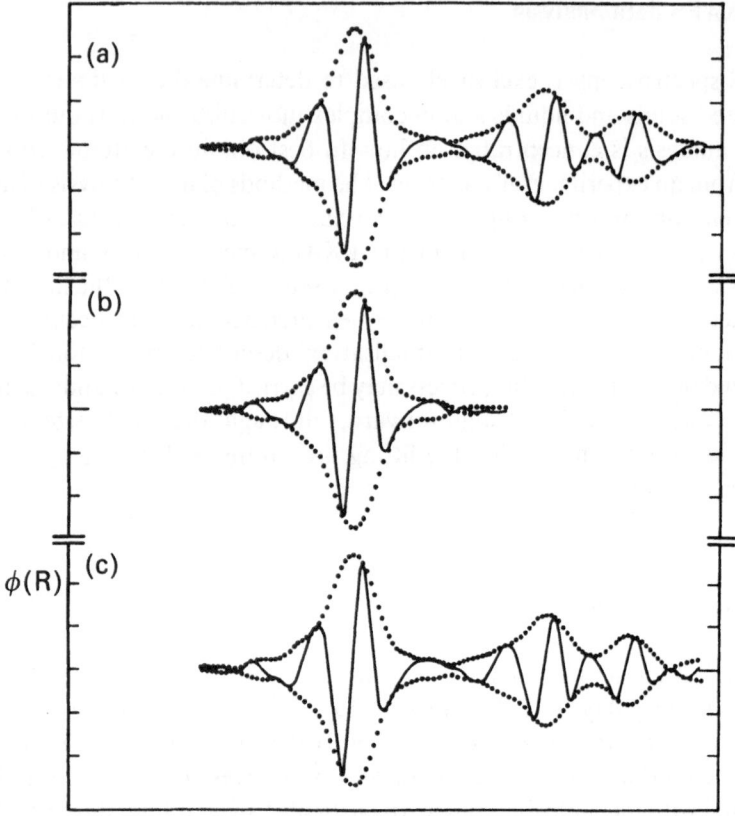

Figure 6.3.   The sine (solid line) and exponential (dotted line) Fourier transforms of the EXAFS on the Ge K edge in crystalline germanium. (a) Full experimental spectrum. (b) Nearest neighbour contribution giving ξ_{Ge-Ge}. (c) Calculated ϕ (r) using ξ_{Ge-Ge} and three guassian peaks. After Hayes and Boyce [1].

The shift in peak position is then $a/2$. The simple Fourier transform is most useful as a preliminary analysis, where its speed and simplicity are major advantages, since it fixes the interatomic distances fairly well and gives a good idea of the amount of information in the spectrum.

To improve the analysis we may weight the experimental spectrum using calculated backscattering amplitudes and take the Fourier transform with respect to $\exp(-2ikR - 2\delta - \psi)$ using calculated phases. In compound systems the scattering parameters we choose are of course those of the chemical type that give the dominant contribution to the spectrum, which is usually that which provides the majority of nearest neighbours. Peaks due to this type of atom will appear at the correct distances in the Fourier transform, whilst those due to other types will be shifted by an amount equal to the difference in the values of $a/2$ between the phase factors of that type and those of the reference type: these shifts are usually small and less than the peak width. This modification gives good RDFs and is extensively used in the presentation of data. An example of a weighted transform is shown in Figure 6.6.

We may also obtain the detailed structural parameters (interatomic distances, coordination numbers, etc.) of the sample from a fit to the Fourier transform of the EXAFS function. If we Fourier transform $k\chi(k)$ with respect to $\exp(-2ikR)$ we obtain a function of R which may be written as

$$\phi(R) = \sum_{\beta} \int_0^{\infty} \frac{dr}{r^2} P_{\alpha\beta}(r)\xi_{\alpha\beta}(R-r) \tag{6.16}$$

where the *peak function* $\xi(R-r)$ is the Fourier transform of the scattering factors which appear in Eqn. 6.13. It is directly analogous to the peak function which appears in X-ray diffraction, where it arises from the Q dependence of the atomic scattering factor. ϕ is therefore the convolution of the partial RDFs with this peak function. It is an array of peaks, one for each peak in $\sum_{\beta} P_{\alpha\beta}(r)$, but these peaks are shifted from the true interatomic distances because $\xi(r)$ does not peak at $r = 0$, due to the k dependence of the scattering factors; this shift is identical to that described above.

The first step in detailed data analysis [1] is to obtain the form of the peak function $\xi_{\alpha\beta}(r)$ for each pair of atoms. These may be extracted from the $\phi(r)$ obtained from the EXAFS measured on a standard of known crystal structure, if such a crystal with non-overlapping peaks in its partial RDF is available. This technique assumes that peak functions are transferable from one material to another: this has been found to be the case to a high degree of accuracy. This condition corresponds to saying that the scattering factors are independent of chemical environment. Alternatively, we may calculate a $\chi(k)$ for a known simple peak shape (such as a δ-function!) using calculated scattering data and the experimental data range. This can then be Fourier transformed to give $\xi_{\alpha\beta}(r)$ for use in the fitting routine.

Whichever method is used to obtain a set of $\xi_{\alpha\beta}(r)$, the final step in data analysis is to calculate a simulated $\phi(r)$ using them and a model for the $P_{\alpha\beta}(r)$. The shape of the peaks in the model $P_{\alpha\beta}(r)$ need not be constrained and in particular may be allowed to show asymmetry. The model parameters in $P_{\alpha\beta}(r)$ are then varied until a best fit is found between the simulation and the experimental $\phi(r)$: a standard least squares routine may be used for this. The stages of this process are illustrated in Figure 6.3. This method of data analysis has been extensively developed by Hayes and co-workers and is well described in the literature [1].

6.3.2 *k-space analysis*

The Fourier transform of the EXAFS function shows us how much information is contained within the data and gives approximate interatomic distances. To obtain more detailed information we need to fit to the data. This may be done in real space, fitting to a Fourier transform of $\chi(k)$ as described above, but is usually more reliably done in k-space, fitting to the spectrum itself. In many ways, k-space analysis is also safer, since we avoid the problems associated with a finite-range Fourier transform, which at the very least smooths away the experimental noise. Fitting in k-space can be done with respect to the raw data.

It is possible to separate out the contributions to the EXAFS function from different shells of atoms, i.e. from different peaks in the partial RDFs, if these are well-defined, and this forms the basis of the simplest method of k-space analysis [3]. To perform this separation, the experimental EXAFS spectrum is first Fourier transformed with respect to $\exp(-2ikR)$. A small part of the resulting real-space spectrum, containing only one peak, is then back-transformed to k-space, the result being the contribution to the EXAFS from atoms lying in a single shell; such atoms are assumed to be all of the same chemical type. This is referred to as a *filtered* spectrum. Such a filtered spectrum taken from a crystalline sample of known structure may be used to obtain an empirical form of the amplitude factor $|f(\pi)|\exp(-2\sigma^2k^2)$ and of the phase $2\delta + \psi$ for the pair of atoms in question since the other parameters in Eqn. 6.13, N_j and R_j, are known exactly. These may then be used in place of the possibly inaccurate calculated scattering data in the k-space fitting method described below.

The empirical scattering data may also be used directly in a simple graphical method to determine the structural parameters of an unknown sample, although the applicability of this method is limited to simple cases. In this method we measure the EXAFS spectra from our system of interest and also from a standard crystal which contains the same pair correlations. We then filter both to isolate the EXAFS from a given atom pair at a given distance. (This is not always possible, hence the limited applicability.) If the interatomic distance is the same in the unknown and standard samples, i.e. if the two peaks in the two Fourier transformed spectra line up, then the two filtered spectra will have the same period. If not then we may determine the interatomic distance in the unknown by comparing the phases of the two spectra as a function of k; this is usually done by plotting the k values at which the filtered spectra go through zero where we have

$$2k_nR + 2\delta + \psi = n\pi, \quad n = 1,2,3,... \qquad (6.17)$$

If R is known for the standard sample then we can obtain $2\delta + \psi$ as a function of k for a particular atom pair (these have also been tabulated [17]) and hence obtain R in the unknown sample. Further, if we measure the amplitudes at maxima and minima and ratio them then we have

$$\frac{\chi(k,\text{sample})}{\chi(k,\text{standard})} = \frac{N\exp(-2\sigma^2k^2)}{N_s\exp(-2\sigma_s^2k^2)} \qquad (6.18)$$

from the expression given as Eqn. 6.13. Taking the natural logarithm of this

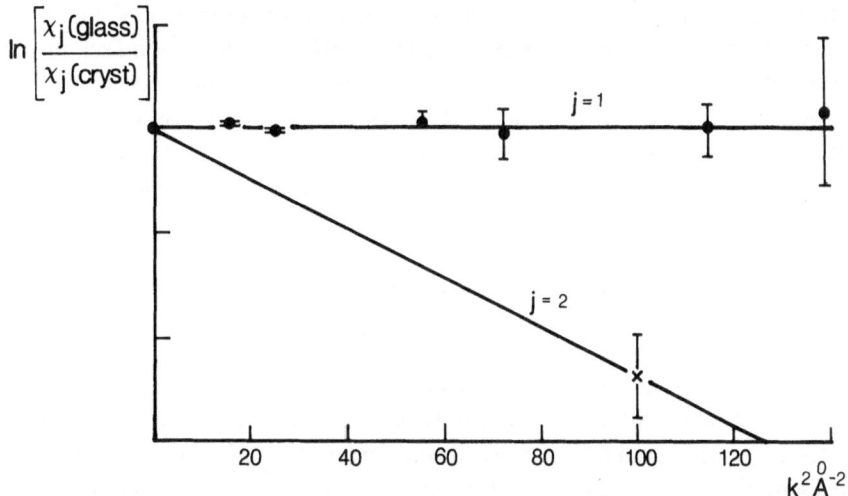

Figure 6.4. Illustration of the graphical method of EXAFS analysis, after Sayers *et al.* [41], for GeO_2. $j = 1$ is the nearest neighbour Ge–O coordination and $j = 2$ is the second neighbour Ge–Ge coordination.

ratio and plotting it against k^2 will give us $(\sigma_s^2 - \sigma^2)$ as a slope and N/N_s as intercept from which the structural parameters of the unknown may be determined. An example is shown in Figure 6.4. It is apparent from this figure that the nearest neighbour oxygen shell in amorphous GeO_2 has the same structural parameters as that in the crystal, whilst the second neighbour germanium shell is much more disordered ($\Delta\sigma^2 = 0.012$ Å^2.) although the coordination is the same. This extra disorder comes from the static disorder characteristic of the amorphous state and corresponds to an RMS distortion of 6° in the Ge—O—Ge bond angle of 130°.

The main problem in the use of this method is the need for two Fourier transforms, each taken over a limited data range, to obtain the filtered spectra. Particular care must be taken with the second transform as regards the range of distances included and the window function used, and serious difficulties obviously arise if the sample contains two shells of atoms at only slightly different distances. Its major advantage lies in its computational simplicity, in that no calculated scattering data are needed if a suitable standard is available; and the simplicity of the graphical technique for extracting structural parameters.

The most commonly used method of data analysis in k-space relies on fitting the experimental EXAFS spectrum itself by calculating a spectrum using an assumed set of structural parameters and adjusting these until a best fit to experiment is obtained. This is done shell-by-shell (remember, a shell of atoms is that set of N_j identical atoms at the same distance R_j from the central atom) using the expressions given as Eqn. 6.14 or the corresponding version derived from Eqn. 6.10 if the curved wave theory is to be used. Calculated scattering factors are used in general, although empirical factors obtained in the manner described above may also be employed. This method requires good scattering information and can be rather time-consuming computationally, but it undoubtedly gives the most detailed structural information. It is also the most suitable method for use when the exact expression for the EXAFS function, Eqn. 6.10, has to be used or when multiple scattering or single crystal effects need to be included. Least squares fitting in k-space has been particularly developed by EXAFS groups using the SRS [9–13] and has resulted in the interactive code EXCURVE. This program utilises the curved wave theory [9] as well as the small atom approximation and also incorporates single and multiple scattering options.

The scattering factor $f_j(\pi)$ and the electron mean free path λ are both aspects of the interaction between the photoelectron and the atoms and other electrons of the system. The scattering properties of atoms are usually represented in terms of energy dependent scattering phase shifts and if we assume that the atomic potential is spherically symmetric we obtain the standard result

$$f(\theta) = \frac{1}{ik} \sum_l (2l + 1)(e^{2i\delta_l} - 1)P_l(\theta) \qquad (6.19)$$

The scattering phaseshifts δ_l are energy-dependent and different for atoms of different Z. If the form of the atomic potential $V(r)$ is known then they may be calculated by solving the Schrodinger equation, as is demonstrated in standard texts on quantum mechanics [5,18]. It is found that the δ_l are independent of the bonding environment of the atom, except for extremely low electron energies; this is the basis of the transferability of scattering factors mentioned above. Thus a data bank containing scattering phase shifts for all the elements of interest suffices for the EXAFS data analysis of all samples. Tabulations of this data have been published [17,19]. Some examples of backscattering factors $f(\pi)$ are shown in Figure 6.5. The different energy dependences of $f(\pi)$ for atoms of different Z are clear in this figure; it is this property which we use to identify the atom type in this method of EXAFS data analysis. In this example, the effects of inelastic scattering are not included in $f(\pi)$; they are treated by the exponential factor which appears in Eqn. 6.14. The form used is a constant imaginary part, V_i, added to the potential seen by the photoelectron. This gives a mean free path for elastic scattering

$$\lambda = - 2k/V_i \qquad (6.20)$$

in atomic units. This has the usual $\sqrt{E}$ variation valid for electron energies well above the plasmon threshold. V_i typically takes the value $- 4\,\text{eV}$ in metals and $- 2\,\text{eV}$ in insulators.

Once we have a set of scattering data, fitting an EXAFS spectrum is straightforward if we use a standard least squares routine. The structural parameters to be varied in obtaining the optimum fit are those which appear in Eqn. 6.14; N_j, R_j, σ_j^2, and the chemical type which fixes $f_j(\pi)$. The fitted values

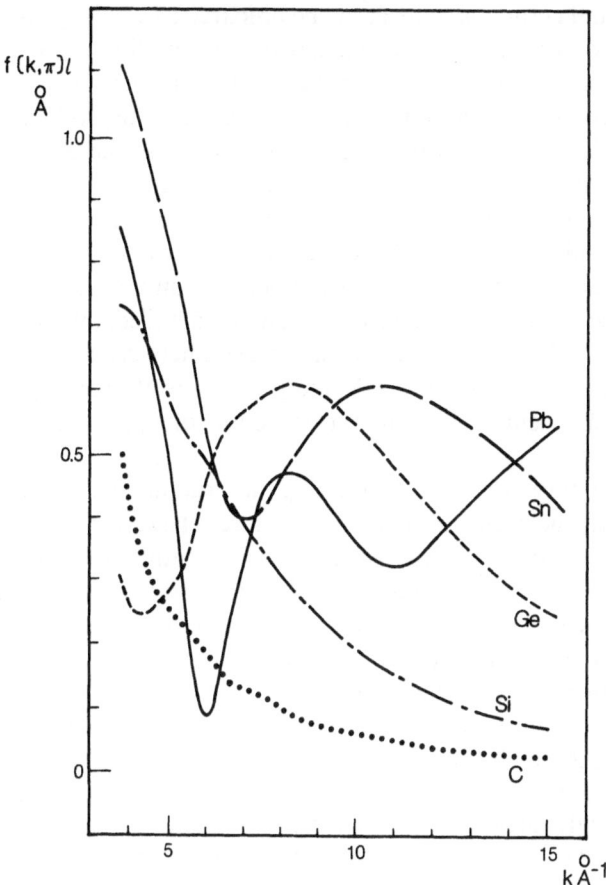

Figure 6.5.   Calculated backscattering factors, after Hayes and Boyce [1].

of these parameters constitute the structural information which we may obtain from EXAFS. In general, we may say that R_j values can be fitted to an accuracy of about 0.02–0.05 Å, better results being obtained if we check and modify our scattering data using spectra from known standard samples. N_j and σ_j^2, which are highly correlated, can be fitted to about 10%. The variation in $f_j(\pi)$ is such that we can expect to be able to separately identify different atoms if their Z values differ by 5 or 10: thus differentiating between transition metals is difficult as is differentiating between C, N and O (on the basis of $f_j(\pi)$ alone) but we can seperate shells of elements occupying different rows of the periodic table fairly easily.

6.4   The structure of oxide glasses

The classic glass-forming system is an oxide glass, based on silica; silicate glasses have of course been known and used for over two thousand years.

Silica itself (and germania) may easily be prepared as a bulk glass but for most applications other materials are mixed with it: the addition of an alkali lowers the melting point and viscosity, making the soda-lime silicates very easy to work with; lead increases the refractive index and is used in the production of optical and crystal glasses; transition metals colour the glass. Therefore the practical oxide glass contains many atomic species and the unique power of EXAFS in the measurement of the local coordination of a particular atomic species is especially useful in elucidating its structure. A study of the extended structure on the X-ray absorption edges of all the different atomic species in the glass, if it were possible, would give a vast amount of detailed information on the arrangement of the atoms which would be extremely useful in understanding the properties of these materials. Rapid progress is at present being made in this area and recent results on the structure of oxide glasses are briefly described here.

As a preliminary step in understanding the structure of complex glasses it is usual to divide the chemical elements into three classes: the network formers, such as silicon, germanium and arsenic which, together with oxygen, form the covalently bonded network; the network modifiers, such as the alkali metals, which are ionic species that tend to disrupt the network (hence their effect on the viscosity of the melt); and the intermediates, such as the transition metals, which may form part of the network or occur interstitially. The class into which a particular atom falls is largely fixed by its ionicity and the structure of network glasses may be understood in terms of the balance between ionic and covalent bonding [20].

Unfortunately most of the network-modifying species, as well as the components of silica itself, are light atoms whose K absorption edges occur at low energies. At such energies the X-ray absorption coefficient is high and so samples have to be made very thin if the traditional method of measuring the absorption coefficient by means of a transmission experiment is to be used (see Chapter 1). It is extremely difficult to make such thin samples of a uniform thickness and this practical problem limits transmission experiments to the study of K edges above about 2 keV (corresponding to $Z > 14$, silicon) except under very favourable circumstances. This problem may be avoided by using the newer techniques of electron yield or fluorescence EXAFS, and good data on sodium and silicon edges have been obtained by the use of these methods. They are also more appropriate for studying the environment of very low concentration species (see Chapter 11). The problems are less severe in the case of germanate glasses, since at least the Ge K edge is accessible to transmission studies. Information may also be obtained on the environment of transition metal atoms in oxide glasses, since their K edges are easily accessible.

In this section we shall briefly review the contribution made by EXAFS studies to our knowledge and understanding of the structure of oxide glasses. To date, more than fifty EXAFS studies of cations in oxide glasses have provided information on the local atomic coordination of network-formers such as Si, Ge and As; network modifiers such as Na, K and Ca; and intermediates such as Ti, Fe and U. This work has given new insight into properties such as cation diffusion, nucleation and corrosion behaviour of glasses.

6.4.1   *Network modifiers in silicate glasses*

Silica and alkali silicate glasses are exceptional in that their ductility allows very thin samples to be prepared by blowing films from a molten blob. Such a technique was used by Greaves *et al.* [21] to prepare glass samples approximately 1 μm thick, the optimum thickness for a transmission measurement of the Si and Na K edges. The EXAFS spectra were analysed in k-space using calculated scattering parameters. The phase-corrected Fourier transforms of

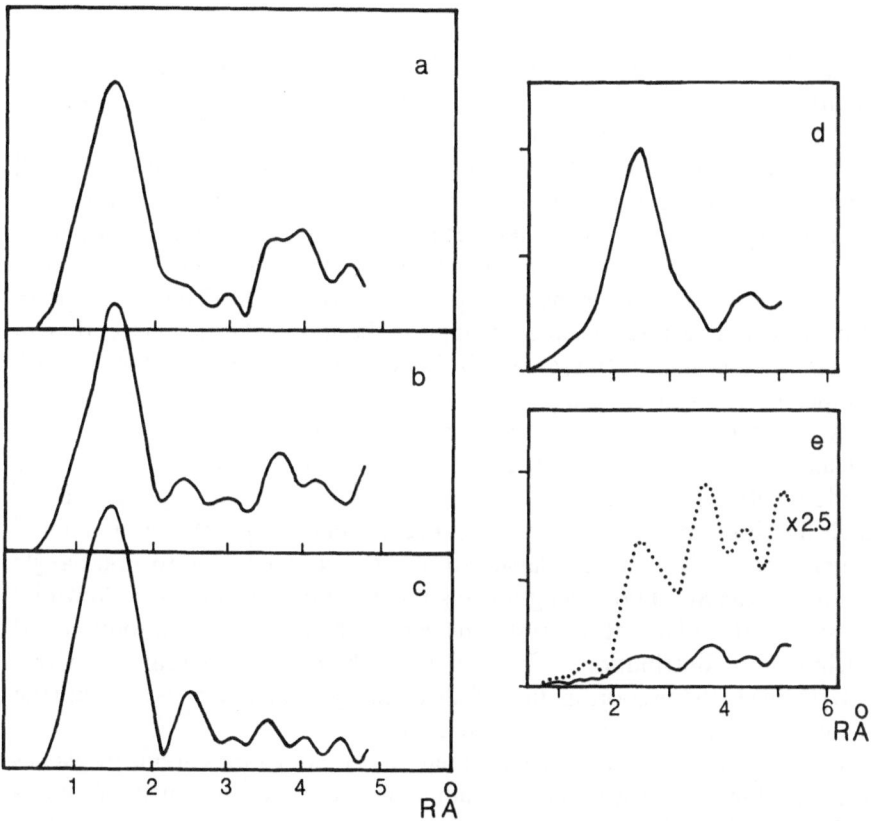

Figure 6.6.   Phase-corrected Fourier transforms of EXAFS in silicate glasses [21]. Si K edge in (a) α-quartz, (b) $Na_2Si_2O_5$ glass, (c) $Na_2CaSi_5O_{12}$ glass. Na K edge in (d) $Na_2Si_2O_5$ glass and (e) $Na_2Ca\ Si_5O_{12}$ glass.

Table 6.1 Structure of silicate glasses

Sample	Edge	Neighbour	$R(\text{Å})$	N	$\sigma^2(10^{-4}\,\text{Å}^2)$
α-quartz	Si	Si–O	1.60	4	10 ± 1
		Si–Si	3.06	4	100 ± 10
		Si–O	3.56	4	100 ± 10
SiO_2 glass	Si	Si–O	1.61	4	10 ± 1
		Si–Si	3.17	4	200 ± 70
$Na_2Si_2O_5$ glass	Si	Si–O	1.61	4	10 ± 1
		Si–Si	3.17	4	200 ± 70
	Na	Na–O	2.30	5	56 ± 5
$Na_2CaSi_5O_{12}$ glass	Si	Si–O	1.61	4	10 ± 1
		Si–Si	3.17	4	210 ± 70
	Na	Na–O	2.43	2	140 ± 10
		Na–O	2.86	1	280 ± 50
		Na–O	3.33	3	200 ± 50
	Ca	Ca–O	2.26	8	1000 ± 100
			± 0.02	± 1	

the EXAFS, corresponding to a radial distribution function, are shown in Figure 6.6 for the three glasses SiO_2, $Na_2Si_2O_5$ and $Na_2CaSi_5O_{12}$, for all of which the oxygen content is stoichiometric.

The near-neighbour silicon environment was found to be the same in all three glasses, even down to the mean square variation in distance. Detailed fits to the EXAFS spectra showed that the SiO_4 unit and the Si–O–Si angle persist in all three glasses. Numerical results are given in Table 6.1. The Si–O distance is the same in the glasses as in α-quartz but the Si–Si distance is larger, indicating that the inter-tetrahedral angle in the glasses is rather larger than in quartz. In the glasses this angle, Si–O–Si, has an average value of 160°, compared to 144° in α-quartz, with a mean variation about this value of 20°. This is rather less than the value of 30° given by molecular dynamics calculations [22] but about the same as that found in model continuous random networks. Molecular dynamics calculations are known to give bond angle distributions which are too wide because of their use of non-directional atomic potentials. In these glasses all of the disorder is due to bond-angle distortion, the Si–O bond length being remarkably constant, as is shown by the small value of σ_1^2, at its crystalline value. The modifier ions, sodium and calcium, give contributions at around 3.6 Å (these were too weak to analyse fully) in line with the coordination of silicon in crystalline silicates and with the results of molecular dynamics calculations.

When investigating the local coordination of modifier ions in silicate glasses it is important to realise that in crystalline silicates the environment of cations such as Na, Ca or Al varies considerably from one mineral to the next [23]. In binary silicates such as $Na_2Si_2O_5$ the sodium environment consists of a trigonal biprism of five oxygens at 2.3 Å whilst in ternary silicates such as

$Na_2CaSi_5O_{12}$ it is a trigonal antiprism of three oxygens at a bonding distance of 2.3Å with a further three at non-bonding distances around 3Å. The environment of calcium in crystalline silicates is, by contrast, extremely distorted: although the local symmetry is roughly octahedral, the oxygen coordination is usually between 7 and 9. Most of these oxygens are at bonding distances around 2.4Å but some are squeezed out to larger distances giving rise to a characteristic asymmetric Ca–O pair distribution function. These asymmetric distributions can usually be modelled reasonably well as a sum of two or three gaussian peaks.

Perhaps surprisingly, the environment of the network modifier cations in alkali silicate glasses is very similar to that in the crystals, even down to the existence of two different sodium sites, and is well-defined. The first demonstration of the well-defined nature of modifier environments came with the EXAFS studies of Greaves *et al.* [21] referred to earlier. They found that in sodium disilicate ($Na_2Si_2O_5$) glass the sodium atom was coordinated by five oxygen atoms at 2.3Å with a small value of σ_1^2 (see Table 6.1), very similar to the environment in the crystal. The same result has been obtained in molecular dynamics simulations [24]. However, in the soda lime silicate glass, $Na_2CaSi_5O_{12}$, the oxygen coordination of sodium was found to be six, these atoms being split evenly between two shells at 2.4Å and 3.5Å. This split shell shows up clearly in the Fourier transform of the Na K edge EXAFS shown in Figure 6.6. Similar results have been obtained for the sodium environment in aluminosilicate glasses [25].

The highly distorted calcium environment in silicate glasses causes problems in data analysis, which has to be done by modelling the distribution by a sum of gaussian peaks [26]. Despite this problem, useful information on the calcium environment in simple silicate and more complex mineral glasses has been obtained. Thus in $CaSiO_3$ the calcium environment, as determined by EXAFS [27] and X-ray diffraction [28] is made up of six oxygen atoms at about 2.3Å with a further two at 2.5Å. A comparison of the two measurements of this environment shows up the advantages EXAFS has over diffraction methods in the investigation of multicomponent glasses. In the mineral glasses [29] formed from anorthite ($CaAl_2Si_2O_8$) and diopside ($CaMgSi_2O_6$) the oxygen coordination of calcium is about eight, with the distances spread between 2.3Å and 3.1Å, being concentrated near 2.4Å and 2.7Å. Anorthite and its glass have very similar calcium sites, both being strongly distorted in a way characteristic of feldspars. Diopside and its glass show very different calcium sites, the calcium environment being much more ordered in the crystal, a pyroxene, where the distribution has a mean Ca–O distance which is about 0.1Å shorter than that in the glass. The calcium environment is very closely correlated to bulk properties: anorthite and its glass have densities of about 2.8g/cm^3, as does diopside glass, but the density of diopside crystal (with the shorter mean Ca–O distance) is 3.2g/cm^3. Since the silicon environment in almost all silicates, crystal or glass, is very much the same, density differences might be expected to correlate strongly with modifier environment.

Potassium is a considerably larger network-modifying cation than is sodium or calcium and typically is coordinated by 6–12 oxygens in crystalline silicates whilst aluminium is a common constituent of geological materials which is thought to act as a network former. The EXAFS on the potassium and aluminium K edges has been studied for a number of mixed alkali silicate glasses formed by quenching molten mixtures of albite ($NaAlSi_3O_8$) and orthoclase ($KAlSi_3O_8$). The aluminium was found [30] to be always four-fold coordinated by oxygen in these glasses, with an Al–O distance of 1.77 Å and very small mean square variation. Thus it is clearly acting as a network former with a very well-defined site. However, in the glasses studied, potassium was found [31] to be coordinated by nine to eleven oxygen atoms at 3.0–3.1 Å, there being a very strong correlation between coordination number and bond length and a well-defined variation of both with the Na/Al ratio. This is shown in Figure 6.7. These results provide the first direct structural constraint on models of the mixed-alkali effect in aluminosilicate glasses.

The implications of the results reviewed here for our view of the structure of oxide glasses are substantial. It is now clear that the modifier ions, such as Na, K or Ca, play a major role in determining the structure of the glasses and do not just occupy occasional sites or holes in the silicate network as the conventional Zachariasen model [47] implies. Rather, the structure of silicate glasses should be seen as a natural extension of the layer or chain crystalline structures where the local coordinations of network-former and network-modifier atoms are retained but where long range order is lost,

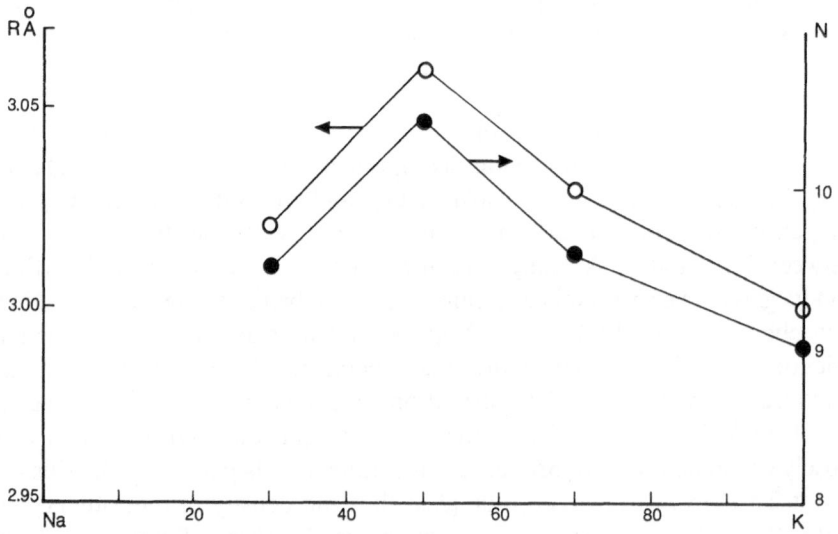

Figure 6.7. Potassium-oxygen coordination in mixed-alkali silicate glasses.

chiefly through random disorder of the bond angles. Such ideas have been developed by Greaves [21, 29] who describes a 'modified random network' model of silicate glasses, partly covalent (network formers) and partly ionic (network modifiers). This contains two types of oxygen atom, bridging and non-bridging, the type depending on the atom's relation to silicon or modifier atoms. The model has the appearance of being made up of 'islands' of network-formers separated by percolation channels of modifiers and may be used to explain the lower viscosity and higher ionic transport coefficients (which are closely connected to corrosion resistance) of alkali silicate glasses when compared to silica.

6.4.2   Intermediates in silicate glasses

Transition metal atoms in oxide glasses have traditionally [20] been thought to behave in a manner intermediate between that of network-formers and network-modifiers. Thus if they are four-fold coordinated by oxygen then they were considered to be substituting for silicon and thus acting as network formers, whilst if more highly (usually six-fold) coordinated then they were considered to be acting like sodium or calcium, as network-modifiers. The coordination reflected the ionisation state of these atoms in the glass and was therefore controlled by their electronegativity. Little direct structural information was available from diffraction techniques since the concentration of the metal atom in the glass was usually so low (< 5%) that its contribution to the total pair distribution function was swamped by other contributions.

EXAFS studies are particularly well suited to investigating the local environment of dilute species, since they measure the environment of a particular atom type immediately. Also the energies of the K edges of transition metal atoms, at 5–10 keV, are particularly conveniently placed for synchrotron sources and crystal spectrometers. Many EXAFS investigations of transition metal environments in oxide glasses have now been published and they comprise almost all the information we have on this subject.

One of the first of these studies was that of titanium in silica glass by Sandstrom et al. [32, 33]; this is a particularly interesting system for structural studies since the thermal expansion coefficient of the glass, which is presumably closely related to the bonding, may be adjusted to near zero by the correct choice of titanium content. Sandstrom et al. found that with a TiO_2 content of less than 7 wt% all of the titanium atoms were four-fold coordinated by oxygen, the Ti–O distance being 1.81 Å, the same as in Ba_2TiO_4, the only crystal with a purely four-fold titanium environment. With higher metal content some of the titanium atoms were found to be six-fold coordinated to oxygen at a distance of 2.15 Å. The proportion of six-fold sites increases approximately linearly with metal content, suggesting that there is a maximum number of available four-fold sites, up to 14 wt% of TiO_2 where 30% of all titanium atoms are six-fold coordinated. With higher metal content phase separation occurs. The energy of the Ti K edge shows that all of the titanium occurs as Ti^{4+} [34] ions, independent of coordination. Both Ti–O distances are in agreement with those found in silicate crystals containing four- and six-fold coordinated titanium sites. Similar results have been found for titanium in a cordierite glass [35], where the proportion of six-fold sites tends to increase with partial crystallisation. Clearly the behaviour of titanium in silicate glasses can be described by the traditional model, the four-fold atoms replacing silicon in the network, the six-fold atoms being interstitial; classic intermediate behaviour.

EXAFS studies which show up the influence of ionisation state are those of Binsted et al. [36] and Waychunas and Brown [37] on Fe in mineral glasses. Like all transition metals, iron is expected to be coordinated by oxygen alone in silicate crystals and glasses and this is indeed the experimentally observed situation. In the crystals iron atoms exist as Fe^{2+} and Fe^{3+} and both may be tetrahedrally or octahedrally coordinated by oxygen [23]. The ionisation state is determined by measuring the energy of the iron K edge and some information on the symmetry may be obtained by looking at pre-edge features, but EXAFS analyses are the only means by which details of the local environment may be obtained. (The difficulties involved in diffraction studies are shown up particularly clearly in a paper by Henderson et al. [38])

The EXAFS results [36] show that iron in a sodium silicate glass of composition $NaFeSi_2O_6$ is almost entirely in the form of Fe^{3+} and the short bond length, Fe–O = 1.86 Å, shows that it is tetrahedrally coordinated. In the crystal the iron atom has a distorted octahedral environment with an average bond length of 2.03 Å. In calcium silicate glasses [39] ($CaFeSi_2O_6$ and $CaFeSiO_4$) iron is found to have a somewhat lower edge energy and a longer bond length of 1.92 Å. These results are interpreted as being due to a proportion, about 30%, of octahedral Fe^{2+} in these glasses. This proportion is

Table 6.2 Transition metals in silicate glasses.

| Metal | Silicate glass | | Crystals | |
	R(Å)	N	R(Å)	N
Ti	1.81	4	∼ 1.7	4
	2.15	6	∼ 1.9	6
Zr	1.89	4	2.065	4
Mn	2.12	6	2.20	6
Fe	1.91	4		
	2.07	6	2.15	6
Co	1.98	4		
	2.20	6	2.13	6
Ni	2.03	6	2.08	6
Ce	1.71	4	1.74	4
	1.84	6	1.88	6

in good agreement with the stoichiometry. Combining these results with other data produces a consistent picture of Fe^{3+} acting as a network-former and Fe^{2+} as a network-modifier in silicate glasses. This information allows us to interpret a body of viscosity and ion diffusion data in a straightforward and consistent way.

The results of the many EXAFS studies of transition metal environments in oxide glasses are summarised in Table 6.2. All of them behave in an intermediate way, always being coordinated by oxygen atoms alone in four-fold tetrahedral or six-fold octahedral sites with bond lengths very similar to those found in silicate crystals. The power of the EXAFS technique in the study of the local environment, combined with edge energy measurements and near edge data (both of which come from the same experiment as the EXAFS) has allowed us to unravel many of the structural properties of complex silicate glasses.

6.4.3 *Alkali germanate glasses*

Germanate glasses based on germania, GeO_2, are more easily studied by means of EXAFS than are the corresponding silicate glasses because the K absorption edge of germanium is conveniently placed (at 9 keV) for transmission measurements using standard X-ray spectrometers. All EXAFS investigations of these glasses to date have used the Ge K edge. Germanate glasses are mainly of interest because of the use of germanium to produce refractive index gradients in optical fibres. This application gives special importance to a knowledge of the local structure around the germanium atoms in multicomponent glasses; as is the case for silicate glasses, alkalis are added to germania to produce a low melting point glass.

Traditionally the germanate glasses have been considered as directly analogous to the silicate glasses but recent structural studies, mostly using EXAFS, have shown that major differences exist between the two. The origin of this difference lies in the more metallic nature of the Ge–O bond when compared to the Si–O bond. The effect of this is seen in crystalline GeO_2 which exists in two forms, one of α-quartz structure with Ge tetrahedrally coordinated to four oxygen atoms as in SiO_2, and one of rutile structure where the germanium is six-fold coordinated. SiO_2 does not exist in the rutile structure at normal pressures.

There have been several studies of amorphous GeO_2 [40, 41] which use the EXAFS on the Ge K edge to investigate the local environment of the germanium atom and its temperature dependence. All show that the Ge–O coordination in the glass is very similar to that in the α-quartz form of the crystal, Ge being coordinated by four oxygen atoms at 1.74 Å with very little static disorder. The mean square variation in the Ge–O bond length is in fact slightly less in the glass than in the crystal and varies but little with temperature [40]. Thus we may say that the GeO$_4$ tetrahedral unit is well-defined in the glass and is somewhat more rigid than in the crystal.

However, although the first shell is very much the same in crystal and glass, the more distant shells in the glass are more disordered due to a static variation in the bond angle [41] (see Figure 6.4). The Ge–Ge distance of 3.15 Å is the same in crystal and glass, showing that the intertetrahedral angle, Ge–O–Ge, is on average the same at 130° [41]. The static mean square variation in the Ge–Ge distance shows that in the glass this angle has an RMS variation of about 6°, much less than the corresponding value in silica glass; nonetheless it is this bond angle variation which gives the flexibility needed to form a continuous random network. The X-ray diffraction data are in general agreement with these results, although of lower resolution. The best data on amorphous GeO$_2$ are in fact a very recent neutron study [42].

The structural effects of adding different amounts of the oxides of the alkali metals to germania to form alkali germanate glasses have been the subject of several EXAFS studies and have also been investigated by means of infrared and Raman spectroscopy and the measurement of X-ray emission line shifts.

Cox and McMillan [43] determined the environment of the germanium atoms by fitting the EXAFS on the Ge K edge in k-space using calculated scattering parameters. Only the nearest neighbour oxygen shell was found to contribute to the EXAFS signal, indicating that the bond angle distortion in these glasses is considerably larger than it is in pure germania; the same effect occurs in silicate glasses as noted above. The detailed analysis showed that this oxygen shell was split in glasses which contained lithium. In pure amorphous germania the Ge atom was found to be four-fold coordinated by oxygen, the bond length being 1.71 Å in agreement with earlier results. In the glasses containing lithium an extra contribution from six-fold coordinated Ge, with a bond length of 1.84 Å, was needed to obtain a good fit to the data. The bond lengths of the two sites were found to be independent of the lithium content and are very close to the values of 1.74 Å (four-fold in α-quartz GeO$_2$) and 1.90 Å (six-fold in rutile GeO$_2$) found in the crystals. The number of six-fold coordinated atoms was found to increase with lithium content up to about 20% LiO$_2$ where 25% of the Ge atoms were in such sites, and to saturate thereafter. The mean square variations in bond lengths are much the same for the two sites and similar to those found in crystalline materials; we may therefore say that both sites have a well-defined oxygen environment.

The other two EXAFS studies [44, 45], which obtained somewhat less precise results than those of Cox and McMillan but which looked at glasses containing Li, Na or K, used an analysis based on the average measured Ge–O bond length to determine the proportion of six-fold sites. In the Fourier transform of the EXAFS spectrum the position of the nearest neighbour peak was found to vary with alkali content, this being interpreted as due to the presence of a proportion of the longer Ge–O bonds of six-fold sites. The proportion of six-fold sites may be determined from the average bond length if the two distinct Ge–O distances are known; in these studies the crystalline values were used. The results obtained are essentially identical to those of Cox and McMillan and show no dependence on alkali type. Sakka and Kamiya [44] also determined the average Ge–O coordination, and hence the proportion of six-fold sites, by the graphical method of EXAFS analysis described previously (see Section 6.3.2 and Figure 6.4); the results closely parallel those obtained from the bond lengths. Since these authors looked at higher alkali content than did Cox and McMillan, they were able to show that the proportion of six-fold sites peaks at about 20% alkali and then falls, reaching zero at about 35% alkali content. Studies using infrared peak shifts [44] and X-ray emission line shifts [46], calibrated against crystals with different proportions of six-fold sites, show very similar results. All of these results are summarised in Figure 6.8.

Lapeyre *et al.* [45] have also measured the Ge–Ge distance in these glasses. This was found to be independent of alkali content and to take the value of

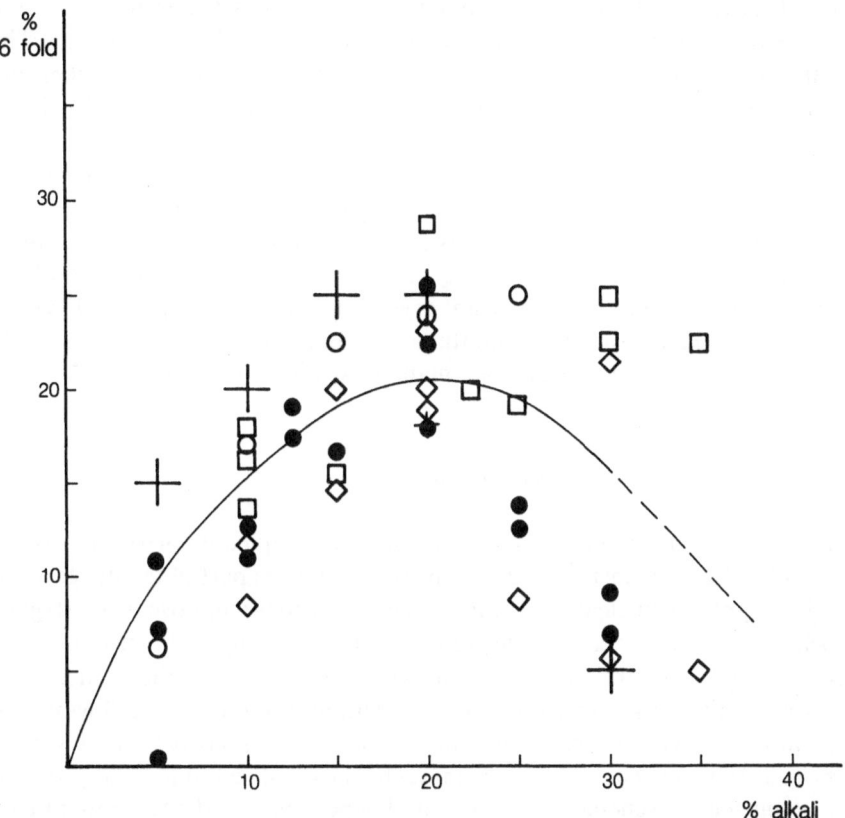

Figure 6.8. Proportion of six-fold coordinated Ge in alkali germanate glasses. The line is drawn to guide the eye. $\bigcirc$, EXAFS [43]; $\square$, EXAFS $\bar{R}$ [44]; $\diamondsuit$, EXAFS $\bar{N}$ [44]; $\bullet$, IR [44]; $+$, X-ray edge shift [46].

3.14 Å, the same as in crystalline and glassy GeO_2. The existence of this peak, weakened though it is by the fairly large mean square variation in bond length, confirms the existence of a three-dimensional random network even at the highest alkali content.

All of these results confirm that the germanium sites are well-defined in alkali germanate glasses and that the disorder is accommodated by bond angle distortion in the network. The presence of six-fold coordinated Ge atoms shows that alkali germanate glasses are not directly analogous to alkali silicate glasses and explains the long-known variations in density and refractive index (hence the optical fibres interest) shown by these glasses as a function of alkali content. Alkali borate glasses behave in a similar way. The changes in the structure of the germanate glasses as alkali atoms are added (in the form of the oxides, X_2O) may be interpreted as a result of the replacement of GeO_4 tetrahedra by GeO_6 octahedra in order to accommodate the changing oxygen proportion without creation of non-bridging oxygens. 'Depolymerization' processes are therefore not of great importance in these glasses. Indeed Raman data [48] suggest that no non-bridging oxygens exist in glasses with less than 20% alkali although they are probably present in more alkali-rich glasses. The proportion of GeO_6 octahedra peaks at about 20% alkali because, according to Zachariasen [47], the octahedra cannot link to each other. Alkali-rich glasses maintain a continuous random network by using the extra oxygens to link GeO_4 tetrahedra. Thus the proportion of six-fold sites falls above 20% alkali content. As is the case with silicate glasses, the structure of the alkali germanate glasses may also be closely related to the corresponding crystalline germanates, where the same linking restrictions apply. For the case of the alkali germanate glasses the relevant crystals are, for example, $2Li_2O.9GeO_2$ and $Na_2O.4GeO_2$; the latter contains the highest proportion of six-fold germanium sites of any ternary germanate, at 20% alkali content!

6.4.4   Corrosion studies of silicate glasses

The structural chemistry underlying the process of aqueous corrosion of oxide glasses has been extensively studied in recent years as part of the programme for the disposal of nuclear waste material in the form of (supposedly inert) glass blocks. For the most part, these studies have used analytical methods which give no *direct* structural information about the ways in which the surface of the glass is modified as corrosion proceeds. Although the precise diffusion and dissolution processes occurring during the aqueous corrosion of silicate glasses are still unclear, evidence from such analytical techniques supports the following general scheme [49]. The initial dependence of the leaching rate on the square root of the leaching time points to the corrosion rate being controlled by a diffusion mechanism. Alkali atoms leave the glass and are replaced by water, resulting in the formation of silanol groups; it is not, however, clear which diffusion process (water in or alkali out) is rate determining or whether H^+, H_3O^+ or molecular water is active in the interdiffusion. Also, it seems certain that more water enters the glass than is required simply to replace depleted alkali cations. As corrosion progresses the leaching rate assumes a linear time dependence, indicating the wholesale dissolution of the glass network at the surface.

Detailed study of the corrosion process requires the use of a technique which gives structural information directly, which is also surface sensitive and which can be used for in situ experiments (i.e. it should not require a high vacuum environment which would alter the composition of the leached layer by, for example, causing the evaporation of water). X-ray absorption spectroscopy has been extensively used to study the bulk structure of oxide glasses, as we have seen from the results discussed in this chapter, but the standard transmission or fluorescence geometries are not surface sensitive. Good surface sensitivity can be achieved by using a glancing angle geometry [50] with detection via either the reflected or fluorescence signals. The surface sensitivity arises from the reflection properties of X-rays at very low angles of incidence. As discussed earlier, in Section 1.3.1, for a monochromatic X-ray beam incident upon a flat surface there exists a well-defined angle of incidence, θ_c, below which total external reflection takes place, the penetration depth of the refracted beam being 10–40 Å for keV X-rays. Above the critical angle the penetration depth is determined by the absorption coefficient and is of the order of microns. Thus by varying the angle of incidence of the X-ray beam, changes in the glass structure may be measured non-destructively as a function of depth and analysed by standard EXAFS techniques.

This method has been used by Greaves et al. [51] to study the aqueous corrosion of a borosilicate glass containing uranium, a model nuclear waste disposal glass. The EXAFS on the uranium L_3 edge at 17.2 keV was measured on the wiggler beam line at the SERC Daresbury Laboratory at glancing angles of incidence above and below the critical angle of 0.1°. The data were analysed by fitting the EXAFS spectrum in k-space using the curved-wave theory described in Section 6.2.

Spectra taken at an angle of $2\theta_c$ (corresponding to a penetration depth of $2 \mu m$) after various leaching times show changes in the uranium environment with the initial formation of uranyl hydroxide groups (the natural outcome of alkali ion exchange with H_3O^+) which convert finally to hydrated uranyl silicates, the end product of the corrosion treatment. This process is consistent with the formation of a gel layer at the surface of leached borosilicate glass and the formation of a surface precipitate, and establishes uranium as a network-modifying cation. As such it is a mobile ion in the modified random network model mentioned above and so might be expected to cluster. This is confirmed by the detection of a significant proportion of U–U correlations, more particularly near the surface, as can be seen in Figure 6.9.

Spectra taken at an angle of incidence of $\theta_c/2$, corresponding to a

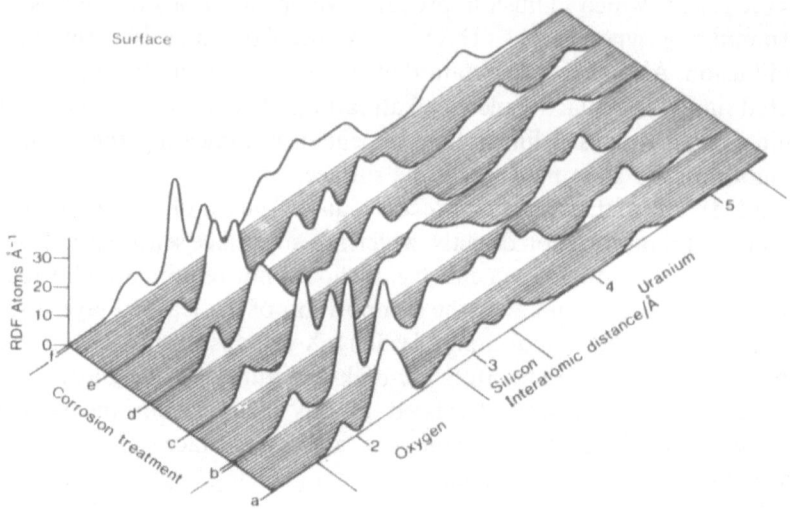

Figure 6.9. Phase-corrected Fourier transforms of the EXAFS on the uranium L_3 edge in uranium silicate glasses as a function of corrosion treatment, after Greaves *et al.* [51]. (a) Bulk structure measured at 45° incidence. At angle of incidence $\phi_c/2$: (b) polished glass as received; (c) after 15 min leaching; (d) after 15 min leaching and subsequent drying; (e) after 30 min leaching; (f) after 90 min leaching.

penetration depth of some tens of angstroms, show significant differences from those taken at $2\theta_c$. Moreover there are progressive changes in structure with leaching. This can be clearly seen in Figure 6.9 which shows a collection of RDFs for an angle of incidence of $\theta_c/2$. On leaching significant changes, characteristic of the formation of uranyl hydroxide-like complexes, were observed for less than 30 min exposure to water, with the formation of a uranyl silicate-like precipitate after 90 min exposure. The corrosion process is now seen to be cyclic, the precipitate forming, washing off and then reforming, leading to a steady leaching away of the glass (compare RDFs b and f in Figure 6.9). This process may be explained in terms of the modified random network model [29] rather successfully as a result of ion exchange of the mobile network-modifying Na and U by H_3O^+ and consequent complexing into uranyl-like units near the surface. This is followed by the removal of Na from the surface and the dissolution of the borosilicate network leading to the recondensation of hydrated uranyl silicate complexes.

This detailed modelling of the corrosion process in silicate glasses requires an experimental technique with three attributes: it must be surface-sensitive; it must give direct structural information; and it must be applicable to in situ experiments using water-covered surfaces. Glancing-angle X-ray absorption spectroscopy uniquely possesses these three attributes.

References

1. T.M. Hayes and J.B. Boyce, *Solid State Phys.* **37** (1983) 173.
2. B.K. Teo and D.C. Joy (eds.), *EXAFS Spectroscopy*, Plenum Press, New York (1981).
3. D.C. Koningsberger and R. Prins (eds.), *X-Ray Absorption*, John Wiley, New York (1988).
4. S.P. Cramer and K.O. Hodgson, *Prog. Inorg. Chem.* **25** (1979).
5. B.H. Bransden and C.J. Joachain, *Physics of Atoms and Molecules*, Longman, London (1983).
6. J.A. Victoreen, *International Tables for X-Ray Crystallography*, Kynoch Press, Birmingham (1962), Vol. 3.
7. R. de L. Kronig, *Z. Phys* **70** (1931) 317; **75** (1932) 190.
8. E.A. Stern, *Phys. Rev. B.* **10** (1974) 3027.
9. P.A. Lee and J.B. Pendry, *Phys. Rev. B* **11** (1975) 2795.
10. S.J. Gurman, N. Binsted and I. Ross, *J. Phys. C: Solid State Phys.* **17** (1984) 143.
11. S.J. Gurman, N. Binsted and I. Ross, *J. Phys. C: Solid State Phys.* **19** (1986) 1845.
12. S.J. Gurman, *J. Phys. C: Solid State Phys.* **21** (1988) 3699.
13. S.J. Gurman, *J. Phys. C: Solid State Phys.* **16** (1983) 2987.
14. P.J. Durham, J.B. Pendry and C.H. Hodges, *Comput. Phys. Commun.* **25** (1982) 193.
15. J.J. Boland, S.E. Crane and J.D. Baldeschweiler, *J. Chem. Phys.* **77** (1982) 142.
16. S.J. Gurman and J.B. Pendry, *Solid State Commun.* **20** (1976) 287.
17. B.K. Teo and P.A. Lee, *J. Am. Chem. Soc.* **101** (1979) 2815.
18. J.B. Pendry, *Low Energy Electron Diffraction*, Academic Press, London (1974).
19. E. Pantos and D. Firth, *Pro EXAFS and Near Edge Structure II*, Springer-Verlag; *Chem. Phys.* **27** (1983) 110.
20. R.F. Pettifer, in *EXAFS for Inorganic Systems*, eds. C.D. Garner and S.S. Hasnain, SERC Daresbury Laboratory (1981) pp. 57–64.
21. G.N. Greaves, A. Fontaine, P. Lagarde, D. Raoux and S.J. Gurman, *Nature* **293** (1981), 611.
22. T.F. Soules, *J. Chem. Phys.* **71** (1979) 4570.
23. R.N.G. Wyckoff, *Crystal Structures*, John Wiley, New York (1964).
24. S.K. Mitra and R.W. Hockney, in *Structure of Non-Crystalline Materials 1982*, eds. P.H. Gaskell, J.M. Parker and E.A. Davis, Taylor & Francis, London, New York (1982) p. 316.
25. D.A. McKeown, G.A. Waychunas and G.E. Brown, *J. Non-Cryst. Solids* **74** (1985) 325.
26. G.N. Greaves, K. Simkiss and M.G. Taylor, *Biochem. J.* **221** (1984) 855.
27. R.G. Geere, P.H. Gaskell, G.N. Greaves, J. Greengrass and N. Binsted, in *EXAFS and Near-Edge Structure III*, Springer, Berlin (1983) p. 256.
28. C.D. Yin, M. Okano, H. Morikawa, F. Masumo and T. Yamanaka, *J. Non-Cryst. Solids* **80** (1986) 167.
29. G.N. Greaves, *J. Non-Cryst. Solids* **71** (1985) 203.
30. D.A. McKeown, G.A. Waychunas and G.E. Brown, *J. Non-Cryst. Solids* **74** (1985) 349.
31. W.E. Jackson, G.E. Brown and C.W. Ponader, *J. Non-Cryst. Solids* **93** (1987) 311.
32. D.R. Sandstrom, F.W. Lytle, P.S.P. Wei, R.B. Greegor, J. Wong and P. Schultze, *J. Non-Cryst. Solids* **41** (1980) 201.
33. R.B. Greegor, F.W. Lytle, D.R. Sandstrom, J. Wong and P. Schultze, *J. Non-Cryst. Solids* **55** (1983) 27.
34. M. Emilio, L. Incoccia, S. Mobilio, M. Guglielmi and G. Fagherazzi, in *EXAFS and Near-Edge Structure III*, Springer, Berlin (1983) p. 331.
35. T. Dumas and J. Petiau, *J. Non-Cryst. Solids* **81** (1986) 201.
36. N. Binsted, G.N. Greaves and C.M.B. Henderson, *Prog. Exp. Petrol (NERC)* **10** (1984).
37. G.A. Waychunas and G.E. Brown, in *EXAFS and Near-Edge Structure III*, Springer, Berlin (1983) p. 336.
38. G.S. Henderson, M.E. Fleet and G.M. Bancroft, *J. Non-Cryst. Solids* **68** (1984) 333.
39. N. Binsted, G.N. Greaves and C.M.B. Henderson, *J. de Phys C* **8** (1986) 837.
40. J. Wong and F.W. Lytle, *J. Non-Cryst. Solids* **37** (1980) 273.
41. M. Okuno, C.D. Yin, H. Morikawa, F. Marumo and H. Oyanagi, *J. Non-Cryst. Solids* **87** (1986) 312; D.E. Sayers, F.W. Lytle, E.A. Stern, *Phys. Rev. Lett* **35** (1975) 584.
42. J.A.E. Desa, A.C. Wright, R.N. Sinclair, *J. Non-Cryst. Solids* **99** (1988) 276.
43. A.D. Cox and P.W. McMillan, *J. Non-Cryst. Solids* **44** (1981) 257.

44. S. Sakka and K. Kamiya, *J. Non-Cryst. Solids* **49** (1982) 103.
45. C. Lapeyre, J. Petiau and G. Calas, in *The Structure of Non-Crystalline Materials*, ed P.H. Gaskell, Taylor & Francis, London (1982).
46. C.D. Yin, K. Morikawa, F. Marumo, Y. Gohshi, Y.Z. Bai and S. Fukashima, *J. Non-Cryst. Solids* **69** (1984) 97.
47. W.H. Zachariasen, *J. Am. Chem. Soc.* **54** (1932) 3841.
48. H. Verweij and J.H.J.M. Buster, *J. Non-Cryst. Solids* **34** (1979) 81.
49. R.H. Doremus, in *Treatise on Materials Science and Technology*, eds. M. Tomozawa and R.H. Doremus, Academic Press, New York (1979) p. 41.
50. R. Fox and S.J. Gurman, *J. Phys C: Solid State Phys.* **13** (1980) L249.
51. G.N. Greaves, N.T. Barrett, G.M. Antonini, F.R. Thornley, B.T.M. Willis and A. Steel, *J. Am. Chem. Soc.* **111** (1989) 4313

7   EXAFS studies of ionically conducting solids

A.V. CHADWICK

7.1   Introduction

Of the wide variety of solid classes that have been successfully studied by EXAFS, this chapter will focus on a group of materials that can be classified together as ionically conducting solids. This classification requires some clarification as the compounds are not simple, normal ionic crystals (e.g. alkali halides) where the structures can be determined by diffraction methods. The characteristic feature of all the materials that will be considered here is that they are all unusually good ionic conductors of electricity. In general, a high ionic conductivity in crystals is associated with a highly defective lattice, due either to intrinsic disorder or the effect of added impurities, thus there are complex problems concerning the local structures around the ions. These problems, to which the EXAFS technique is ideally matched, have to be resolved if the detailed mechanisms of conduction are to be understood. A range of crystalline ionic materials is discussed, including pure, heavily doped and mixed compounds, where EXAFS studies have played a major role in relating the structure and ionic transport. High ionic conductivity is not limited to crystalline solids and materials of contemporary interest are the solvent-free complexes formed between salts and polyethers. The EXAFS technique has provided unique information on these systems and a description of that work will be included in this chapter. We begin, however, by outlining the general phenomenon of ionic conduction in solids. This will help to emphasise the kind of structural problem that needs to be resolved and the points where the EXAFS technique will be used to special advantage. Reference to experimentation and data analysis will be brief, as these matters are covered in more detail in Chapters 1 and 6. For further information, the reader is referred to several useful experimental and theoretical texts [1–5].

7.2   Ionic conductivity in solids

In an ionic crystal with a wide band gap the principal contribution to the electrical conductivity arises from the motion of ions in the applied electric field and this is termed ionic conductivity. For simple, binary compounds (e.g.

the alkali and silver halides) the mechanisms of ionic conduction are understood in great detail [6–8] and involve the migration of ions via point defects. Like all crystalline solids, an ionic crystal at a temperature above absolute zero will contain point defects, the simplest examples being unoccupied lattice sites (vacancies) and ions in the lattice interstices (interstitials). However, these defects will have effective charges with respect to the perfect lattice. Thus in a 1:1 univalent ionic crystal the cation vacancy, anion vacancy, cation interstitial and anion interstitial will have effective formal charges $-1e$, $+1e$, $+1e$ and $-1e$, respectively, where e is the protonic charge. As a result the condition of electroneutrality imposes restrictions on the exact nature of the disorder. In a pure NaCl crystal the dominant defects are Schottky pairs, i.e. equal numbers of cation and anion vacancies. Cation Frenkel pairs, i.e. equal numbers of cation vacancies and interstitial cations, dominate in pure AgCl. The concentration of a particular type of defect r is usually expressed in terms of a site fraction, c_r, which is the ratio of the number of defects n_r, and the total number of sites in the crystal the defect could occupy, N_r. Provided the defect concentrations are low ($c_r < 10^{-2}$) mass-action expressions can be derived [6] to relate the various defect site fractions in a crystal and for NaCl the site fractions of cation vacancies, c_+, and anion vacancies, c_-, are given by

$$c_+ c_- = K_S = \exp(-g_S/kT) \qquad (7.1)$$

Here K_S is the Schottky equilibrium constant, g_S is the Gibbs free energy of formation of the Schottky pair, T is the temperature (K) and k is the Boltzmann constant. An analogous expression can be written for cation Frenkel disorder in terms of cation vacancy and cation interstitial concentrations (c_+ and c_i, respectively) and g_F, the Gibbs free energy of formation of the Frenkel pair. The relative magnitudes of g_S and g_F prescribe the nature of the dominant disorder in any crystal. Defect energies of the dominant disorder are typically $\sim 2\,\text{eV}$ and therefore defect concentrations are low even at the melting point, e.g. $c_r \sim 10^{-3}$.

In a pure crystal the condition of electroneutrality leads to equal concentrations of the two components of the defect pair, for example, in NaCl

$$c_+ = c_- = K_S^{1/2} = \exp(-g_S/2kT) \qquad (7.2)$$

The deliberate addition (doping) of impurities with a different valency (allovalent) to the host crystal will perturb the defect concentrations. This can be visualised by considering the effective charges of the species. Thus doping NaCl substitutionally with divalent cations like Sr^{2+} (effective charge $+1e$) leads to the creation of cation vacancies. If the concentration of Sr^{2+} is $c_{Sr^{2+}}$ then applying the conditions of charge and site balance yields [6]

$$c_+ = c_{Sr^{2+}} + c_- \qquad (7.3)$$

Combining Eqns. 7.1 and 7.3 allows the concentration of defects in doped NaCl to be evaluated. Again, analogous expressions can be derived for other types of intrinsic disorder and dopants. However, it should be noted that the above expressions relate to isolated defects. A dopant and the corresponding defect it creates will tend to associate due to their opposite effective charges to form near neighbour pairs (which are neutral but have an electric dipole moment) and higher aggregates [6]. At low dopant levels this process can also be treated in terms of mass-action relationships. It is not necessary to give the details here and it only needs stating that the fraction of dopant in aggregates will increase with increasing dopant concentration and with decreasing temperature.

The diffusion of atoms in crystals occurs via the migration of the point defects and several mechanisms have been established [6–10]. For example, the vacancy mechanism involves an atom jumping into a neighbouring vacancy on the same sub-lattice. An applied electric field across an ionic crystal will bias the otherwise random diffusive jumps of the defects and give rise to ionic conduction. The specific conductivity, σ, of an ionic crystal is given by [6–8]

$$\sigma = \sum N_r c_r |q_r| \mu_r \tag{7.4}$$

where N_r is the number of sites per unit volume available to defect r and q_r is the effective charge of the defect. The defect mobility, μ_r, is directly related to the jump frequency of the defect, w_r, and from a transition-state theory treatment [6] this can be written as

$$w_r = v \exp(-\Delta g_r / kT) \tag{7.5}$$

Here, v is a jump attempt frequency (usually taken as the Debye frequency) and Δg_r is the free energy barrier to the jump of the defect. In simple crystals typical values of the Δg_r are 0.5–1.0 eV. In a real crystal, one defect of the pair is usually the more mobile and Eqn. 7.4 is dominated by one contribution. For example, the mobile defects in NaCl are cation vacancies and in this case

$$\sigma = N_+ e c_+ \mu_+ \tag{7.6}$$

Since defect concentrations and mobilities are low in most crystals it follows from Eqns. 7.4–7.6 that σ is low, typically $10^{-3} \, \Omega^{-1} \, cm^{-1}$ at the melting point.

A number of solids are known which have unusually high ionic conductivities, comparable to those found in molten salts and aqueous solutions of strong electrolytes, and these are referred to as 'superionic conductors' or 'fast-ion conductors' or 'solid electrolytes' [11–18]. These have evoked considerable interest for their potential technological applications in devices such as solid-state batteries, electrochromic displays, coulometers, sensors, etc. In qualitative terms the origins of the unusual behaviour can be explained by examination of the above equations for simple crystals; clearly σ will be increased by increasing c_r and/or μ_r. These conditions can be achieved by various mechanisms which depend upon the particular type of fast-ion conductor. A number of simple, stoichiometric materials undergo a solid state transition to a highly defective structure on one of the sub-lattices and examples are δ-Bi_2O_3 [19, 20] and α-AgI [21, 22]. Non-stoichiometric materials are also highly defective and can exhibit fast-ion behaviour. The best example of this type is Na β-alumina [23] which has an ionic conductivity of $1.4 \times 10^{-2}\,\Omega^{-1}\,cm^{-1}$ at room temperature due to the motion of Na^+ ions. In this material the Na^+ ions are located on planes between the spinel blocks, and the high conductivity arises from there being more sites than there are Na^+ ions to occupy them and low energy pathways between the sites. Another type of fast-ion conductor comprises solids which have been heavily doped with allovalent impurities. A number of materials with the fluorite structure, both halides [24] and oxides [25], fall into this class as the open structure will accommodate high impurity and defect concentrations. A more recently discovered class of fast-ion conductors is the polymer electrolytes [26, 27] where an alkali metal salt is dissolved in an elastomer and the flexing of the polymer chain allows the ions to move relatively freely.

The classical concepts outlined above for expressing defect concentrations and ionic conductivity in simple, low defect crystals are not quantitatively valid for fast-ion conductors. In highly disordered solids, $c_r > 10^{-2}$, defect–defect interactions are difficult to include in the thermodynamic treatments and the defect concentrations cannot be calculated with any accuracy. The validity of the jump diffusion model has also been questioned when the ions are highly mobile (see, for example, [14]). Thus the quantitative theoretical modelling of fast-ion conduction has proved to be a very difficult and long-standing problem. At the present time the most promising approach appears to be that of computer modelling using molecular dynamics simulations and this method has had some successes with the structurally simple fast-ion conductors [28].

A wide range of experimental techniques [14] has been employed with varying degrees of success to resolve the problems posed by fast-ion conductors. Many of these problems involve a detailed knowledge of the structure, not simply the long-range order which can be determined by diffraction methods, but the local structure around specific ions. It is this specific information that can be provided by the EXAFS technique.

In the materials described below there are examples of the diversity of questions concerning fast-ion conductors that have been resolved by EXAFS studies, including:

(i) the changes of environment of the mobile ions in metal halides (e.g. α-AgI)
(ii) the structure of defect aggregates in heavily doped crystals (e.g. rare earth ions in CaF_2);
(iii) the nature of sites occupied by dopant ions in stabilised oxide fluorite crystals, Y^{3+} doped ZrO_2 and Y^{3+} doped Bi_2O_3;
(iv) the defect structures in mixed ionic crystals (e.g. $RbBiF_4$);
(v) the environments of ions in salt-containing polymers (e.g. RbX complexes with polyethylene oxide).

7.3   Silver iodide

EXAFS studies of fast cation conductors were pioneered by Boyce *et al.* [30] whose study of AgI revealed the power of the technique for this class of materials.

Below 147°C, at atmospheric pressure, AgI exists in the β-form in which the Ag^+ ions reside in the regular tetrahedra of I^- ions in the hexagonal wurtzite structure [31]. The ionic conductivity which is due to the motion of Ag^+ ions, is low in magnitude ($< 10^{-4}\,\Omega^{-1}\,cm^{-1}$) and has an activation energy of 0.5 eV. In other words, in this phase AgI is a normal ionic crystal. At 147°C β-AgI undergoes a first-order phase transition to the α-AgI form. The ionic conductivity, still due to the Ag^+ ions, increases to $1\,\Omega^{-1}\,cm^{-1}$ (i.e. by a factor of 10^4) at the transition and in the α-phase the activation energy is $\sim 0.05\,eV$. Since the discovery of this unusual behaviour of AgI a number of analogous materials have been reported in which fast-ion conduction is the result of a phase transition, e.g. CuI, CuBr and CuCl. However, AgI remains the most outstanding and thoroughly investigated example.

The $\beta \to \alpha$ transition at 147°C involves the transformation to a structure in which the I^- ions form a body-centred cubic structure and a highly disordered Ag^+ sub-lattice. The structure proposed by Ströck [31] for this phase is shown in Figure 7.1 and the unit cell contains 42 available sites for the two Ag^+ ions. These sites are 12 equivalent tetrahedrally coordinated sites (d), 24 three-fold

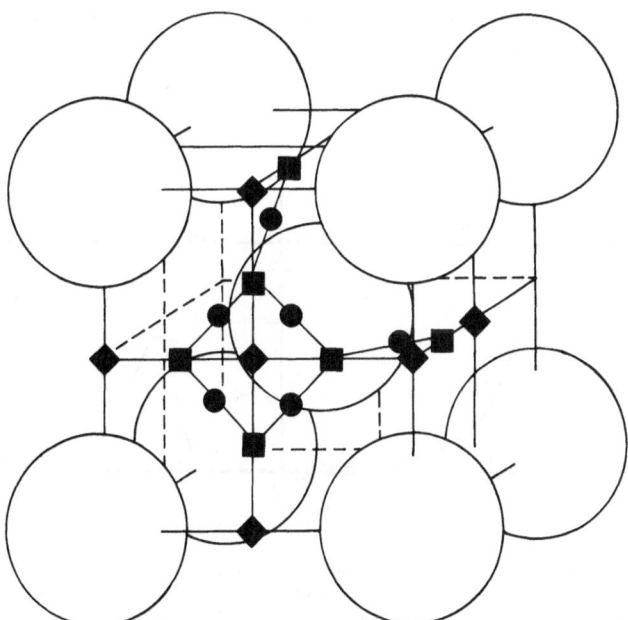

Figure 7.1.   The structure proposed by Ströck for α-AgI. The large circles represent I^- ions. The Ag^+ ions are distributed over the 6 b sites (♦), 12 d sites (■) and 24 h sites (●).

coordinated sites (h) and 6 octahedrally coordinated sites (b). Thus the simplest model of the ionic conductivity is that the Ag^+ sub-lattice has 'melted' and the ions are distributed in a liquid-like fashion around the I^- ions. Various attempts have been made to provide a quantitative model for this system (see, for example [3, 32, 33] and references therein). A problem was the lack of good structural data, a deficiency caused in part by the difficulties in preparing single crystals of α-AgI. A major refinement of the structural information for α-AgI was provided by the EXAFS study.

EXAFS data were collected at the Ag K edge in AgI at $-197, 20, 98, 198$ and $302°C$ at SPEAR [30]. The details of the data analysis can be found in the original work, and in a later paper [32]. Further papers describe the interpretation of the data [32–35]. At the elevated temperatures the EXAFS were dominated by the contribution from the I^- ions which are nearest neighbours to the Ag^+ ion. Thus the spectrum at $-197°C$ was fitted with a narrow gaussian centred at 2.82 Å (the $Ag^+–I^-$ nearest neighbour distance obtained from diffraction data). The fitting parameters were then used in the analysis of the higher temperature data. Significant changes were found in the RDF plots, shown in Figure 7.2, as the sample was heated; the main peak shifted slightly to lower r and became asymmetric, and the peak amplitude decreased. This indicated that the Ag^+ ions were retaining tetrahedral coordination in the α-phase as the changes were small, which is consistent with

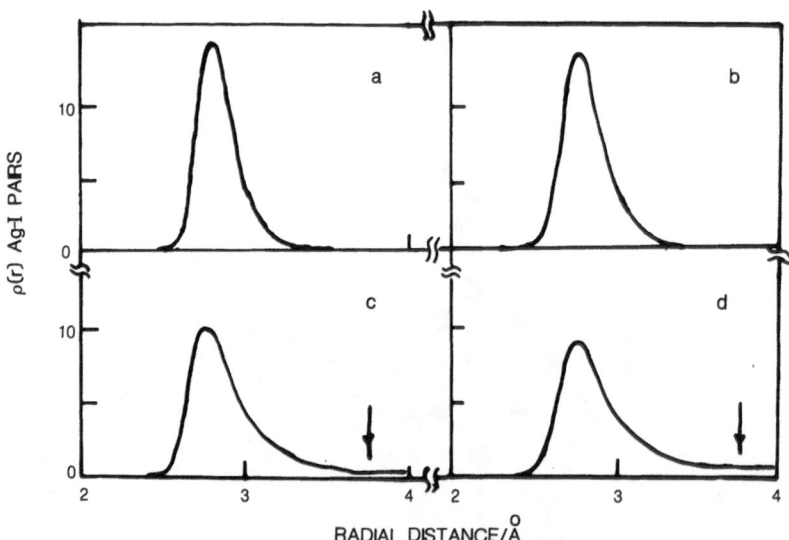

Figure 7.2. The radial distribution function of I^- ions around Ag^+ in AgI. (a), (b) β-phase at 20 and 98°C, respectively; (c), (d) α-phase at 198 and 302°C, respectively. Each distribution function has been normalised to contain four I ions in the first peak. The arrows in (c) and (d) indicate the location of the contribution from a Ag^+ ion occupying the centre of a face between adjacent tetrahedra (after [32]).

the fact that the d sites have the largest space for the cations. The data were inconsistent with there being significant occupation of the octahedral b sites.

A variety of models of the disorder in α-AgI were quantitatively tested with the EXAFS data [32]. The best fit was obtained with the excluded volume model which approximates the anion-cation interaction to a hard-sphere potential [36]. In this model the I$^-$ ions are fixed to a b.c.c. lattice and the Ag$^+$ ions uniformly distributed over the remaining volume. The location of the Ag$^+$ ions is then dominated by one parameter, $r_{excluded}$, which is roughly the sum of the effective hard-sphere radii of Ag$^+$ and I$^-$. Slight softness of the potential and thermal vibrations of the anions was accounted for by convoluting a gaussian with the pair distribution function. Thus the model contains only two parameters; $r_{excluded}$ and the width of the gaussian. An excellent fit was obtained to both β- and α-AgI and it was possible to generate the cation density map shown in Figure 7.3. This shows the density peaks on the tetrahedral sites and an easy conduction pathway is along $\langle 110 \rangle$ directions through the three-fold coordinated sites.

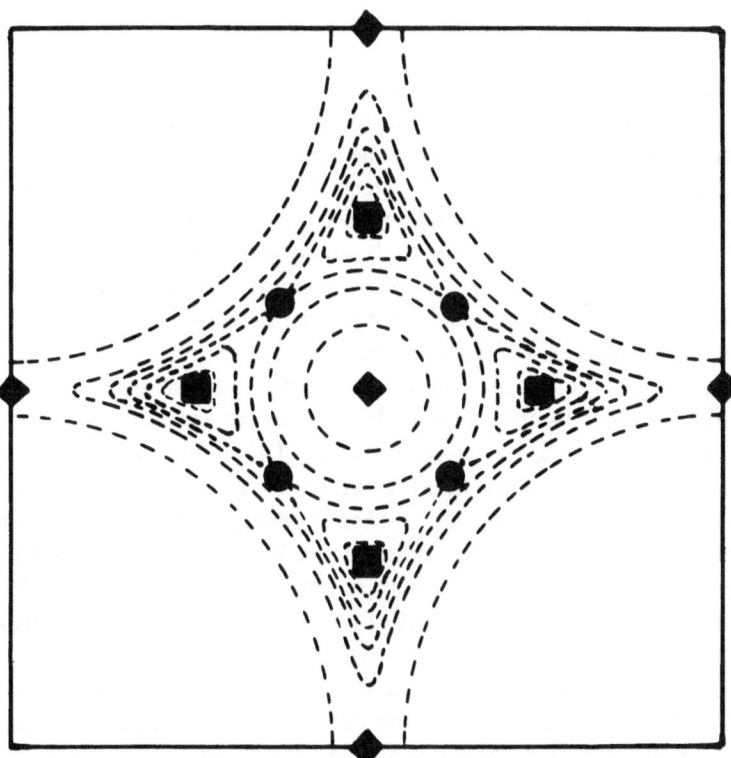

Figure 7.3.   A contour plot of the Ag$^+$ ion density in the (100) plane of AgI at 302°C. The symbols for sites have the same notation as in Figure 7.1 (after [35]).

At the time the EXAFS work was first published it appeared inconsistent with existing neutron diffraction data [32]. However, later neutron data for single crystal AgI [37] yielded ion densities consistent with the EXAFS work. The difference between the two techniques in the studies of fast-ion conductors has been stressed by Hayes and Boyce [3]; diffraction data give precise information of the immobile ions but little on the mobile ions as their distribution is disordered. In contrast, EXAFS will locate the mobile ion with respect to the immobile ions. Thus the two techniques should be regarded as complementary.

Finally, it is worth noting that the EXAFS data for Cu halides have been successfully fitted with the excluded volume [38, 39] and parameters for AgI and CuI have been compared with molecular dynamics simulations [39–41].

7.4   Rare-earth doped alkaline earth fluorides

A number of halides (e.g. CaF_2, SrF_2, BaF_2, PbF_2 and $SrCl_2$) and technologically important oxides (e.g. UO_2, ThO_2, CeO_2 and cubic stabilised ZrO_2) adopt the fluorite structure shown in Figure 7.4(a). It consists of a simple cube of anions with alternate cubes occupied by cations. At low to moderate temperatures, the point defects and ionic conductivity in lowly doped crystals of the halides are well understood [24, 42] and are typical of normal ionic solids, as described in Section 7.2. The predominant defects are anion Frenkel pairs, anion vacancies and anions in the centre of the unoccupied cubes, and the vacancies are the more mobile defects. However, the 'open' nature of the structure gives rise to unusual and interesting behaviour. The pure halides exhibit a broad thermal anomaly at about 0.8 of the melting temperature and the high temperature regime exhibits fast-anion conduction. These are excellent models of fast-ion behaviour due to the simplicity of the structure and they have been extensively studied (see, for example [24, 43–45], and references therein). A second consequence of the 'open' structure is the ability of the alkaline earth fluorides, e.g. CaF_2, to dissolve very high concentrations of rare earth fluorides, LnF_3, (Up to 50 mol% [46]); the Ln^{3+} ions substitute for the host cations with the formation of F^- interstitials. At Ln^{3+} concentrations in excess of 1 mol% there is evidence from a wide range of experiments, including neutron diffraction [47], dielectric relaxation spectroscopy [48], electron-spin resonance [49], laser spectroscopy [50] and computer modelling techniques [51] for the formation of defect clusters. A great deal of attention has been focussed on these systems, partly for their intrinsic interest and also as models for non-stoichiometric materials. However, there has been considerable debate over the cluster structures and the EXAFS studies described below have provided invaluable insight into this long-standing problem.

The early neutron diffraction data for Y/CaF_2 [47] showed the presence of

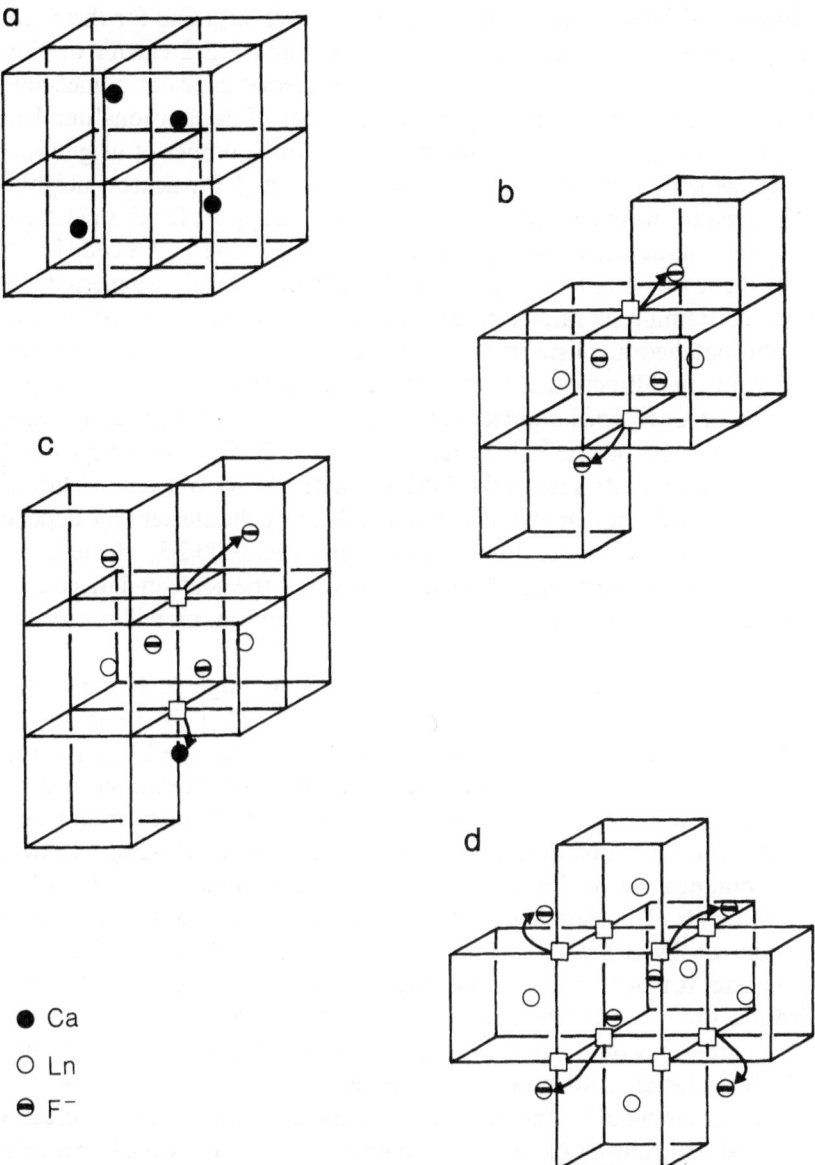

● Ca
○ Ln
⊖ F⁻

Figure 7.4.   (a) The fluorite structure. (b) The $2|0|2|2_1$ dopant dimer cluster. (c) The $2|0|2|3_1$ dopant dimer cluster. The arrows show the postulated relaxation pathways of the cations. (d) The $6|0|8|6_1$ $\langle 111 \rangle$ dopant hexamer. Five of the six lanthanide dopants and four of the F^- interstitials, sited opposite the edges of the central cube, are shown. The arrows shown have a 45° rotation of the central cube faces, followed by an outward relaxation, will generate 8 of the 14 F^- interstitials.

F^- interstitial ions displaced from the cube-centre sites and for the 5 mol% doping this was interpreted in terms of the small 2:2:2 cluster originally proposed for UO_{2+x} by Willis [52]. In the systematic cluster nomenclature [51] this is termed the $2|0|2|2|_1$ cluster (number of dopant ions|number of anion vacancies|number of relaxed normal anions|number of neighbouring interstitials|subscript denoting interstitial position (1 for nearest neighbour site to dopant)) which is shown in Figure 7.4(b). Support for this cluster was provided by static-defect energy calculations using the HADES code [53, 54] whereas X-ray studies [55–57] of more heavily doped CaF_2 suggested larger clusters containing six rare-earth cations packed around an interstitial site— the cubo-octahedral cluster based on the unit shown in Figure 7.4(d). More recent neutron diffraction studies provided support for the cubo-octahedral cluster in Er/CaF_2 [58] and clear evidence for $\langle 111 \rangle$ interstitials (inconsistent with the cubo-octahedral cluster) in La/CaF_2 [59]. It was proposed that the $\langle 111 \rangle$ interstitials were in the $2|0|2|3_1$ cluster shown in Figure 7.4(c). This work agreed with the calculations [51] in indicating the cluster type depended on the dopant ion radius, i.e. large dopants, small $2|0|2|3_1$ clusters; small dopants, large cubo-octahedral clusters. However, the neutron data could not be unambiguously interpreted as all atom-atom correlations will contribute to the data.

EXAFS spectra were collected at the Ln L III edge in 10 mol% LnF_3 doped CaF_2 for a wide range of rare earths [60–62]. This specific concentration was chosen as the spectroscopic data [63, 64] for the Er/CaF_2 system indicated the presence of only one type of predominant cluster. The normalised EXAFS, $\chi(E)$ versus E, are reproduced in Figure 7.5 and clearly show the Ln environment is varying across the rare-earth series. There are two dominant series within which the features are similar, (i) La to Nd, and (ii) Gd to Yb. The detailed analysis [61, 62] of the data associated the variations of EXAFS with a shift from small to large cluster with increasing dopant size. It is worth noting the origins in the trends shown in Figure 7.5 which are predicted by the computer simulations [51]. Firstly, the smaller $2|0|2|3_1$ cluster would have a larger average first shell than the cubo-octahedral cluster, 2.40 and 2.35 Å, respectively. The second shell of the $2|0|2|3_1$ cluster would contain $11\,Ca^{2+}$ ions and $1\,Ln^{3+}$ ion, whereas the cubo-octahedral cluster has $8\,Ca^{2+}$ and $4\,Ln^{3+}$ ions. Finally the larger cluster would significantly distort the lattice and create large static displacements. Thus the contributions from neighbouring shells would be diminished. Representative examples of the two types were analysed in depth; the Nd/CaF_2 EXAFS were fitted on the basis of a $2|0|2|3_1$ cluster and the data for Er/CaF_2 with a $6|0|8|6_1$ cluster. Model compounds were used to assign the phase shifts. The quality of the fits can be judged from Figure 7.6 and the comparison between EXAFS and computer simulation parameters is given in Tables 7.1 and 7.2. The agreement between the two techniques is very good and differences are probably within the errors; in the EXAFS case

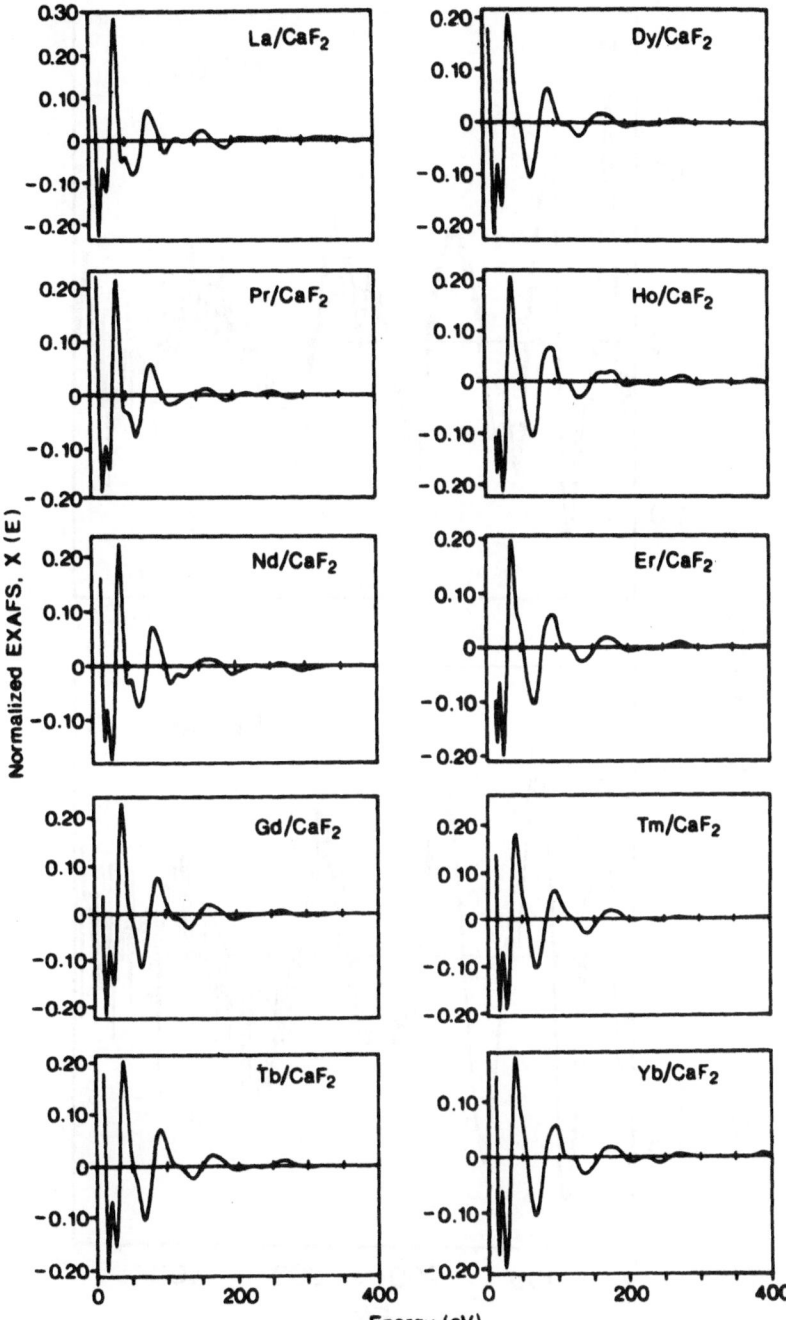

Figure 7.5. The normalised lanthanide Ln L III edge EXAFS for the series $Ca_{0.9} Ln_{0.1} F_{2.1}$ (after [62]).

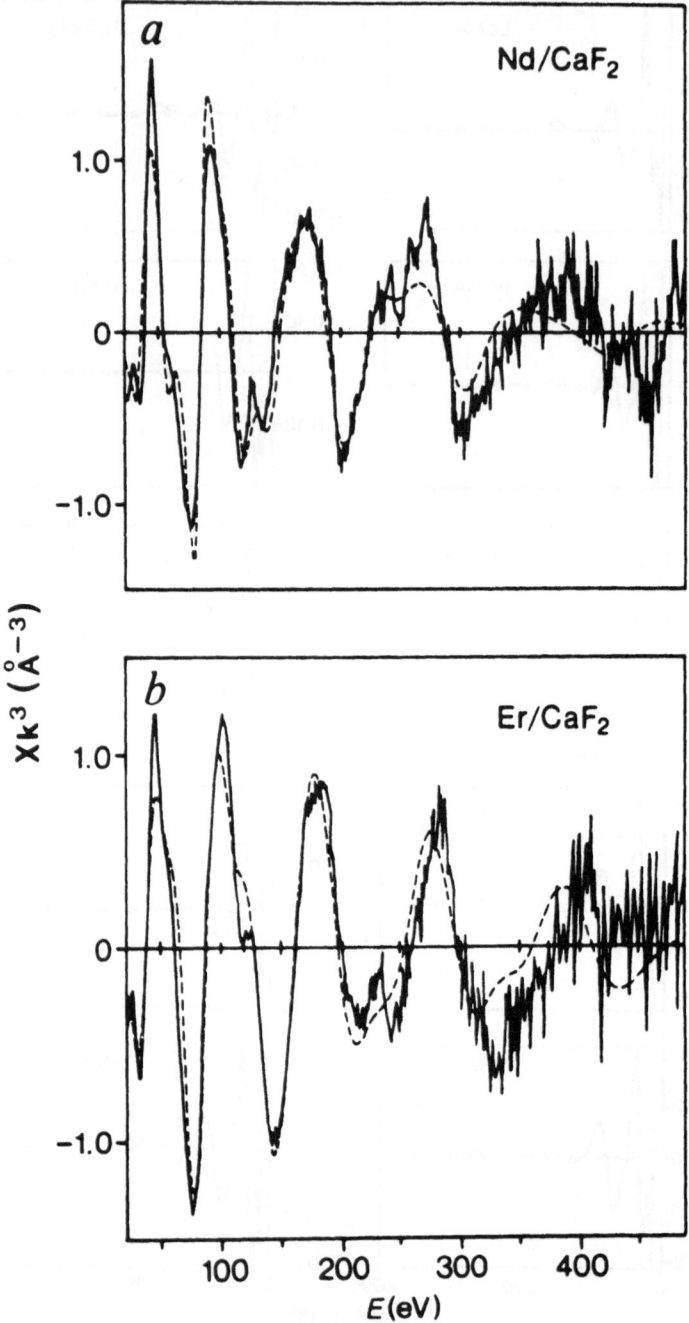

Figure 7.6. The k^3-weighted Ln L III edge EXAFS of Nd/CaF_2 and Er/CaF_2 (full line) and the EXAFS calculated from the parameters in Tables 7.1 and 7.2 (after [62]).

Table 7.1 Comparison of radial distances (RD) obtained from fitting the Nd/CaF$_2$ data to the $2|0|2|3_1$ cluster configuration with the radial distances predicted by computer simulations.

No. atoms	Type	RD from EXAFS (Å)	Debye-Waller factor σ^2(Å^2)	Simulation RD (Å)
10	F	2.40	0.014	2.40
1	Nd	3.97	0.013	3.79
11	Ca	3.91	0.024	3.87
1	F	4.18	0.005	4.05
24	F	4.50	0.028	4.53
6	Ca	5.45	0.026	5.46
24	F	5.95	0.030	5.95
24	Ca	6.69	0.035	6.69

the experimental errors and in the calculations, the accuracy of the interatomic potentials.

It was noted later by Welch [65] that the $6|0|8|6_1$ would not be consistent with the spectroscopic data of the Ln/CaF$_2$ systems with smaller rare-earths. He proposed that a pentamer would be a more appropriate cluster for these materials. The binding energy of the $5|0|8|5_1$ cluster was calculated [66] and it was found to be slightly more stable than the $6|0|8|6_1$ cluster, providing some support for the view. However, the calculated EXAFS spectra of these two clusters were virtually indistinguishable and the basic conclusions outlined above remain unaffected [66].

Finally, the EXAFS technique has been used to provide an understanding of the rapid increase at high temperatures in the ionic conductivity of the heavily rare-earth doped alkaline earth fluorides [66]. In 10 mol% Er/CaF$_2$ this occurs at $\sim$ 870K [67] and EXAFS were collected at the Er L III edge for this system over the range 298–1070K. These experiments show that the basic cluster structure remained constant throughout this temperature range and the increased conductivity was due to the loss of a loosely bound interstitial rather than a complete fragmentation of the cluster.

Table 7.2 Comparison of radial distances (RD) obtained from fitting the Er/CaF$_2$ data to the $6|0|8|6_1$ cluster configuration with the radial distances predicted by computer simulations.

No. atoms	Type	RD from EXAFS (Å)	Debye-Waller factor σ^2(Å^2)	Simulation RD (Å)
9	F	2.35	0.011	2.35
1	F	3.32	0.007	3.45
8	Ca	3.93	0.016	3.83
2	F	4.12	0.007	4.05
4	Er	4.21	0.023	4.05
22	F	4.54	0.034	4.55

7.5   Cubic stabilised zirconia and bismuth oxide

7.5.1   *Cubic stabilised ZrO$_2$*

Pure zirconia, ZrO_2, has a monoclinic structure at normal temperatures. However, the high temperature cubic fluorite phase (stable above $2370°C$) can be stabilised at room temperature by the addition of $\sim 9\,mol\%$ lower valence oxides, such as Y_2O_3 and CaO. The calcium- and yttrium-stabilised zirconias (referred to as CSZ and YSZ, respectively) are good anion conductors at high temperatures and have technological applications as solid electrolytes in sensors, fuel cells and electrocatalysts [15, 17, 68–71]. The good conductivity arises in both cases from the formation of charge-compensating vacancies; one anion vacancy is created by two Y^{3+} ions in YSZ and by one Ca^{2+} ion in CSZ. At these high dopant levels the defect structure must be complex, with some association between the dopant and anion vacancies. The exact nature of the defect structure has been controversial and several models have been proposed on the basis of small defect clusters [72]. Central to the resolution of the controversy is the relative location of the dopant and the anion vacancies, i.e. are vacancies nearest neighbours or are they accommodated elsewhere in the structure. Microdomain models have also been proposed [73]. Neutron diffuse scattering data [74] for YSZ suggest a model of a vacancy between a pair of Y^{3+} ions, whereas site-selective spectroscopy [75] indicates that the Y^{3+} ion is a full eight-fold coordination site at low dopant concentrations. Similarly, for CSZ there is conflicting evidence for the relative locations of Ca^{2+} ions and anion vacancies. At first sight this is an ideal situation where the EXAFS technique might be used to advantage. However, Zr^{4+} and Y^{3+} are isoelectronic and have virtually identical backscattering functions making it impossible to distinguish the two ions in coordination shells. Nevertheless, by careful choice of experiment and data analysis it has been possible [76, 77] to provide information on the ion sites in YSZ and how these change with increasing temperature. In fact there have been two EXAFS studies of YSZ, one at the Daresbury SRS [76, 77] and one at the Cornell High Energy Synchrotron Source (CHESS) [78]. The experiments and the raw data were similar in both studies; however, the interpretation of the results differed. The former of these studies will be considered first.

Catlow *et al.* [76] collected EXAFS data at the Y K edge and Zr K edge over the temperature range -125 to $770°C$ for a powder sample of YSZ (19 mol% Y_2O_3 in ZrO_2, $Zr_{0.81}Y_{0.19}O_{1.90}$). In the same series of experiments, data were obtained for pure Y_2O_3 at room temperature and pure monoclinic ZrO_2 over the temperature range -125 to $770°C$. Data from these model compounds were used to assign the phase shifts. The Fourier transforms of the $k^3\chi(k)$ versus k normalised EXAFS for the Y and Zr edge of YSZ at $-125°C$ are shown in Figure 7.7. The notable differences are (i) the first shell distances; the mean Zr–O distance is less than the average Y–O distance, and (ii) the magnitude of

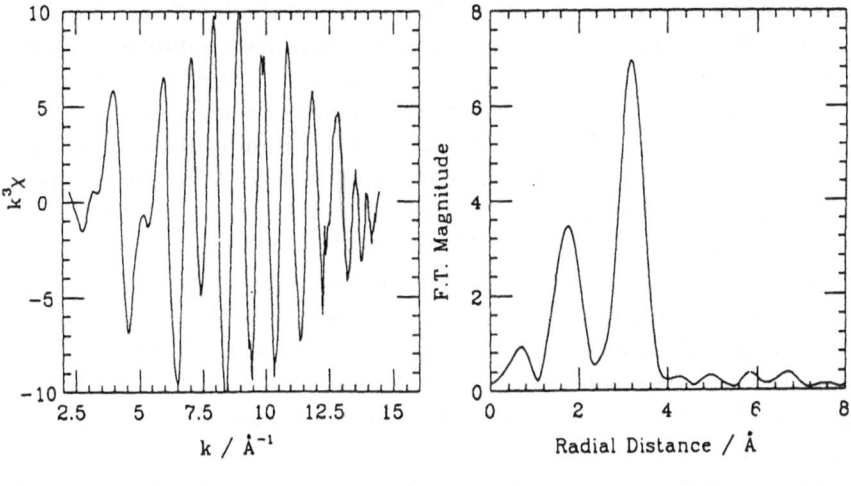

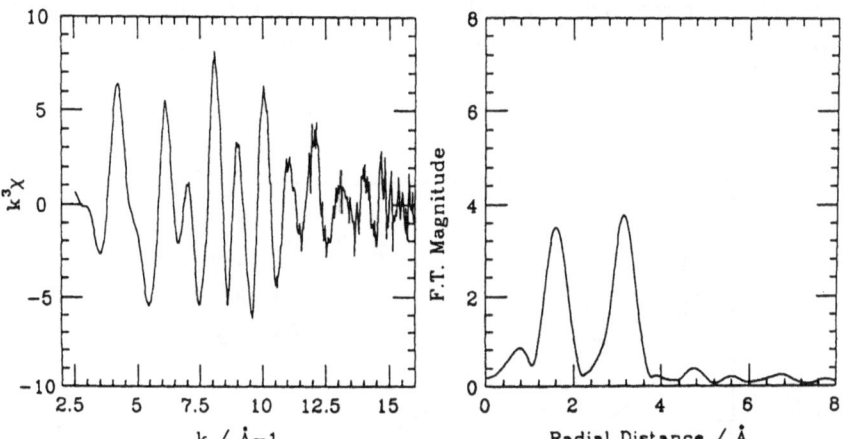

Figure 7.7. (a) The k^3-weighted Y K edge EXAFS for YSZ at $-125°$C. (b) The Fourier transform of the data in (a). (c) The k^3-weighted Zr K edge EXAFS for YSZ at $-125°$C. (d) The Fourier transform of the data in (c) (after [76]).

the second shell, due to cations, is smaller by nearly a factor of 2 in the Zr EXAFS. A feature in both EXAFS is the lack of structure beyond the second shell indicating a high degree of static disorder.

The detailed fitting of these data with the parameters from the model compounds yielded the mean cation-oxygen distances in YSZ as 2.11 Å for Zr–O and 2.28 Å for Y–O. These are very similar to the average cation-oxygen distances in the parent oxides, i.e. 2.16 ± 0.08 Å for monoclinic ZrO_2 and 2.28 ± 0.03 Å for Y_2O_3. As pointed out in Section 6.3.2, it is difficult to determine the coordination number, N_j, precisely, as a result of the strong correlation with

Debye-Waller factors, σ_j^2. However, by a comparison of both spectra it was clear that the Y–O EXAFS required a larger coordination number and/or a smaller amount of disorder to obtain a good fit compared with the Zr–O EXAFS. This would be consistent with the anion vacancies being in the Zr^{4+} coordination shell, a view supported by consideration of the cation-cation EXAFS. The reduced magnitude of the second shell EXAFS must be due to large static variations in the Zr-cation distance. Since the same cation neighbours are shared by both Y^{3+} and Zr^{4+} and the ions are equivalent backscatterers, these variations cannot be due to a general displacement of all cations and were therefore attributed to displacements of only Zr^{4+} ions. It was concluded that this was disorder caused by Zr^{4+} ions relaxing off centro-symmetric sites, the most likely cause being the presence of a nearest-neighbour O^{2-} vacancy. By contrast, electrostatic arguments would place the vacancy adjacent to the Y^{3+} cation.

At high temperatures the anions are likely to execute anisotropic thermal motions. As it would be difficult to incorporate this into a detailed fitting of the EXAFS for YSZ, a very simple approach was taken which involved comparing the areas under the first two peaks of the Fourier transformed Y and Zr EXAFS. These are plotted in Figure 7.8 and it can be seen that they converge

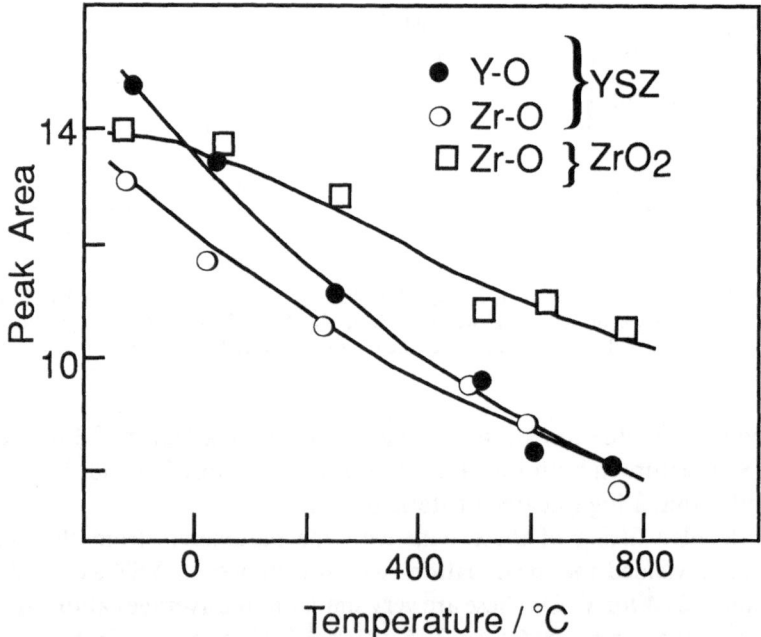

Figure 7.8. Variation with temperature of the area of nearest-neighbour Y–O and Zr–O peaks in magnitude of Fourier transformed EXAFS data for YSZ and monoclinic ZrO_2 (after [76]).

at high temperatures. It was inferred that this was due to the increased mobility of the O^{2-} ions giving rise to a statistical distribution of anion vacancies between Y and Zr nearest-neighbour sites, i.e. the local environments of both cations become more similar as the temperature increases.

Roth et al. [78] also collected EXAFS data for both the Y K edge and the Zr K edge in YSZ samples with a range of Y_2O_3 concentrations (9.4–24 mol%). They interpreted the first major peaks in the Fourier transforms from both sets of spectra as poorly resolved double maxima or asymmetrically shaped distributions suggesting both Y^{3+} and Zr^{4+} are bonded to O^{2-} at several distances. Two-shell models were used to analyse the first peak with the Zr–O distances being shorter than the Y–O distances. The average values in each case were close to those found by Catlow et al. [76]; in contrast, though Roth et al. concluded the O^{2-} vacancy was adjacent to the Y^{3+} cation rather than the Zr^{4+} cation. However, the data of Roth et al. also show that the relative magnitude of the second peak is much smaller in the Zr EXAFS. This feature was not discussed in their paper. It is quite clear though, from this observation that the Zr^{4+} ions must be statistically disordered in YSZ. The same conclusion has been reached in an anomalous dispersion diffraction study of YSZ powder [79]. The simplest explanation for the origin of this disorder is that the O^{2-} vacancy is located adjacent to the Zr^{4+} cation. Further support for this picture of the disorder in YSZ comes from computer simulations [80] which show that Zr^{4+}–O^{2-} vacancy pairs are energetically favoured with respect to Y^{3+}–O^{2-} vacancy pairs. It can also be justified in chemical terms from a consideration of ionic radii ($Zr^{4+} = 0.98\text{ Å}$, $Y^{3+} = 1.16\text{ Å}$) and the preference of Zr^{4+} for a 7-coordinate environment, as found in monoclinic ZrO_2. Recent EXAFS studies [81] of Gd^{3+} stabilized ZrO_2, which is expected to be similar to YSZ due to the same charge and comparable ionic radii of the dopant ($Gd^{3+} = 1.20\text{ Å}$, $Y^{3+} = 1.16\text{ Å}$), showed that as the Gd^{3+} concentration increased the coordination number of the Zr^{4+} ion decreased. The coordination number of the Gd^{3+} remained constant suggesting that the O^{2-} vacancy was located preferentially adjacent to the Zr^{4+} ion.

The recent EXAFS study of CSZ [77] provides an interesting comparison with the data for YSZ. The difference in dopant cation charge means that the same concentration of Ca^{2+} creates twice the number of anion vacancies as Y^{3+} and a more disordered structure is expected. Data were collected for dopant levels of 8 mol% ($Zr_{0.9}Ca_{0.08}O_{1.92}$) and 19 mol% ($Zr_{0.81}Ca_{0.19}O_{1.90}$), i.e. 4% and 9.5% anion vacancy concentrations, respectively, compared to 5% in the YSZ sample [76] discussed above. The Zr K edge EXAFS for the CSZ samples were very similar to the equivalent spectra for YSZ. In particular, the second shell peak height was small and correlated with the vacancy concentration. The peak height for the $Zr_{0.92}Ca_{0.08}O_{1.92}$ sample resembled more closely that of the $Zr_{0.81}Y_{0.19}O_{1.90}$ sample whose anion vacancy concentration was similar. The Zr–O distance in the CSZ samples was 2.12 Å which, within the experimental error, is the same as that in YSZ. It is therefore reasonable to invoke similar arguments to those applied for YSZ and conclude that the anion vacancy is also nearest-neighbour to the Zr^{4+} ion in CSZ.

Ca K edge EXAFS were also taken [77] for the two CSZ samples. The Ca–O shell is more disordered than the Zr–O shell and it was not possible to determine the Ca^{2+} coordination number. This suggests that the large Ca^{2+} ion (1.26 Å radius) causes large displacements of the O^{2-} neighbours. These studies suggest that EXAFS could be used to detect cation ordering in CSZ, which would explain the appearance of microdomains in cubic stabilised zirconia. Such a possibility arises from the difference in photoelectron back-scattering powers, $f_j(\pi)$, of Ca and Zr (unlike Y and Zr which are similar), and would result in changes in the appearance of the second shell with ordering. Thus it should be feasible to investigate the evolution of cation ordering, at least qualitatively, with composition and thermal treatment.

7.5.2   Rare-earth doped bismuth oxide

Pure bismuth oxide, Bi_2O_3, has a phase transition at 729°C from the low temperature monoclinic α-phase to the fluorite-structured δ-phase which is stable up to the melting point (824°C). In the δ-phase 25% of the normal anion sites are unoccupied and this is clearly associated with the fact that it has the highest known oxygen ion conductivity, $\sim 1\Omega^{-1}cm^{-1}$ [20]. There have been several models [70, 82] proposed for the conduction mechanism in this material, although it is still a point of controversy. The δ-phase can be stabilised at room temperature by the addition of certain trivalent cations, such as Y^{3+} and the smaller rare-earths, and there has been considerable interest in these materials as solid electrolytes [70]. The very high oxygen ion conductivity of the δ-phase of pure Bi_2O_3, however, is not maintained on doping. Nevertheless, it is still good, e.g. $\sim 10^{-2}\Omega^{-1}cm^{-1}$ at 500°C compared with $\sim 10^{-3}\Omega^{-1}cm^{-1}$ for YSZ at the same temperature. Even less is known about the conduction process in doped Bi_2O_3 than in the pure material. Recent neutron scattering studies [83, 84] of Ln_2O_3/Bi_2O_3 showed there is considerable anion disorder with $\sim 30\%$ of the normal anions considerably displaced along $\langle 111 \rangle$ directions, whereas the cations exhibit small $\langle 100 \rangle$ displacements. The number of $\langle 111 \rangle$ displaced anions decreases with increased concentration of dopant and there is evidence for microdomains at high dopant concentrations. The question arises as to whether the disorder is centred around the dopant or the host cation. Thus the problem is analogous to that described earlier for YSZ and is one where EXAFS studies offer considerable assistance.

Catlow and co-workers performed EXAFS studies for Bi_2O_3 doped with Y_2O_3 [85, 77], Er_2O_3 [86, 77] and Yb_2O_3 [86, 77] with dopant concentrations in the range 20–40 mol%. EXAFS spectra were collected at $-195°C$ and room temperature at the rare-earth and Bi L III edges, and the Y K edge. Additional measurements were made at temperatures up to 750°C for the 40 mol% Y doped Bi_2O_3 and the 25 mol% Er doped Bi_2O_3. The pure oxides were used as model compounds to assign the phaseshifts. The EXAFS showed no major variation with the nature or concentration of the dopant and therefore the discussion can be confined to the 40 mol% Y^{3+} doped Bi_2O_3.

The Fourier transforms of the $k^3\chi(k)$ versus k normalised EXAFS for the Y and Bi of 40 mol% Y^{3+} doped Bi_2O_3 at 80K obtained by Catlow and co-workers are shown in Figure 7.9. It is interesting to compare the data with that obtained for YSZ shown in Figure 7.7. In the present system, examination of the first shells shows the Bi–O contribution is much smaller than the Y–O contribution, whereas in YSZ the Zr–O and Y–O contributions have very similar magnitudes. In addition, the second shell EXAFS for both Y and Bi is considerably reduced in Y_2O_3/Bi_2O_3. The detailed analysis revealed considerable static disorder around the Bi^{3+} cations. It was not possible to obtain a reasonable fit to the Bi–O shell due to the very large asymmetry of this peak. The much reduced Bi-cation second shell compared to the Y-cation second shell was taken as further evidence for the disorder being around the Bi^{3+} cations, i.e. the same reasoning used to explain the YSZ data. Thus the $\langle 100 \rangle$ cation displacements observed in the diffraction data are most likely to be associated with the displacement of Bi^{3+} ions from centro-symmetric sites. Heating the sample had the effect of reducing the amplitude of Y EXAFS compared to the Bi EXAFS suggesting the local environments were becoming more similar and the mobile O^{2-} ions were those sited around the host cations.

Kamijo *et al.* [87] have reported an EXAFS study of Y_2O_3/Bi_2O_3 with the dopant concentrations at 25 and 40 mol%. The experimental procedures employed by these workers using the synchrotron at the Japanese Photon Factory were similar to those of Catlow and co-workers and in many respects the experimental data were similar in both studies. However, in this study both the Y–O and Bi–O peaks were treated as asymmetric and were fitted using two shell models. Those workers concentrated on the analysis of these first peaks and concluded that the oxygen vacancies in the structure were localised around the dopant. Their data, however, show a much reduced Bi-cation peak with respect to the Y-cation peak, as Catlow and co-workers found. This was not discussed by Kamijo and co-workers although, as we have seen, it is difficult to reconcile a disordered cation shell with a fully coordinated anion shell.

There are clear similarities in these EXAFS experiments between Y doped Bi_2O_3 and ZrO_2. As Catlow and co-workers have argued, the most consistent interpretation of both anion and cation shells in the two stabilised cubic systems is that static disorder at low temperatures is chiefly confined to the

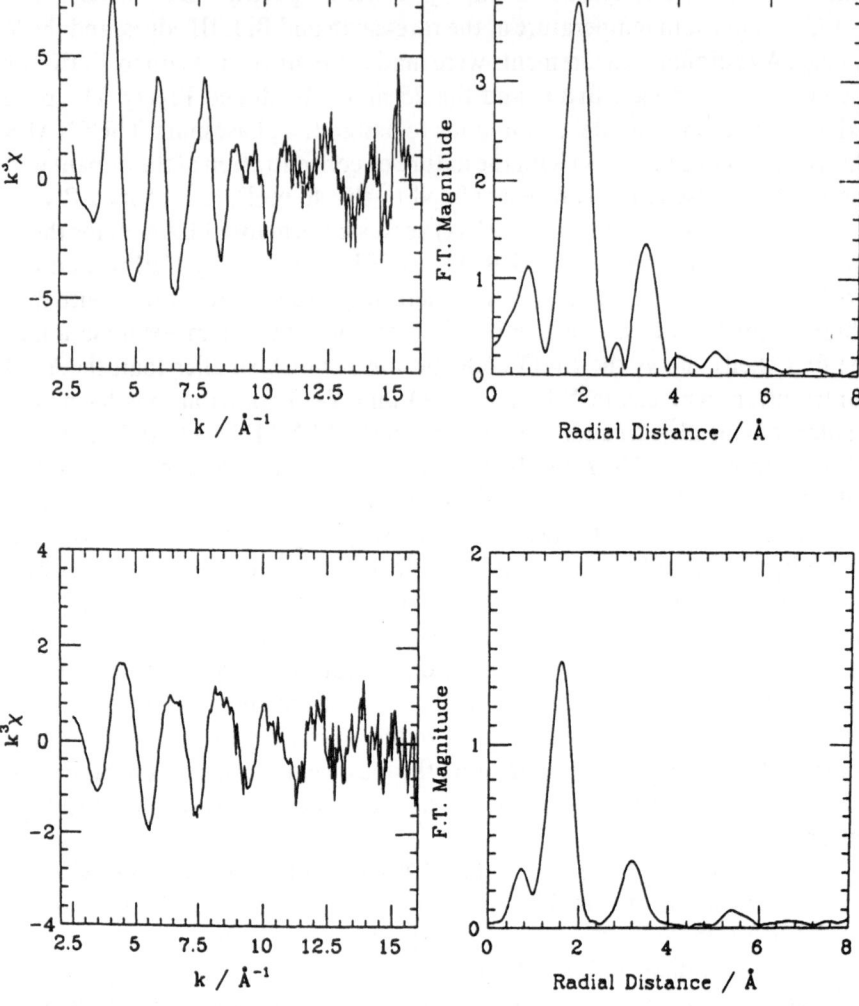

Figure 7.9.   (a) The k^3-weighted Y K edge EXAFS for 40% Y_2O_3/Bi_2O_3 at 80K. (b) The Fourier transform of the data in (a) (c) The k^3-weighted Bi L III edge EXAFS for 40% Y_2O_3/Bi_2O_3 at 80K. (d) The Fourier transform of the data in (c) (after [85]).

host cation. However, static disorders is more pronounced in doped Bi_2O_3. In terms of ionic radii Zr^{4+} is smaller than Y^{3+}, whereas Bi^{3+} is nominally larger than Y^{3+}. In both systems the mean distance between the host cation and the nearest oxygens is the same, 2.11 Å, less than the sum of ionic radii for O^{2-} and Bi^{3+}. However, Bi^{3+} has a stereochemically active lone-pair and this has been given as the reason why it adopts a highly anisotropic coordination geometry. In Y_2O_3/Bi_2O_3 then, the role of the dopant appears to be as an 'absorber' of the disorder generated around the Bi^{3+} ions as well as a stabiliser of the fluorite structure.

7.6   Mixed fluorides

The fast-ion conductivity of the fluorite-structured halides has been outlined in Section 7.4. There has been considerable interest in mixed fluorides based on SnF_2 (e.g. $PbSnF_4$) and BiF_3 (e.g. $RbBiF_4$) which have a fluorite phase and a high ionic conductivity at moderate temperatures [88]. For example, the conductivity of $RbBiF_4$ is $5 \times 10^{-3} \Omega^{-1} cm^{-1}$ at $100°C$ [89]. The high conductivity can be attributed to a highly disordered anion sub-lattice and neutron diffraction studies [89] have shown that there is a substantial occupancy of the cube-centre sites by interstitial F^- ions. In these mixed systems the distribution of the two cations is random and the obvious question arises, as with the doped oxides, as to the possible preferential location of anion disorder around a particular cation.

EXAFS studies have been performed for both $RbBiF_4$ [90, 91] and $PbSnF_4$ [92]. The results for both systems are analogous and attention will be focussed on the investigation of $Rb_{1-x}Bi_xF_{1+2x}$ (where $0.5 \leqslant x \leqslant 0.75$) which involved a study [90] of the EXAFS of the Rb K edge and Bi L III edge over the temperature range 80–473 K. The k^2 weighted data for the $x = 0.5$ samples are displayed in Figure 7.10. Both series of spectra show only one well-resolved frequency which was attributed to the nearest-neighbour anions. No cation order was detectable, which was consistent with the large Debye-Waller factor

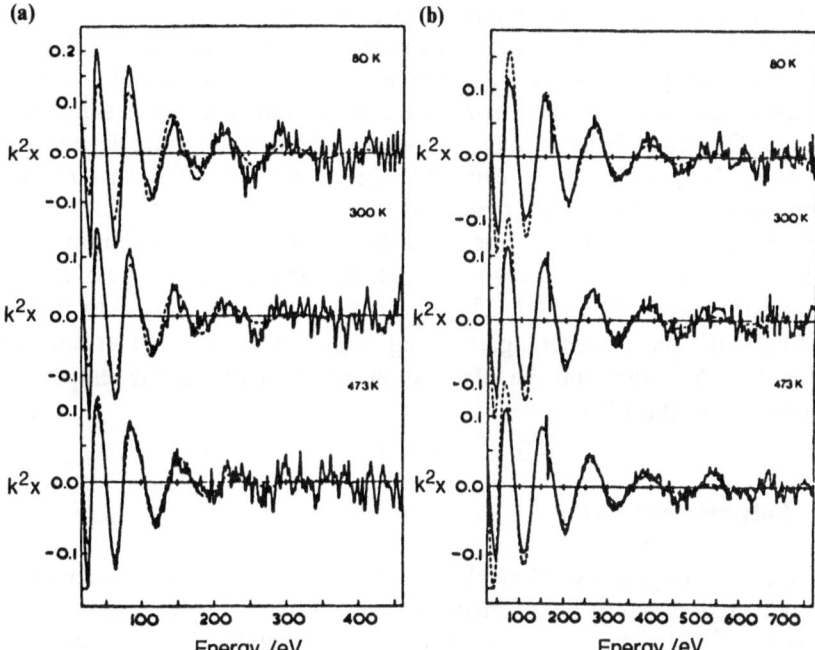

Figure 7.10.   The k^2-weighted EXAFS data for (a) the Rb K edge and (b) Bi L III edge in $RbBiF_4$ at 80, 300 and 473K (after [90]).

found for cations in the neutron study. The major difference in the two sets of spectra is the temperature dependence; for Bi the spectra show virtually no change in amplitude and frequency whereas for Rb both amplitude and frequency decrease with increasing temperature. The data were fitted on the assumption that for both cations the nearest-neighbour F^- ions were situated at the mean cation-anion distance of 2.62 Å at 80 K. Not surprisingly the Bi–F distance was hardly changed at 473 K, i.e. 2.60 Å with a Debye-Waller factor, σ_1^2, of 0.013 Å^2. In contrast, at 473 K the Rb–F distance had reduced to 2.54 Å with a Debye-Waller factor, of σ_1^2, 0.025 Å^2. These changes suggest that the static disorder is in the local environment of the Rb^+ ion and arises from an increase in the number of F^- vacancies in the first shell. The relaxation caused by these vacancies would affect the observed decrease in the Rb–F distance with increasing temperature. The absence of similar changes in the Bi–F distance suggests that the vacancies are principally formed in Rb 'rich' regions of the lattice. It was argued in the original paper that the creation of vacancies would be electrostatically more favourable around the singly charged Rb^+ ion than the triply charged Bi^{3+} ion.

The spectra showed increasing order as x was increased. This would be consistent with the above model as increasing the Bi^{3+} concentration would lower the amount of Rb rich regions. It also explains the fact that the conductivity of these systems decreases with increasing x, even though this increases the number of F^- charge carrying ions.

The interpretation of the data for $RbBiF_4$ was necessarily simple as a more complex model would have been inconsistent with the experimental errors. However, more progress has been possible in the understanding of this system by combining the EXAFS work with molecular dynamics simulations [91, 92]. The radial distribution functions and their temperature dependences for both the cations show excellent agreement from the two different techniques, confirming the structural predictions from the EXAFS. In addition, the molecular dynamics modelling has shown that the migration pathway for the F^- ions involves a non-collinear interstitial type of motion.

The situation in $PbSnF_4$ [92] now seems to be very similar to that for $RbBiF_4$, with the disorder again being preferentially located around one cation (the Pb^{2+} ion) and the effects showing up mainly in the temperature dependence of the EXAFS.

7.7 Polymer electrolytes

Polymer electrolytes are relatively new materials that are characterised by a good, predominantly ionic, electrical conductivity in the absence of any solvent [26, 27]. The archetypal polymer electrolytes are the complexes formed between high molecular weight polyethylene oxide (PEO) and alkali metal salts (MX) which were first studied by Wright and co-workers [93, 94].

Armand *et al.* [95] made a wider investigation of these materials and the analogous complexes containing polypropylene oxide (PPO). It was these authors who proposed the application of these materials as electrolytes for re-chargeable solid state batteries which led to the current intense research interest in polymer electrolytes. In comparison to inorganic electrolytes they offer several advantages for battery fabrication. The specific conductivity of the MX.PEO complexes is not particularly outstanding ($\sim 10^{-4}$ to $10^{-3}\,\Omega^{-1}\,cm^{-1}$ at 400 K); however, this is achieved at moderate temperatures and the materials are readily prepared as thin films (a few tens of micrometres thickness) by solvent-casting techniques. Thus the internal resistance of the battery could be low. In addition, the high flexibility of the films avoids the problem of electrolyte fracture and offers the possibility of a compact, lightweight battery design.

The vast majority of the polymer electrolytes are based on the ethylene oxide monomer unit ($-CH_2-CH_2-O-$) incorporated in the polymer backbone or in a side-chain. It is assumed that the incorporation of the salt involves the coordination of the ether oxygens to the cation in a similar manner to the crown-ether salt complexes. A common nomenclature for the PEO complexes is $MX.PEO_x$, where x is the number of the ethylene oxide monomer units per salt molecule. It has been assumed that the fully stoichiometric complex would have four ether oxygens around each cation, i.e. $MX.PEO_4$; however, this is not a general rule [26, 27]. The optimal conductivity is found in materials with $x > 4$, i.e. excess polymer and the phase diagrams of these materials can be very complicated.

High molecular weight PEO is a highly crystalline polymer ($\sim 80\%$ crystallinity) which melts at $\sim 65°C$ (338 K). At room temperature a typical film of $MX.PEO_x$ ($x > 4$) contains three phases; crystalline PEO, crystalline stoichiometric complex and a small amount of amorphous material. At 338 K the pure PEO crystallites melt to give an amorphous 'elastomer' phase which contains the higher melting point crystallites of the stoichiometric complex. As the temperature is increased beyond 338 K these crystallites dissolve into the elastomer phase. It is now realised that the ionic conductivity is due to the migration of ions in the amorphous region of the polymer and therefore a technological goal has been a fully amorphous polymer electrolyte at room temperature. The methyl side-group in PPO prevents the polymer crystallising; however, MX-PPO complexes are poor conductors. An additional condition for good conductivity is that the polymer chain is flexible, i.e. has a low glass-transition temperature.

X-ray diffraction techniques [96] have been used to study the structure of crystalline, near-stoichiometric complexes of $MX.PEO_x$ and have shown that the polymer adopts a helical conformation with small cations (Na^+ and K^+) enveloped by the helix and the larger cations coordinated to the outside of the helices. Whilst the resolution of the structure of the conducting, amorphous phase is clearly not amenable to diffraction methods, valuable information on

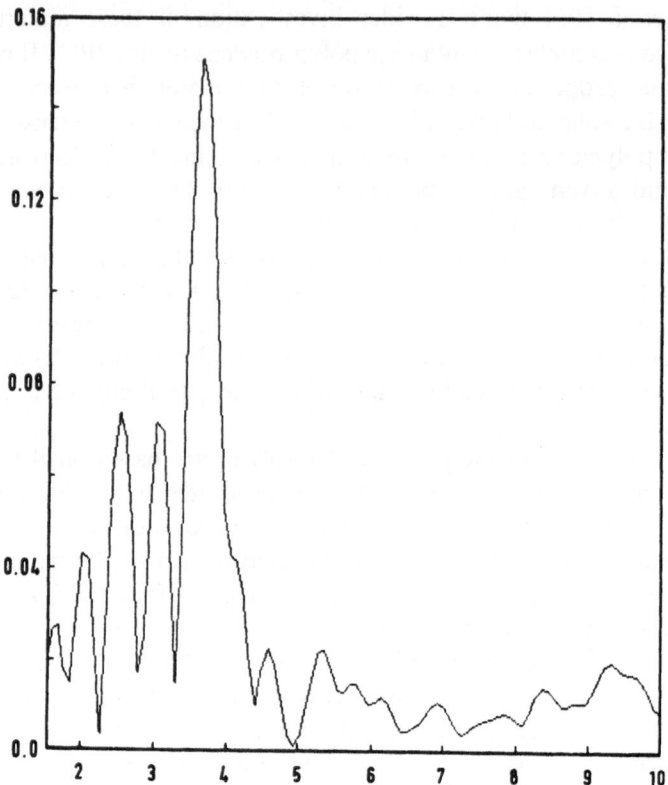

Figure 7.11.   The Fourier transform of the Rb K edge EXAFS for RbI.PEO$_8$ at 80K (after [97]).

the cation environments, particularly their location with respect to the anions, has been obtained from EXAFS studies.

EXAFS spectra [97] were collected at the Rb K edge for RbSCN.PEO$_x$ and RI.PEO$_x$ ($x = $ 4 and 8) over the temperature range 80–453 K. The samples, about 2 mm thick, were cast as films from acetonitrile and thoroughly dried. The RbSCN-18-crown-6-ether complex was used as the model compound for the evaluation of the phase shifts. The spectra showed no strong dependence on salt concentration in both systems and no nearest neighbour Rb–Rb pairs were detected suggesting that the variations in Rb–Rb distances is so large that their contribution to the EXAFS is smeared out. The Fourier transforms for RbI.PEO$_8$ and RbSCN.PEO$_8$ at 80 K are shown in Figures 7.11 and 7.12.

The Fourier transform for RbI. PEO$_8$ at 80 K has a large peak at 3.8 Å which was attributed to the presence of the I$^-$ ion, which is a strong back-scatterer. The pair of peaks between 2.5 and 3.8 Å was assigned to the ether oxygens. The data were fitted assuming that there were four oxygens as near neighbours. The best fit yielded two short Rb–O distances of 2.65 Å and two long Rb–O distances of 3.03 Å. The carbons were centred at 3.50 Å (four at 3.46 Å and four at 3.54 Å) and the I$^-$ at 3.75 Å. These distances did not change,

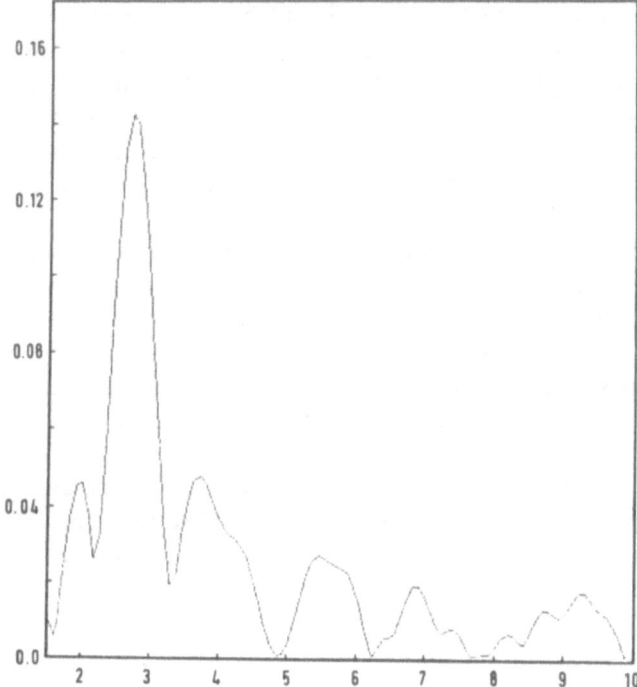

Figure 7.12.   The Fourier transform of the Rb K edge EXAFS for RbSCN.PEO$_8$ at 80K (after [97]).

within the experimental error, between 80 K and 453 K. Static disorder was the dominant contribution to the Debye-Waller factor at 80 K and accounted for 50% of the Debye-Waller factor at 453 K. Over this temperature range the Debye-Waller factor increased by a factor of 3.

The Fourier transform of the EXAFS for RbSCN.PEO$_8$ shows structure up to 6 Å and is more difficult to fit as there are no strong backscatterers to help discriminate different shells. The larger intensity of the first peak in comparison to the Fourier transform for RbI.PEO$_8$ suggests there is an extra atom in the first shell. The data were again fitted assuming that there were four oxygens in the first shell. The best fit yields four oxygens at 2.72 Å, a nitrogen of the SCN$^-$ ion at $\sim$ 3.0 Å, four carbons at 3.24 Å and four carbons at 3.55 Å. It is difficult to precisely locate the nitrogen as it is a weak backscatterer. The relative increase in the Debye-Waller factor over the range 80–453 K was slightly larger than that found in RbI.PEO$_8$.

The major conclusion of this work was that the bulk of the Rb$^+$ ions occupy a site of fixed geometry throughout the temperature range investigated. These sites are surrounded by a constant number of ether oxygens, assumed to be four in the modelling. The anions are clearly close to the cations and it is probably this proximity which causes the perturbations responsible for the differences in local structures of the two systems. The I^- ion splits the oxygen shell into two long and two short Rb–O bands, whereas the SCN^- ion disrupts this structure to give equal Rb–O distances but with an overall increase in the static disorder. Thus the nature of the anion can affect the migration of the cation via its control of the local structure. Aggregation of the ions to form ion pairs, triplets, etc. has been proposed as an important process in the transport of ions in polymer electrolytes. However, it would be misleading to use the EXAFS results to support this argument as they represent the average structure and provide no proof that the ions migrate as aggregates. There was no evidence for anomalous thermal motions in the polyether electrolytes, in contrast to the EXAFS for α-AgI, discussed earlier in Section 7.3, which suggests a picture in which the ions spend the bulk of the time in stable sites and occasionally undergo rapid jumps between these sites. These jumps would be facilitated by the segmental motion of the polymer chain.

Recently, EXAFS studies have been made on PEO complexes containing salts with divalent cations [98] and the RbI and RbSCN complexes with PPO [99]. The latter studies show that the Rb^+ ion local environments are similar to those in the PEO analogues.

7.8   Summary

Many of the studies discussed in this chapter were initiated to test the viability of EXAFS as a structural probe for ionically conducting solids. There is little doubt that they have been successful and the technique is emerging as a major tool in the study of these materials. It has been possible to draw very specific conclusions concerning the structure of particular materials. In addition, clear strategies have evolved, both experimental and in data analysis, for the most effective application of EXAFS.

The preferential location of disorder around particular ions in doped and mixed ionic crystals at low temperatures is a fundamental feature of these materials. EXAFS provides a method for probing the details of this disorder. In the case of rare-earth doped CaF_2 it has been possible to identify the specific structure of the defect clusters of dopant ions and hence resolve a long-standing problem in crystal chemistry. The EXAFS data for cubic stabilised zirconia suggest that the simple assumption that defects aggregate around allovalent dopants is not always correct. The most consistent interpretation for this system is that the excess vacancies are nearest neighbour to the host Zr^{4+} ion which is not expected on electrostatic arguments. The EXAFS study of the fast-ion conductor AgI was particularly successful at several levels, notably in identifying the pathway of the migrating ion. The final class of materials discussed, the polymer electrolytes, are amorphous materials and EXAFS is therefore the unique structural probe for these systems.

Not all of the systems discussed would be regarded as ideal for EXAFS investigations. For example, in YSZ the two cations are isoelectronic and their backscattering factors are equivalent. However, the technique has been extremely informative for this material and it is worth considering strategies for EXAFS work. It is always useful to have some structural model at the outset on which to test the EXAFS results. In some of the cases discussed this was available from other structural techniques (e.g. neutron diffraction for Ln/CaF_2) or from computer modelling (e.g. Ln/CaF_2 and $RbBiF_4$). In the future the combination of EXAFS and computer simulations should be very powerful for studying ionic materials due to the recent developments in the latter methods, for example improved interatomic potentials and faster computers. An important experimental condition is to collect EXAFS spectra over a wide range of temperature. The temperature dependence of the spectra and the fitting parameters can provide vital clues to the local structure. The best example of this in the work that has been discussed here is that of YSZ.

Finally, it is worth considering briefly potential developments and future experiments in this general area. In the work that has been discussed the concentration of the dopant ions was high and transmission experiments were appropriate. However, there are a number of problems that would involve very low dopant concentrations. For example, there is evidence in many systems that the structure of defect aggregates changes with dopant concentration (e.g. Ln/CaF_2). The increased sensitivity of fluorescence EXAFS should allow quite dilute systems (around 0.01 mol%, depending on the system) to be investigated. Similarly, it is possible with soft X-ray absorption fine structure (SOXAFS) to study the local environments of lighter elements than those discussed here. A good candidate would be the study of polyethers doped with sodium salts as the smaller cations can be located inside the polymer helix. Thus there is considerable scope to extend the range of ionically conducting solids accessible to EXAFS.

References

1. H. Winick and S. Doniach, eds., *Synchrotron Radiation Research*, Plenum Press, New York (1980).
2. B.K. Teo and D.C. Joy, eds., *EXAFS Spectroscopy: Techniques and Applications*, Plenum Press, New York (1981).
3. T.M. Hayes and J.B. Boyce, *Solid State Phys.* **37** (1982) 173.
4. B.K. Teo, EXAFS: basic principles and data analysis, in *Inorganic Chemistry Concepts* **9**, Springer-Verlag, New York (1986).
5. D.C. Koningsberger and R. Prins, eds., *X-Ray Absorption*, John Wiley, New York (1987).
6. A.B. Lidiard, in *Handbuch der Physik*, Vol. XX, Springer-Verlag, Berlin (1957) p. 246.
7. L.W. Barr and A.B. Lidiard, in *Physical Chemistry—An Advanced Treatise*, Vol. 10, eds. H. Eyring, D. Henderson and W. Jost, Academic Press, New York (1970) p. 151.
8. J. Corish and P.W.M. Jacobs, in *Surface and Defect Properties of Solids*, Vol. 2, eds. M.W. Roberts and J.M. Thomas, Chemical Society, London (1973) p. 160.
9. C.P. Flynn, *Point Defects and Diffusion*, Oxford University Press, London (1972).

10. J. Philibert, *Diffusion et Transport de Matiere Dans les Solides*, Les Editions de Physique, Paris (1985).
11. W. van Gool, ed.., *Fast Ion Transport in Solids*, North-Holland, Amsterdam (1973).
12. J. Hladik, ed., *Solid Electrolytes*, Academic Press, New York (1972).
13. S. Geller, ed., Solid electrolytes, in *Topics in Applied Physics* 21, Springer-Verlag, Berlin (1977).
14. M.B. Salamon, ed., Physics of superionic conductors, in *Topics in Current Physics* 15, Springer-Verlag, Berlin (1979).
15. P. Hagenmüller and W. van Gool, eds., *Solid Electrolytes*, Academic Press, New York (1978).
16. P. Vashishta, J.N. Mundy and G.K. Shenoy, eds., *Fast Ion Transport in Solids*, North-Holland, New York (1979).
17. S. Chandra, *Superionic Solids; Principles and Applications*, North-Holland, Amsterdam (1981).
18. G.D. Mahan and W.L. Roth, eds., *Superionic Conductors*, Plenum Press, New York (1976).
19. T.A. Wheat, A. Ahmad and A.K. Kuriakose, eds., *Progress in Solid Electrolytes*, Publication ERP/MSL 83–94, Energy, Mines and Resources, Ottawa, Canada (1983).
20. T. Takahashi and H. Iwahara, *Mater. Res. Bull.* **13** (1978) 1447.
21. K. Funke, *Prog. Solid State Chem.* **11** (1976) 345.
22. S. Geller, ed., in *Topics in Applied Physics* 21, Springer-Verlag, Berlin (1977) p. 41.
23. J.H. Kennedy, in *Topics in Applied Physics* 21, ed. S. Geller, Springer-Verlag, Berlin (1977) p. 105.
24. A.V. Chadwick, in *Defects in Solids—Modern Techniques*, NATO-ASI Series B, Vol. 147, eds. A.V. Chadwick and M. Terenzi, Plenum Press, New York (1986) p. 37.
25. J.A. Kilner and B.C.H. Steele, in *Non-Stoichiometric Oxides*, ed. O.T. Sorenson, Academic Press, New York (1981) p. 233.
26. M.A. Ratner and D.F. Shriver, *Chem. Rev.* **88** (1988) p. 109.
27. J.A. MacCallum and C.A. Vincent, eds., *Polymer Electrolyte Reviews—1*, Elsevier North-Holland, New York (1987).
28. C.R.A. Catlow, in *Defects in Solids—Modern Techniques*, NATO-ASI, Series B, Vol. 147, eds. A.V. Chadwick and M. Terenzi, Plenum Press, New York (1986) p. 269.
29. G.N. Greaves, G.P. Diakun, P.D. Quinn, M. Hart and D.P. Siddons, *Nucl. Instrum. Methods* **A 208** (1983) 335.
30. J.B. Boyce, T.M. Hayes, W. Stutius and J.C. Mikkelsen, Jr., *Phys. Rev. Lett.* **38** (1977) 1362.
31. L.W. Ströck, *Z. Phys. Chem. B* **25** (1934) 411; **31** (1936) 132.
32. J.B. Boyce and T.M. Hayes, in *Topics in Applied Physics* 21, ed. S. Geller, Springer-Verlag, Berlin (1977) p. 5.
33. J.B. Boyce and T.M. Hayes, in *EXAFS Spectroscopy: Techniques and Applications*, eds. B.K. Teo and D.C. Joy, Plenum Press, New York (1981) p. 103.
34. T.M. Hayes, J.B. Boyce and J.L. Beeby, *J. Phys C* **11** (1978) 2931.
35. J.B. Boyce, T.M. Hayes and J.C. Mikkelsen, Jr., *Phys. Rev. B.* **23** (1981) 2876.
36. T.M. Hayes and P.N. Sen, *Phys. Rev. Lett.* **34** (1975) 956.
37. R.J. Cava, F. Reidinger and B.J. Wuensch, *Solid State Commun.* **24** (1977) 411.
38. J.B. Boyce, T.M. Hayes, J.C. Mikkelsen, Jr. and W. Stutius, *Solid State Commun.* **33** (1980) 183.
39. J.B. Boyce, T.M. Hayes and J.C. Mikkelsen, Jr., *Solid State Commun.* **35** (1980) 237.
40. P. Vashishta and A. Rahman, *Phys. Rev. Lett.* **40** (1978) 1337.
41. P. Vashishta and A. Rahman, in *Fast Ion Transport in Solids*, eds. P. Vashishta, J.N. Mundy and G.K. Shenoy, North-Holland, New York (1979) p. 527.
42. A.B. Lidiard, in *Crystals with the Fluorite Structure*, ed. W. Hayes, Clarendon Press, Oxford (1974) p. 101.
43. A.V. Chadwick, *Solid State Ionics* **8** (1983) 209.
44. M.T. Hutchings, K. Clausen, M.H. Dickens, W. Hayes, J.K. Kjems, P.G. Schnabel and C. Smith, *J. Phys. C.* **17** (1984) 3903.
45. A.R. Allnatt, A.V. Chadwick and P.W.M. Jacobs, *Proc. R. Soc. (London) A* **410** (1987) 385.
46. E.G. Ippolitov, L.S. Garashina and A.G. Maklaskov, *Russ. Inorg. Mater.* **3** (1967) 59.
47. A.K. Cheetham, B.E.F. Fender and M.J. Cooper, *J. Phys. C* **4** (1971) 3107.
48. J.J. Fontanella, D.J. Treacy and C.G. Andeen, *J. Chem. Phys.* **72** (1980) 2235.
49. J.M. Baker, E.R. Davies and J.P. Hurrell, *Proc. R. Soc. (London) A* **308** (1968) 403.
50. J.C. Wright, *Cryst. Lattice Defects Amorphous Mater.* **12** (1985) 505.

51. P.J. Bendall, C.R.A. Catlow, J. Corish, P.W.M. Jacobs, *J. Solid State Chem.* **51** (1984) 159.
52. B.T.M. Willis, *Proc. Br. Ceram. Soc.* **1** (1964) 9.
53. C.R.A. Catlow, *J. Phys. C* **6** (1973) L24.
54. C.R.A. Catlow, *J. Phys. C* **9** (1976) 1845.
55. P. Gettman and O. Greis, *J. Solid State Chem.* **26** (1978) 255.
56. O. Greis and D.J.M. Bevan, *J. Solid State Chem.* **24** (1978) 113.
57. D.J.M. Bevan, J. Strahle and O. Greis, *J. Solid State Chem.* **44** (1982) 75.
58. C.R.A. Catlow, A.V. Chadwick and J. Corish, *Radiation Effects* **75** (1983) 61.
59. C.R.A. Catlow, A.V. Chadwick and J. Corish, *J. Solid State Chem.* **48** (1983) 65.
60. C.R.A. Catlow, A.V. Chadwick, G.N. Greaves and L.M. Moroney, *Radiation Effects* **75** (1983) 159.
61. C.R.A. Catlow, A.V. Chadwick, G.N. Greaves and L.M. Moroney, *Nature* **312** (1984) 601.
62. C.R.A. Catlow, A.V. Chadwick, G.N. Greaves and L.M. Moroney, *Cryst. Lattice Defects Amorphous Mater.* **12** (1985) 193.
63. J.J. Fontanella and C.G. Andeen, *J. Phys. C.* **9** (1976) 1055.
64. D.S. Moore and J.C. Wright, *J. Chem. Phys.* **74** (1981) 1626.
65. H.K. Welch, *J. Phys. C* **18** (1985) 5637.
66. C.R.A. Catlow, A.V. Chadwick, J. Corish, L.M. Moroney and A.N. O'Reilly, *Phys. Rev. B* (in press).
67. J.A. Archer, A.V. Chadwick, I.R. Jack and B. Zeqiri, *Solid State Ionics* **9,10** (1983) 505.
68. J.A. Kilner and J.D. Faktor, in *Progress in Solid Electrolytes*, eds. T.A. Wheat, A. Ahmad and A.K. Kuriakose, Publication ERP/MSL 83–94, Energy, Mines and Resources, Ottawa, Canada (1983).
69. Proceedings of Conference on High Temperature Solid Oxide Electrolytes, August 16–17, 1983, compiled by F.J. Salzano, Publication BNL-51728, Brookhaven Nat. Lab., New York (1983).
70. B.C.H. Steele, in *High Conductivity Solid Ionic Conductors*, ed. T. Takahashi, World Scientific Pub. (1988).
71. *Science and Technology of Zirconia*, American Ceramic Society.
72. M. Morinaga, J.N. Cohen and J. Faber, *Acta Crystallogr. A* **36** (1980) 520.
73. J.G. Allpress and J.H. Rossell, *J. Solid State Chem.* **15** (1975) 68.
74. D. Steele and B.E. Fender, *J. Phys. C* **7** (1974) 1.
75. J. Dexpert-Ghys, M. Faucher and P. Caro, *J. Solid State Chem.* **54** (1984) 179.
76. C.R.A. Catlow, A.V. Chadwick, G.N. Greaves and L.M. Moroney, *J. Am. Ceram. Soc.* **69** (1986) 272.
77. L.M. Moroney, *Adv. Ceram.* **23** (1987) 649.
78. W.L. Roth, R. Wong, A.I. Goldman, E. Canova, Y.H. Kao and B. Dunn, *Solid State Ionics* **18, 19** (1986) 1115.
79. L.M. Moroney, P. Thompson and P.E. Cox, *J. Appl. Crystallogr.* (in press).
80. A.N. Cormack, *Mater. Sci. Forum* **7** (1986) 177.
81. T. Vehara, K. Koto, S. Emura and F. Kanamaru, *Solid State Ionics* **23** (1987) 331.
82. P.W.M. Jacobs and D.A. MacDonaill, *Solid State Ionics* **23** (1987) 279 and 295.
83. P.D. Battle, C.R.A. Catlow, J.W. Heap and L.M. Moroney, *J. Solid State Chem.* **63** (1986) 8.
84. P.D. Battle, C.R.A. Catlow and L.M. Moroney, *J. Solid State Chem.* **67** (1987) 42.
85. P.D. Battle, C.R.A. Catlow, A.V. Chadwick, G.N. Greaves and L.M. Moroney, *J. Phys.* **47** (1986) C8–669.
86. P.D. Battle, C.R.A. Catlow, A.V. Chadwick, P. Cox, G.N. Greaves and L.M. Moroney, unpublished results.
87. N. Kamijo, H. Kageyama, K. Koto, H. Maeda, M. Hida, T. Ishida and H. Terauchi, *J. Phys. Soc. Jpn* **55** (1986) 2217.
88. J.-M. Reau and J. Grannec in *Inorganic Solid Fluorides*, ed. P. Hagenmuller, Academic Press, New York (1985) p. 423.
89. J.-M. Reau, S. Matar, G. Villeneuve and J.-L. Soubeyroux, *Solid State Ionics* **9, 10** (1983) 563.
90. C.R.A. Catlow, L.M. Moroney, S.M. Tomlinson, A.V. Chadwick and G.N. Greaves, in *EXAFS and Near Edge Structure III*, ed. K.O. Hodgson, B. Hedman and J.E. Penner-Hahn, Springer-Verlag, Proc. Phys. **2** (1984) 435.
91. P.A. Cox and C.R.A. Catlow, *Abstracts of the International Conference on Lattice Defects in Insulating Crystals*, Parma, Italy (1988) p. 355.
92. P.A. Cox, Ph.D. Thesis, University of Keele (1988).

93. D.E. Fenton, J.M. Parker and P.V. Wright, *Polymer J.* **14** (1973) 589.
94. P.V. Wright, *Br. Polymer J.* **7** (1975) 319.
95. M.B. Armand, J.M. Chabagno and M.J. Duclot, in *Fast Ion Transport in Solids,* eds. P. Vashista, J.N. Mundy and G.K. Shenoy, North-Holland, New York (1979).
96. J.M. Parker, P.V. Wright and C.C. Lee, *Polymer J* **22** (1981) 1305; T. Hibma, *Solid State Ionics* **9, 10** (1983) 1101.
97. C.R.A. Catlow, A.V. Chadwick, G.N. Greaves, L.M. Moroney and M.R. Worboys, *Solid State Ionics* **9, 10** (1983) 1107.
98. K.C. Andrews, M. Cole, R.J. Latham, R.G. Linford, H.M. Williams and B.R. Dobson, *Solid State Ionics* **28–30** (1988) 929.
99. C. Bridges, P.A. Cox, C.R.A. Catlow and A.V. Chadwick, unpublished results; C. Bridges, Ph.d. Thesis, University of Kent (1988).

8 Applications of EXAFS to the study of metal catalysts

J. EVANS

8.1 Introduction

Metal catalysts can be sub-divided into several classes. The most obvious division is between *homogeneous* ones, which are generally discrete transition metal complexes in solution, and *heterogeneous* ones in which the metal centres in question are on the surface of a solid support. In both cases the active species is generally a metal centre present in low concentration in a (generally) disordered medium. EXAFS is a uniquely appropriate technique for establishing structural parameters in such circumstances, with its element specificity allowing the metal centres to be probed directly. For the transition series, the matrix around the metal (typically either an organic solvent or an inorganic oxide support for these two classes of catalyst) does not interfere with the experimental observations, excepting that it contributes to the spectrum background absorption. Unlike other techniques often applied in qualitative analysis of catalysts, such as infrared spectroscopy and nuclear magnetic resonance, the *total* metal content is probed in X-ray absorption spectroscopy and there is less chance of a qualitative analysis being biased by the technique employed. Its weakness is in the lack of resolution between chemically inequivalent sites of the same element (as compared to nuclear magnetic resonance (NMR) for example). As we shall see in the examples below, EXAFS is best utilised in combination with such techniques. When this has been carried out, EXAFS has provided the kind of descriptions about the local structure, particularly in heterogeneous catalysts, that has not been possible hitherto. This new information has proven to be a crucial test of some of the suppositions held previously.

Application of EXAFS in the study of catalysts has been reviewed previously in monographs [1, 2], a compilation of lectures on catalyst characterisation [3], reviews [4, 5] and in a monograph concerned with mixed-metal catalysts [6]. In this chapter, following a brief description of experimental methods, the applicability of EXAFS for investigating specific types of catalyst is illustrated. Many preliminary accounts of such research can be found in the publications accompanying the international EXAFS

Conferences held in Italy [7], United States [8] and France [9]. The physics and data analysis of EXAFS have been introduced in Chapter 6.

8.2 Sampling methods

Although there have been reports of the use of laboratory X-ray sources in the field of oxide supported metal catalysts [10–12], these have utilised rather high metal loadings (4–10%). For more dilute samples, synchrotron based experiments have been much preferred to give spectra of acceptable quality in a reasonable time-span. Loadings of approx. 0.3% have been typical of studies using these more intense light sources, and these could generally be lowered to 0.1% [13, 14].

With the exception of the early 3d metals (Ti–Cr), absorption of X-radiation at the energies of the metal absorption edges by typical gaseous catalyst feed-stocks (e.g. CO, N_2 and H_2) is quite low. Absorption due to organic liquids is also acceptable for EXAFS studies of elements from about cobalt on through the periodic table. Thus it is relatively straightforward to investigate heterogeneous and homogeneous catalysts under operating conditions using fluorescence detection methods. This requires a large aperture X-ray transparent window (typically Be or Mylar). Smaller apertures are needed for transmission experiments. For the heavier transition series (using the 4d metals' K edges and 5d L III edges) such experiments are more favourable since absorption by catalyst matrices is still less. Fluorescence and transmission detection geometries and other experimental details are described in Chapter 1.

These varying circumstances have led to a variety of sample chamber designs which can be applied with advantage to different problems. Relatively simple cells may be employed to provide environmental control, but some of those with higher specifications are due to Lytle [15] (for heterogeneous catalysts, with a temperature range of 100–800K and fluorescence detection), to Koningsberger [16] (heterogenous catalysts, temperature range 77–773K, transmission detection), Goulon et al. [17] (homogeneous catalysts, 77K and above, transmission detection) and Leach and Street [18] (homogeneous catalysts, 15 bar, transmission detection).

8.3 Homogeneous transition metal catalysts

Among the earliest examples of the application of X-ray absorption spectroscopy was the use of the rhodium K edge to probe the Monsanto process for asymmetric alkene hydrogenation which is used in the manufacture of l-DOPA used in treating Parkinson's disease [19]. The complexes studied involved the chiral phosphines which have provided the highest enantiomeric excess of the required l-isomer, viz. the monodentate CAMP (cyclohexylanisylmethylphosphine) and the bidentate diPAMP (bis[phenylanisylphosphino]-ethane), both of which possess chiral centres at phosphorus. In both cases, the catalyst precursor environment $[Rh(P)_2(MeOH)_2]^+$ could be identified. The spectra were also acquired with a substrate, an acylaminocinnamic acid (Ac), added. Attempts to fit the data on the CAMP system on the basis of the simple displacement of methanol by the substrate were unsucessful, probably due to the presence of mixture of species in solution. The results on the bidentate ligand system, which is apparently the preferred one for the process having the higher stereo-selectivity, provided strong evidence for the coordination of both the C=C bond (Rh–C 2.28 Å) and a carbonyl oxygen (Rh–O 2.01 Å), as well as the retention of the two phosphorus donors (Rh–P 2.26 and 2.29 Å) in the species [Rh(diPAMP)(Ac)]$^+$. Indeed a key to its selectivity, which involves the preferential binding of one face of the prochiral alkene and the subsequent faster hydrogenation the less favoured diastereoisomer, may be the relative simplicity of the solution equilibria.

Very different examples of homogeneous catalysts are provided by the Ziegler type catalysts derived from the reduction of 3d metal complexes by AlEt$_3$ [17, 20]. In the earlier report [20], nickel and cobalt octanoate were reduced and the EXAFS of the resulting solutions were dominated by metal-metal backscattering. In the more thoroughly studied nickel system, there was evidence for disordered clusters having a Ni–Ni bond length of 2.48 Å and a second metal shell at about 3.5 Å $[R_{M...M}/R_{M-M} \simeq 1.41]$, consistent with a close-packed structure. At high Al/Ni ratios, a shell at approx. 2.0 Å, attributable to carbon, increased at the expense of the nickel backscattering, indicative of a smaller mean cluster size. This change could be correlated with a decrease in catalytic activity. It seems that these catalysts should be considered microheterogeneous; comparisons between the reduced Fe(acac)$_3$ solutions and iron carbides suggested that these clusters could also be carbides.

Alkene polymerisation catalysts have been prepared from the solvolysis of the dinitrosyl complexes $[M(NO)_2Cl]_2$ in acetonitrile [21]. A particular difficulty arises in these species due to the strong multiple scattering from these 'slightly bent' nitrosyl groups (see Section 6.2.1). Binsted et al. [22] have noted that double and triple scattering dominate the single scattering contributions in linear M–C–O units. This is shown in Figure 8.1. The dramatic phase and amplitude variation that occurs with bond angle over a 150–180° range allows bond angles to be estimated to approx. 2° in favourable cases. In the nitrosyl containing catalyst, the difficulty is compounded by the linear heavy atom chain in the coordinated acetonitrile ligands in the active species, considered to be $[M(NO)_2(CNMe)_p]^+$; for M = Fe $p = 3$ and M = Co $p = 2$. However the M–CNMe bond length is substantially longer (0.3–0.4 Å) than that of the strong M–NO linkage, so resolution is possible. The problem was treated by

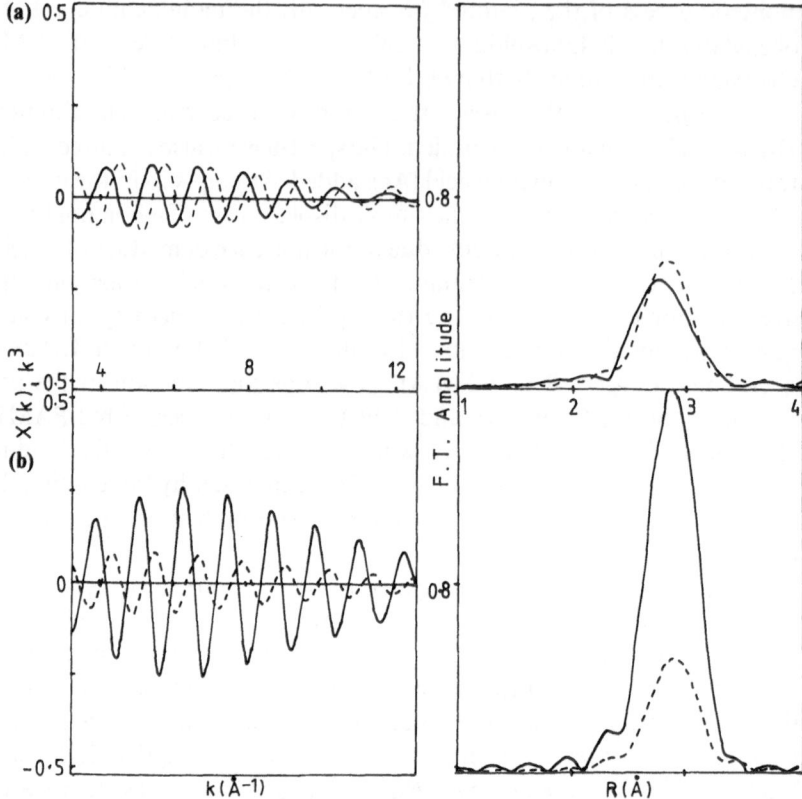

Figure 8.1. Calculated single scattering (---) and (——) total EXAFS contributions of the oxygen shell in a Co–C–O unit of fixed Co–C and C–O bond lengths. (a) Co–C–O = 150°; (b) Co–C–O = 180°.

comparison of the imaginary parts of the Fourier transform of the active catalytic species with two model complexes of differing M–N–O angle, viz. $[Fe(NO)_2(PPh_3)Cl]$ (166°) and $[Fe(NO)_2(PPh_3)_2]$ (178°). The phase and amplitude in the M...O shell region matched the former more closely, indicating an approximate M–N–O angle of 165° in the active species. In nitrosyl complexes this angle is an important indicator of the metal-ligand bonding, varying as it does between 120 and 180°.

Difficulties with this approach may well arise when there is no model compound available with the required bond angle; the marked sensitivity of the multiple scattering contributions to the bond angle and lengths in the unit in question means that the tolerance is rather small (within approx. 2°). In these situations an alternative approach is to utilise the full spherical wave multiple scattering formalism of Lee and Pendry [23] which has been successfully coded [24, 25] and tested on a series of cobalt [22] and iron [26] carbonyl complexes with differing mean M–C–O bond angles. This has recently been employed to study the formation of the cluster species $[Ru_3(\mu$-

H)(μ-CO)(CO)$_{10}$]$^-$ in an ethylene hydroformylation catalyst system [18]. Multiple scattering effects had to be explicitly considered to provide an accurate analysis of the Ru K edge EXAFS which was recorded on a 4.5 mM solution of [Ru$_4$(μ-H)$_3$(CO)$_{12}$]$^-$ (which is converted into the trinuclear anion under the reaction conditions) extracted from an autoclave at 150°C and 225 lb/inch^2 gauge pressure. This was the same concentration as was employed in both the catalysis studies and an FTIR study concentrating on the intense C–O stretching modes. The low absorption of the solvent (tetrahydrofuran) at the energy of this edge was exploited and a long path length (1 cm) was adopted in order to obtain good quality transmission spectra on this low concentration (corresponding to approx. 0.2% Ru). The extension of this approach to in situ studies of homogeneous catalysis seems to be entirely viable.

8.4   Surface organometallic species

In recent years organometallic chemistry has been used to provide a more controlled means of generating specific species on the surface of, generally, an oxide support in an attempt to combine the high specificity of molecular, homogeneous catalysts with operational and activity advantages of some heterogeneous catalysts [27]. A fundamental advantage is that identifying the metal site responsible for a given catalytic reaction opens up the possibility of achieving an understanding of the reaction mechanism on a molecular level [28]. EXAFS has been used to characterise a few of these materials [5]. One particular series of materials serves to illustrate this chemistry and also how the improvements in analysis procedures in recent years have added to the precision of the EXAFS-determined structures.

In 1980 Besson *et al.* [29] reported that the trinuclear cluster [Os$_3$(CO)$_{12}$] reacted with the surface of sil-type silica to yield a species formulated as [Os$_3$(μ-H)(μ-O-SIL)(CO)$_{10}$] from gas evolution and infrared studies (comparing the C–O stretching absorptions with those of model compounds) (Figure 8.2). This included Os L III edge EXAFS data on this material. An estimate of the Os–Os distance was made by comparison of the shell radius at the peak maximum in the Fourier transform with that of the model compound [Os$_3$(CO)$_{12}$]. A relatively short metal-metal bond of approx. 2.68 Å was proposed. However, this peak in the Fourier transform is a composite of both the Os–Os and carbonyl Os...O shells. So the apparent shell radius is a function of parameters pertaining to both shells and such a simple procedure is subject to some potential error.

A later analysis was carried out on the analogous material on alumina [30]. This used a spherical wave analysis procedure, but encountered difficulty in fitting the carbonyl oxygen shell, multiple scattering calculations being unavailable at that time. The procedure adopted was to utilise modified phase

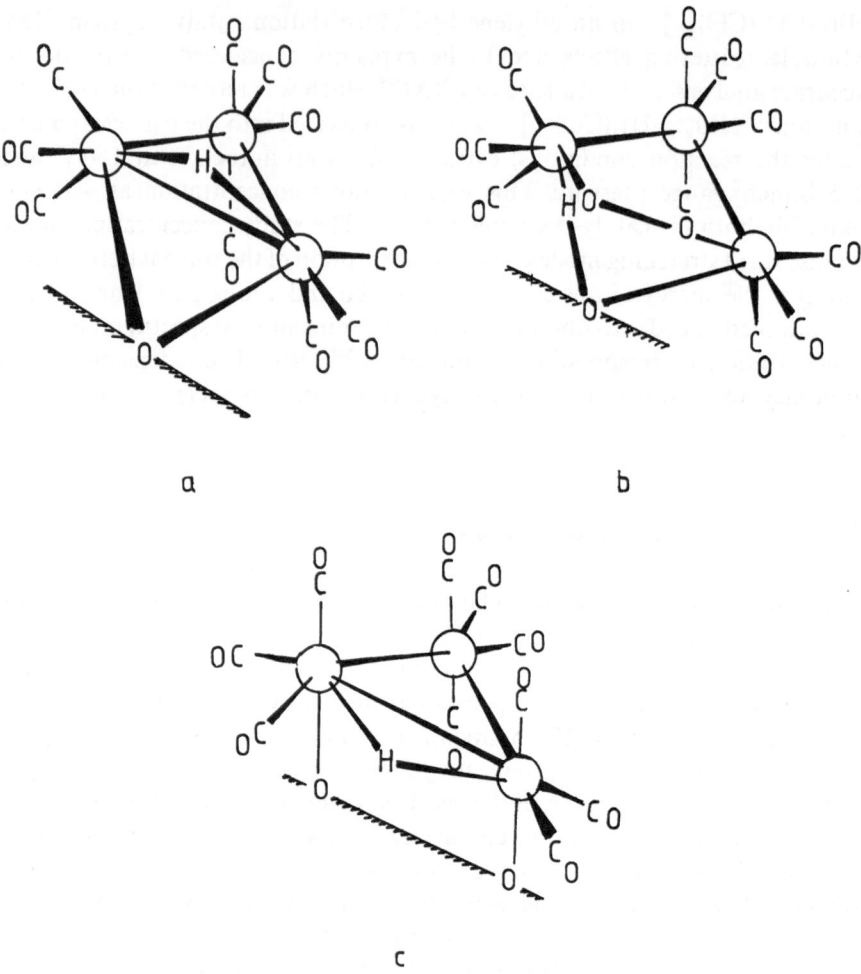

a   b

c

Figure 8.2. Structures proposed for the first chemisorption product of $[M_3(CO)_{12}]$ on silica (M = Ru and Os) and alumina (M = Os).

shifts which replicated the backscattering amplitude envelope of the carbonyl groups, but yielded a systematically short estimate (by about 0.1 Å) of the Os...O distance. Two alternative structures ((a) and (b) in Figure 8.2), differing in the coordination number of the minor Os–O shell of the cluster to surface link, and, perforce from model complexes, having an isosceles cluster triangle in the case of the second example. Both models were found to fit the data equally well; the EXAFS was dominated by the 10 carbonyl groups and two shorter M–M bonds. However, extra detail about the cluster structure was now apparent, including the Os–O bond length, the mean Os–C distance and an estimate of the Os–Os distance more consistent with model compounds of structure (a) of 2.84 Å. This material has been re-investigated by Koningsberger *et al.* [31] using an empirical method of deriving the carbonyl phase shifts and backscattering factors. Quantitative differences were reported in this analysis from the earlier one of this author. In particular, a firm estimation of the mean Os–O coordination number of 0.67 was derived, consistent with structures (a) and (c) in Figure 8.2, and a longer Os–Os distance of 2.88 Å. The closest model complex characterised by X-ray diffraction is $[Os_3(\mu\text{-H})(\mu\text{-}OSiEt_3)(CO)_{10}]$, for which the mean metal-metal bond length is 2.80 Å [32]. Some error may have accrued in this analysis from the means of deriving the empirical phase shifts from $[Os_3(CO)_{12}]$. The composite shell in the Fourier transform due to the Os–Os and Os...O peak was resolved by considering the data at higher photoelectron energy than 11 Å^{-1} to be due to the heavy metal only; normally backscattering from such light elements as oxygen is relatively minor in this region. However this represents a worst case for such a differentiation, there being four CO groups per metal versus two Os–Os bonds, and the energy range of backscattering from the former is considerably extended by multiple scattering [22]. So the transfer of these backscattering parameters to another complex differing in the CO/Os ratio and other structural changes will be subject to error.

The most recent report on such materials provided strong evidence for the bridging hydride sites in structures (a) or (c) by the observation of the hydride deformation mode in the infrared [33]. A ruthenium K edge EXAFS analysis was made on the $[Ru_3(CO)_{12}]$/silica product which utilised the spherical wave multiple scattering formalism (see Section 6.2.1 and [25]). This indicated an essentially equilateral triangle with Ru–Ru separation 2.79 Å, and Ru–C and Ru–O bond lengths of 1.90 and 2.06 Å, respectively.

It is doubtful whether this is the end of this particular story, especially since there have been persuasive reports of the $[Os_3(CO)_{12}]$/silica species acting as the active species in the catalysis of ethylene hydrogenation and the cluster-oxide link being involved in accommodating an incoming alkene [34]. But so far it serves to illustrate two generalities about the role of EXAFS in catalysis research. Firstly, over the period of time covering this research (8 years), the detail that can be obtained about the structure of metal sites on the surfaces of amorphous oxides has increased very considerably. In particular, the multiple scattering intrinsic to carbonyl groups has ceased to become an intractable problem of analysis and is now a means of estimating the mean M–C–O bond angle. Any analysis which has not taken proper account of this will inevitably be potentially error-prone. Secondly, it must also be clear that there are confidence limits to any data analysis, and other complementary characteris-ation methods are necessary. For example, the strongest evidence for structures (a) or (c) (which are not differentiable by EXAFS) against structure (b) in Figure 8.2 is probably not the coordination number of the metal-oxygen surface link; it is arguable that the difference between 1.33 and 0.67 can be distinguished for this minor shell. Rather the original gas evolution analysis and the more recent $\delta(M_2H)$ infrared observation are more persuasive.

Arguably one of the best cases for structural characterisation in which

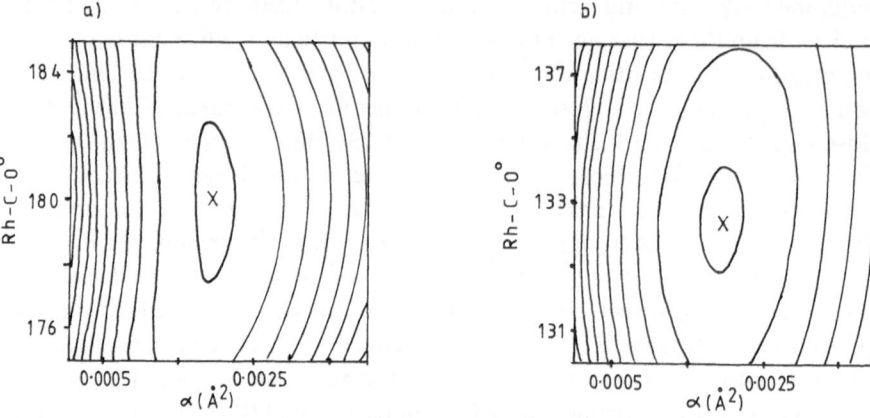

Figure 8.3. Contour maps of the fit index of the Rh K edge EXAFS of $[Rh_6(CO)_{16}]$ plotted for the Rh–C–O bond angles for (a) the terminal and (b) the μ_3-CO groups against the Debye-Waller factor, α. Lower contour 0.905, increment 0.04 for (a) and 0.05 for (b).

EXAFS is a key part is for $[Rh_6(CO)_{16}]$ and its derivatives. In that octahedral cluster, each metal is coordinated to two terminal and two μ_3-CO units of substantially different Rh–C bond length (by 0.3 Å) and Rh–C–O angle (132 against 180°), but essentially identical Rh...O distance. In spite of this latter fact, the Rh...O shell may be successfully split into part of the two multiple scattering units represented by the two types of carbonyl ligand [35]. This is most graphically demonstrated by contour maps of the fit index for the two Rh–C–O bond angles, showing minima at the crystallographic values (Figure 8.3). Most cluster carbonyls only seem to allow the estimation of a mean M–C–O angle, but the favourable populations and wide angle difference at the rhodium centres make this a best-case example. EXAFS also provides good estimates of the Rh–Rh and Rh...Rh distances that make up the octahedral cluster unit. This cluster can be tethered to silica as the derivative $[Rh_6(CO)_{15}\{PPh_2CH_2CH_2SIL\}]$, and finger-printed by infrared and ^{31}P–CP MAS NMR spectroscopy [36]. The rhodium EXAFS of this material can be fitted to essentially the same parameters as the parent complex, thus providing a very detailed structural description of this supported cluster.

8.5 Oxide supported metal ion sites

The identification of metal ion sites in catalysis is illustrated by a series of reports on nickel in zeolite Y [37–39]. In the first two, nickel was introduced into the zeolite by ion exchange and the initial Ni^{2+} site identified as $[Ni(OH_2)_6]^{2+}$ (Ni–O 2.05 Å). In one study [37] the sample was calcined in air and then back-exchanged with Ca^{2+}, which removed about two-thirds of the original nickel. The nickel K edge EXAFS now indicated a Ni–O coordination number of 4 (Ni–O 2.07 Å) and contributions to backscattering atoms in a 3–4 Å range. This was considered as evidence of binding of the nickel to the zeolite cage, these extra shells being attributed to framework atoms. An alternative explanation of similar evidence was presented in the second report [38]. The sample history was rather different in this case with samples being treated with NaOH and also calcined at 643K in a stream of oxygen (10%) in helium. However, the conclusions drawn were that these treatments formed oligomeric Ni–O–Ni units (by comparison with Ni EXAFS data on NiO) which were too little disordered to be located by electron microscopy and X-ray diffraction (possibly residing within the super-cage). On reduction with hydrogen (643K, 6 h), the alkali treated sample showed incomplete reduction to the metal, unlike the calcined material.

An alternative nickel precursor was adopted in the third report, namely $Ni(CO)_4$ [39]. The volatile carbonyl was absorbed into NaY, LiY and a de-aluminated version, labelled Y*. Considerable changes in the infrared absorptions of the carbonyl stretching modes were observed in the two ion exchanged samples. Ni K edge XANES and EXAFS showed that there was essentially no change at the nickel centres. Thus the infrared changes are clearly not due to any oxidation of the metal or ligand substitution but a Lewis acid-base interaction between the alkali metal cation and the oxygen of a carbonyl group; this evidently causes little perturbation of the $[Ni(CO)_4]$ bond angles and bond lengths, but must modify the C–O stretching and interaction force constants. Other Ni(0) complexes were generated in these super-cages. For example treatment of NaY with $Ni(CO)_4$ and PBu^t_3 afforded a species identified as $[Ni(CO)_3(PBu^t_3)]$ by ^{31}P–CP MAS NMR and Ni X-ray absorption spectroscopy. It is interesting that this complex is sterically too large to pass through the channels in the faujasite structure. Evidently the smaller reagents migrate into the super-cages, form the complex in situ which is then trapped. A complex infrared spectrum was obtained for this material, presumably due to $M^+...OC$-Ni interactions, and it would have been very difficult to qualitatively characterise this intra-zeolite complex by that technique. A similar in situ synthesis of $[Ni(PMe_3)_4]$ was achieved from $Ni(allyl)_2$ and trimethylphosphine. In this case, there was some evidence for higher disorder in the Ni–P shell in the super-cage as compared with the pure complex.

The adsorbed nickel carbonyl is reactive under both reducing and oxidising atmospheres, forming the metal and NiO, respectively. When used in conjunction with electron microscopy, EXAFS provided evidence for a bimodal particle size distribution for the metal crystallites. This seems quite plausible with electron microscopy locating the larger particles, migrated onto the outside of the zeolite, and EXAFS also detecting some very small clusters which may still be within the zeolite pores. In contrast, the NiO particles were substantially still within the zeolite cages, indicating that sintering is more difficult than for the metal.

8.6   Oxide supported metallic catalysts

As has been implicit in the discussion above, EXAFS generally provides complementary information to electron microscopy in metal particle characterisation. Microscopy will locate the larger particles and indicate their morphology. EXAFS studies of large particles, being a population weighted measurement, will normally be biased towards the atoms in the bulk, but will provide a more accurate measurement of inter-nuclear distances than electron microscopy. Considerations of the radius ratios of outer shells can indicate the structure type of these particles [40], with the three normal bulk metal structures being readily distinguished if shells are identified over a 5 Å range. Further than that the mean growth pattern and particle sizes (i.e. two- or three-dimensional particles) can also be estimated from the coordination numbers of these shells [41]. For example, the $\sqrt{2}R$ shell in a close packed structure will be absent in a monatomic layer. This procedure has been used to suggest that an Os/silica sample consisted of 7 ± 2 Å discs, whereas the less oxyphilic iridium formed small (6 ± 2 Å) spheres on the same support [41]. The coordination number test was shown to be reasonably sensitive for particles of up to about 20 Å in diameter. In that size regime, the proportion of surface metal atoms is higher and so one would expect that it would be just in these circumstances that EXAFS can then be used to investigate the structures at those metal atoms that will be involved in chemisorption and catalysis. However, this assumes that these chemical changes preserve the particle size. As we shall see, this may not be apposite, giving EXAFS a wider application.

Studies on supported rhodium catalysts have provided a new understanding of the chemistry at small metal particles on oxide supports. Two samples were investigated in an early study, namely 1% Rh/silica and 0.5% Rh/alumina [42]. Hydrogen chemisorption measurements indicated that the rhodium atoms were principally surface ones. However, the mean Rh–Rh coordination number was much lower for the alumina catalyst (1.5 as opposed to 9), indicating the possibility of extremely small (even mononuclear) metal units. The metal-metal bond lengths were essentially the same as bulk rhodium under the measurement conditions (purging with helium after H_2 reduction at 400°C). Rather more detailed analyses have been carried out on the first coordination shell for Rh/alumina samples under a hydrogen atmosphere [43, 44]. Even though coordinated chlorine was detected in the impregnated and dried catalysts (prepared from $RhCl_3.3H_2O$), there was no evidence for its retention following hydrogen reduction. One new proposal was for a long Rh^0-O_{oxide} interaction of approx. 2.7 Å, which is about 0.55 and 0.6 Å longer than Rh^I-O and $Rh^{III}-O$ bonds, respectively. The evidence for this was derived from the imaginary part of the Fourier transform of the feature near 2.8 Å which contained a more complex phase signature than that of the metal alone. This possibility has important implications for other studies that have relied on simple Fourier transform amplitude comparisons, assuming that a feature

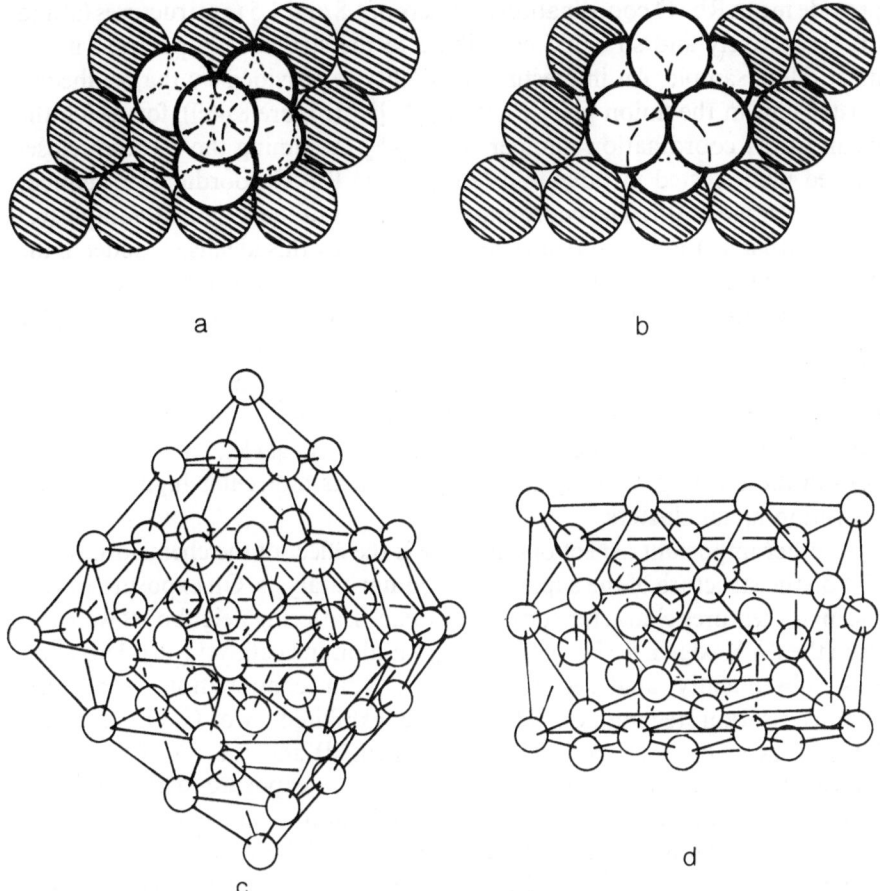

Figure 8.4.   Possible structures of small rhodium crystallites on an alumina support. (a) and (b) as contributors to the 0.57% Rh sample and (c) and (d) to the 2.0% one. In (a) and (b) hatched circles are oxygen, open circles are rhodium.

at this distance was purely metal backscattering. The Rh–Rh bond length (2.68 Å) was essentially the same as for the bulk metal (see below). The Rh–Rh coordination number increased with metal loading (from 3.7 for 0.57% Rh to 6.5 for 2.0% Rh) offsetting a concomitant reduction in the average number of Rh^{0}–O bonds (from 2 for 0.57% Rh to 1 for 2.0% metal).

It is interesting to consider the possible clusters that might contribute to such mean coordination values. Concentrating on fragments of close packed arrays, two geometries are the trigonal bipyramid and octahedron which have 3.6 and 4.0 nearest neighbours metals, respectively (Figure 8.4(a) and (b)). Both these values are within experimental error of the estimation for the 0.57% Rh sample. If a typical close-packed oxide plane is considered as the oxide binding site, then placing the metal atoms such that they are in three-fold hollows will provide mean Rh–O coordination numbers of 1.8 and 1.5 for structures (a) and (b), respectively and so could contribute to the structure population in the most dilute sample. An interesting possibility is provided by the octahedral cluster unit in the anion $[Ni_{38}Pt_6(CO)_{48}H]^{5-}$ (Figure 8.4(c)), for which the mean M–M coordination number is 7.6 [45]. Assuming the 10 atom close-packed plane rested in three-fold hollows, the Rh–O coordination number would average as 0.7. The coordination numbers of the 2% metal sample (6.5 for Rh–Rh and 1 for Rh–O) deviate from those for this idealised model in the direction of a larger footprint on the oxide for a given number of atoms, moving in the direction of hemispherical growth. A modification of this structure is to remove the 10 atoms of one face of the cluster, exposing a 12 atoms surface (giving Rh–Rh and Rh–O coordination numbers of 6.2 and 1.1), as in Figure 8.4(d). Clearly all of these values are in the region of the experimental values for the 2.0% Rh catalyst. It seems clear that the particle sizes in these catalysts are in the same regime as molecular high nuclearity transition metal clusters.

Another important aspect of this work is the effect of the chemisorption of CO on the structure of the supported rhodium catalysts. On exposure to CO, the high k features characteristic of heavy backscattering neighbours were lost from the Rh K edge EXAFS of the 0.57% Rh/alumina catalyst [46]. This effect was subsequently analysed in detail [47], in combination with other characterisation techniques. The sample exhibited two C–O stretching frequencies in the infrared (2095 and 2023 cm^{-1}) and the EXAFS analysis indicated total disruption of the metal particles on exposure to CO to form a $Rh(CO)_2(O\text{-alumina})_3$ unit. This change was accompained by a dramatic shortening of the Rh–O bond length from that of the reduced catalyst to a value of 2.12 Å, typical of Rh^{I}–O distances. The implication is that exposure to CO induces oxidation of the metal centres; this has recently been shown by infrared spectroscopy to be effected by isolated (rather than hydrogen-bonded) hydroxyl groups on the oxide surface [48]. The key aspects of these results hinge on the EXAFS measurements being carried out under carefully controlled conditions. The results provided a unifying explanation for apparently conflicting evidence. This explanation required that CO chemi-sorption be a corrosive process, a matter of some import since that is often used as a means of estimating the surface area of the metal component in oxide supported catalysts!

This behaviour may well not be limited to this one system. There is now evidence of a similar pattern of behaviour for a highly dispersed ruthenium metal catalyst supported on alumina prepared from $[Ru_3(CO)_{12}]$ [49]. Although detailed analysis was not presented of the Ru K edge EXAFS of a 2% Ru sample, it seems that following hydrogen reduction, a metal catalyst was formed with a mean Ru–Ru coordination number of approx 4. On exposure to CO there was a marked change in the EXAFS and Fourier transform consistent with the loss of virtually all the metal-metal bonds and the formation of isolated $Ru(CO)_n$ units. Similar units, with a local environment of $[cis\text{-}Ru^{II}(CO)_2(O^-)_4]$, have been identified by EXAFS and infrared studies on the aerial oxidation of $[Ru_2(\mu\text{-}H)(\mu\text{-}O\text{-}SIL)(CO)_{10}]$ [35]. Indeed the $Rh^I(CO)_2$ site described above may be formed readily from the chemisorption product of $[Rh_4(CO)_{12}]$ on alumina.

EXAFS is now having a key role in understanding chemisorption on supported metal catalysts. Hydrogen chemisorption measurements are also widely employed to estimate metal surface areas, with a H/M stoichiometry of unity generally assumed. However, comparisons of surface area estimates from the EXAFS derived M–M coordination numbers with values from hydrogen chemisorption, indicate that this assumption is not necessarily valid [50–52]. The evidence indicated that the actual stoichiometry was a characteristic of the metal, and was not strongly dependent upon either the support or the mode of preparation. The elemental order noted for H/M stoichiometry was Ir > Rh > Pt. For iridium H/M ratios of > 2 were determined, with the existence of polyhydride sites on the surface being the probable explanation.

This change in the structure of the surface metal sites might be anticipated to cause some change in other metric parameters considering that the incorporation of a μ-H site into a metal cluster complex normally results in the dilation of the bridged bond by approx. 0.1 Å. A preliminary report of a Ir L III edge study has indeed shown some contraction in the metal-metal distance once chemisorbed hydrogen is removed from a reduced Ir/alumina catalyst [53]. It would be interesting to see if the observed contraction of bond length with decreased particle size observed on weakly interacting supports such as carbon [54] became manifest once hydrogen is removed from catalyst surfaces. Oxygen chemisorption at 77K appeared to occur without changing the metal-metal bonding. However, even warming to 100K caused some disruption.

A controversial phenomenon in recent years has been termed SMSI (strong metal support interaction). This involves modification of the chemisorption properties of a late transition element supported on a reducible oxide when treated with hydrogen at relatively high temperatures (700–800K). Thus, for example, Rh/titania treated in this way loses its capacity to chemisorb H_2 or CO. One common explanation has been the coverage of the noble metal by the reduced oxide. This has been studied by two groups using X-ray absorption spectroscopy. Haller and co-workers used EXAFS [55] and near edge structure (XANES) [56] to provide evidence of direct Rh–Ti bonding in catalysts reduced at 773K, there being a peak to the low R side of the Rh–Rh peak in the Fourier transform of the Rh K edge EXAFS. Further support for this proposition is gained by an entirely empirical analysis of the XANES structure (up to 90 eV from the edge). This involved taking combinations of the spectra of four model species: Rh powder, the alloys RhTi and Rh_3Ti, and a highly dispersed Rh/titania catalyst which did not display SMSI. Analysis in this way of samples reduced at 623 and 773K both contained significant proportions of the features of the components ascribed to Rh metal, dispersed 'Rh' and Rh_3Ti. The conclusion drawn therefore, is that in the SMSI state, the rhodium approaches the environment in these inter-metallics. It is not yet clear that such a procedure is unambiguous. Other models may have been adopted, including rhodium oxides or even the dimer $Rh_2(O_2CMe)_4$ as a small unit with oxygen donors, which might also provide XANES patterns that could be found to contribute to the features observed in the SMSI sample. Without any consideration of the structural sensitivity of the near edge structure in these examples, assessment of this data is difficult.

Indeed this problem is a difficult one for EXAFS to address, since the discrimination of small contributions of the two rather light elements, oxygen and titanium, would require excellent experimental data and analysis methods. It is therefore not surprising that opposing conclusions have been drawn about the structural cause of SMSI [57]. Consideration was made of both the conventional k^3- and k-weighted EXAFS, which are sensitive to high and lower Z elements, respectively, and this indicated that the lower R-space components in the Rh K edge EXAFS should be better ascribed to oxygen rather than titanium. Some further evidence for this was obtained by imaging samples sintered in a high resolution electron microscope (HREM); these could be ascribed to fcc rhodium, and not its alloys with titanium. Thus the analysis of the SMSI sample reduced at 723K indicated a Rh^0–O link at 2.60 Å (CN 1.9), a Rh–Rh bond at 2.63 Å (CN 3.4) and also 2.8 non-bonded Ti atoms at 3.41 and 4.39 Å. The nearest neighbour values are very similar to those discussed above for the 0.57% Rh/alumina catalyst, suggesting also a mean metal nuclearity of 5–6 atoms. The HREM studies indicated that the predominant exposed face of the support was [101] anatase. While some

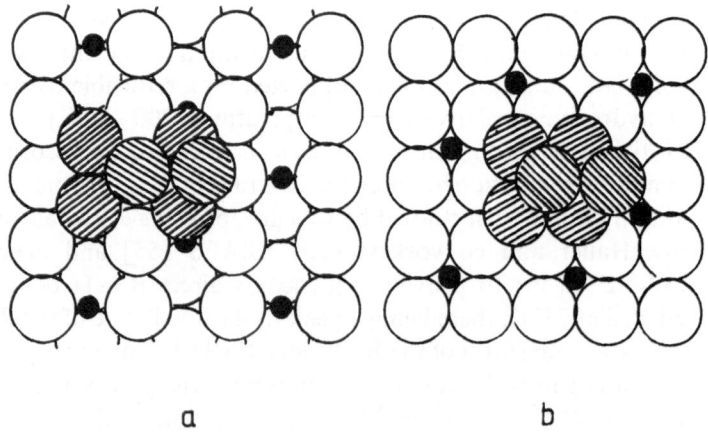

a   b

Figure 8.5. Representative structure for a Rh/titania site in an SMSI sample. (a) on anatase [101], (b) on [001]. Open circles are oxygen, hatched are rhodium, shaded are titanium.

rhodium crystallites were observed on this face, the majority were observed on the edges of the crystallites. It is expected that these will be the [001] face, which has the lowest surface energy. A pyramidal five-atom cluster with registration on these faces would have Rh–Rh coordination numbers of 2.4 and 3.2, respectively and these would be increased to 3.0 and 3.7 by adding a sixth atom in clusters similar to the capped square pyramidal skeleton in $[Os_6(\mu\text{-}H)_2(CO)_{18}]$. A presentation of this unit on these two faces is given in Figure 8.5 in which the non-nearest neighbour Ti sites may also be identified. In these structures the mean Rh–O coordination number is 2.7. It seems clear from these values that there can be little encapsulation of the rhodium by the support, as required by one of the SMSI models. That would increase this coordination number to a value greatly in excess of the experimental one. The SMSI chemisorption effects were ascribed to be electronic in nature involving the reduced titanium sites.

8.7 Oxide supported alloy catalysts

Mixed metal catalysts have received considerable attention, due in no small part to their use as petroleum reforming catalysts [6]. X-ray absorption spectroscopy in principle can be valuably applied in this area, particularly when the components differ greatly in atomic number; both elements may be probed independently, and also distinguished in data analysis. This can be illustrated by one of the series of papers from Sinfelt et al. [13] concerned with Os–Cu/silica. Incorporation of small proportions of copper modifies the selectivity of the catalyst causing a suppression of cyclohexane hydrogenolysis but maintaining activity for dehydrogenation to benzene. A 2% Os–0.6% Cu/silica catalyst was investigated that was estimated to have a mean particle size of approx. 15 Å. For this, the total coordination numbers around copper and osmium were fitted to the respective K and L III edge EXAFS as 9.5 and 12.5. Interestingly, the percentage osmium neighbours around osmium (83%) and copper (51%) also differed. Even though these elements are virtually immisicible in the bulk, evidently there are mixed metal particles. A similar study on Ru–Cu/silica found that the more volatile element had segregated entirely to the surface of these particles, thus accounting for the marked effect of copper on catalytic activity [58]. In the 1:1 molar Os–Cu sample with such a small mean diameter, there are insufficient copper atoms to cover the interstitial osmium atoms, and so a model was evolved in which the core of the clusters is osmium, with a mixed metal periphery. There were interesting changes in the 'white line' intensity at the Os L III edge. This is a resonance feature and corresponds to dipole allowed 2p–5d transitions and its intensity has been related to the number of vacancies in the 5d sub-shell. An Os/silica catalyst was found to have a substantially more intense white line than the bulk metal. Addition of copper to the supported catalyst partially reversed this trend, suggesting that part of the action of the copper is to contribute electrons to the empty 5d states of the osmium atoms; this may also perturb the catalytic activity of these materials.

The investigation of the Pt–Ir reforming catalysts is one of the most difficult alloy catalyst systems to provide definitive results. Since the components are neighbours in the periodic table, they possess virtually identical 'fingerprints' (phase shifts, ψ_j, and backscattering factors, $f_j(\pi)$, see Section 6.2.2). In addition the Ir L III edge extends for only 350 eV before there is overlap with its neighbour's absorption edge; so the Pt L III edge is also contaminated by iridium EXAFS. The problem was approached by careful choice of reference materials to derive empirical scattering parameters and to separate the EXAFS components from the two edges in the region of overlap [59]. Therefore the analysis was restricted to a single shell. However, for a 10% Pt–10% Ir/silica catalyst, the first shell radii at the two edges differed. These were 2.753 and 2.724 Å at the platinum and irridium edge, respectively, indicating some partitioning of the elements since these radii differ from those derived for the single catalysts (Pt–Pt 2.774 Å; Ir–Ir 2.712 Å). The values for the alloy catalyst are intermediate between these homonuclear bond lengths, indicating the presence of heteronuclear bonds.

EXAFS studies of the Pt–Re reforming catalysts are only slightly easier. The Re L III edge can be probed over all its useful energy range, but the Group 7 element's L II absorption edge truncates the Pt L III data at approx. 350 eV. One way around this is to utilise the less intense Pt L I edge [60]. After EXAFS analysis using empirical phase shifts, and observations of the 'white line' intensity at the Re L III edge, it was concluded that reduction of a 1% Pt–1% Re/alumina catalyst at 775K afforded both metals in a fully reduced state. The bond lengths obtained were: Pt–Pt 2.75 Å; Pt–Re 2.63 Å; Re–Re 2.73 Å; very little coordination to oxygen was detected. Similar Re L III edge studies by another group have been interpreted differently [61]. While there was agreement about the reduction of the platinum, this group found evidence for two oxygens on average per rhenium atom. The model proposed consisted of the more oxyphilic rhenium adopting the sites next to the alumina support, with small platinum clusters (which exhibit smaller coordination numbers of 3 and 5 at reduction temperatures of 300 and 500°C) dispersed on this base. Exposure to H_2S (0.5 mol%) in hydrogen at 725K was found to have little effect on the structure of the alloy catalyst, even though on average each rhenium atom was considered to be coordinated to one sulphur atom [60]. Other work has shown that the form of the sulphiding treatment has a marked effect upon the structures of the catalysts [62]. The poor performance of one catalyst could be correlated with the conversion of platinum to PtS, the rhenium being unaffected. An effective procedure caused partial sulphiding at both metal centres. Indeed, the plantium now seemed to be protected from oxidation by a layer of PtS. Reduction of the sulphided catalyst caused the PtS to be completely converted to fcc metal crystallites, while the rhenium was still partially in an ReS_2 environment.

Analysis of Pt–Sn reforming catalysts presents a potentially easier problem, with the two elements having substantially differing atomic numbers. Comparison has been made between a preferred catalyst on alumina, and a silica supported version [63]. For the alumina based material, the tin was introduced prior to gel formation, and the platinum added subsequently. Analysis of the Pt EXAFS indicated approx. 3.5 Pt and 0.7 oxygen neighbours, the latter being at a distance of 2.06 Å. The tin on the other hand only had light elements in the coordination sphere (1.9 O and 0.5 Cl), with near edge features consistent with a Sn^{2+} type site. Although there is no direct Sn–Pt interaction, the presence of the tin in the support effectively reduced the mean size of the platinum crystallites from a regime like the representations (c) or (d) in Figure 8.4 to one including (a) or (b). The Pt–Sn/silica preparation was different with the two metals being introduced onto a preformed silica. In this case analysis of the EXAFS at the two elements' edges supported the presence of bimetallic clusters, and this was confirmed by power X-ray diffraction measurements which showed lines due to PtSn, Pt and Pt_3Sn. There is a clear structural difference to correlate with the varied catalytic properties on the two supports. Perhaps surface oxygen atoms bound to the small proportion of tin in the alumina act as nucleation sites and also prevent sintering to the larger platinum particles observed for Pt/alumina.

8.8   Concluding comments

In this selective review there have been examples to illustrate how EXAFS has been used to provide new information about metal catalysts in disordered media. In many homogeneous catalysts and surface organometallic centres, there is one dominant species and the molecular structure can be probed in detail by well-performed EXAFS data analysis. Conventional heterogeneous catalysts are more complex materials, and X-ray absorption spectroscopy is one of the most powerful ways of probing these materials. Their chemical complexity generally makes interpretation rather less precise, but the evidence is that particle sizes in many active catalysts are similar to those in high nuclearity cluster complexes [64].

All of the experiments described above have been conventional X-ray absorption spectroscopy. Other developments clearly impinge on catalytic science. Three examples suffice to illustrate this.

(1) X-ray absorption is one of the techniques that can be applied to both single crystal surface science experiments and conventional catalysts. It may therefore be used as a linking technique like TPD and FTIR. One may cite the recent study of the chemistry of $[Ru_3(CO)_{12}]$ on Cu(111) [65] as of possible relevance to the supported Ru–Cu catalysts.

(2) Energy dispersive X-ray absorption spectroscopy can provide time resolution of 10^{-3} s, and so can be used to monitor dynamic processes such as catalyst reduction [66].

(3) Totally reflected EXAFS (REFLEXAFS) has been reported to provide sensitivity approaching one monolayer on a flat surface. Unlike most SEXAFS measurements, the fluorescence detection used allows interfaces with condensed phases to be investigated. An example is the investigation of a layer of lead on an Ag(111) electrode under an electrolyte solution monitored with the Pb L III edge [67].

All of these new experiments, in conjunction with the recent advances in data analysis would suggest that X-ray absorption spectroscopy will have an increasingly important role in catalytic science.

References

1. D.C. Koningsberger and R. Prins, eds., *X-Ray Absorption Spectroscopy*, John Wiley, New York (1988).
2. H. Winich and S. Doniach, eds., *Synchrotron Radiation Research*, Plenum Press, New York (1980).
3. J.M. Thomas and R.M. Lambert, eds., *Characterisation of Catalysts*, John Wiley, Chichester (1980).
4. J.H. Sinfelt, *Acc. Chem. Res.* **20** (1987) 134.
5. J. Evans, in *Catalysis*, eds. G. Bond and G. Webb, Specialist Periodical Report, Royal Society of Chemistry, London, **8** (1989) 1.
6. J.H. Sinfelt, *Bimetallic Catalysts: Discoveries, Concepts and Applications*, John Wiley, New York (1983).
7. A. Bianconi, L. Incoccia and S. Stipcich, eds., *EXAFS and Near Edge Structures*, Springer-Verlag, Berlin (1983).
8. K.O. Hodgson, B. Hedman and J.E. Penner-Hahn, eds., *EXAFS and Near Edge Structure III*, Springer-Verlag, Berlin (1984).
9. *J. Physique.* **47** (1986) Colloque C8.
10. S. Khalid, R. Emrich, R. Dujari, J. Shultz and J.R. Katser, *Rev. Sci. Instrum.* **53** (1982) 22.
11. K. Tohji, Y. Udagawa, S. Tanabe and A. Ueno, *J. Am. Chem. Soc.* **106** (1984) 612.
12. N. Kakuta, K. Tohji and Y. Udagawa, *J. Phys. Chem.* **92** (1988) 2583.
13. J.H. Sinfelt, G.H. Via, F.W. Lytle and R.B. Greegor, *J. Chem. Phys.* **75** (1981) 5527.
14. D. Bazin, H. Dexpert, P. Lagarde and J.P. Bournonville, *J. Catalysis* **110** (1988) 209.
15. F.W. Lytle, G.H. Via and J.H. Sinfelt, in *Synchrotron Radiation Research*, eds. H. Winich and S. Doniach, Plenum Press, New York (1980) p. 401.
16. D.C. Koningsberger and R. Prins, in *X-Ray Absorption Spectroscopy*, eds. D.C. Koningsberger and R. Prins, John Wiley, New York (1988) p. 321.
17. C. Esselin, E. Bauer-Grosse, J. Goulon, C. Williams, Y. Chauvin, D. Commereuc and E. Freund, *J. Physique.* **47** (1986) C8-243.
18. J. Evans, Gao Jingxing, H. Leach and A.C. Street, *J. Organomet. Chem.* **372** (1989) 61.
19. B.R. Stults, R.M. Friedman, K. Koenig, W. Knowles, R.B. Greegor and F.W. Lytle, *J. Am. Chem. Soc.* **103** (1981) 3235.
20. J. Goulon, E. Georges, C. Goulon-Ginet, Y. Chauvin, D. Commereuc, H. Dexpert and E. Freund, *Chem. Phys.* **83** (1984) 357.
21. D. Ballivet-Tkatchenko, C. Esselin and J. Goulon, *J. Physique.* **47** (1986) C8-343.
22. N. Binsted, S.L. Cook, J. Evans, G.N. Greaves and R.J. Price, *J. Am. Chem. Soc.* **109** (1987) 3669.
23. P.A. Lee and J.B. Pendry, *Phys. Rev. B* **11** (1975) 2795.
24. S.J. Gurman, N. Binsted and I. Ross, *J. Phys. C* **17** (1984) 143.
25. S.J. Gurman, N. Binsted and I. Ross, *J. Phys. C* **19** (1986) 1845.
26. N. Binsted, J. Evans, G.N. Greaves and R.J. Price, *J. Chem. Soc., Chem. Commun.* (1987) 1330.

27. J. Evans in *Surface Organometallic Chemistry: Molecular Approaches to Surface Catalysis*, eds. J.-M. Basset et al., Kluwer, Dordrecht (1988) p. 47.
28. J.-M. Basset et al., eds., *Surface Organometallic Chemistry: Molecular Approaches to Surface Catalysis*, Kluwer, Dordrecht (1988).
29. B. Besson, B. Morawek, A.K. Smith, J.M. Basset, R. Psaro, A. Fusi and R. Ugo, *J. Chem. Soc., Chem. Commun.* (1980) 569.
30. S.L. Cook, J. Evans and G.N. Greaves, *J. Chem. Soc., Chem. Commun.* (1983) 1287; S.L. Cook, J. Evans, G.S. McNulty and G.N. Greaves, *J. Chem. Soc., Dalton Trans.* (1986) 7.
31. F.B.M. Duivenvoorden, D.C. Koningsberger, Y.S. Uh, and B.C. Gates, *J. Am. Chem. Soc.* **108** (1986) 6254.
32. L. D'Ornelas, A. Choplin, J.-M. Basset, L.-Y. Hsu, and S.G. Shore, *Nouv. J. Chim.* **9** (1985) 155.
33. V.D. Alexiev, N. Binsted, J. Evans, G.N. Greaves, and R.J. Price, *J. Chem. Soc., Chem. Commun.* (1987) 395.
34. A. Choplin, B. Besson, L. D'Ornelas, R. Sanchez-Delgardo, and J.-M. Basset, *J. Am. Chem. Soc.* **110** (1988) 2783.
35. N. Binsted, J. Evans, G.N. Greaves and R.J. Price, *Organometallics* **8** (1989) 613.
36. V.D. Alexiev, N.J. Clayden, R.J. Crowte, C.M. Dobson, J. Evans, D.J. Smith and P.S. Western, unpublished results.
37. G. Woolery, G. Kuehl, A. Chester, T. Bein, G. Stucky and D.E. Sayers, *J. Physique.* **47** (1986) C8-281.
38. M. Sano, T. Maruo, H. Yamatera, M. Suzuki and Y. Saito, *J. Am. Chem. Soc.* **109** (1987) 52.
39. T. Bein, S.J. McLain, D.R. Corbin, R.D. Farlee, K. Moller, G.D. Stucky, G. Woolery and D.E. Sayers, *J. Am. Chem. Soc.* **110**, (1988) 1801.
40. J. Evans, *Chem. Brit.* **22** (1986) 813.
41. R.B. Greegor and F.W. Lytle, *J. Catalysis* **63** (1980) 476.
42. G.H. Via, G. Meitzner and J.H. Sinfelt, *J. Chem. Phys.* **79** (1983) 1527.
43. J.B.A.D. van Zon, D.C. Koningsberger, H.F.J. van't Blik, R. Prins and D. Sayers, *J. Chem. Phys.* **80** (1984) 3914.
44. J.B.A.D. van Zon, D.C. Koningsberger, H.F.J. van't Blik and D.E. Sayers, *J. Chem. Phys.* **82** (1985) 5742.
45. A. Ceriotti, F. Demartin, G. Longoni, M. Manassero, M. Marchionna, G. Piva and M. Sansoni, *Angew. Chem., Int. Ed. Engl.* **24** (1985) 696.
46. H.F.J. van't Blik, J.B.A.D. van zon, T. Huizinga, D.C. Koningsberger and R. Prins, *J. Phys. Chem.* **87** (1983) 2264.
47. H.F.J. van't Blik, J.B.A.D. van Zon, T. Huizinga, D.C. Koningsberger and R. Prins, *J. Am. Chem. Soc.* **107** (1985) 3139.
48. P. Basu, D. Panayotov and J.T. Yates, Jr., *J. Am. Chem. Soc.* **110** (1988) 2074.
49. T. Mizushima, K. Tohji and Y. Udagawa, *J. Am. Chem. Soc.* **110** (1988) 4459.
50. F.B.H. Duivenvoorden, B.J. Kip, D.C. Koningsberger and R. Prins, *J. Physique.* **47** (1986) C8-227.
51. B.J. Kip, F.B.M. Duivenvoorden, D.C. Koningsberger and R. Prins, *J. Am. Chem. Soc.* **108** (1986) 5633.
52. B.J. Kip, F.B.M. Duivenvoorden, D.C. Koningsberger and R. Prins, *J. Catalysis* **105** (1987) 26.
53. D.C. Koningsberger, F.B.M. Duivenvoorden, B.J. Kip and D.E. Sayers, *J. Physique.* **47** (1986) C8-255.
54. G. Apai, J.F. Hamilton, J. Stöhr and A. Thompson, *Phys. Rev. Lett.* **43** (1979) 165.
55. S. Sakellson, M. McMillan and G.L. Haller, *J. Phys. Chem.* **90** (1986) 1733.
56. D.E. Resasco, R.S. Weber, S. Sakellson, M. McMillan and G.L. Haller, *J. Phys. Chem.* **92** (1988) 189.
57. J.H.A. Martens, R. Prins, H. Zandbergen and D.C. Koningsberger, *J. Phys. Chem.* **92** (1988) 1903.
58. J.H. Sinfelt, G.H. Via and F.W. Lytle, *J. Chem. Phys.* **72** (1980) 4832.
59. J.H. Sinfelt, G.H. Via and F.W. Lytle *J. Chem. Phys.* **76** (1982) 2779.
60. G. Meitzner, G.H. Via, F.W. Lytle and L.H. Sinfelt, *J. Chem. Phys.* **87** (1987) 6354.
61. D. Bazin, H. Dexpert, P. Lagarde and J.P. Bournonville, *J. Physique.* **47** (1986) C8-293.

62. R.J. Oldman, *J. Phys.* **47** (1986) C8-321.
63. G. Meitzner, G.H. Via, F.W. Lytle, S.C. Fung and J.H. Sinfelt, *J. Phys. Chem.* **92** (1988) 2925.
64. G. Longoni, A. Ceriotti, M. Marchionna and G. Piro, in *Surface Organometallic Chemistry: Molecular Approaches to Surface Catalysts*, eds. J.-M. Basset *et al.*, Kluwer, Dordrecht (1988) p. 157.
65. T.K. Sham, T. Ohta, Y. Yokoyama, Y. Kitajima, M. Funabashi, N. Kosugi and H. Kuroda *J. Chem. Phys.* **88** (1988) 475.
66. H. Dexpert, *J. Phys.* **47** (1986) C8-219.
67. M.G. Samant, G.L. Borges, J.G. Gorden II, O.R. Melroy and L. Blum, *J. Am. Chem. Soc.* **109** (1987) 5970.

9 Looking at solid surfaces with synchrotron radiation

D. NORMAN and D.A. KING

9.1 Introduction

For many years, the study of solid surfaces has lagged behind that of the solid, liquid and gas phases. The revolution that occurred in physical chemistry in the late 1930s, heralded by the accurate determination of structures by X-ray crystallography and infrared and, later, NMR spectroscopies, almost completely bypassed solid surfaces until the 1960s. These methods were not sensitive enough for the small number of molecules at a surface. One solution was to make surfaces with a high surface area to increase the density of surface species in, for example, an infrared probe. However, this approach is limited by the heterogeneous nature of the surfaces exposed. The alternative is to prepare a mirror-finish surface of the desired material from single crystals, cut and carefully polished to expose a given crystal plane, with an area of about $1 \, cm^2$. The crystal is cleaned in a UHV environment (better than 10^{-8} Pa) to avoid its subsequent contamination. Almost every technique, both spectroscopic and crystallographic, that can be used to study bulk materials has now been made sensitive enough to study surfaces prepared in this way. In addition, some special methods have been developed that are unique to surface science [1].

9.2 Surface science: the tools

Since electrons in solids have a mean free path of only a few atomic diameters in the useful energy range of 50–3000 eV, a number of particle-based techniques are surface-sensitive. Thus only those electrons that are generated within the surface will escape into the vacuum with a well-defined energy. Most commonly used techniques are low energy electron diffraction (LEED), yielding the surface crystallography, and Auger electron spectroscopy (AES), which gives the complementary elemental analysis of the surface; both of these techniques involve electrons as the incident probe and the scattered particle. Surface vibrations can be measured, with high sensitivity, by high resolution electron energy loss spectroscopy (HREELS), another electrons in/electrons out technique. Backscattering techniques involving the impact of ions, in the energy range 300–3000 eV, or neutral atoms on solids are sensitive to the

Table 9.1 Surface sensitive spectroscopies based on incident photons.

Particle in	Particle out	Technique	Comment
hv, IR	hv, IR	RAIRS: reflection-absorption infra red spectroscopy	Brightness of synchrotron radiation below $1000\,cm^{-1}$ provides advantages; otherwise better to use conventional source
hv, UV	hv, IR	SERS: surface-enhanced Raman spectroscopy	Little advantage over laboratory laser source
hv, UV-VUV, X-ray	electron	UPS/XPS: ultra-violet/X-ray photo-electron spectroscopy	Tunability, polarisation and intensity of synchrotron radiation give some clear advantages
hv, X-ray	hv or electron	SEXAFS: surface extended X-ray absorption fine structure	Currently only possible with synchrotron radiation
hv, X-ray	hv, X-ray	Grazing incidence X-ray diffraction	Very difficult without synchrotron radiation

surface layer only, and provide useful structural information, and elemental analysis.

Certain photon-based techniques are also surface sensitive and can often be enhanced using synchrotron radiation. The most important are listed in Table 9.1. The incident probe used may penetrate deep into the solid, but photoemitted electrons, for example, will only be ejected back into the vacuum, where they can be counted and their momentum analysed, without suffering inelastic collisions if the photoelectron excitation process occurred within the top few atomic layers of the solid. Quantised features in the spectrum are thus related to processes in the surface layer.

Reflection-absorption infrared spectroscopy is also surface sensitive because grazing incidence infrared photons are reflected from the surface region. Absorption bands from vibrations in adsorbed species with a finite component to the dynamic dipole, associated with a vibration normal to the surface, may be observed with this technique with better resolution than with HREELS. Recent results suggest that sensitivity can be almost as good. Finally, at near-grazing incidence X-ray photons can be totally reflected from the surface, giving X-ray diffraction from surface layers.

9.3   Advantages of the synchrotron source

The use of synchrotron radiation for infrared spectroscopy offers no advantage in the region above about $1000\,cm^{-1}$ ($>125\,meV$) but at longer wavelengths, synchrotron radiation should give improved brightness and polarisation compared to a conventional black-body source. Attempts are now being made at a number of synchrotron radiation centres to exploit the infrared range of the synchrotron radiation output for surface studies. However, for Raman spectroscopy, it appears that fixed wavelength lasers provide the necessary intensity and synchrotron radiation gives no help.

It is at higher energies that the use of synchrotron radiation for surface science really comes into its own [2]. For photoelectron spectroscopy, major advantages arise from the tunability, polarisation and brightness of the synchrotron light source. Tunability provides the opportunity to optimise surface sensitivity by maximising cross-sections and by gearing the kinetic energy of the photoemitted electron to the minimum escape depth. Constant initial state and constant final state experiments can also be performed and shape resonances can be directly measured. These are revealed in plots of photoemission intensity versus frequency of the incident radiation. They are important in establishing exactly what transition is occurring and probing the predicted bond orbitals at the surface. The polarised nature of the source allows the symmetry of electron states in the surface to be determined. There is normally no advantage in intensity for photoemission measurements with synchrotron radiation however, and very high resolution work is probably best performed with, for instance, a He resonance lamp. Likewise, unless one has access to unlimited amounts of time at a synchrotron source, routine measurements should always be carried out with laboratory sources.

X-ray diffraction studies of surfaces are possible with a high-intensity X-ray generator (e.g. rotating anode), but only for surface structures which remain stable and free from contamination for a long time. In practice, most interesting surfaces become contaminated within an hour or two, even under UHV conditions, making synchrotron radiation almost essential for these low count rate experiments.

Finally, there are several techniques that simply could not be performed without synchrotron radiation. For example, surface extended X-ray absorption fine structure (SEXAFS) spectroscopy, which produces interatomic spacings for absorbate systems with more precision than any other method; and the related technique of near-edge X-ray absorption fine structure (NEXAFS) spectroscopy, which provides a direct means of determining the orientation of molecular absorbates, with respect to the surface plane. These techniques demand the tunability over a wide energy range, the brightness and the polarisation characteristics that are, for the time being, unique to synchrotron radiation.

9.4 Photoemission

9.4.1 *Valence level photoemission*

Photoemission is by far the most widely used technique for studying the electronic structure of surfaces and near-surface regions of solids. Here we give

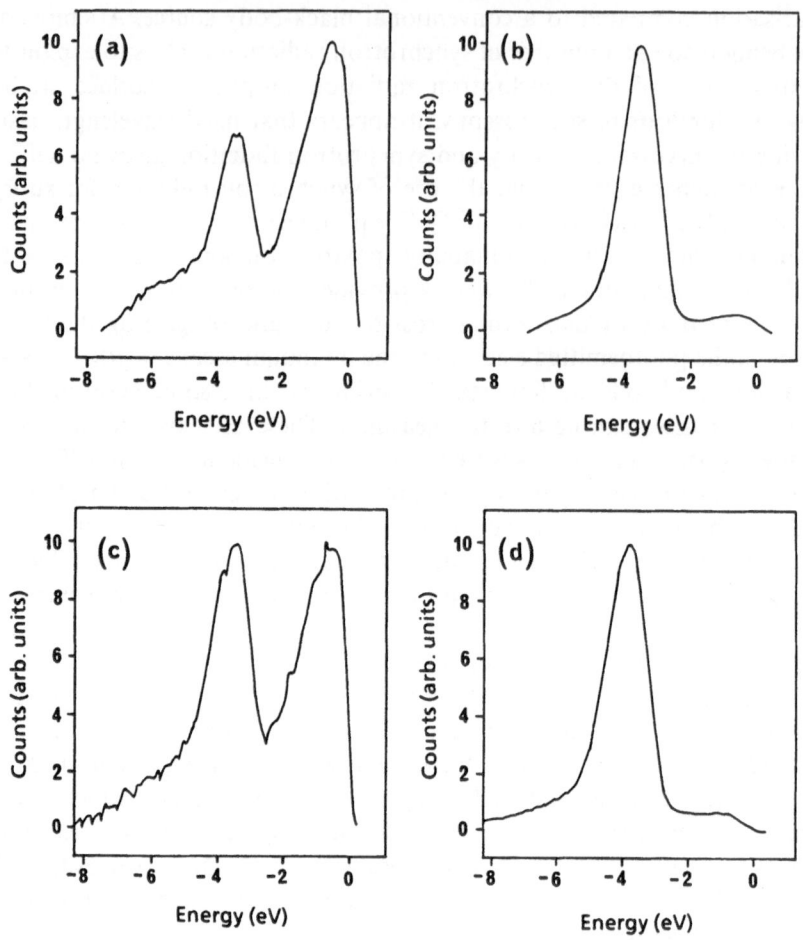

Figure 9.1. Photoemission spectra of (a), (b) $Cu_{30}Zr_{70}$ and (c), (d) $Cu_{40}Zr_{60}$ at photon energies of (a), (c) 40 eV and (b), (d) 120 eV.

a few examples to illustrate the variety of ways in which the experimental variable may be manipulated to yield interesting information.

The tunable photon energy is of great importance in helping to assign valence band features to states of different symmetry. For instance, Figure 9.1 shows photoemission yield from amorphous metallic glasses of $Cu_{30}Zr_{70}$ and $Cu_{40}Zr_{60}$ measured at two different photon energies [3]. The valence band comprises a mixture of states derived from Cu 3d and Zr 4d orbitals. The former exhibit a fairly smooth variation in cross-section as the energy is varied but the 4d states, with a node in their wavefunction, change in intensity by about two orders of magnitude in the range chosen, allowing easy separation of the contributions to the valence band, with the Zr-derived states found to be near the Fermi level ($E = 0$) and thus dominating the conductivity. This is a particularly simple example since the main parts of the Cu-derived and Zr-derived bands are almost completely separated from each other. However, a similar method has been used for crystalline alloys of Cu–Pd where the bands overlap, and allowed empirical derivation of the Cu and Pd partial density of states [4]. Both of these examples are of systems where the real interest lies in the bulk properties, rather than the surface, but photoemission is so powerful a

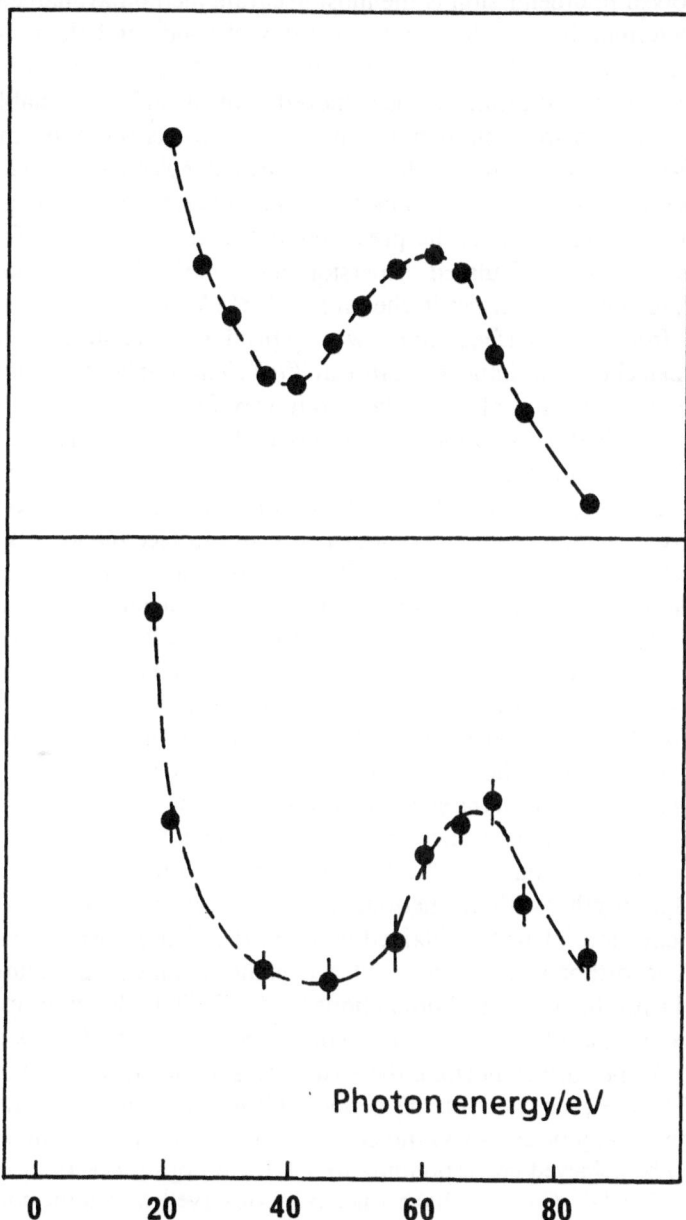

Figure 9.2. Photoemission intensity variation from a surface state on Cu 6.8% Pd (111) as a function of photon energy: theory (top) and experiment (bottom).

technique for determining electronic structure that the experimenters were prepared to tolerate its surface sensitivity!

9.4.2 Angle-resolved photoemission

Angle-resolved photoemission is the most accurate method available for the detailed determination of the band structures of solids and their surfaces. Further work on alloys, this time on purely surface properties, illustrates the degree of sophistication that can be achieved with the aid of a tunable light source. A first-principles photoemission theory predicts the appearance of Schockley-type surface states on alloys, i.e. a particular electronic surface state that has no correspondence with bulk electronic states. On the [111] plane of several Cu–Pd alloy crystals the predicted states were observed [5] giving close agreement with calculated dispersion and polarisation dependence. In particular, for photon energies in the range 20–85 eV, the predicted intensity variation from the surface state was reproduced remarkably closely (Figure 9.2) including a shape resonance at 70 eV. This implies that the theory gives a good description of the surface state wavefunction.

Another application is to look at the mode of bonding of molecules to a surface. Carbon monoxide, which used to rate as the surface scientist's favourite adsorbate, is invariably bound to surfaces through the carbon atom, and usually with its molecular axis normal to the surface plane. It has been shown that on one group of surfaces it is not: the face centred cubic [110] planes. Photoemission spectra, taken with a He II resonance lamp, show the 1π, 5σ and 4σ orbitals of adsorbed CO on Pt clearly, with a strong bonding shift of the 5σ orbital on the carbon atom (Figure 9.3(a)). The 4σ orbital (on the O atom) at 11.7 eV below the Fermi level is fully resolved in the spectrum. The polar angular intensity dependence of the 4σ emission provided a means of accurately determining the orientation of the C–O axis.

A theoretical analysis of the spectra requires assumptions that can be tested on the CO 4σ shape resonance, at $hv \simeq 40$ eV. This shape resonance was experimentally determined on the Daresbury Storage Ring, and is shown in Figure 9.3(b) together with the theoretical prediction (solid line). The variation of the 4σ emission intensity (obtained with He II radiation) with polar angle in the plane orthogonal to the incidence plane is shown in Figure 9.3(c), compared with the theoretical predictions for C–O tilt angles with respect to the surface normal of 0° and 25°. It is concluded that the C–O axis is tilted $25 \pm 2°$ from the surface normal; the structure is shown in Figure 9.3(d). As indicated in the plan view of the surface, tilting arises from the repulsive interaction of neighbouring CO molecules along the atom rows, and allows a packing to be achieved corresponding to a 1:1 ratio of surface Pt atoms: CO molecules. Similar structures have since been observed with a monolayer of CO on Ni[110], Cu[110] and Pd[110], except that on Ni and Pd the bridge sites are preferred. This study nicely illustrates the successful combination

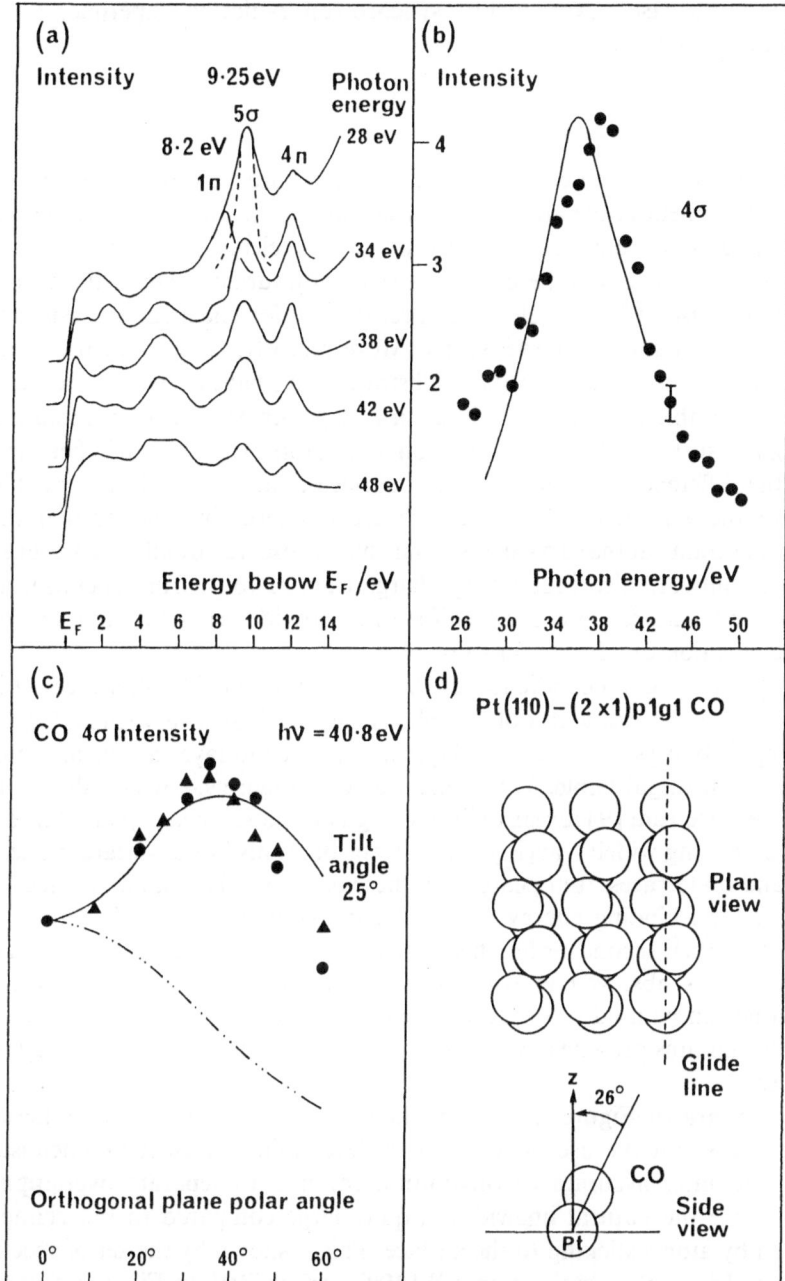

Figure 9.3.   (a) Photoelectron spectra taken from a Pt[110] surface with a monolayer of CO. The positions of the CO molecular orbitals are shown. (b) Photoemission intensity from the CO 4σ orbital as a function of photon energy. (c) The dependence of the CO 4σ intensity on polar emission angle. (d) The structure of CO on Pt[110].

of laboratory-based experiments, synchrotron radiation experiments and theoretical calculations.

9.4.3   Core level photoemission

The binding energies of core level electron states are sensitive to the valence level environment of the atom. The corresponding chemical shifts of the core level energies of adatoms, typically by several electronvolts, are used to monitor the presence of different valence states on surfaces. This is the basis of the familiar use of ESCA (electron spectroscopy for chemical analysis) as a routine tool in surface spectroscopy. More recently, using high resolution instrumentation, core level shifts of substrate atoms have been measured, and, under favourable circumstances, it has proved possible to use this technique to distinguish surface substrate atoms from bulk atoms [8]. The shifts here arise from the difference in coordination of surface atoms and bulk atoms. This narrows the valence band, which is shifted in energy in order to maintain charge neutrality at the surface. A similar shift is observed by all the core levels. Further shifts are also induced by charge transfer to or from chemisorbed atoms, and surface core level shifts thus provide a useful tool for the characterisation of adsorption sites.

Tungsten is a good candidate for these studies due to the sharpness of its $4f_{7/2}$ level, i.e. the half width of this photoemission line is very small. On the more open W crystal planes, top layer atoms, second layer atoms and bulk atoms are distinguishable. This is illustrated in Figure 9.4, which shows the core level spectrum [9] for six different singular and stepped crystal planes of tungsten having widely varying proportions of atoms in the surface plane at steps and on terraces. It appears from these data that the dominant effect on the core level binding energy is the effective atomic coordination number. There are other, second-order effects, such as the small lateral translation of surface atoms observed in some surface reconstructions, but clearly this technique can be used to distinguish substrate atoms at steps from those in terraces and thus provide a way of following details of adsorption on stepped surfaces.

The spectra of Figure 9.4 clearly show very distinct peaks for different crystal faces, but the use of surface core-level shift in adsorbate chemistry usually requires accurate deconvolution routines to separate overlapping peaks, since the natural linewidth is quite large compared to the changes induced by atoms sticking to the surface. This is shown by the set of spectra shown in Figure 9.5 obtained on a W[100] surface [10]. In Figure 9.5(a) the peaks arising from the surface layer (S), the underlayer (U), and the bulk (B) are illustrated. The remaining spectra, (b)–(h), were taken successively during nitrogen adsorption, up to 0.5 monolayers. In these spectra the underlayer (U) and bulk (B) peaks are still distinguishable but the surface peaks are shifted to S_1, S_2, S_3; the subscript denotes the number of N atoms bonded adjacent to the

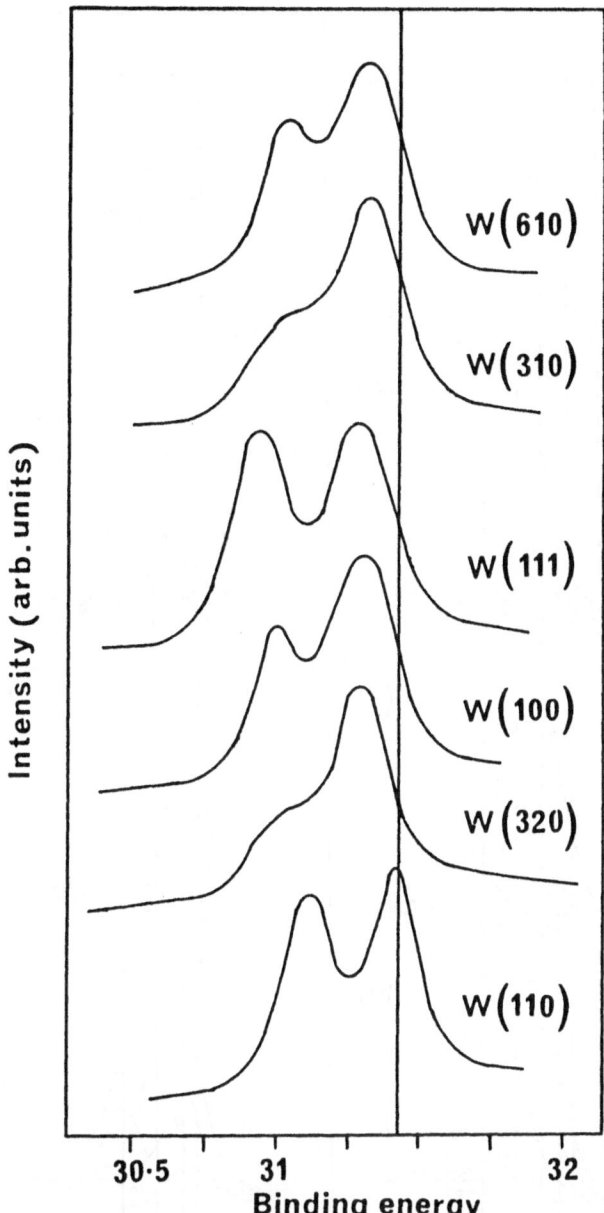

Figure 9.4. W $4f_{7/2}$ spectra at $hv = 75\,eV$ from 6 different crystal planes of tungsten; the vertical line indicates the position of the peak for bulk atoms.

surface W atom. The direction of the shift is due to net charge transfer to the N adatoms. The fine details of these spectra are revealed only after the curve-fitting depicted under each peak.

An analysis of this set of spectra provided a remarkable amount of detail concerning the adsorption process. Previously, using LEED, a contracted

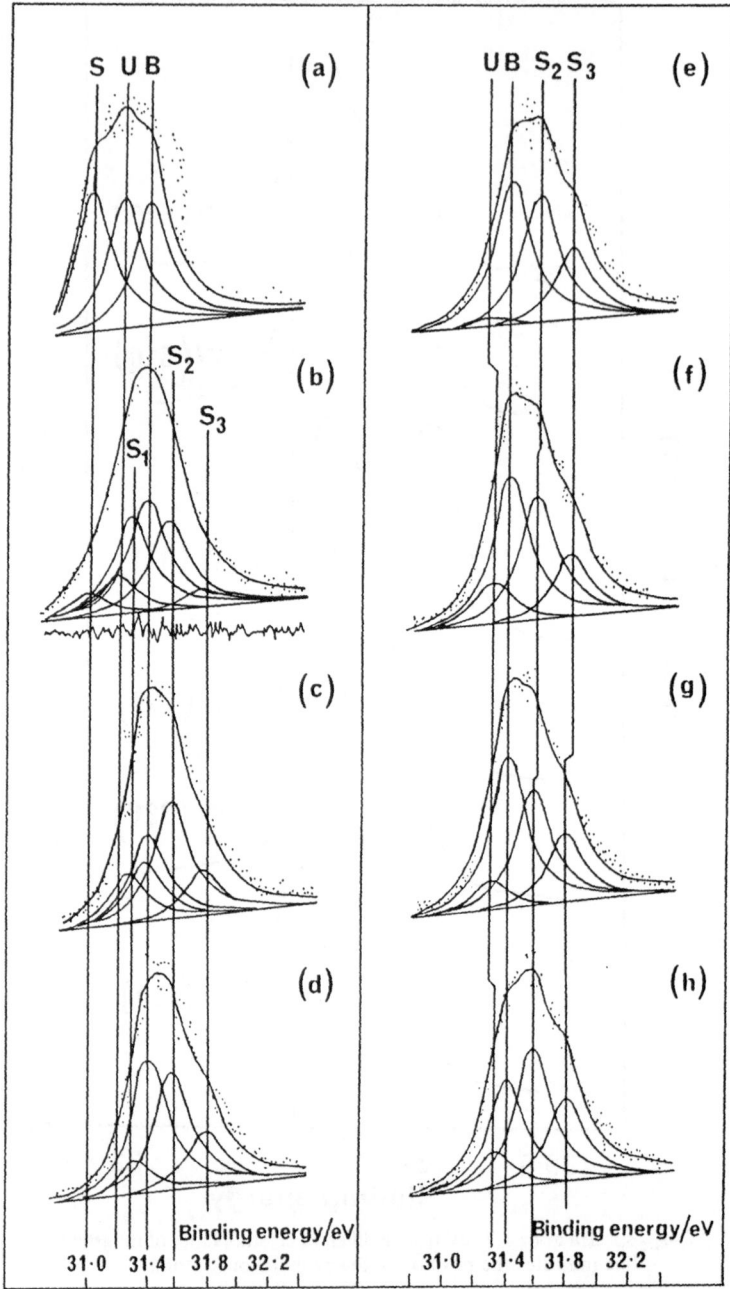

Figure 9.5. The W $4f_{7/2}$ photoemission peak, at high resolution, from W[100] at increasing coverages θ of nitrogen. (a) Clean surface, $\theta = 0$; for remainder, $\theta =$ (b) 0.19, (c) 0.30, (d) 0.34, (e) 0.40, (f) 0.46, (g) 0.56 and (h) 0.59.

domain structure had been observed at a fractional coverage $\theta = 0.4$, involving lateral displacements of W surface atoms towards N adatoms in four-fold hollow sites [11]. In this structure, the surface density of W atoms within each 2D domain or island, containing about 16 N adatoms, is higher than in the normal W[100] surface due to the contraction or inward collapse of the structure. At $\theta = 0.5$, the W atoms return to their bulk positions. The core level shift work confirms the displacive structure at $\theta = 0.4$, but also demonstrates that even at lower coverages small contracted islands are formed.

Although, as mentioned earlier, conventional X-ray sources (usually Al Kα at $hv = 1486.6$ eV or Mg Kα at $hv = 1253.6$ eV) may be used for core level photoemission studies, considerable advantages can accrue from synchrotron radiation, particularly in enhanced sensitivity or better resolution. An example of the latter is the spectrum of Figure 9.6, which depicts the Al$2p_{1/2, 3/2}$ core levels from an Al (111) surface after exposure to 100 L (1 L (Langmuir) = 1×10^{-6} Torr s) of oxygen [12]. A spectral resolution of 0.18 eV allows the separation of five peaks, indicated by the dashed lines in the figure, which are interpreted to correspond to metallic aluminium and an oxide state at the extremes, with in between, three distinct chemically shifted states associated with surface Al atoms bonded to one, two or three O nearest neighbours.

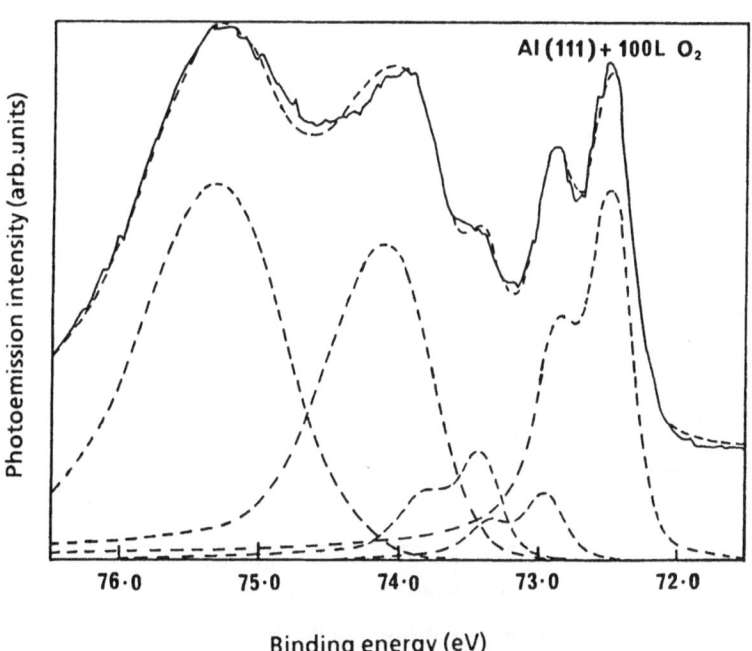

Figure 9.6. Al2p photoemission spectrum from Al[111] exposed to 100 L oxygen (Figure 9.1) and the fit to this (dashed curve) achieved by a sum of the individual contributions shown below. These individual contributions are from right to left (a) metallic state, (b), (c), (d) chemisorption states 1, 2 and 3 and (e) oxide state.

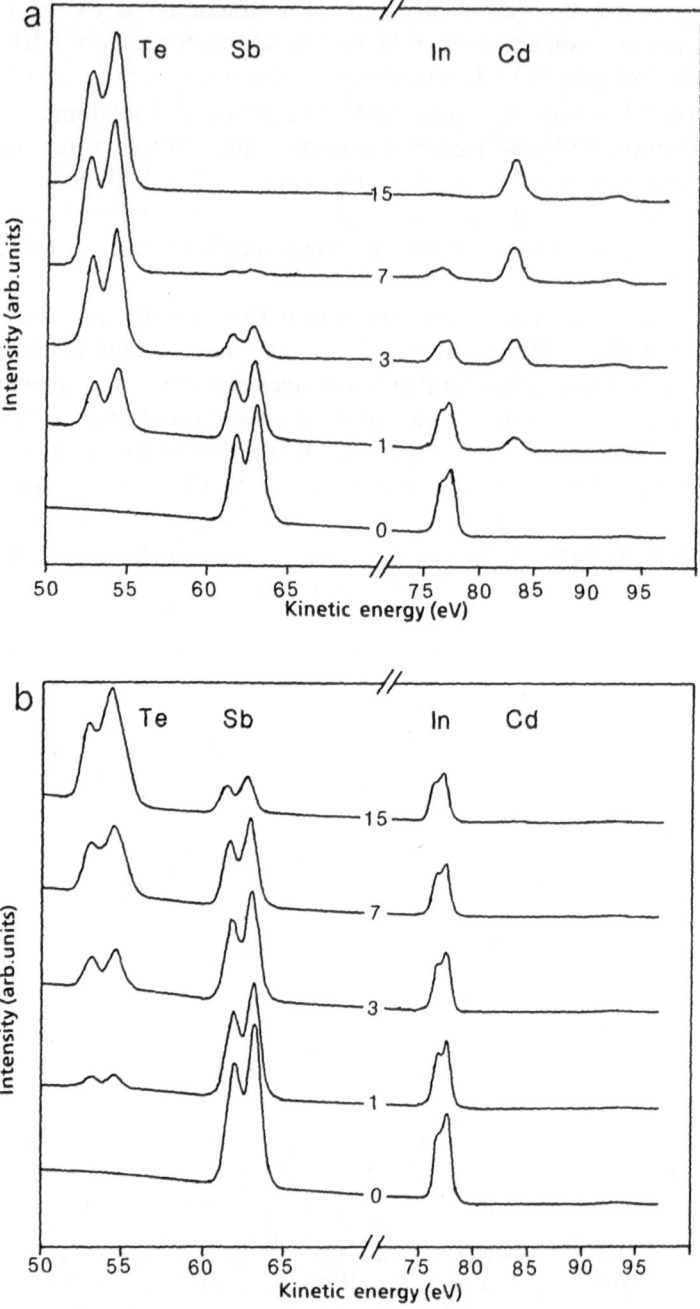

Figure 9.7. (a) Soft X-ray photoelectron spectra ($hv = 100\,eV$) of clean InSb (100) and following deposition of CdTe layers onto a room temperature substrate, showing core levels of Te, Sb, In and Cd with the valence bands just visible. Each set is labelled with the CdTe layer thickness in angstroms. (b) Photoelectron spectra ($hv = 100\,eV$) of clean InSb (100) and following the deposition of CdTe layers onto a substrate at 500K. Each set is labelled with the CdTe layer thickness in angstroms.

One final example of photoemission core level spectroscopy concerns the reactivity of the interface between the III–V semiconductor InSb and an overlayer of the II–VI material CdTe [13]. The 4d core levels of Te, Sb, In and Cd can all be observed with a photon energy of 100 eV, as shown in Figure 9.7. The top panel gives the effects of deposition of CdTe onto an InSb substrate at room temperature, with the CdTe layer thickness given in angstroms. The In and Sb peaks are attenuated at a constant rate and CdTe grows stoichiometrically; the difference in intensity of the Cd and Te peaks in the uppermost curve, for instance, is just caused by differences in photoemission cross-section. Although stoichiometric, this room temperature deposit is not ordered and epitaxial growth requires the substrate to be held at an elevated temperature. The lower panel in Figure 9.7 shows the results of the same experiment with the InSb held at 500 K. Under these conditions, the reactivity is completely different and a 'chemical soup' is found in which the two compounds appear to split into their four constituent atoms; the In signal does not decrease with the Sb peak, the Te level is broadened and Cd is barely visible. The interface is neither abrupt nor stoichiometric. Further measurements indicate that an interfacial layer, largely indium telluride, is formed with a thickness of several angstroms.

9.5   X-ray absorption spectroscopy

9.5.1   *Surface EXAFS*

The technique of EXAFS described in Chapter 6, has been successfully applied to a number of surface structural determinations, yielding adsorbate-substrate distances with a precision unrivalled by any other method [14]. The key to its usefulness is that it is a short-range probe, requiring no long-range order. Thus a reaction may be followed through its various stages, as with the study of oxygen on aluminium [15], starting from initial chemisorption of isolated adatoms on a single crystal, through the formation of ordered overlayers, and the growth of an amorphous bulk oxide on the surface. No other technique is available to follow the detailed surface geometry through all these phases.

In another series of surface EXAFS (SEXAFS) experiments, on chlorine atoms adsorbed on silver [16], the enhanced precision in bond length determination over other surface techniques has been exploited together with the ability to examine both ordered and disordered phases. In each case, the data reveal the adatom site, the nearest-neighbour spacing, and, since all atoms within a radius of 5.5 Å of the absorber atom (Cl) could be detected, the overlayer structure. Thus for Cl on Ag[111], at both 1/3 and 2/3 of a monolayer the Cl atoms are found to occupy three-fold hollow sites on the surface, with a Cl–Ag bond length of 2.70 ± 0.01 Å, independent of coverage. This site provides three-fold coordination to surface Ag atoms. The structures determined at these two coverages are shown in Figure 9.8. At 2/3 monolayer

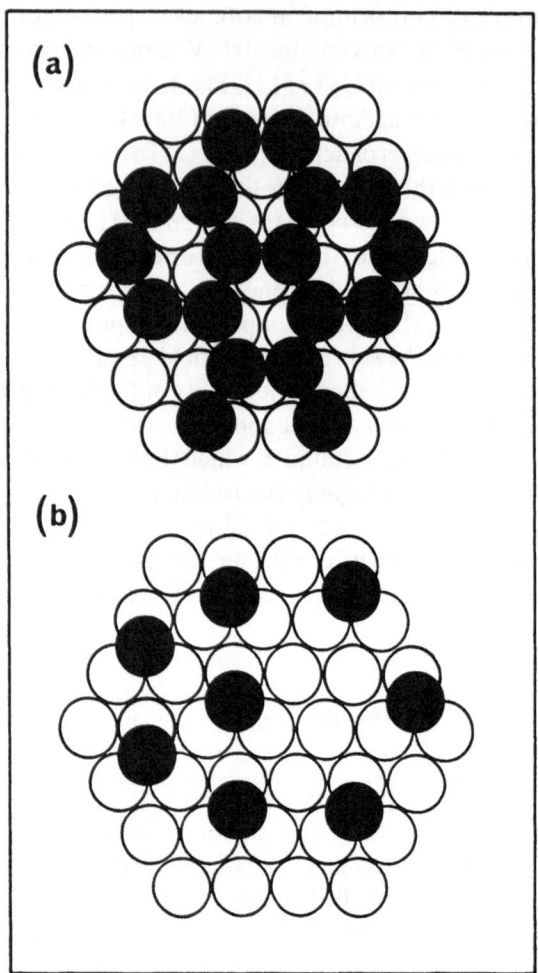

Figure 9.8. The structures of Cl on Ag[111] at (a) 2/3 monolayer and (b) 1/3 monolayer, determined by SEXAFS.

the Cl adatoms form a vacancy honeycomb structure, which shows weak LEED diffraction beams. At 1/3 monolayer, Cl–Cl nearest neighbour sites are avoided, and no long-range order is apparent from LEED patterns. Neither structure could have been determined by LEED. The 2/3 monolayer honeycomb structure (Figure 9.8(a)), may have implications for the ethylene epoxidation reaction; Ag is the preferred catalyst, and chlorine is used in the feedstock to prevent complete oxidation to CO_2.

The invariance of the adsorbate-substrate bond length in the Cl–Ag system is in strong contrast to the results found for Cs on Ag[111] [17]. At 0.15 monolayers Cs, the interatomic spacing is found to be 3.20 Å while at 0.3 monolayers, corresponding almost to a close-packed layer of Cs atoms on the

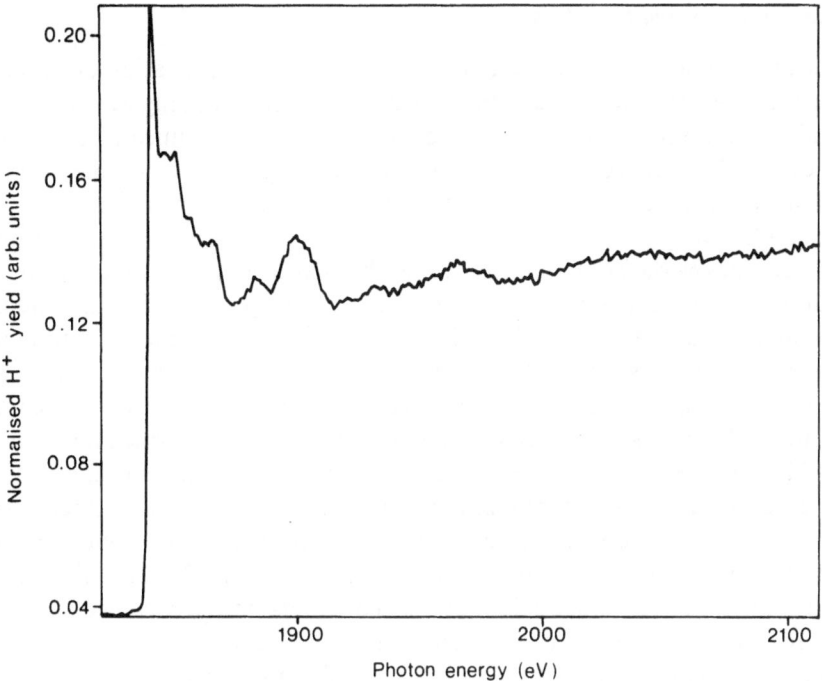

Figure 9.9.   The yield of H$^+$ ions as a function of photon energy above the Si K-edge from a Si(100) surface which had been exposed to water.

surface, the bond length is 3.50 Å. This difference is due to a bonding transition in the overlayer, from an ionic state at low coverages to a largely covalent state in the close-packed layer. Again, the three-fold site is preferred, even in the densely packed layer. Alkali metal adatoms are used as promoters in many catalytic reactions, and the objective of this work is to provide a sound basis for understanding promoter action.

SEXAFS may also be seen in the spectrum of the yield of ions photo-desorbed from a surface, as in Figure 9.9, which depicts the intensity of H$^+$ ions collected as a function of photon energy from a silicon (100) surface after exposure to water [18]. H$_2$O adsorption is dissociative, but the (2 × 1) structure is maintained, with H and OH attached to either end of Si$_2$ surface dimer pairs. The mechanism for desorption of H$^+$ is initiated by the adsorption of photons in the range above the Si K absorption edges, followed by interatomic Auger transitions which lead to the breaking of the Si–H surface bond. The yield of protons thus is directly related to the surface absorption coefficient of silicon atoms. With adsorbate present, the Si–Si bond distance in the dimer was found to be 2.35 Å, which is the same as the bulk interatomic spacing. This result is consistent with models used to interpret surface vibrational data.

9.5.2   X-ray standing waves

A further application of tunable soft X-rays to the determination of surface structure is via the use of interference to produce X-ray standing waves (XSW) parallel to a crystal surface. When the photon energy is scanned through the region of a Bragg reflection in the crystal, the standing wave outside the surface will move. This is equivalent to rotating the sample, as in a measurement of a crystal's 'rocking curve' (see Eqn. 1.13). As a node passes through the position of an atomic overlayer of impurity atoms, the characteristic Auger electron or fluorescence signal from these overlayer atoms will go through a minimum. Similarly a standing wave antinode will produce a maximum. Combining experimental results with rather simple calculations of the effect can yield the vertical spacing of an adsorbate layer to a precision of about 0.05 Å. Figure 9.10 demonstrates this for a regular overlayer of chlorine atoms on a Cu[111] surface [19]. This technique actually measures the spacing of the overlayer atoms from a continuation of the perfect bulk lattice rather than from the real surface layer, which may be relaxed or reconstructed in some way. This may be turned to an advantage since the use of XSW in conjunction with some other technique might allow deductions on the presence and the nature of such a surface rearrangement. Since SEXAFS and XSW can use the same experimental apparatus, this is a natural combination of methods for surface adsorbate geometry.

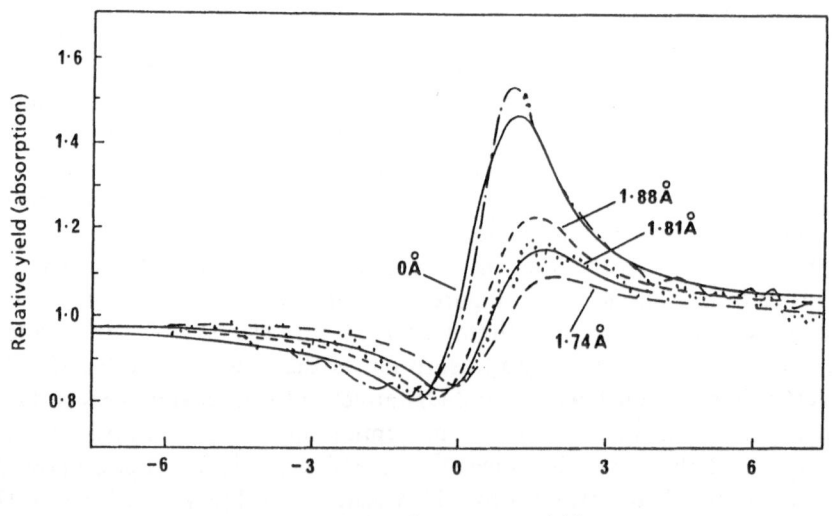

Figure 9.10.   Experimental CuL_3VV (–.–.–) and $ClKVV$ (......) Auger electron yields as a function of photon energy around the normal incidence [111] Bragg reflection from a Cu[111] ($\sqrt{3} \times \sqrt{3}$) R30° − Cl surface, compared with theoretical adsorption profiles for different adsorber locations, Δz. The theoretical curves include the effect of angular energy broadening as well as finite temperature and incoherent contributions.

9.5.3 *NEXAFS*

The near-edge structure in the adsorption spectrum (NEXAFS) from atomic adsorbates is dominated by multiple scattering of electrons between adsorbate and substrate atoms [14], and is therefore relatively difficult to analyse for structural detail. However, much interest has been generated by NEXAFS

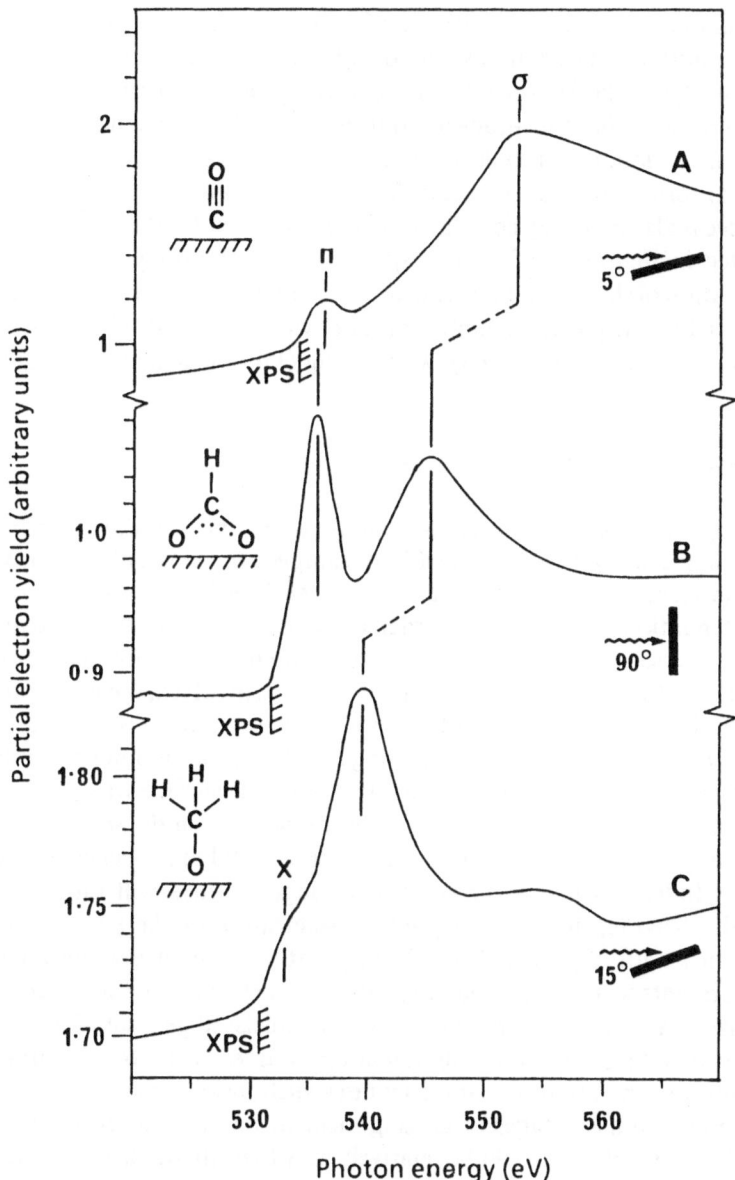

Figure 9.11.   NEXAFS spectra of the O K edge region for (A) CO, (B) formate (HCOO), and (C) methoxy (CH_3O) on Cu[100].

studies of molecules, particularly involving excitation from the K edges of C, N and O, which show very pronounced near-edge structure. This structure is dominated by intra-molecular resonances; for molecules with π-bonding, the spectrum exhibits a strong resonance close to the absorption edge, corresponding to the excitation of a 1s electron into the antibonding π^* orbital. This is strongest when the polarisation of the incident radiation is parallel to the π orbital, and is zero when it is perpendicular to π [20].

For all molecules, with or without π bonding, excitation of the 1s electron into the antibonding σ^* molecular orbital produces a resonance which is strongest when polarisation is along the inter-nuclear axis between two atoms in the molecule. The transitions are thus governed by simple dipole selection rules and variations in polarisation provide a way of determining the molecular orientation at surfaces. The energy above threshold of the σ resonance is also inversely correlated with the intra-molecular bond length, as illustrated in Figure 9.11 for three species with carbon-oxygen bonds [21]. The variation of shape resonance position is shown for carbon monoxide, with a short (1.13 Å) triple C–O bond, formate, with a quasi-double C–O bond about 1.25 Å long, and methoxy with a long (1.43 Å) single C–O bond.

9.6   X-ray diffraction from surfaces

Scattering of X-rays has been for many years the standard method of determining the crystallography of bulk samples, but it is only recently that the same techniques have been applied to surface structures. X-rays have bulk attenuation lengths of the order of microns, so that Bragg scattering involves thousands of atoms; since the peak intensity of a diffracted beam is proportional to the square of the number of atoms involved, the peak intensity of X-rays diffracted from a monolayer is lower than from the bulk by at least a factor of 10^7. It is therefore not surprising that X-ray diffraction has not found routine use as a surface structural tool. However, the surface sensitivity can be greatly enhanced by making the X-ray beam incident on the sample at a low grazing angle and this is the key to the experimental approaches now being adopted. Indeed, below a certain critical angle, X-rays will undergo total external reflection, limiting the photon penetration of the surface to their extinction length, typically 20 Å. Working at very small grazing angles of course presents severe experimental problems, particularly with respect to the beam size and the flatness and the necessarily precise alignment of the sample. The use of intense synchrotron radiation sources, with the natural high collimation of the photon beam, facilitates such measurements.

The potential advantages of grazing-incidence X-ray scattering lie in two areas. Theoretically, it should be relatively easy to evaluate surface structures, since, unlike electrons, X-rays are only weakly scattered and it is relatively simple to calculate the expected intensities for any surface structure from kinematic diffraction theory. Experimentally, X-ray scattering is sensitive to long-range order across the surface, and, most importantly, may be used to study clean surfaces, or, since the photons will penetrate a less dense overlayer, solid–solid interfaces (such as semiconductor heterojunctions) or solid-liquid interfaces.

The first of these advantages is well illustrated in a study of the adsorption of tin atoms onto a Si(111) surface [22]. Scans of the intensity of scattered X-rays, using a wavelength of 1.38 Å, yield a series of crystallographic structure factors which are then compared with calculations for a variety of possible surface models. It is found that Sn atoms occupy sites above second-layer Si atoms, but the fit is not as good as expected until displacements of sub-surface silicon atoms are included in the calculation. To avoid having too many arbitrarily variable parameters, the Keating model for strain energy is incorporated into the fitting routine, leading to quite precise displacements down to the sixth plane of Si, 8.64 Å below the adsorbed Sn atoms.

The technique has also been used to examine the perfection of growth of an epitaxial layer of germanium grown onto a crystalline germanium surface

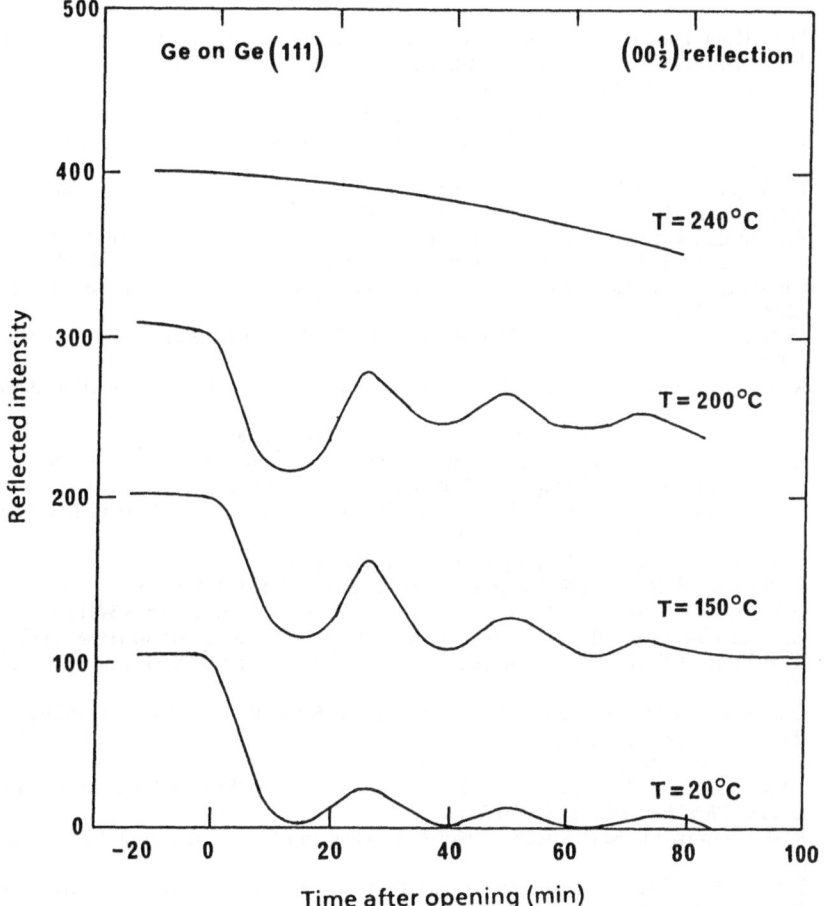

Figure 9.12.   The reflected signal from Ge[111] during Ge deposition, shown as a set of curves corresponding to different substrate temperatures. The curves are normalised to a starting signal of 100 and are given vertical offsets of 100 units.

[23]. The original surface, although cut very close to a (111) plane, inevitably contains steps. An angle of X-ray incidence of as much as 6° was used, which corresponds to destructive interference between an island one atomic layer thick and the underlying plane, thus giving maximum experimental sensitivity to island growth. The reflected signal is shown in Figure 9.12 as a function of time during deposition of Ge at a constant rate and at different substrate temperatures. When half the surface is covered by islands, the reflected intensity goes to zero. As deposition continues, the intensity increases again, with the period of this oscillation corresponding to the growth of one bilayer of the (111) surface. At a certain temperature (240°C with the deposition rate used in this experiment), the diffusion length is sufficiently high that the arriving Ge atoms move to steps on the surface and growth proceeds by step flow; X-ray intensity oscillations are no longer seen.

References

1. D.A. King and D.P. Woodruff, eds., *The Chemical Physics of Solid Surfaces and Heterogeneous Catalysis*, Vols. 1–4, Elsevier, Amsterdam (1982–1985).
2. I.T. McGovern, D. Norman and R.H. Williams, in *Handbook of Synchrotron Radiation*, Vol. 2, ed. G.V. Marr, Elsevier, Amsterdam (1987) Chap. 7.
3. D. Grieg, B.L. Gallagher, M.A. Howson, D.S-L. Law, D. Norman and F.M. Quinn, *Mater. Sci. Eng.* **99** (1988) 265.
4. H. Wright, P. Weightman, P.T. Andrews. W. Folkerts, C.F.J. Flipse, G.A. Sawatzky, D. Norman and H. Padmore, *Phys. Rev. B* **35** (1987) 519.
5. R.G. Jordan, G.S. Sohal and P.J. Durham, *J. Phys. F: Metal Phys.* **16** (1986) L135.
6. P. Hofmann, S.R. Bare, N.V. Richardson and D.A. King, *Solid State Commun.* **42** (1982) 645.
7. S.R. Bare, K. Griffiths, P. Hofmann, D.A. King, G.L. Nyberg and N. V. Richardson, *Surf. Sci.* **120** (1982) 367.
8. D. Spanjaard, C. Guillot, M.C. Desjonquéres, G. Tréglia and J. Lecante, *Surf. Sci. Rep.* **5** (1985) 1.
9. K.G. Purcell, J. Jupille and D.A. King, in *Surface Sciences*, Vol. 14, Springer-Verlag, Berlin (1988) p. 477.
10. J. Jupille, K.G. Purcell and D.A. King, *Solid State Commun.* **58** (1986) 529.
11. K. Griffiths, C. Kendon, D.A. King and J.B. Pendry, *Phys. Rev. Lett.* **46** (1981) 1584.
12. C.F. McConville, D.L. Seymour, D.P. Woodruff and S. Bao, *Surf. Sci.* **188** (1987) 1.
13. K.J. Mackey, P.M.G. Allen, W.G. Herrenden-Harker, R.H. Williams, C.R. Whitehouse and G.M. Williams, *Appl. Phys. Lett.* **49** (1986) 354.
14. D. Norman, *J. Phys. C: Solid State Phys.* **19** (1986) 3273.
15. D. Norman, S. Brennan, R. Jaeger and J. Stöhr, *Surf. Sci.* **105** (1981) L297.
16. G.M. Lamble, R. Brooks, S. Ferrer, D.A. King and D. Norman, *Phys. Rev. B* **34** (1986) 2975.
17. G.M. Lamble, R. Brooks, D.A. King and D. Norman, *Phys. Rev. Lett.* **61** (1988) 1112.
18. R. McGrath, I.T. McGovern, D.R. Warburton, G. Thornton and D. Norman, *Surf. Sci.* **178** (1986) 101.
19. D.P. Woodruff, D.L. Seymour, C.F. McConville, C.E. Riley, M.D. Crapper, N.P. Prince and R.G. Jones, *Surf. Sci.* **195** (1988) 237.
20. J. Stöhr and R. Jaeger, *Phys. Rev. B* **26** (1982) 4111.
21. J. Stöhr, J.L. Gland, W. Eberhardt, D. Outka, R.J. Madix, F. Sette, R.J. Koestner and U. Döbler, *Phys. Rev. Lett.* **51** (1983) 2414.
22. K.M. Conway, J.E. MacDonald, C. Norris, E. Vlieg and J.F. van der Veen, *Surf. Sci.* **215** (1989) 555.
23. E. Vlieg, A.W. Denier van der Gon, J.F. van der Veen, J.E. MacDonald and C. Norris, *Phys. Rev. Lett.* **61** (1988) 2241.

10 Protein crystallography

I.D. GLOVER and J.R. HELLIWELL

10.1 Introduction

This chapter will outline the benefits to macromolecular crystallography of exploiting synchrotron radiation. The use of central synchrotron radiation facilities is now an intrinsic part of the research projects in a large number of protein crystallographic groups. The exploitation of the characteristics of synchrotron radiation—very high intensity, fine natural collimation and an almost continuous spectrum in the hard X-ray region—have led to significant progress in many areas of macromolecular crystallography which may be summarised as follows:

(1) Protein crystallographic data collection, including
   (a) the reduction of radiation damage,
   (b) the study of large unit cells,
   (c) the use of small crystal volumes.
(2) Anomalous scattering and phase determination.
(3) Time-resolved crystallography e.g. of enzyme intermediate states.
(4) The study of molecular motion via diffuse scattering.

All the above applications can be studied using incident monochromatic radiation, usually focussed in one or two dimensions. However in some cases, for example the reduction of radiation damage and more particularly in time-resolved crystallography, the use of polychromatic data collection is yielding promising results. This technique utilises the intensity and collimation of the synchrotron radiation beam combined with the large wavelength spread necessary for Laue data collection.

10.2 Instrumentation for protein crystallography

In general, the macromolecular crystallographer is concerned with the collection of monochromatic data from small crystals, usually of dimensions less than $(0.5\,mm)^3$. Synchrotron radiation is characteristically emitted as a very intense broad fan of radiation with a continuous wavelength spectrum in the X-ray region. X-ray optics, particularly monochromators for wavelength selection and mirrors for focussing and/or harmonic wavelength rejection, are therefore important considerations in the design of facilities for macro-molecular crystallography. The data collection medium, or detector, is also a major design consideration. X-ray sensitive film has been the mainstay detector for many years and still has favourable characteristics although electronic area detector systems are now increasingly used.

10.2.1   X-ray optics

The fundamental optical requirements for a doubly focussed X-ray diffraction system have already been outlined in Chapter 1. In this section we consider the particular needs for macromolecular crystallography. These include require-ments for anomalous dispersion and also for Laue diffraction.

The simplest optical arrangement is that of a bent plane mirror with bent single crystal (see Sections 1.4 and 4.2). Despite its simplicity, the combination provides a point focussed monochromatic beam at the sample with reasonable spectral resolution. This mode is very suitable when substantial demagnific-ation of the horizontal source is needed to focus the beam to a point of the order of the dimensions of a typical crystal [1–3]. When short wavelengths are available, as on a wiggler beam line, this arrangement can be optimised to use 0.9 Å radiation. This wavelength has the advantage of low sample absorption and air scatter. It also enhances absorption sensitivity due to Br in the photographic film.

Tunability can also be provided in this arrangement but this requires refocussing of the crystal and, for a substantial change in wavelength, a diffe-rent crystal. For the rapid tunability and high spectral resolution required for optimised anomalous dispersion, the single crystal geometry must be aban-doned in favour of a double crystal monochromator [4]. Mirror focussing may be used but at the expense of spectral resolution, unless the mirror is placed 'downstream' of the monochromator. In conjunction with plane double crystals, a bent plane mirror will give a line focus whilst toroidal mirrors offer point focussing. As discussed in Chapter 1, double crystal monochromators have, however, been designed with a saggital bending mechanism, usually on the second crystal to give a horizontal focussing option [5].

Polychromatic Laue diffraction is a significantly different application and has required different approaches [6,7]. In its purest form the full white beam is incident on the sample with no optical elements inserted in the synchrotron radiation beam. Point focussing of the white beam may, however, be achieved using a toroidal mirror. In the narrow band pass Laue method, two approaches are used, both giving a beam with a band pass, $d\lambda/\lambda$, of ~0.2 at the sample. Firstly in a 'coventional' synchrotron radiation beam, a broad band pass monochromator based on a multilayer synthetic microstructure device (see Section 1.3.4) may be used. More recently the spectral structure of an undulator insertion device has been used [7].

10.2.2 Detectors

We turn now to the detectors suitable for synchrotron radiation macro-molecular diffraction. These have been briefly discussed in Chapter 1. The following account highlights the special requirements for protein crystallography.

X-ray sensitive film (see Section 1.5.4) has traditionally been used at synchrotron radiation centres as the simplest and most flexible area detection system. Films with small grain size but with reasonable speed and low intrinsic chemical fog levels are available, which make them suitable for even high resolution data collection where signal-to-noise considerations are paramount. Photographic film acts as an integrating detector but is off-line, that is all data analysis is carried out remotely and gives no real time control over the diffraction experiment. In order to match the dynamic range of the diffraction pattern at different X-ray wavelengths, several films in a pack may be required, from two or three films at 1.5 Å to three films with interleaving metal foils at wavelengths below 1 Å.

Electronic area detection systems have been developed to overcome the shortcomings of off-line detectors, both in terms of information per incident photon and in real time control of the diffraction experiment. They comprise multiwire proportional counters (MWPC), TV-based detector systems and CCD detectors. Another development is the image plate.

MWPC systems have been reviewed in Section 1.5.4. The electronic noise level is generally low but both global and local count rates limit peak intensities that may be recorded. Low intrinsic noise and counting statistics make the devices more efficient than film. The spatial resolution is limited by a number of factors which include wire spacing or interpolation accuracy and parallax. This limits either the unit cell size or Bragg resolution obtainable in a diffraction experiment. This can be improved, though, if multiple detector arrays are available [9]. Parallax becomes a problem on flat detectors at high scattering angles. This may be overcome by pressurising the detector or by incorporating a spherical drift chamber [5, 8].

Despite these limitations, MWPC detector systems are extremely well suited to optimised anomalous dispersion experiments (see later) where intensity is sacrificed in favour of high spectral resolution and where counting statistics are needed.

TV detector systems are integrating devices and therefore do not offer counting statistics. However, they operate on-line offering real time control of a diffraction experiment. One particular example, the Enraf-Nonius FAST [10] is illustrated in Figure 10.1. The detector combines an aperture and

Figure 10.1.   View of the area detector diffractometer (the FAST) mounted on the SRS wiggler protein crystallography workstation at Daresbury. The detector is mounted on its side.

spatial resolution well matched to protein diffraction patterns particularly when hard X-rays are used. They exhibit very good count rates and a detective quantum efficiency (DQE) far better than photographic film and have proved useful in the collection of high resolution data and with very large unit cells [11].

CCD detectors are likely to become important in the future as imaging devices for diffraction. Their size is currently limited to $(10 \times 10 \, \text{mm})^2$ although larger devices can be made. Typical pixel sizes are of the order of $20 \, \mu\text{m}$. Strauss *et al.* [12] and Naday *et al.* [13] are developing a CCD-based system for protein crystal data acquisition. Allinson *et al.* [14] have used a CCD to look at a portion of a Laue protein diffraction pattern for on-line measurements.

Image plates are attracting considerable interest in the crystallographic community as a two-dimensional detector system. They are an analogue integrating device similar to film in some ways but have some very specific advantages. The plates have a far wider dynamic range of 10^5, do not suffer from the Wooster effect, have higher absorption efficiency throughout the X-ray region and lack background fog which results in a higher DQE [15]. They are also re-usable.

For a full discussion and review of instrumentation employed at synchrotron radiation centres see [16] and the references therein. Arndt [17] has reviewed the characteristics and performances of detectors commonly used in macromolecular crystallography.

10.3 Use of the high intensity and collimation of synchrotron radiation in protein crystallography

10.3.1 *Fundamentals of the protein crystallographic technique*

Bragg's law predicts the angle of reflection of any diffracted ray from specific atomic planes whereby

$$n\lambda = 2d \sin \theta$$

where d is the interplanar spacing of that set of planes, λ, is the wavelength of the X-rays and n is an integer. The closer the separation of the planes then the larger the value of θ, the diffraction angle for a given wavelength; this corresponds to the higher resolution data.

The intensity measured for a given reflection is proportional to $F(hkl)$ where

$$F(hkl) = \sum_j f_j \exp\{2\pi i(hx_j + ky_j + lz_j)\}$$

where f_j is the atomic scattering factor for X-rays from the jth atom of coordinate (x_j, y_j, z_j) expressed as fractions of the cell a, b, c. This is the structure factor equation and the Fourier inverse of $F(hkl)$ is the electron density $\rho(xyz)$

$$\rho(xyz) = \frac{1}{V}\sum_h \sum_k \sum_l F(hkl) \exp \alpha(hkl) \exp\{-2\pi i(hx + ky + lz)\}$$

If the amplitude and phase of the structure factor are known for all hkl planes or reflections, then the electron density can be calculated for all points (x, y, z) in the cell and so the crystal structure is then solved. Of course it is impossible to measure all h, k, l reflections so the summation is usually terminated with a finite number of terms at a certain resolution limit known as the Bragg resolution, i.e. $\lambda/2 \sin \theta_{max}$.

For an ideally mosaic crystal rotating with constant angular velocity, ω, through the reflecting position, the total diffracted energy $E(hkl)$ may be written:

$$E(hkl) = \frac{e^4}{m^2 c^4 \omega} I_0 \lambda^3 PLA \frac{V_x}{V} |F(hkl)|^2$$

where λ is the wavelength of the incident beam of intensity I_0 the crystal volume is V_x, the unit cell volume is V, P is a polarisation factor which depends on the state of polarisation of the incident beam and L is the Lorentz factor which takes into account the relative time each reflection spends in the reflecting position; e, m and c have their usual meanings. From Darwin's equation the value of synchrotron radiation in crystallographic data collection can now be clearly seen. The extremely high intensity available can be used to overcome the inherent weakness of individual reflections and the large time required to collect the voluminous datasets from proteins. These considerations are especially serious in the case of crystals with large unit cells such as virus crystals. Small samples give weak diffraction patterns. Moreover, all proteins suffer to a greater or lesser extent from radiation damage. High resolution data is affected the most. The fall-off in the atomic scattering factor and atomic thermal vibration makes the high angle reflections weaker. In addition radiation damage disrupts the diffraction pattern at atomic levels of resolution most severely.

10.3.2   Reduction of radiation damage

Radiation damage is usually manifest in a time-dependent degradation of the crystal diffraction pattern. This affects the high resolution data first and progresses to lower resolution with time. The exact mechanisms of damage are unclear. It has been found that the high intensity of synchrotron radiation beams and hence short exposure times allow significant quantities of data to be collected before radiation damage becomes limiting.

The reduction of radiation damage has several benefits. Firstly the initial resolution, or maximum Bragg angle, that can be observed is usually improved with synchrotron radiation. It is not that this data is absent when a conventional source is used, but that with synchrotron radiation statistically significant data can often be collected before radiation decay occurs. The extension of data resolution may also be due, in part, to the well collimated geometry of the synchrotron radiation beam. Because of this, the high angle reflections collected as spots on a film can be smaller for synchrotron radiation than for a conventional source, thus making the average optical density recorded stronger and so more statistically significant. Finally, more data per sample crystal can often be collected with synchrotron radiation to the extent that a complete dataset may be obtained where this was not previously routine with a conventional X-ray source.

The benefit of high resolution data is obvious once electron density maps are calculated. The amount of detail increases dramatically, the accuracy of fitting and refinement improves and a detailed analysis of the macromolecular geometry and interactions can be made more confidently as atomic resolution is approached. Figure 10.2 shows a portion of the electron density map from the haem unit of haemoglobin calculated at 1.5 Å resolution from data collected at the Daresbury SRS using the short wavelength (0.9 Å) and high intensity of station 9.6 (Liddington and Dodson, unpublished data, 1988).

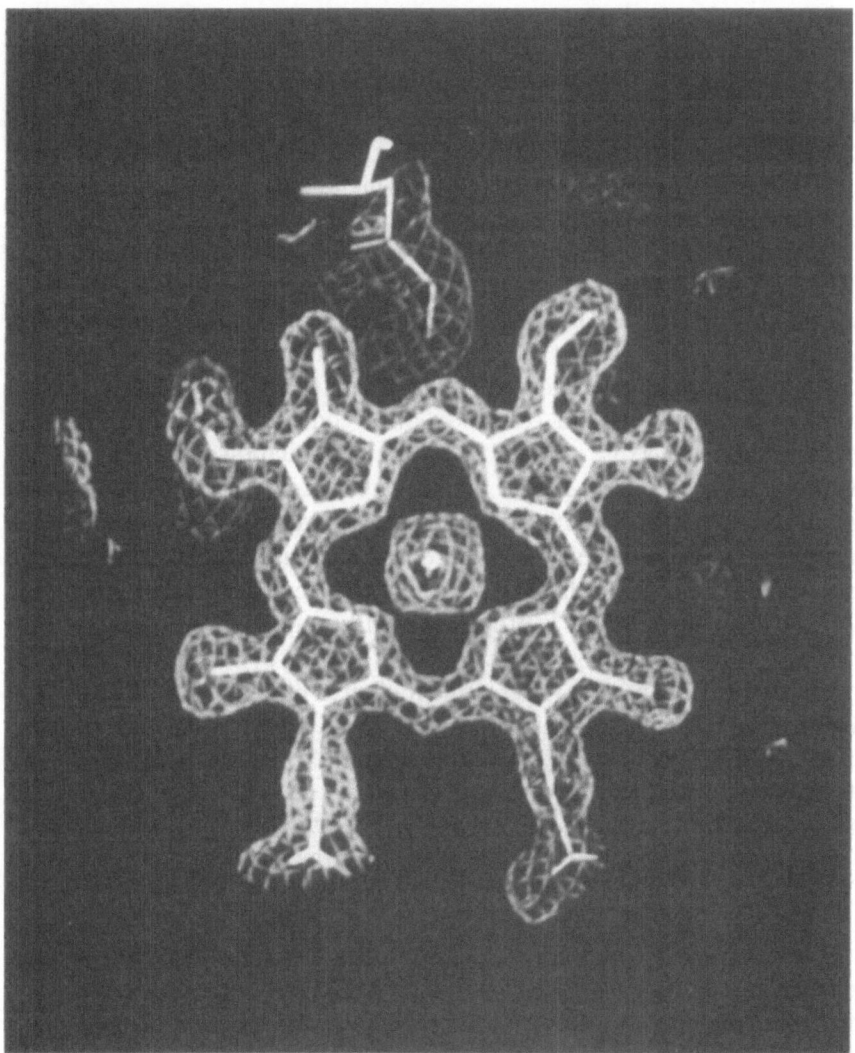

Figure 10.2. A portion of the 1.5Å electron density map of the haem unit in haemoglobin (Liddington *et al.*, unpublished).

10.3.3 *Large unit cells: virus crystallography*

The combination of all the advantages of synchrotron radiation listed above is especially needed in virus crystallography where the unit cells are very large and the crystals often small. Data collection from very large unit cell crystals benefits from the fine natural collimation of synchrotron radiation. Despite this, exposure times can still be fairly long and in some cases only one data photograph can be collected per sample crystal due to radiation damage problems.

Plant and animal viruses are very large macromolecular assemblies, consisting of a protein coat, the capsid, enclosing a core of nucleic acid. Molecular weights of intact viruses are often greater than 10^6 and structure determination would appear to be a daunting task. The number of virus structures solved to date is, however, growing due to major advances in the handling and processing of data and the impact of synchrotron radiation facilities.

Recently Rossmann et al. [18] solved the structure of human rhinovirus (a picorna virus responsible for the common cold) in little more than 12 months. Nearly a million reflections were collected on the protein crystallography facility at the Cornell Synchrotron source in a matter of days. This conveyed a speed advantage over data collection on a conventional source and also ameliorated an otherwise impossible problem of radiation damage when long exposure times were used. The far greater rate of radiation damage of picorna virus crystals compared with crystals of plant viruses is symptomatic of an inherently less stable protein capsid. The capsid consists of an icosahedral shell made up of 60 copies each of four proteins, enclosing the genetic material, RNA, which constitutes 30% of the molecular weight.

The processing of the synchrotron radiation data by Rossmann et al. [19] was specifically adapted to the treatment of virus diffraction data. They were also able to use the intrinsic symmetry of the virus to provide phasing information in the final stages of the analysis. Low resolution phasing was provided via two heavy atom derivative datasets again collected using synchrotron radiation. The high resolution phasing was carried out using real space molecular replacement whilst extending the resolution in small steps. The final skew-averaged map allowed the location of 811 out of 855 residues in four distinct polypeptide chains [18].

Another important technical development has been the use of short wavelengths at the SRS wiggler workstation [3] for the study of foot and mouth disease virus (FMDV). The combination of large unit cell and small crystal volume caused problems of signal-to-noise which were resolved by using a short wavelength of 0.9 Å, the weaker diffraction being tolerated in order to reduce the background with long crystal to film distances. The short wavelength also reduced radiation damage. Even so only one exposure per FMDV crystal was possible. With no assessment of crystal orientation possible, data photographs had to be taken in a largely random setting and a dataset slowly built up from a large series of exposures. Once complete, however, the data were used to solve the structure, phasing initially by molecular replacement using the rhinovirus model averaged with the mengo virus model at low resolution. This was followed by phase extension and symmetry averaging to obtain the final electron density map. Figure 10.3

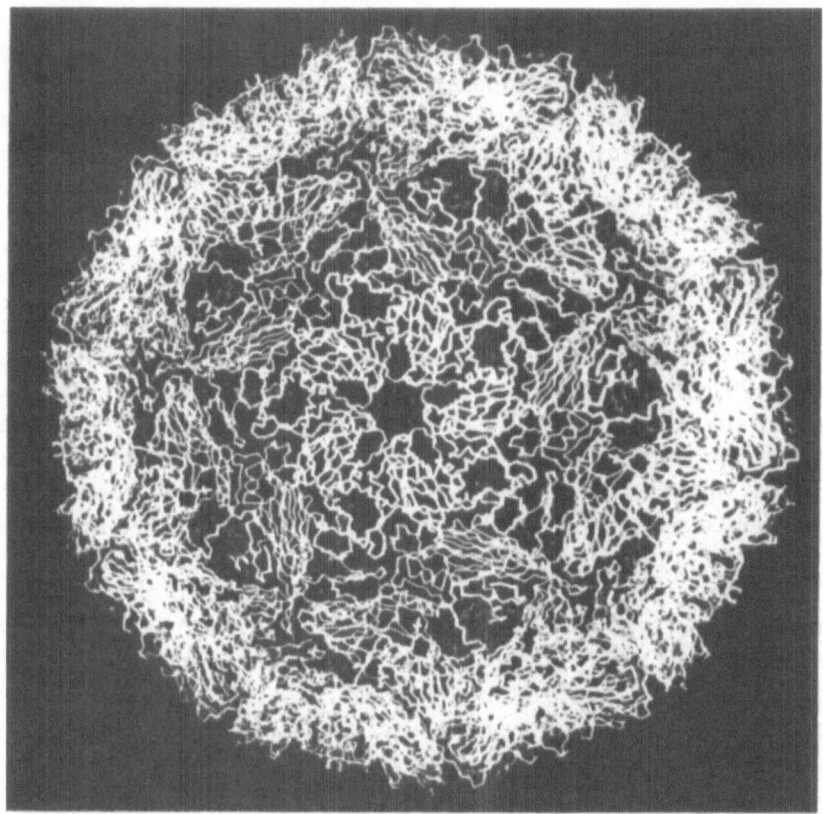

Figure 10.3.   A view of the foot and mouth disease virus structure [20].

shows a view of the viral capsid down the five-fold symmetry axis at a resolution limit of 2.8 Å [20].

The success of these new methods allows rapid structure determination of viruses, particularly the improvement of crystal lifetime using short wavelengths [20, Liddington *et al.* unpublished, Rossmann *et al.* unpublished]. Moreover, since an intact animal virus can now be crystallised, the surface proteins involved in immunogenic interactions can be viewed directly at the site of binding. This is very exciting and future studies will include other important pathogenic animal viruses and even more complex structure determinations of exceptional biological interest.

The new generation of synchrotron sources, e.g. the ESRF at Grenoble, offer the opportunity of larger unit cell data collection using X-ray undulator beam lines.

10.3.4 *Small sample volume*

Small sample volume (Vx) affects the strength of the diffraction pattern as seen above from Darwin's formula. High intensity can be used to overcome this problem. Additionally to obtain a reasonable signal-to-noise ratio the X-ray background must be minimised. This is especially true when the scattering is weak. The fine collimation of the beam is the only solution to this problem. (See also Chapter 4 where the analogous problem in SAXS is examined.)

Many proteins currently selected for structural study crystallise only with small sample volume. This is quite often true in the case of novel engineered mutants or otherwise commercially interesting proteins. Sample volumes as small as $(20 \, \mu m)^3$ have yielded very strong diffraction patterns with the white synchrotron radiation beam [21]. In this application area efforts to reduce the X-ray background at the detector include replacement of traditional glass capillaries with thin film capillaries as the sample holder. The lifetime of such tiny protein crystals is poor and experiments need to be carried out at low temperatures to prolong sample lifetime.

Andrews *et al.* [22] have exploited the advantages of synchrotron radiation using the wiggler protein crystallography workstation at the SRS [3] to solve the structure of a poorly ordered silicate crystal using monochromatic data collected from a crystal of just $18 \times 8 \times 175 \, \mu m^3$. The crystal was mounted dry, in air. The data collected using a beam of 0.2 mm diameter and a wavelength of 0.88 Å to improve signal-to-noise. The Enraf-Nonius FAST television area detector was used. The refined mosaic spread was high (2–3°) and the processed data of relatively low quality. The structure, however, was solved using direct methods and refined to crystallographic R value of 10.3%. Although this was a small molecule, the result is an encouraging example of what can be achieved in the field of data collection from extremely small volumes, albeit of a poorly ordered crystal. Some very small crystals have a high mosaic spread which limits the resolution limit unless deconvolution is done (Rizkallah and Harding, pers. commun.).

10.4 Anomalous scattering and phase determination

The solution of the phase problem is central to protein crystallography. Classically this has been solved using the technique of multiple isomorphous replacement (MIR), whereby a series of heavy atom derivative crystals are prepared and the intensity differences between native and derivative diffraction patterns used to derive phases. The method does, however, depend critically on the preparation of derivative crystals which are isomorphous with the native form and ideally with the heavy atom bound to a small number of high occupancy sites. Lack of strict isomorphism, the existence of multiple or disordered heavy atom sites limits the resolution and accuracy of phase determination [23].

A separate approach to phasing is provided by the technique of anomalous scattering. The phenomenon (discussed also in Chapters 2 and 6) arises from resonance effects due to the fact that electrons, particularly from K and L shells, are tightly bound in the atom and scatter differently from free electrons (Thompson scattering). Core level electrons are excited by the incident X-radiation and as the resonant frequency (elemental absorption edge) is approached the atomic scattering factor becomes complex, i.e.

$$f = f + f' + if''$$

where f' and f'' are the dispersion (real) and absorption (imaginary) components of the resonance induced scattering. The exact energy for resonance, the absorption edge, clearly depends on the type of atom and as shown in Figure 10.4, the magnitudes of both f' and f'' vary greatly in the region of a K or L shell absorption edge. The presence of an f'' component in the total Bragg scattering leads to a breakdown of Friedel's law, i.e.

$$I(hkl) \neq I(\overline{hkl})$$

which can be used not only to determine absolute configuration, but also phase angles [24]. The anomalous scattering of metals in protein heavy atom derivatives has been used to determine heavy atom positions by use of anomalous difference Pattersons [25], or to supplement isomorphous derivative information in phase determination [26–28]. When only a single heavy atom derivative may be prepared, the anomalous scattering information may be used to break the inherent two-fold ambiguity in the phases calculated from a single derivative. Without recourse to phase improvement procedures such as that of Wang [29], however, this approach has been limited to small proteins [30]. Anomalous scattering information alone may also be used in phase determination particularly where an anomalously scattering atom is naturally present in the protein molecule. The method was first discussed by Herzenberg and Lau [31], who suggested the use of naturally occurring sulphur within proteins; it has since been used successfully in phasing the structure of crambin [32]. Smith and Hendrickson [33] also used the method for phase determination of trimeric haemerythrin, where they took advantage of the anomalous scattering of the iron atoms, resolving the two-fold ambiguity in the phases, in both cases, by taking the phase closest to that of the anomalously scattering atoms. In all these cases data were collected at the characteristic wavelength of a conventional X-ray source and as such are far removed from the elemental absorption edge. The residual f'' effect was extracted by painstaking data collection and subsequent processing. f' effects would be inaccessible to the majority of such experiments. If, however, the incident X-ray energy is selected or tuned to a particular absorption edge anomalous scattering (absorption) effects may be maximised. This is becoming

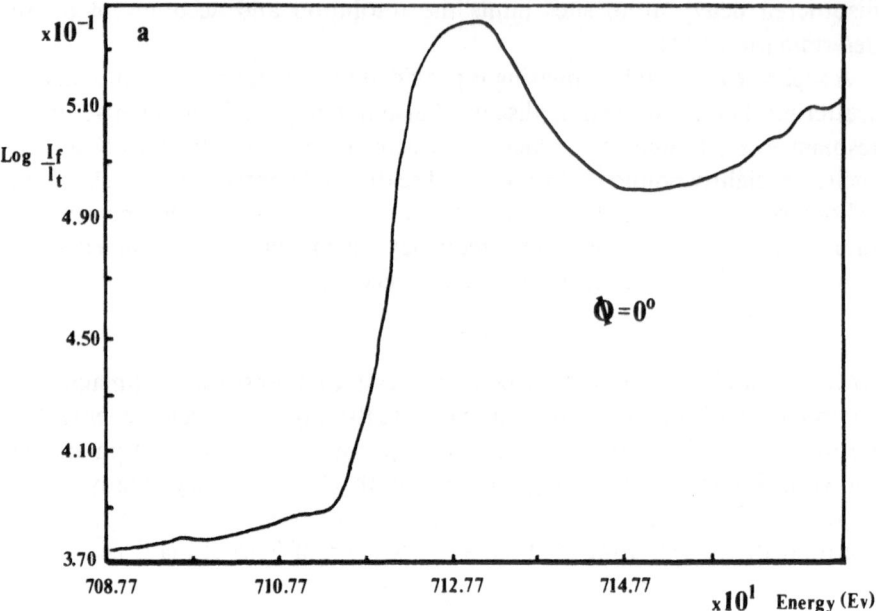

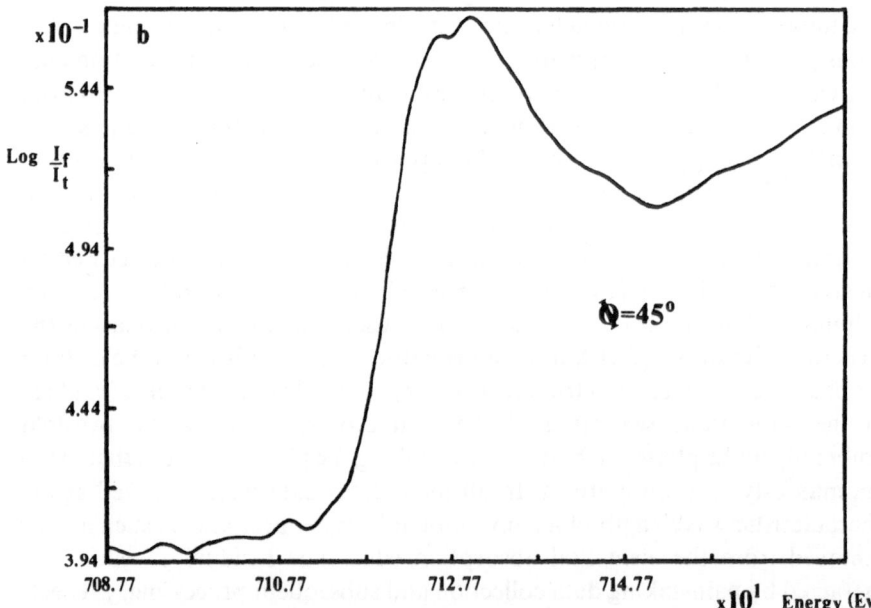

Figure 10.4. (a) Recorded XANES spectrum from a single crystal of an Fe metalloprotein using the very high energy resolution optics on station 8.1 at Daresbury. (b) As in (a) but at a different crystal orientation indicating dichroism.

an increasingly useful technique in the collection of heavy atom derivative data with significant anomalous scattering recorded on the very hard X-ray wiggler beam line at the Daresbury SRS. This approach is given the acronym MIROAS, multiple isomorphous replacement with optimised anomalous dispersion.

10.4.1 *Multiple wavelength anomalous diffraction methods*

A distinct approach to the problem of phase determination using anomalous scattering was suggested by Okaya and Pepinsky [34], and discussed by Ramaseshan [35], who pointed out that data collected at several wavelengths close to an absorption edge could be used to derive an unambiguous phase estimate. The principle of the so-called multiple wavelength anomalous diffraction (MAD) approach lies in being able to select and collect data at a series of well-defined wavelengths which optimise the anomalous scattering contributions and their relative values. For a crystal containing an anomalous scatterer, the net intensity of each Bragg reflection will be energy or wavelength dependent and this variation may be used to solve the phase problem in a manner analogous to, but without the inherent problems of, the isomorphous replacement method. The availability of synchrotron radiation sources with an almost continuous spectrum in the X-ray region (~ 2–0.5 Å) has recently stimulated a great deal of interest in the technique, allowing the sampling of the K shell absorption edges of metals such as Cu, Fe, and Zn often naturally occurring in proteins as well as heavy atoms commonly used in derivative preparation, e.g. Hg, Au and Pt.

Previously two wavelength experiments using conventional Ni and Cu target X-ray sources had been shown to be useful for the Fe edge in the pioneering work of Hoppe and Jakubowski [36]. Cascarano *et al.* [37] extended the approach of Singh and Ramaseshan [38] to show that with two wavelengths, either both on the same side or with each on either side of an absorption edge, direct methods may be used to determine the position of the anomalous scatterer. Karle [39, 40] has set out an algebraic analysis of the multiple wavelength approach. Woolfson [41] has set out a procedure for phase determination using anomalous scattering and Kahn *et al.* [42] have obtained by an independent method an electron density map for a terbium derivative of parvalbumin using data collected at three wavelengths about the terbium absorption edge. A similar experiment carried out by Harada *et al.* [43] used the anomalous scattering from the native haem iron atom in cytochrome *c'*. In all cases the implicit assumption was made that the wavelength independent component of the structure amplitude was known. Hendrickson [44], Fourme *et al.* [45] and Tickle and Glover [46] have developed the method of Karle [39, 40] to obtain phase information. This has involved a reformulation of the structure factor equation involving the separation of wavelength dependent and wavelength independent factors. Once protein phases are known, anomalous scattering information may be used to obtain accurate metal–metal distances in proteins or with multiple wavelength data to distinguish between metal atoms of similar atomic weight within a protein [47, 48].

The MAD experiment is technically and experimentally very demanding and has required innovations in data collection and analysis. In principle the experiment consists of the collection of a series of datasets (normally 3–6), including Bijvoet pairs, in the vicinity of an absorption edge. X-ray wavelengths are selected so that they optimise not only Bijvoet (f'') but also dispersion (f') difference. Even when maximised or optimised the anomalous contributions to the total scattering are small, typically 5–10 electrons (f') in the case of K shell elements, giving changes of only a few percent in the observed Bragg intensities. Any approach to the determination of phases using the MAD technique must aim at accuracy and the reduction of systematic errors. Success depends on the accurate estimation of small intensity differences. There are several important experimental pre-requisites. High quality X-ray optics are required particularly monochromator systems with high spectral (wavelength) resolution, rapid tunability with high and reproducible setting accuracy. Systematic errors will arise principally from sample to sample variation, radiation damage (due to its effect on data recorded at different times), sample absorption and the lack of sufficient energy resolution in the monochromator system as well as non-uniformity of response and instabilities in the area detector. The need for accuracy in the recorded intensities has prompted the widespread use of electronic detection systems and the development of sophisticated data collection strategies aimed at the reduction of systematic errors.

Optimisation of the wavelengths at which data are collected is crucial; the phasing power will depend on maximising both the f' and f'' contributions to various data sets. The absorption edge defines the points of maximum $|f'|$ and f''. The absolute position of the absorption edge will depend on both the ionisation state and environment in which the anomalous scatterer is found. Hence measurement of the absorption edge is carried out using a fluorescence detection system for the crystal in situ on the diffractometer (see Figure 10.4(a)). Once wavelengths are selected, data collection is carried out using a strategy whereby a small segment of data is recorded, with Bijvoet mates, at each wavelength before proceeding to the next segment of data. Strategies such as this minimise the effects of radiation damage and absorption differences on equivalent data recorded at the different wavelengths. In principle, data collected at three wavelengths will give a unique phase determination but in practice, more normally 4–5 wavelengths are collected.

Considerable success has been achieved recently such as the determination of the structure of lamprey haemoglobin by MAD methods at 3 Å resolution [50] and in the application of the technique to the ab initio structure determination of protein structures (reviewed by Moffat [49]). Besides the obvious candidates for such an approach, the metalloproteins, the technique may also be applied in the case of a protein where only a single derivative may be prepared. A particularly exciting area is that developed by Cowie [51] and exploited by Hendrickson involving the engineering and expression of seleno-methionyl proteins. The use of selenium as an anomalous scattering centre has potentially widespread application as a general vehicle for protein structure determination. Hendrickson *et al.* [52] have used the Se anomalous scattering from a seleno-methionyl modified streptavidin using data collected at five wavelengths about the selenium absorption edge. Data were collected at the Photon Factory, Japan using a four circle diffractometer with single counter. The initial MAD phased electron density maps were of sufficient quality to be interpreted in terms of the protein sequence [52]. Guss *et al.* [53] have used the MWPC protein crystallographic facility at SSRL, Stanford, in the determination of the structure of a basic blue copper protein (CBP) from cucumber seedlings, a 12 500 molecular weight protein containing a single copper atom, using the MAD technique. They collected data at four wavelengths about the Cu absorption edge and using the software package of Hendrickson, produced electron density maps at 2.5 Å resolution. These maps have been successfully interpreted. This structure solution illustrates the power of the MAD technique over MIR methods. Although the crystallisation of CBP had been carried out several years before, all attempts at the production of heavy atom derivatives had failed and until the MAD method was attempted no progress could be made with structure solution.

When the technique is applied to larger proteins the relative contribution of the anomalous scattering contribution decreases leading to more stringent experimental requirements. In certain cases the absorption edge itself may exhibit dichroism, with the absorbtion edge structure changing as the orientation of the crystal is varied (see Figure 10.4(b)). These effects lead to the edge position changing due to the strong orientation of ligands. Templeton *et al.* [54] show this effect clearly in their high resolution investigation of absorption edges in uranyl compounds. These factors and their effects on the subsequent phase calculation will not be detected unless very high resolution monochromator systems are exploited.

10.5 Time-resolved crystallography

Structural studies of intermediate states in dynamic processes are one of the more important general applications of synchrotron radiation. Examples in biological studies include the visualisation of a fully productive enzyme–substrate complex or the pathways involved in a structural transition in a molecule. Such time-resolved information would fundamentally improve our understanding of biological structure function relationships. An experimental obstacle to be overcome is the ability to collect crystallographic data in time scales short enough to monitor various stages in such a dynamic process without recourse to stopping the action by flash freezing or chemical shock. The extremely high intensity of synchrotron radiation sources, particularly radiation from insertion devices, now allow data collection in the millisecond or even very much lower time regimes. Two general approaches have developed: the first is aimed at the collection of representative monochromatic data in short time scales, as pioneered by Bartunik (see below) and exploited in different forms by several groups; and the second uses polychromatic Laue diffraction methods to collect complete datasets or significant fractions of datasets in the subsecond time regime.

Bartunik *et al.* [55] collected monochromatic test data on a millisecond time scale on carbonmonoxy myoglobin, where structural changes were induced by the debinding of the CO ligand by a laser. A small subset of reflections sensitive to the structural changes was collected on an electronic linear detector and the relative intensities during the course of CO debinding analysed.

As noted, the alternative approach is the polychromatic and narrow band pass synchroton radiation Laue method of data collection. This method is yielding exciting results. In its purest form, the technique allows the complete synchrotron radiation spectrum to be incident on the sample crystal. The incident intensity of a polychromatic synchrotron radiation beam is several orders of magnitude higher than the monochromatic beam and allows a concomitant reduction in exposure times.

10.5.1   *Laue crystallography*

The availability of synchroton radiation X-ray sources has renewed interest in the Laue diffraction method. This method directly exploits the polychromatic nature of synchrotron radiation whereby the integration of reflected intensities is done over wavelength. Preliminary studies of Laue diffraction from protein crystals [3, 16, 21, 56, 57, 58, 59, 60] and small inorganic crystals [59, 61] suggested that the Laue method possessed advantages over conventional monochromatic data collection strategies for certain experiments. It makes optimum use of the entire synchrotron radiation spectrum and the very high intensity of the white beam affords a reduction in exposure times of several orders of magnitude. The Laue method thus permits very brief exposure times in the millisecond time scale from a strongly scattering protein crystal [55, 56, 74] and the examination of microcrystals [21]. The stationary crystal yields integrated intensities directly that are relatively insensitive to small transient changes in unit cell dimensions. A typical Laue diffraction pattern contains many more reflections than a typical monochromatic photograph corresponding to a much larger volume of reciprocal space being sampled. In favourable cases of high space group symmetry and careful choice of beam direction with respect to the crystal axes a single exposure may yield the bulk of the unique data [63]. These advantages are particularly appropriate to dynamic experiments in which the diffracted intensities change rapidly with time in response to a structural perturbation.

Moffat *et al.* [56] used a modified, narrow band pass Laue technique $(d\lambda/\lambda < 0.2)$ with a synchrotron radiation beam from CHESS with successful results from myoglobin and haemoglobin crystals. Helliwell [16, 60, 64] exposed a pea lectin crystal in the full $(0.2 \text{ Å} < \lambda < 2.5 \text{ Å})$ SRS wiggler white beam and obtained a high signal-to-noise Laue pattern. It seems to be the case that the more robust protein crystals (where robust here is defined as a property of relatively good lifetime in the monochromatic beam) survive several (3–10) exposures to the white beam whereas less robust crystals only withstand a beam of limited wavelength range (narrow band pass Laue).

A fundamental complexity of the Laue method is the overlapping of various orders of a Bragg reflection which may be stimulated by the white X-ray beam. However Cruickshank *et al.* [65] show that even with an infinite wavelength spread, 72.8% of all Bragg reflections occur as singlets (spots consisting of a single reciprocal lattice point stimulated by a given wavelength in the available band pass). With more realistic experimental wavelength ranges this proportion increases to more than 83% and depends only on the ratio of the maximum to minimum wavelength incident at the sample and not on space group or limiting resolution. Hence multiple orders are not a serious limitation. However, lower resolution reciprocal lattice points do tend to occur in multiplet spots (Figure 10.5).

A further complexity is the need to avoid or minimise spatial overlap of neighbouring reflections in what are often very dense diffraction patterns. This is usually minimised by using very small collimated beam sizes of the order of 0.2 mm diameter or less. Despite reducing the incident flux this requirement does, however, improve the observed signal-to-noise characteristics of the Laue patterns in the following way: the amount of extraneous material (e.g. mother liquor, glass capillary) in the beam is limited and so the background scatter is reduced. Very short exposure times are preserved even with such a small incident beam and associated sample volumes because of the very high intensity. For unfocussed wiggler radiation this can be as high as 10^{14} polychromatic photons/s per mm^2 compared with a monochromatic focussed wiggler beam of 10^{12} photons/s per mm^2 or a rotating anode beam of 10^9 CuK_α photons/s per mm^2. Computational approaches involving deconvolution of spatial overlaps have been successful [66].

The interpretation of Laue patterns has required the development of new software for prediction, spot integration, film-to-film scale factor determination, harmonic spot unscrambling, the derivation of wavelength dependent correction factors and wavelength normalisation [57, 67, Szebenyi, Schildkamp and Moffat, unpublished]. The software is based on the already successful packages used for oscillation data processing [68, 69]. The

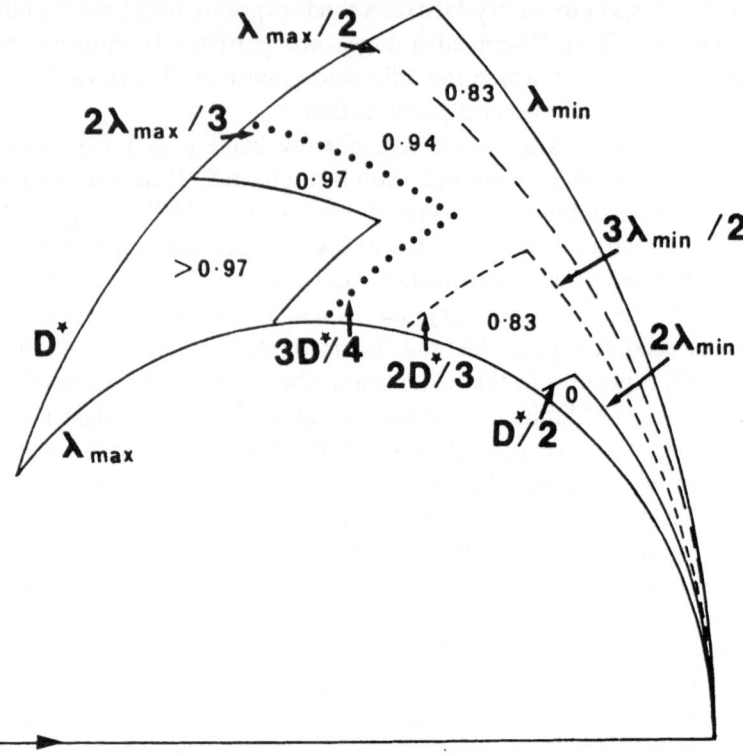

Figure 10.5. Ewald sphere construction in the Laue method. The numbers marked represent the probability that a given reciprocal lattice point would be measured as a single component spot. For full details see [65].

processed pea lectin Laue data compare favourably with monochromatic oscillation film data [70]. Hence, data of useful statistical quality can be collected in very short times. This is of potential use for the study of protein crystal intermediate states. Laue methods were exploited by Farber *et al.* [71] in the determination of the positions of the heavy atom binding sites in crystals of xylose isomerase. Helliwell *et al.* [6] have used Laue data, recorded with a focussing mirror, of a Hg protein crystal derivative to derive phases from isomorphous and anomalous data.

10.5.2   *Time-resolved studies of the phosphorylase enzyme*

The structural study of enzyme function in general would, ideally, involve the direct imaging in an electron density map of the enzyme substrate complexes. This approach has been used for the first time by Johnson, Hajdu and co-workers studying the enzyme glycogen phosphorylase b. The experiments used both monochromatic methods, aimed at minimising data collection times using the Daresbury SRS [62, 73], and more recently exploited the advantages of full white beam Laue data collection methods.

The use of full white beam data collection enabled this group to extend their structural investigation of the mechanism of phosphorylase b into a time regime where enzyme action in the crystal can be easily monitored. The geometry of the Laue method had additional advantages since a single exposure yielded 30% of all of the data to 3 Å resolution for the tetragonal crystal form of phosphorylase and the exposure times were very short.

In the experiment, the binding of maltoheptose to the glycogen binding site of phosphorylase was monitored. The crystal was kept in a thermostated flow cell and Laue datasets of three exposures (six films in a pack) of 1 s each were taken at different angular settings before, during and after the addition of maltoheptose. Each film in a film pack was scaled to its counterpart in the native Laue dataset and the measurements kept unmerged. The initial photographs related to the native protein. Fractional intensity changes detected in subsequent photographs with the same crystal orientation could then be used to calculate a difference Fourier map. Taking phases from the monochromatic data, the coefficients used were as follows:

$$\frac{(\text{F-Laue-deri} - \text{F-Laue-native})}{\text{F-Laue-native}} \times \text{F-mono-native}$$

where F-mono-native is the structure factor amplitude from a reference native dataset, F-laue-native is the corresponding amplitude measured on the starting native Laue photograph and F-Laue-deri the Laue measurement following the addition of the ligand. The difference Fourier map showed malthoheptose at the glycogen binding site and was of better quality than the equivalent monochromatic map (calculated with identical subsets of data). It clearly showed four of the seven sugar units (three presumably protrude into solvent and show increased mobility) and difference density due to side chain movements in the active site [73]. This is the first example of an electron density map being obtained from a protein crystal using the Laue method of data collection. The total data collection time per dataset was only 3 s.

The advantages of the difference Laue technique is that wavelength and position dependent corrections are not needed [55, 57] except in the case of significant change in the mass absorption coefficient during the course of the experiment (e.g. the binding of heavy atoms).

10.5.3 Structural transitions in insulin

Reynolds et al. [74] have used the polychromatic Laue diffraction method to investigate the structural transition of crystalline insulin from the 4Zn to 2Zn form. The start and end points of the transition are well characterised, the structures of both the 4Zn and 2Zn forms have been solved and the induction of the transition by halide ion concentration has been investigated in solution. Reynolds and co-workers have now extended this investigation to the crystalline state. In their experiments a 4Zn insulin crystal was removed from its mother liquor (containing 0.6 M NaCl) and mounted in a tube surrounded by, but not in contact with, the (halide free) 2Zn insulin mother liquor. A series of Laue exposures was then recorded over a period of 2–3 h. Their preliminary results showed first a sharp well-ordered 4Zn insulin Laue diffraction pattern for up to 40 min, then a loss in order of the crystal occurred with a much weaker and more diffuse Laue pattern accompanied by a series of 2 or 3 rings, indicating 2 or 3 distinct crystal cells or orientations. Finally, after about 135 min, the crystal re-annealed exhibiting a sharp well-ordered pattern which could only be interpreted in terms of a 2Zn insulin crystal. Although the results are still preliminary, they indicate that the transition from the 4Zn to the 2Zn form may be induced and observed in the crystal. The transition is between two similar structures but involves the change of the B1–B8 region from an extended chain conformation to an α-helix, in which the B1 residue moves by up to 25 Å. The single Laue patterns, recorded with an exposure time of typically 3 s, extend to 2.3 Å resolution for both well-ordered forms.

10.5.4   *100 ps data collection*

The ultimate time resolution accessible from a synchrotron radiation source is the time width of a single bunch (see Chapter 1). Moffat and co-workers have succeeded in collecting Laue diffraction data from strongly diffracting samples using radiation produced by a single electron bunch circulating in the CHESS ring. Due to the low intensity of bending magnet radiation, Moffatt *et al.* exploited instead the very high power undulator beam line at CHESS. The natural spectral characteristics of the first harmonic of the undulator provided a very intense beam with a band width $(d\lambda/\lambda)$ of approximately 0.2, well suited to the narrow band pass Laue technique. A series of shutters was used, synchronised to the circulating electron frequency to permit exposures to be recorded with the very brief pulse of radiation from a single bunch. A series of exposures was recorded on small molecules which showed good diffraction and finally they successfully recorded the diffraction pattern from a single crystal of lysozyme using an exposure time of only 100 ps [75].

10.6   Synchrotron radiation and diffuse scattering from protein crystals

Of great interest to the molecular biologist is the relationship between protein form and function. In recent years it has been realised that, although accurate static structural information is necessary, some appreciation of molecular flexibility and dynamics is essential. Classically this information has been derived from the crystallographic atomic thermal parameters and more recently from molecular dynamics simulations (see for example, [76]) which yield independent atomic trajectories. A characteristic of protein crystals, however, is that their diffraction patterns only extend to quite limited resolution, even employing synchrotron radiation. This lack of resolution is especially apparent in medium to large proteins where diffraction data may only extend to 2 Å or worse, thus limiting any analysis of the protein conformational flexibility from refined atomic thermal parameters. It is precisely these crystals where flexibility is likely to be important in the protein function.

The diffuse scattering background which nearly always occurs in a protein diffraction pattern may arise from several sources including thermal diffuse scattering, static disorder scattering, solvent disorder, Compton scattering, fluorescence, scattering from mounting tubes, air scattering and intrinsic film fog. The static or dynamic displacement of atoms in crystals causes a breakdown in translational symmetry, leading to a reduction in the Bragg intensities at high resolution and the concomitant appearance of diffuse scattering at and between the reciprocal lattice positions. In macromolecular crystals the diffuse scattering is often quite strong, rich in detail and apparently quite distinct to the crystal form. It represents a potentially rich source of information regarding atomic displacements.

Static disorder arises when unit cells exist with different arrangements of the time-averaged positions. Orientational disorder occurs in molecular crystals where molecules or flexible domains or side groups may take up different positions breaking down the translational symmetry. Dynamic disorder arises from thermal vibrations and is present in crystalline materials. Two types of lattice vibrations may be distinguished, acoustic modes due to the propagation of ultrasonic waves in the crystal and optic modes of vibration such as are observed in infrared and Raman spectra. Ultrasonic vibrations give rise to thermal diffuse scattering which peaks at the reciprocal lattice positions and is observed characteristically as haloes around the Bragg peak. Optic mode vibrations along with other disorder modes give rise to continuous diffuse scattering which is distributed continuously but non-uniformly throughout reciprocal space and is often observed as lines or streaks in a diffraction pattern and is often strongly oriented.

The measurement of diffuse scattering impose stringent experimental conditions for data collection. Firstly the separation of Bragg from thermal diffuse scattering requires extremely fine collimation, producing a very fine Bragg peak against the broader shoulder of diffuse scattering. The very fine collimation and minimal beam divergence necessary requires the use of synchrotron radiation.

The continuous diffuse scattering has its origins in inelastic events and as a result is of very much lower intensity than the Bragg data. For it to be collected in reasonable time scales with good statistics, a very intense source is required, again with the constraints of very fine collimation. The diffuse scattering events are also energy dependent imposing further conditions on the spectral divergence. All these conditions make synchrotron radiation sources a pre-requisite in the measurement of inelastic scattering events in protein or other macromoleclular crystals.

The form of the thermal diffuse scattering has been investigated using low resolution reflections from samples of bovine ribonuclease [77]. Diffraction patterns were recorded using the bending magnet beam line at the Daresbury SRS with a wavelength of 1.488 Å and 0.2 mm diameter collimation. The incident horizontal divergence was severely limited using slits. Under these experimental conditions, the Bragg reflection spatial and angular widths were minimised to allow investigation of the form of the diffuse scattering haloes at the reciprocal lattice positions. Experimental plots clearly show the broad diffuse scattering shoulders to the Bragg peaks. Exposure times needed for still diffraction patterns were of the order of 20 min compared to 20 s for a normally recorded still pattern. The close agreement between calculated and observed plots showed that at low resolution, at least the acoustic scattering may be explained in terms of scattering from a single phonon or ultrasonic wave propagated in the crystal. At progressively higher resolutions, however, single phonon interactions will be less well adhered to and multiphonon interactions become more significant [78].

The thermal diffuse scattering peaking at the Bragg positions may in fact constitute a significant source of error in measurements of the integrated intensity of a Bragg reflection. The fine collimation and low beam divergence may be used to collect data in which the acoustic scattering contributions are minimised and its effect on model refinement assessed (Glover, Harris, Helliwell and Moss, unpublished). In both the above cases, synchrotron radiation is essential if data is to be recorded in realistic times.

Continuous diffuse scattering, due to its origins, is a potentially good source of dynamic information regarding correlated atomic displacements in the crystal. The ability to record well-exposed diffuse scattering patterns from macromolecular crystals in reasonable time scales has prompted considerable interest in both the qualitative and quantitative interpretation of such features. These features are different in both degree and detail from crystal to crystal and are apparently independent of radiation damage [77, and references below]. The diffuse scattering from γII crystallin, a 20 kDa protein which diffracts to 1.4 Å resolution on a synchrotron radiation source, has been recorded under a variety of chemical perturbations. Figure 10.6(a) shows a native still diffraction pattern which gives evidence of distinct thermal but little continuous diffuse scattering. Cross-linking and the preparation of heavy atom derivatives significantly alter the appearance of the diffraction pattern.

Figure 10.6. Monochromatic still photograph from γ II crystallin: (a) Native; (b) Hg derivative showing distinct increase in diffuse scattering.

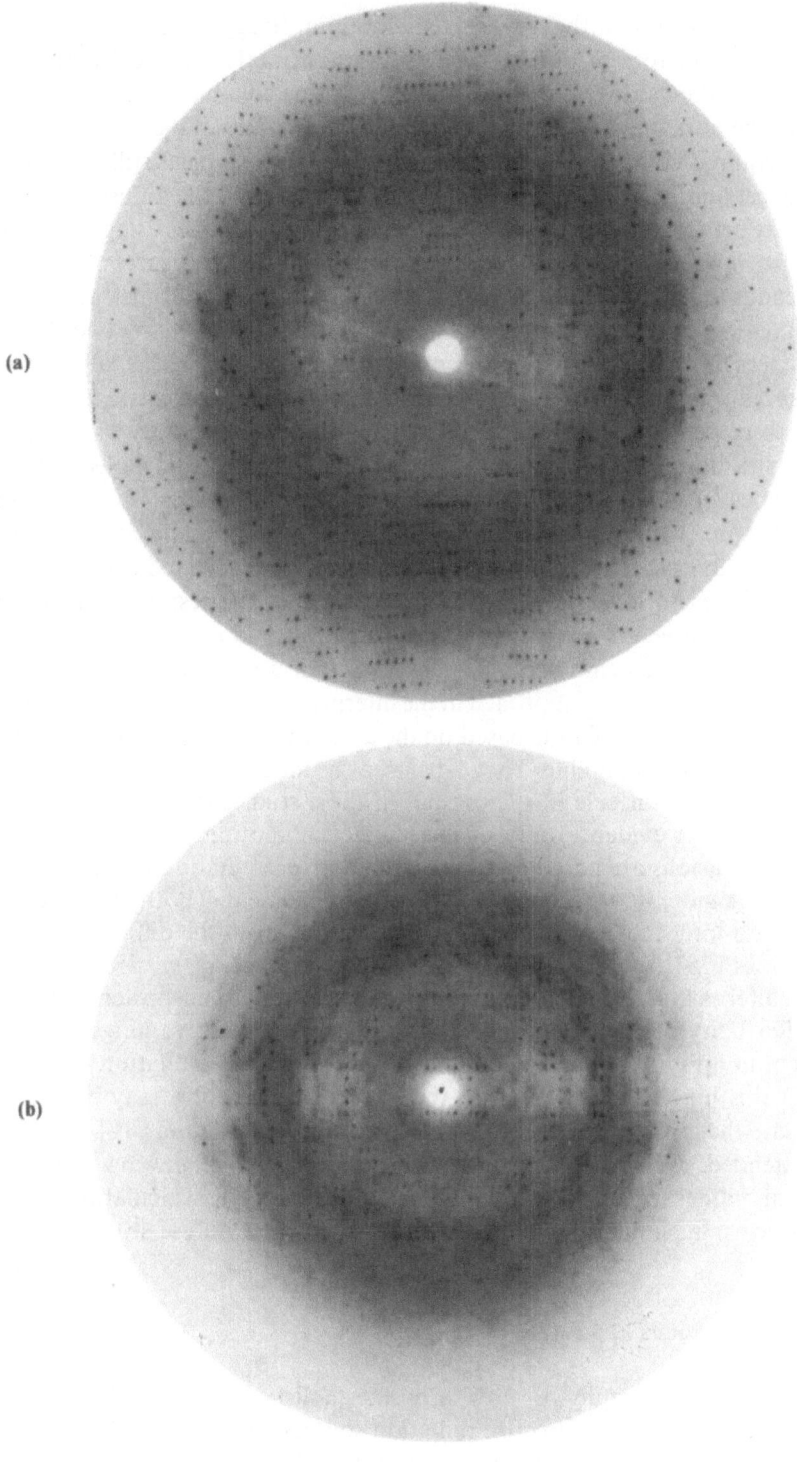

(a)

(b)

Figure 10.6(b) shows the severe reduction in Bragg resolution and the concomitant increase in diffuse scattering as the crystal is chemically modified. In all cases the data were collected using very fine collimation and slitting to reduce the horizontal beam divergence. Exposure times of the order of 1000 s were required compared to approximately 30 s for a typical still pattern.

Doucet and Benoit [79] have collected and measured the diffuse scattering from the orthorhombic form of lysozyme. The exposure times using the LURE synchrotron radiation source were 30 min compared to exposures of the order of seconds to record the Bragg peaks alone. The diffuse scattering features they observed were predominantly rows of modulated intensity along two families of planes (100) and (001) with other minor contributions. Their analysis of the measured intensities of diffuse scattering demonstrated the existence of rigid body displacements which are correlated within short periodic rows of aligned molecules along the a and c axes of the unit cell. The correlated molecules form two types of super-unit with amplitudes of motion of 5×10^{-4} nm and 7×10^{-4} nm, within which the molecules are sterically coupled via intermolecular interactions. The physical origins of the correlated displacements were assigned to local disorder in one or more of the residues involved in contact with a neighbouring molecule. This produced a displacement in the latter, all displacements being correlated within a super-unit. Caspar *et al.* [80] have used a similar approach in the interpretation of the diffuse scattering from insulin crystals. This is based on the estimation of atomic displacements and their range of coupling. They advanced the idea of liquid-like movements of parts of the molecule in the crystal. In these studies the diffuse scattering provides direct evidence only of intermolecular motions and contacts and these conclusions are based on mainly qualitative simulation of the observed patterns. A more rigorous approach requires refinement of both the correlated intra- and inter-molecular displacements directly from the diffuse scattering intensities [78].

An alternative approach uses intuitive arguments about protein structure and flexibility derived from molecular graphics. A putative hinge region is present in 6-PGDH relating domains of the structure. Simulation of motion about this hinge and its effect on the molecular transform may be used to simulate the gross features of the molecular diffuse scattering [81]. This may be extended, for smaller molecules, to a calculation of the low frequency normal modes of vibration which may be used, via calculation of the transforms, in simulations of the recorded molecular diffuse scattering.

10.7   Conclusions and future directions

Protein crystallography is now a well-established part of the research programmes of the synchrotron radiation sources whose spectra extend into the X-ray region. Synchrotron radiation considerably increases the rate at which data are collected. The higher resolution data that are measured improve the refinement of the macromolecular structures, which is most important in many cases in relating structure to function. The development of intense radiation sources by means of wiggler insertion devices permits the investigation of weakly diffracting crystals, whether this is due to the size of the crystal (microcrystals) or due to the large size of the unit cell or some disorder or large solvent content of the crystal. The technique of optimised anomalous dispersion coupled to an intense source with tunable wavelengths in the range 0.5–2.0 Å allows a more direct method of solving the crystallographic phase problem for small and medium sized protein single crystal structures.

The synchrotron radiation spectrum of an intense energy continuum of radiation makes possible rapid data collection in the sub-second to nano-second regime by the Laue diffraction method. This rate of data collection is sufficient to investigate most of the crystal kinetics and catalytic intermediate reaction states that can be produced.

X-ray undulator devices and new high brightness multipole wiggler magnets are very important for the further development of the use of synchrotron radiation in macromolecular crystallography [82]. Dedicated sources of this type are being constructed at Grenoble and at Argonne.

References

1. J.R. Helliwell, T.J. Greenhough, P.D. Carr, S.A. Rule, P.R. Moore, A.W. Thompson and J.S. Worgan, *J. Phys. E* **15** (1982) 1363.
2. C. Nave, J.R. Helliwell, P.R. Moore, A.W. Thompson, J.S. Worgan, R. Greenall, A. Miller, S.K. Bentley, J. Bradshaw, W.J. Pigram, W. Fuller, D.P. Siddons, M. Deutsh and R.T. Tregear, *J. Appl. Crystallogr.* **18** (1985) 396.
3. J.R. Helliwell, M.Z. Papiz, I.D. Glover, J. Habash, A.W. Thompson, P.R. Moore, N. Harris, D. Croft and E. Pantos. *Nucl. Instrum. Methods A* **246** (1986) 617.
4. R.P. Phizackerly, C.W. Cork and E.A. Merritt, *Nucl. Instrum. Methods* (1986) 579–595.
5. R. Kahn, R. Fourme, R. Bosshard and V. Saintage, *Nucl. Instrum. Methods* (1986) 596.
6. J.R. Helliwell, S. Harrop, J. Habash, B.G. Magorrian, N.M. Allinson, D. Gomez de Anderez, M. Helliwell, Z. Derewenda and D.W.J. Cruickshank, *Rev. Sci. Instrum.* In press (1989).
7. D. Bilderback, *Rev. Sci. Instrum.* In press (1989).
8. R. Kahn, R. Fourme, R. Bosshard and V. Saintagne, *Nucl. Instrum. Methods* **201** (1982) 203.
9. N.H. Xuong, D. Sullivan C. Nielsen and R. Hamlin, *Acta Crystallogr. B* **41** (1985) 267.
10. U.W. Arndt and D.J. Gilmore, *J. Appl. Crystallogr.* **12** (1979) 1.
11. M.Z. Papiz and S.J. Andrews, in *Computational Aspects of Protein Crystal Data Analysis*, eds. J.R. Helliwell, P.A. Machin and M.Z. Papiz, Daresbury Laboratory (1987).
12. M.G. Strauss, I. Naday, I.S. Sherman, M.R. Kraimer and E.M. Westbrook, *New Scientist.* **34** (1987) 389.
13. I. Naday, M.G. Strauss, I.S. Sherman, M.R. Kraimer and E.M. Westbrook, *Opt. Eng.* **26** (1987) 788.
14. N.M. Allinson, R. Brammer, J.R. Helliwell, S. Harrop. B.G. Magorrian, T. Wan et al., *J. X-Ray Sci. Technol.* In press (1989).
15. J. Miyahara, K. Takahashi, Y. Amemiya, N. Kamiya and Y. Satow, *Nucl. Instrum. Methods A* **246** (1986) 572.
16. J.R. Helliwell, *Rep. Prog. Phys.* **47** (1984) 1403.
17. U.W. Arndt, *J. Appl. Crystallogr.* **19** (1987) 145–163.

18. E. Arnold, G. Vriend, M. Luo, J.P. Griffith, G. Kamer, J.W. Erickson, J.E. Johnson and M.G. Rossmann, *Acta Crystallogr. A* **43** (1987) 346.
19. G. Vreind, M.G. Rossmann, E. Arnold, M. Luo, J.P. Griffith and K. Moffat, *J. Appl. Crystallogr.* **19** (1986) 134.
20. R. Acharya, E. Fry, D. Stuart, G. Fox, D. Rowlands and F. Brown, *Nature* **337** (1989) 709.
21. B. Hedman, K.O. Hodgson, J.R. Helliwell, R.C. Liddington and M.Z. Papiz, *Proc. Natl. Acad. Sci. USA* **82** (1985) 7604.
22. S.J. Andrews, M.Z. Papiz, R. McMeeking, A.J. Blake, B.M. Lowe, K.R. Franklin, J.R. Helliwell and M.M. Harding, *Acta Crystallogr. B* **44** (1988) 73.
23. T.L. Blundell and L.N. Johnson *Protein Crystallography*, Academic Press, New York (1976).
24. J.M. Bijvoet, *Proc. Acad. Sci. Amsterdam* **52** (1949) 313.
25. M.G. Rossmann, *Acta Crystallogr.* **14** (1961) 383.
26. A.C.T. North, *Acta Crystallogr.* **18** (1965) 212.
27. B.W. Matthews, *Acta Crystallogr.* **20** (1966) 82.
28. P. Argos and F.S. Mathews *Acta Crystallogr. B* **29** (1973) 1604.
29. B.C. Wang, *Methods Enzymol.* **115** (1985) 106.
30. T.L. Blundell, J.E. Pitts, I.J. Tickle, S.P. Wood and C.-W. Wu, *Proc. Natl. Acad. Sci. USA* **78** (1981) 4175.
31. A. Herzenberg and M.S.M. Lau, *Acta Crystallogr.* **22** (1967) 24.
32. W.A. Hendrickson and M.M. Teeter, *Nature* **290** (1981) 107.
33. J.L. Smith and W.A. Hendrickson, in *Computational Crystallography*, ed. D. Sayre, Clarendon Press, Oxford (1982) pp. 209–222.
34. Y. Okaya and R. Pepinsky, *Proc. Natl. Acad. Sci. USA* **42** (1957) 286.
35. S. Ramaseshan, in *Advanced Methods in X-Ray Crystallography*, ed. G.N. Ramachandran, Academic Press, New York (1961).
36. W. Hoppe and U. Jakubowski in *Anomalous Scattering*, eds. S. Ramaseshan and S.C. Abrahams, Munksgaard (1975).
37. G. Cascarano, G. Giacovazzo, A.F. Peerdeman and J. Kroon, *Acta Crystallogr. A* **38** (1982) 710.
38. A.K. Singh and S. Ramaseshan, *Acta Crystallogr. B* **24** (1968) 35.
39. J. Karle, *Int. J. Quantum. Chem.* **7** (1980) 357.
40. J. Karle, in *Computational Crystallography*, ed. D. Sayre, Munksgaard (1982).
41. M.M. Woolfson, *Acta Crystallogr. A* **40** (1984) 32.
42. R. Kahn, R. Fourme, R. Bosshard, M. Chiadmi, J.L. Risler, O. Dideberg and J.P. Wery *FEBS Lett.* **179** (1985) 133.
43. S. Harada, Y. Masanori, Y. Murakawa, N. Kasai and Y. Satow, *J. Appl. Crystallogr.* **19** (1986) 448.
44. W.A. Hendrickson, in *Crystallographic Computing* 3, eds. G.M. Sheldrick, G. Kruger, and R. Goddard, Clarendon Press, Oxford (1985) pp. 277–285.
45. R. Fourme, in *Abstracts of the 2nd International Conference on Biophysics and Synchrotron Radiation*, Chester, U.K. (1988).
46. I.D. Glover, PhD Thesis, London University (1984).
47. H. Einspahr, K. Suguna, F.L. Suddath, G. Ellis, J.R. Helliwell and M.Z. Papiz *Acta Crystallogr. B* **41** (1985) 335.
48. Y. Kitigawa, N. Tanaka, Y. Hata, Y. Katsube and Y. Satow, *Acta Crystallogr. B* **43** (1987) 272.
49. K. Moffat, *Nature* **336** (1989) 422.
50. W.A. Hendrickson, J.L. Smith, R.P. Phizackerly and E.A. Merrit, *Proteins* **4** (1988) 77.
51. D.B. Cowie and G.N. Cohen, *Biochem. Biophys. Acta* **26** (1957) 252.
52. Hendrickson et al., *Proc. Natl. Acad. Sci. USA* In press (1989).
53. J.M. Guss, E.A. Merritt, R.P. Phizackerley, B. Hedman, M. Murata, K.O. Hodgson and H.C. Freeman, *Science* **241** (1988) 806.
54. L.K. Templeton and D.H. Templeton, *Acta Crystallogr. A* **45** (1989) 39.
55. H-D. Bartunik, R. Fourme and J.C. Phillips, in *Uses of Synchrotron Radiation in Biology*, ed H.B. Sturhmann Academic Press, New York (1982).
56. K. Moffat, D.M.E. Szebenyi and D. Bilderback, *Science* **223** (1984) 1423.
57. K. Moffat, W. Schildkamp, D. Bilderback and K. Volz, *Nucl. Instrum. Methods A* **246** (1986) 627.

58. D.H. Bilderback, K. Moffat and D.M.E. Szebenyi, *Nucl. Instrum. Methods* **222** (1984) 245.
59. J. Hails, M.M. Harding, J.R. Helliwell, R.C. Liddington and M.Z. Papiz, *Daresbury Laboratory Preprint* DL/SCI/P428E (1984).
60. J.R. Helliwell, *J. Mol. Struct.* **130** (1985) 63.
61 I.G. Wood, P. Thompson and J.C. Matthewman, *Acta Crystallogr. B* **39** (1983) 543.
62. J. Hajdu, R. Acharya, D.I. Stuart, P.J. McLaughlin, D. Barford, N.G. Oikonomakos, H. Klein and L.N. Johnson, *EMBO J.* **6** (1987) 539.
63. M. Elder *Inf. Quart. Protein Crystallogr. at Daresbury Laboratory* **19** (1986) 31.
64. I.J. Clifton, D.W.J. Cruickshank, G. Diakun, M. Elder, J. Habash, J.R. Helliwell, R.C. Liddington, P.A. Machin and M.Z. Papiz, *J. Appl. Crystallogr.* **18** (1985) 296.
65. D.W.J. Cruickshank, J.R. Helliwell and K. Moffat, *Acta Crystallogr. A* **43** (1987) 656.
66. T.J. Greenhough and A.K. Shrive, In preparation.
67. J.W. Campbell, J. Habash, J.R. Helliwell and K. Moffat, *Inf. Quart. Protein Crystallogr. at Daresbury Laboratory* **18** (1986) 35.
68. U.W. Arndt and A. Wonacott, in *The Rotation Method in Protein Crystallography*, eds. U.W. Arndt and A. Wonacott, North-Holland, Amsterdam (1977).
69. M.G. Rossmann, *J. Appl. Crystallogr.* **12** (1979) 225.
70. J.R. Helliwell, J. Habash, D.W.J. Cruickshank, M.M. Harding, T.J. Greenhough, J.W. Campbell, I.J. Clifton, M. Elder, P.A. Machin, M.Z. Papiz and S. Zurek, Daresbury Lab Preprint DL/SCI/P558E; *J. Appl. Crystallogr.* submitted (1988).
71. G.K. Farber, P.A. Machin, S. Almo, J. Hajdu and G. Petsko, *Proc. Natl. Acad. Sci. USA* (1988).
72. P.J. Mclaughlin, D.I. Stuart, H.W. Klein, N.G. Oikonomakos and L.N. Johnson, *Biochemistry* **23** (1984) 5862.
73. J. Hajdu, P.A. Machin, J.W. Campbell, T.J. Greenhough, I.J. Clifton, S. Zurek, S. Gover, L.N. Johnson and M. Elder, *Nature* **329** (1987) 178.
74. C.D. Reynolds, B. Stowell, K.K. Joshi, M.M. Harding, S.J. Maginn and G.G. Dodson, *Acta Crystallogr. B* **44** (1988) 512.
75. K. Moffatt et al., to be published.
76. J.A. McCammon, *Rep. Prog. Phys.* **47** (1984) 1.
77. I.D. Glover, G. Harris, J.R. Helliwell and D.S. Moss, in preparation (1989).
78. D.S. Moss and G. Harris, submitted to *Acta Crystallogr* (1989).
79. J. Doucet and J.P. Benoit, *Nature* **325** (1987) 643.
80. D.L.D. Caspar, J. Clarage, D.M. Salunke and M. Clarage, *Nature* **332** (1988) 659.
81. J.R. Helliwell, I.D. Glover, A. Jones, E. Pantos and D.S. Moss, *Biochem. Soc. Trans.* **14** (1986) 653.
82. J.R. Helliwell, *ESRF Red Book* (1988) 329.

11 X-ray absorption spectroscopy of biological molecules

C.D. GARNER

11.1 Introduction

During the last decade, X-ray absorption spectroscopy (XAS) has become a routine and vital technique for investigating the local environment of atoms in biological systems. This advance has been made possible by the availability of synchrotron radiation facilities. Thus, the availability of electro-magnetic radiation with high intensity throughout an extensive range of the X-ray region has permitted reliable data to be obtained for a wide range of elements in a relatively short time. XAS is ideally suited to probe the immediate environment of specific atoms, particularly the d-transition metal atoms, in biological systems [1–3]. A major attraction is the ability to probe the nature of a catalytic site of a metallo-enzyme, directly and selectively. This is especially true since the technique is not limited by the physical state of the sample and successful investigations are possible with present instrumentation for elements down to millimolar concentrations. Furthermore, changes in local structure can be monitored as a function of the state of the system, thus providing direct information on the relationship between chemical structure and biological function.

This chapter briefly outlines the experimental and analytical aspects of XAS relevant to studying biological systems and then presents a selective review of the research accomplished.

11.2 Experimental aspects

The X-ray absorption spectra associated with the copper and zinc K edges of Cu, Zn-metallothionein are shown in Figure 11.1, revealing the clear element specificity of the technique, determined by the core level binding energy. An X-ray absorption spectrum can usually be divided, for convenience of interpretation, into three regions: the pre-edge and edge; the X-ray absorption near edge structure (XANES); the extended X-ray absorption fine structure (EXAFS).

The XANES and EXAFS signals are small in comparison to the atomic absorption resulting from excitation of the core electron. Thus, good quality XANES and EXAFS data require an intense and stable X-ray source. So far

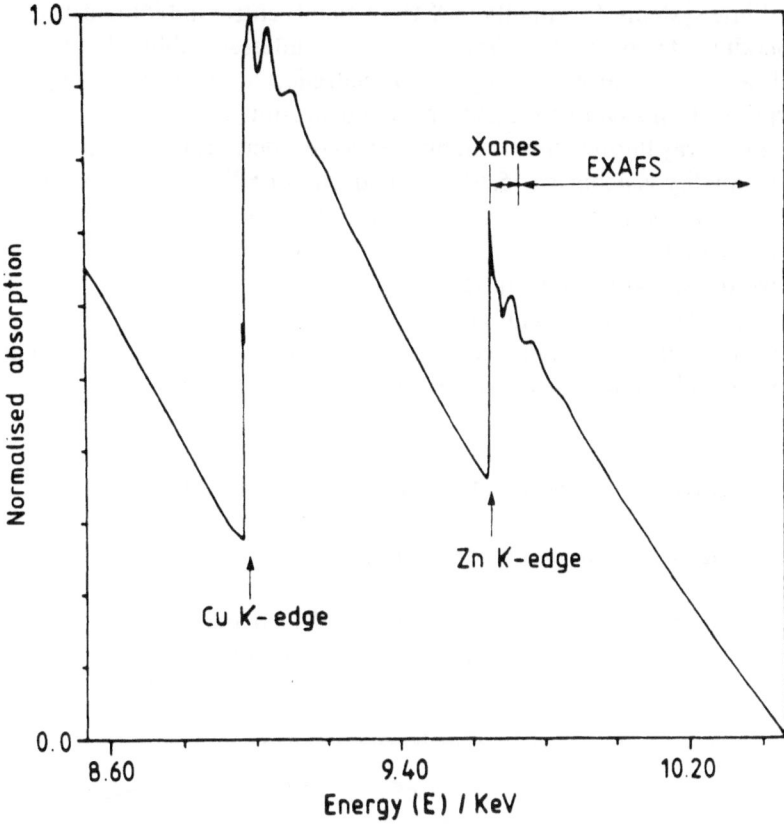

Figure 11.1.   X-ray absorption spectrum of Cu, Zn-metallothionein.

there has been no EXAFS study of a metalloprotein using a laboratory X-ray source such as a rotating anode; all such investigations have required synchrotron radiation sources. A recent and valuable development at many such synchrotron radiation facilities has been the inclusion of insertion devices, especially wavelength shifters and multipole wiggler magnets, to increase the intensity available particularly at shorter wavelengths. The spectral purity, high brightness, and stability of the beam are essential for probing metal centres in biological systems by XAS. Thus, samples of interest are often available in small quantities with a low (millimolar) concentration of the metal and good quality spectra usually require that several scans, each of approx. 45 min duration, be recorded. Furthermore, radiation damage of proteins is usually time dependent, rather than just proportional to the radiation dose; therefore, it is advantageous to record data in as short a measuring time as possible.

The experimental arrangements for measuring XAS are described in Chapter 1. For biological systems focussed beams are generally required. The use of wide aperture monochromators [4] can also be advantageous in maximising the intensity of X-rays at the specimen. However, as with studying other dilute systems, the quality of XAS of metal centres in biological systems is generally detector limited. Since the atom of interest is diluted in a host of protein and solvent atoms, the signal is a small effect above a large background absorption. If, instead of measuring the absorption directly, the secondary process of X-ray fluorescence is monitored, a considerable improvement in the spectral quality is achieved [5, 6]. The majority of EXAFS data for metallo-proteins have been recorded in the fluorescence mode using an array of Na(Tl)I scintillation counters with several (say upto 8) scans being averaged to improve the signal-to-noise ratio. The advent of a multi-element solid state device represents a further and significant improvement in detector sensitivity. Such devices will greatly enhance the quality of spectra and extend the range of systems for which successful studies can be accomplished.

11.3 Information content of an X-ray absorption spectrum

11.3.1 *The position of an absorption edge*

Provided proper calibration has been accomplished and an edge position can be expressed without ambiguity, any 'chemical' shift in the absorption edge is a

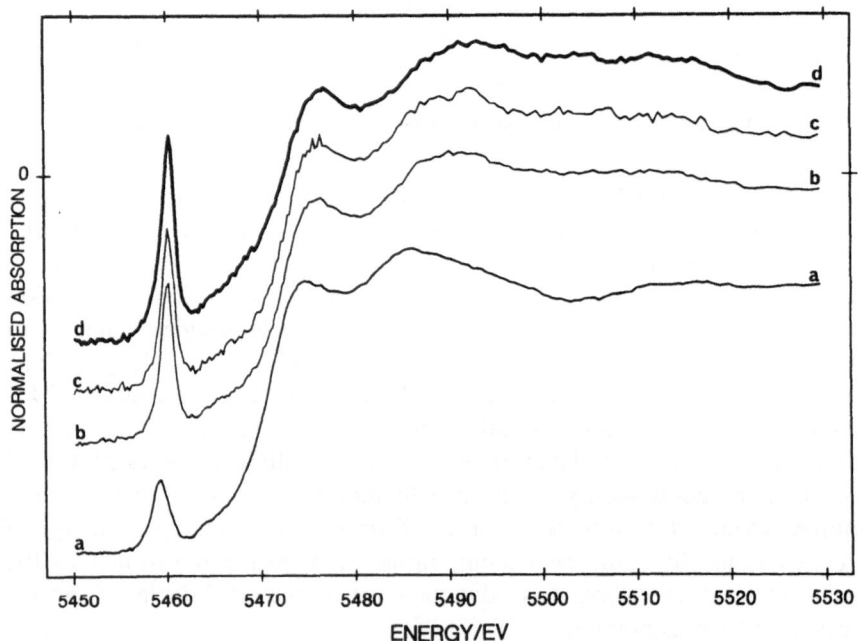

Figure 11.2. Vanadium K edge and XANES spectra of samples of bromoperoxidase: (a) dithionite reduced bromoperoxidase; (b) native bromoperoxidase; (c) native bromoperoxidase plus 10 mM H_2O_2; (d) native bromoperoxidase plus 250 mM KBr.

measure of the net charge on the primary absorber and, therefore, can serve as an indicator of the element's oxidation state. This is apparent from Figure 11.2 [7], where the position of the vanadium K edge is some 2 eV lower in energy for the reduced sample (a) of bromoperoxidase from *Ascophyllum nodosum*, which contains V(IV), than for the oxidised samples (b, c, and d), which contain V(V).

11.3.2 *Pre-edge and/or edge features*

The excitation of a core electron into the continuum may be convoluted with transitions from the core level to outer bound states. A clear pre-edge transition is evident in each of the spectra of Figure 11.2. This corresponds to a 1s to 3d transition of the vanadium centre. The position and intensity of such features are dependent upon the electronic structure and the local symmetry at the primary absorber as the atomic transition is forbidden by the $\Delta l = \pm 1$ selection rule. The pre-edge feature of the vanadium centre of reduced bromoperoxidase (Figure 11.2(a)) is *c.* 1 eV to lower energy than that of the oxidised forms of the enzyme (Figures 11.2(b,c,d)). In addition, the peak has a different profile for V(IV) as compared to V(V); the height is approximately halved and it is significantly broadened. These observations indicate a significant change in geometry upon reduction and, especially for vanadium systems [7, 8]; valuable structural insights may be obtained from the nature of such changes.

11.3.3 *XANES and EXAFS*

These spectral features arise as a consequence of local electron diffraction as shown schematically in Figure 11.3. The principal distinction between XANES and EXAFS is that the former invariably involves multiple scattering of the photoelectron within the cluster of atoms surrounding the primary absorber, whereas the latter usually does not (although imidazole [9] and carbonyl [10] groups represent notable exceptions). Interpretation of EXAFS has progressed [11] from the plane wave, single scattering approximation [12], to a full spherical wave treatment [13] which allows the inclusion [9] of multiple scattering pathways. EXAFS analysis proceeds by removal of the smooth background due to the atomic absorption to produce a damped sine wave (Figure 11.3). Analytical procedures in *k*-space involve simulations of EXAFS profiles and refinement of structural and other parameters to produce the optimum agreement between the experimental and theoretical data as described in Chapter 6.

The structural parameters immediately available from EXAFS analysis are the distance (R), the occupation number (N), and Debye-Waller parameter (σ^2)

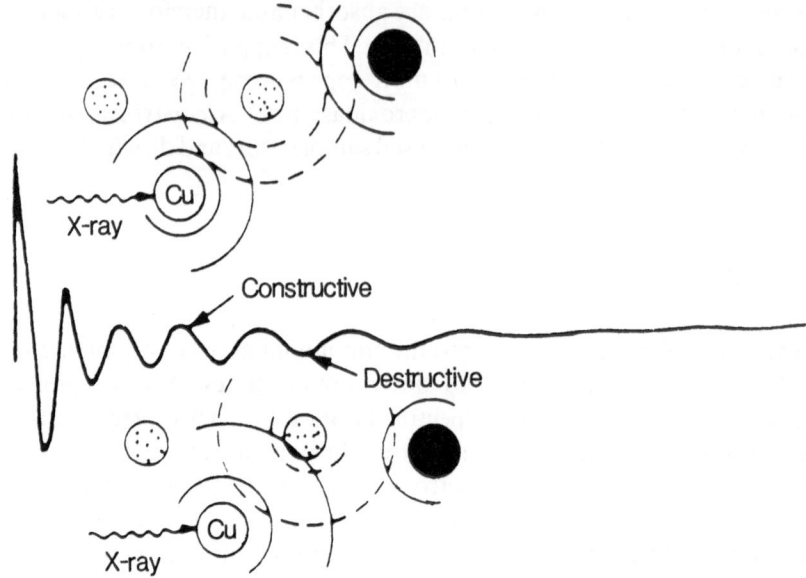

Figure 11.3. Diagrammatic representation of the origin of EXAFS features associated with a copper K edge. The emanating spherical wave of the excited photoelectron is backscattered by the surrounding atoms and the resulting interference modifies the X-ray absorption coefficient as shown.

of each shell of atoms located near (usually within 3.5 Å) the primary absorber. Provided that absorber and backscatterer phase shifts are carefully checked by use of suitable model compounds, R can be determined much more accurately ($\leqslant 1\%$) than N ($\geqslant 10\%$). There is a further important uncertainty, in that backscattering contributions from atoms of a similar atomic number (e.g. C, N, and O, or P, S, and Cl) can rarely be distinguished. This present limitation is especially frustrating for metal centres in proteins and emphasises the need to integrate the information from other spectroscopic and structural techniques into the EXAFS analysis.

XANES is a very sensitive, if empirical, fingerprint of the immediate environment about the primary absorber and direct comparisons can prove extremely useful. This is apparent in Figure 11.2 where, in agreement with the comments made above, the XANES profiles clearly manifest a structural change at vanadium upon reduction.

11.4　Applications

Excellent reviews of the application of XAS to characterise metal centres in biological systems have been published [1–3] and readers are referred to these. This review concentrates on some of the more recent advances.

11.4.1 *Metallothioneins*

Metallothioneins are a unique and widely distributed group of proteins. They are characterised by their low molecular weight (approx. 6000), high cysteinyl content, and the ability to bind substantial numbers of metal ions [14]. The proteins bind Cu and Zn as part of the normal metabolism of these elements and offer protection from the invasion of inorganic forms of the toxic elements Cd, Pb and Hg; in addition other metals such as Fe and Co can be induced to bind. EXAFS is ideally suited to probe the environment of these different metal atoms (see Figure 11.1) and the results of several such studies [15] are summarised in Table 11.1. Thus, in each case, the data are consistent with the primary coordination of the metal deriving from the cysteinyl residues. As indicated earlier, uncertainties of approx. 10% in N are inherent in EXAFS and, to reduce confusion, appeal should be made to other information. For systems as simple as those in Table 11.1, correlations of bond length with coordination number are helpful. Thus, for Cu in pig liver metallothionein, reference may be made to crystallographic information for $Cu(SR)_2$, $Cu(SR)_3$, and $Cu(SR)_4$ centres for which the Cu–S bond lengths are approx. 2.15, 2.24–2.29, and 2.35 Å, respectively. Thus, on this basis, copper would appear to be three coordinate in the Cu, Zn-metallothioneins [16].

The first EXAFS study of a Cd environment in a protein was achieved for Cd_5Zn_2- and Cd_7-metallothionein from rat liver [15, 17]. The two samples manifest identical EXAFS and the data are consistent with a shell of four sulphur atoms at 2.51 Å. These results demonstrate that the Cd atom inequivalence observed by [113]Cd-NMR studies [18] of metallothioneins does not arise from marked variations in atom type, coordination number, or metal ligand distances within the metal's first coordination sphere.

Table 11.1 Structural parameters deduced from EXAFS of metallothioneins [15].

Absorber	Scatterer	$R(\text{Å})$
Zn in Zn_7-MT	4S	2.33
	1S	4.10
	1Zn	5.00
	1Zn	5.20
Cu in Cu_6Zn_3-MT	3S	2.25
Cu in Cu_5Zn_2-MT	3S	2.25
Pb in Pb_7-MT	2S	2.65
Hg in Hg_7-MT	3S	2.42
Cd in Cd_7-MT	4S	2.51
Cd in Cd_4Co_3-MT	4S	2.51
Co in Cd_4Co_3-MT	4S	2.30
Co in Co_7-MT	4S	2.31
Co in Co_3-MT	4S	2.31
Fe in Fe_7-MT	4S	2.32
Fe in Fe_3-MT	4S	2.32

Clear evidence for the close approach of metal atoms in Cd_7-metallothionein has been obtained from ^{113}Cd-NMR spectroscopy [18] and indications of the distances involved have been sought from EXAFS. In the case of the Zn_7-metallothionein from rabbit liver some evidence was obtained for Zn...Zn separations of approx. 5 Å [16]. In contrast, Cu...Cu approaches of 2.74 Å were reported [19] for canine liver $Cu_{7,8}$-metallothionein. However, the principal conclusion is that these metal-metal separations are not coherent and, therefore, the backscattered waves, especially with their high frequency at distances $\geqslant 3$ Å, engage in destructive interference which effectively renders them 'silent' in the EXAFS.

11.4.2 Zinc centres in proteins

Zinc is known to be associated with a large variety of proteins, and is present in enzymes in each of the six categories defined by the International Union of Biochemistry. In enzymes, zinc is often thought to play a catalytic role, i.e. to be essential for and directly involved in the catalysis, but also purely non-catalytic (i.e. structural or regulatory) roles have been implied, and in some cases its role is undefined [20].

Some crystal structures of zinc proteins are available [21], however, information concerning zinc centres in proteins is rather limited. Because of its filled 3d-shell, Zn is not accessible by spectroscopic techniques like optical absorption and electron paramagnetic resonance (EPR). Zinc sites in proteins may be studied indirectly by spectroscopic techniques; Cd substitution and monitoring by ^{113}Cd-NMR spectroscopy or Co substitution and monitoring by ultraviolet visible or circular dichroism are popular procedures. Nevertheless, it should be borne in mind that these surrogates may not be faithful reporters of zinc sites.

XAS in general and EXAFS in particular offers a unique, direct probe of zinc in proteins. So far, only three types of amino acid residues have been identified as ligands of zinc in a protein; an imidazole nitrogen of histidine, the carboxylate oxygen of glutamic or aspartic acid, and the sulphur of cysteine. A tentative classification of zinc enzymes, based on features distinguishable in EXAFS, the only spectroscopic technique applicable to all zinc proteins, may be suggested:

Type-A zinc: coordination by sulphur exclusively;
Type-B zinc: coordination by sulphur and nitrogen and/or oxygen;
Type-C zinc: coordination by nitrogen and/or oxygen.

The potential of EXAFS to distinguish between classes, A, B and C derives from the facts that: (i) the backscattering amplitude of sulphur and nitrogen (or oxygen) are approximately π out of phase when placed at a similar distance from the absorber atom, and (ii) Zn–S bonds are typically about 2.3 Å in length whereas Zn–N/O bonds are approx. 0.3 Å shorter. Therefore, Type-A and

Type-C centres can be readily distinguished. Type-B centres are usually readily identified, but it may be difficult to determine the respective occupancies of the two shells precisely, as the contributions are typically not well resolved in the Fourier transform. Also, it is unfortunate that EXAFS cannot distinguish between carboxylate oxygens and water as ligands of zinc. Coordination of imidazole is usually clearly evident because of the multiple scattering pathways involving all atoms of the 5-membered ring [9].

Type-A zinc sites have been identified in metallothioneins [15, 16] (Table 11.1) and aspartate transcarbamylase [22]. Both involve four sulphur atoms at approx. 2.33 Å from the zinc. Such sites will not be catalytically active as the coordination sphere of the metal is saturated. A function for the zinc as a rivet, holding together the otherwise disordered bonding region of the regulatory chains of the enzyme in a flexible manner, which permits conformational change, is proposed [22].

Although Type-B and Type-C zinc sites are known to be catalytically active, in these zinc can also play an important structural role. This is true for 'zinc finger proteins'. The transcription factor IIIA of eukaryotic cells is a protein with a distinctive repeat sequence consisting of two cysteine and two histidine residues. These bind Zn in a tetrahedral site causing a loop or 'finger' to be formed The finger binds in the wide groove of DNA in a sequence specific manner. Thus, the DNA binding capability is controlled by the coordination at the zinc and a similar structural motif has been identified in numerous proteins. Diakun *et al.* [23] showed that zinc in transcription factor IIIA from *Xenopus laevis* is coordinated by 2S and 2N atoms (at 2.30 and 2.00 Å, respectively) consistent with ligation by two cysteine and two histidine residues.

5-Aminolaevulinate dehydratase catalyses the synthesis of the pyrrole porphobilinogen from two molecules of 5-aminolaevulinic acid. The enzyme from bovine liver consists of eight subunits and is capable of binding eight atoms of zinc, although only four seem necessary for full activity. Dent *et al.* [24] studied the enzyme with all eight sites occupied ($4Zn_A + 4Zn_B$) and with the four zinc atoms necessary for full activity ($4Zn_A$), thus permitting structural information to be obtained for both sites. These workers concluded that Zn_A is bound to two histidines (Zn–N = 2.00 Å), an oxygen (possibly from aspartate, Zn–O = 1.85 Å) and a sulphur from cysteine (Zn–S = 2.28 Å), whereas the Zn_B site involves coordination from three sulphurs (Zn–S = 2.25 Å) and one nitrogen/oxygen ligand (Zn–O = 2.50 Å). Possibly Zn_A plays a functional role and Zn_B a structural role.

11.4.3 *Manganese in photosystem II*

The most important role yet recognised for maganese in nature is its direct involvement in the photocatalytic, four-electron oxidation of two H_2O

molecules to produce O_2 within the photosynthetic apparatus of green plants and cyanobacteria. The Mn atoms appear to be bound at or near the lumenal surface of the thylakoid membrane by membrane-bound polypeptides. This enzyme is crucial for life on this planet as it is the provider of atmospheric oxygen. The active site of the enzyme is thought to possess four Mn atoms but its structure is unknown. George and Prince have accomplished polarised XAS at the Mn K edge for different orientations of the thylakoid membrane. The data obtained indicate that Mn is coordinated by oxygen or nitrogen atoms at a mean distance of 1.9 Å from the metal. Two different Mn–Mn interactions can be discerned at 2.7 Å and 3.3 Å. The orientation dependence of the latter indicates that the 3.3 Å Mn–Mn vector lies approximately along the membrane normal. The results are consistent with the active site involving a tetranuclear manganese cluster [25].

11.4.4   Oxomolybdoenzymes

Molybdenum is an important element within the biosphere, since it is essential for the activity of a large group of enzymes [26]. For many of these, clear evidence has been obtained that molybdenum is the site of substrate binding and conversion. These enzymes are found in organisms which range from bacteria to man; in the former, molybdenum is involved in nitrogen fixation and in the latter, xanthine and sulphite oxidation. There is a clear biochemical distinction between the nitrogenases on the one hand and oxomolybdoenzymes, such as xanthine oxidase and dehydrogenase, sulphite and aldehyde oxidase, and nitrate reductase on the other. Such a distinction was initially apparent from EPR spectroscopy [27] and proven by the isolation of an iron-molybdenum-cofactor (FeMoco) from the nitrogenases [28] and a molybdopterin cofactor (Moco) from the oxomolybdoenzymes [29, 30].

XAS has played a vital role in defining the chemical nature of these molybdenum centres and how they respond to changes in the oxidation level of the protein and/or to the presence of substrates, substrate analogues or inhibitors of enzymic activity [31]. The prefix oxo for this latter group of enzymes is appropriate. Thus, not only does each enzyme catalyse a conversion, the net result of which can be represented as oxygen atom transfer, but also XAS studies have confirmed the presence of at least one terminal oxo ligand (Mo=O) of molybdenum in each of the enzymes.

The simplest of the oxomolybdoenzymes is sulphite oxidase which is responsible for the physiologically vital oxidation of sulphite to sulphate. The molybdenum K edge EXAFS studies achieved [32] for this enzyme are the clearest such data and, thus, the interpretation represents a prototype for other oxomolybdoenzymes. Molybdenum has been investigated in its three accessible oxidation states, Mo(VI), Mo(V), and Mo(IV), as a function of pH and chloride concentration. The Mo(VI) coordination sphere of two Mo=O groups (1.70 Å), one Mo–O/N (2.06 Å) and three Mo–S ligands (2.42 Å) is apparently unaffected by a change in pH (from 6 to 9) and chloride concentration. The reduced molybdenum centres each have a single oxo-ligand (at 1.69 Å), one Mo–O/N and three Mo–S ligands (at 2.00 and 2.37 Å, respectively) and both of these centres appear to bind a chloride ligand at pH 6 in 0.3 M KCl. The results for Mo(V) have a special significance, in that they permit a direct comparison of the EXAFS results with EPR data. EPR spectroscopy [27] shows that the Mo(V) centre exists in two different forms that are in a pH- and anion-dependent equilibrium. George et al. [32] concluded that their EXAFS data were consistent with one chloride ligand binding to the low pH form of the Mo(V) (and Mo(IV)) centre and that the number of oxo-groups remains the same upon transition from the high pH to the low pH molybdenum (V) form.

Thus, for sulphite oxidase, a clear picture emerges of the Mo(VI) centre existing as an MoO_2^{2+} moiety ligated to approx. three sulphur atoms; two of these would be expected to originate from the molybdopterin and the other would arise from a cysteinyl residue of the protein. Reduction to the Mo(V) and Mo(IV) forms does not appear to affect the sulphur ligation but results in the loss of one oxo-group, presumably due to protonation, and the generation of an anion binding site at the molybdenum. The behaviour is consistent with the chemistry of molybdenum in its higher oxidation states. A cis-dioxomolybdenum(VI) centre is generally converted to a mono-oxomolybdenum(V) or (IV) centre upon reduction, provided that steric restrictions prevent dimerisation—as would be expected for the enzymes. Also, the inherent ability of Mo(VI) and Mo(IV) states to interconvert by a change in the number of oxo-groups, viz. $Mo^{VI}O_2^{2+} \leftrightarrow Mo^{IV}O^{2+}$, would seem crucial for the mediation of oxygen atom transfer. This is formally a two-electron process and is difficult to accomplish, except by atom transfer.

Xanthine oxidase is the most available of the oxomolybdoenzymes and is readily extracted from cow's milk. A special feature of this enzyme is its existence in two forms, active and an inactive form caused by loss of a sulphur atom (desulpho). The environments of Mo(VI) and Mo(IV) in desulpho-xanthine oxidase closely resemble [33] that of the corresponding oxidation state for sulphite oxidase. The principal difference between the centre of oxidase active, as compared to desulpho-xanthine oxidase is the presence of one sulphido-group (at 2.18 Å) plus one oxo-group rather than two oxo-groups.

The molybdenum centre of xanthine oxidase is very reactive and both EXAFS and EPR data indicate that the centre of this reactivity is the Mo=S bond. The terminal sulphido-group is lost upon reduction, presumably protonated to form an Mo–SH moiety. Arsenite is a potent inhibitor of xanthine oxidase and the nature of the species formed has been probed by XAS at both the Mo and As K edges. Clear evidence for an Mo–S–As interaction was observed [34].

11.4.5 *Molybdenum nitrogenases*

Nitrogen is an essential element for life and is abundant in the Earth's atmosphere in the form of nitrogen gas (N_2). However, N_2 is not metabolised by most organisms. Consequently, they must obtain their nitrogen in a combined form, from ammonia, nitrate or an organic molecule. Because organic nitrogen is incompletely recycled in living ecosystems and the available ammonia and nitrate are continually metabolised to N_2 through nitrification and denitrification, all life ultimately depends upon the biological fixation of N_2.

Nitrogen fixation occurs only in certain prokaryotes and is catalysed by the complex enzyme system, nitrogenase, which has two protein components. These are the Fe-protein, which acts as a specific reductant of the larger MoFe-protein. Within the MoFe-protein are at least six metal-containing prosthetic groups, including the iron-molybdenum cofactor (FeMoco) centres which are the catalytic centres of the enzyme.

One of the earliest successful applications of EXAFS to probe a metalloenzyme was the study of the Mo site of nitrogenase. Studies were made on both the *Clostridium pasteurianum* and *Azotobacter vinelandii* MoFe-proteins and on FeMoco [35]. These studies definitively showed the involvement of Mo in a polynuclear cluster containing S and Fe, with Mo–S and Mo–Fe distances of 2.36 and 2.72 Å, respectively. This work inspired the successful development of chemical systems containing Mo–Fe–S clusters. XANES and EXAFS studies of these systems have strengthened the basis for the interpretation of corresponding data for the natural system. The most accurate picture of the Mo site currently available involves a coordination sphere of three S (at 2.37 Å), three O (or N) (at 2.10 Å) and three Fe (at 2.70 Å) [36].

The average environment of Fe in FeMoco from *Klebsiella pneumoniae* has also been investigated by XAS [37]. These data provided clear evidence for a longer range structural order (3.68 Å) than observed from molybdenum, in addition to Fe–S (2.20 Å) and Fe–Fe(Mo) (2.64 Å) consistent with the molybdenum K edge studies. Also, the iron atoms do not appear to be bound to light (O/N) atoms in the intact cofactor.

Despite the data obtained from XAS, and a plethora of other spectroscopic studies, the overall structure of FeMoco has remained elusive. Furthermore, it has proved extremely difficult to identify the site at which substrates, or inhibitors, of the enzyme bind. In particular, no changes in the Mo K edge EXAFS could be observed on addition of N_2, C_2H_2, CN^-, MeNC, N_3^-, or CO to the dithionite-reduced enzyme [38]. This suggests that the binding of these ligands does not occur at molybdenum.

11.4.6 Vanadoenzymes

Recently, vanadium has been shown to be essential for the activity of two classes of enzymes, nitrogenases and bromoperoxidases.

Genetic suppression of the 'normal', molybdenum-dependent-nitrogenase of certain classes of *Azotobacter* allows expression of the vanadium-dependent enzyme. This provides the explanation for the early indication of a vanadium requirement for nitrogen fixation observed by Bortels in 1936 [39]. The vanadium and molybdeum nitrogenase systems show many similarities and, in particular, an iron-vanadium cofactor (FeVaco), analogous to FeMoco, has been isolated. Clear evidence of a strong similarity between active sites in the MoFe- and VFe-proteins have been provided by vanadium K edge XAS studies. Two VFe-proteins have been investigated, one from *Azotobacter chroococcum* [40] and one from *Azotobacter vinelandii* [41]. The edge structure shows a weak, single $1s \rightarrow 3d$ transition, the intensity of which precludes the presence of terminal $V=O$ bonds and implies an octahedral coordination around vanadium. The edge and XANES structure is very similar to that of a VFe_3S_4 cubane-like cluster $[NMe_4] [VFe_3S_4Cl_3(DMF)_3]$. The EXAFS results [40] are consistent with vanadium in the VFe-protein of *A. chroococcum* being ligated by three O(N), three S and three Fe atoms at 2.13, 2.33, and 2.75 Å, respectively.

Most haloperoxidases are haemoproteins, however, certain marine algae possess a vanadium-dependent bromoperoxidase. One such species is *Ascophyllum nodosum* and the pre-edge, edge, and XANES recorded for four forms of the enzyme are shown in Figure 11.2; these results have already been discussed in a qualitative manner. The EXAFS data for the native enzyme are consistent with the presence of an oxo-group ($V=O$ 1.61 Å) together with additional coordination by other light atoms, approx. three at 1.72 Å and two at 2.11 Å. No change is apparent at the vanadium centre upon the addition of H_2O_2 or KBr (see Figure 11.2). Although the reduced (V(IV)) state does not appear to participate in the catalytic cycle of the enzyme, there is value in characterising this species. Vanadium in reduced bromoperoxidase is also suggested to be present bound to an oxo-group ($V=O$ 1.63 Å) and ligated by similar groups to those found for the native enzyme, approx. three light atoms at 1.91 Å and two at 2.11 Å. The 2.11 Å distance is suggested to correspond to coordination by the imidazole group of histidine in each case, the outer C and N shells being clearly apparent for the reduced enzyme. The other ligands are suggested to be the phenols of tyrosines since the significant reduction in the V–O distance from 1.91 Å to 1.72 Å upon oxidation clearly parallels known vanadium chemistry and would correspond to an increase in π-donation from the phenolic oxygen to vanadium upon oxidation

The XAS studies presently accomplished suggest that molybdenum and vanadium sites in enzymes may be classified in a similar manner: (a) as part of an Fe_3MS_4 (M = Mo, V) cubane-like cluster which forms a sub-unit of the cofactor of the nitrogenases; (b) bound to one (or more) oxo-groups plus sulphur (in the case of molybdenum) or oxygen/nitrogen (in the case of vanadium) ligands to form a catalytic centre for oxygen atom transfer.

11.5 XAS and protein crystallography

The value of combining XAS and protein crystallography when studying metalloproteins [42] has been amply demonstrated for rubredoxin [43] and Cu, Zn-superoxide dismutase [44]. The principal advantage of this approach is that the crystallographic data provide global details of the protein molecule, including the position of the metal centre(s) and their ligation. This information is of immense value for interpretation of XAS data which should help improve the definition of the metal-ligand bond lengths and permit clear interpretations of structural changes at the metal centre following reactions of the metalloprotein.

Few studies of this kind have been accomplished. However, a promising start has been made in this respect for diferric chicken ovotransferrin, an important protein for sequestering and mobilising iron atoms in serum. Protein crystallographic and EXAFS data have been combined to help define the molecular structure and the nature of the iron sites. The latter information is consistent with a first coordination shell of six light (O/N) atoms, two at 1.85 Å and four at 2.04 Å [45]. However, further aspects need to be addressed, including the reasons for the concomitant binding of a synergistic anion (HCO_3^- and the intramolecular allosteric effects of iron uptake and loss. In these respects, NMR and small angle scattering studies may well be necessary to provide definitive information. With diferric ovotransferrin as a particular example, it is clear that XAS has much to contribute to the characterisation of metal centres in proteins particularly when set in the context of other spectroscopic and structural studies. However, considerable challenges remain, especially to extend the applicability of the technique and to improve the precision of data interpretation.

References

1. S.P. Cramer and K.O. Hodgson, *Prog. Inorg. Chem.* **25** (1979) 1.
2. L. Powers, *Biochim. Biophys. Acta.* (1982) 683.
3. S.S. Hasnain and C.D. Garner, *Prog. Biophys. Mol. Biol.* **50** (1987) 47.

4. M.J. van der Hoek, W. Werner, P. van Zuyler, B.R. Dobson, S.S. Hasnain, J.S. Worgan and G. Luijckx, *Nucl. Instrum. Methods* A **246** (1986) 380.
5. S.P. Cramer and R.A. Scott, *Rev. Sci. Instrum.* **52** (1981) 395.
6. S.S. Hasnain, P.D. Quinn, G.P. Diakun, E.M. Wardell, and C.D. Garner, *J. Phys. E. Sci. Instrum.* **17** (1984) 40.
7. J.M. Arber, E. de Boer, C.D. Garner, S.S. Hasnain and R. Wever, *Biochemistry* (1989).
8. J. Wong, F.W. Lytle, R.P. Messmer and D.H. Maylotte, *Phys. Rev.* B **30** (1984) 5596.
9. R.W. Strange, N.J. Blackburn, P.F. Knowles and S.S. Hasnain, *J. Am. Chem. Soc.* **109** (1987) 7157.
10. N. Binsted, S.L. Cook, J. Evans, G.N. Greaves and R.J. Price, *J. Am. Chem. Soc.* **109** (1987) 3669.
11. D.C. Koningsberger and R. Prins, eds., *X-Ray Absorption Spectroscopy*, John Wiley, New York (1988).
12. P.A. Lee and J.B. Pendry, *Phys. Rev.* B **11** (1975) 2795.
13. S.J. Gurman, N. Binsted and I. Ross, *J. Phys. C.* **17** (1984) 143; **19** (1986) 1845.
14. M. Vasak and J.H.R. Kagi, *Met. Ions Biol. Syst.* **15** (1983) 213.
15. J.M. Charnock, C.D. Garner, I.L. Abrahams, J.M. Arber, S.S. Hasnain, C. Henehan and M. Vasak, *Physica B* **158** (1989) 93.
16. I.L. Abrahams, I. Bremner, G.P. Diakun, C.D. Garner, S.S. Hasnain, I. Ross and M. Vasak, *Biochem. J.* **236** (1986) 585.
17. I.L. Abrahams, C.D. Garner, I. Bremner, G.P. Diakun and S.S. Hasnain, *J. Am. Chem. Soc.* **107** (1985) 4596.
18. J.D. Otvos and I. Armitage, *Proc. Natl. Acad. Sci. USA* **77** (1980) 7094.
19. J.H. Freeman, L. Powers and J. Peisach, *Biochemistry* **25** (1986) 2342.
20. B.L. Vallee, in *Metal Ions in Biology*, vol. 5, ed. T.G. Spiro, John Wiley, New York (1983) 1.
21. e.g. P.M. Colman, J.N. Jansonius and B.W. Matthews, *J. Mol. Biol.* **70** (1972) 701; J.S. Richardson, K.A. Thomas, B.H. Rubin and D.C. Richardson, *Proc. Natl. Acad. Sci. USA* **72** (1975) 1349; D.C. Rees, M. Lewis, R.B. Konzato, W.N. Lipscomb and K.D. Hardman, *Proc. Natl. Acad. Sci. USA* **78** (1981) 3408.
22. J.C. Phillips, J. Bordas, A.M. Foote, M.H.J. Koch and M.F. Moody, *Biochemistry* **21** (1982) 830.
23. G.P. Diakun, L. Fairall and A. Klug, *Nature* **324** (1986) 698.
24. A.J. Dent, D. Beyersmann, C. Block and S.S. Hasnain, *Physica B* **158** (1989) 95.
25. G.N. George and R.C. Prince, *SSRL Activity Report* (1986) 142.
26. T.G. Spiro, *Molybdenum Enzymes*, Wiley-Interscience, New York (1985).
27. R.C. Bray, *Q. Rev. Biophys.* **21** (1988) 99.
28. V.K. Shah and W.J. Brill, *Proc. Natl. Acad. Sci. USA* **74** (1976) 3249.
29. P.T. Pienkos, V.K. Shah and W.J. Brill, *Proc. Natl. Acad. Sci. USA* **74** (1977) 5468.
30. J.L. Johnson, B.E. Hainline and K.V. Rajagopalan, *J. Biol. Chem.* **255** (1980) 1783.
31. S.P. Cramer, in *Advances in Inorganic and Bioinorganic Mechanisms*, vol. 2, ed. A.G. Sykes, Academic Press, London (1983) p. 259.
32. G.N. George, C.A. Kipke, R.C. Prince, R.A. Sunde, J.H. Enemark and S.P. Cramer, *Biochemistry* **28** (1989) 2075.
33. N.A. Turner, R.C. Bray and G.P. Diakun, *Biochem. J.* **260** (1989) 563.
34. S.P. Cramer and R. Hille, *J. Am. Chem. Soc.* **107** (1985) 8164.
35. S.P. Cramer, K.O. Hodgson, W.O. Gillum and L.E. Mortenson, *J. Am. Chem. Soc.* **100** (1978) 3398; S.P. Cramer, W.O. Gillum, K.O. Hodgson, L.E. Mortenson, E.I. Stiefel, J.R. Chisnell, W.J. Brill and V.K. Shah, *J. Am. Chem. Soc.* **100** (1978) 3814.
36. S.D. Conradson, B.K. Burgess, W.E. Newton, L.E. Mortenson and K.O. Hodgson, *J. Am. Chem. Soc.* **109** (1987) 7507.
37. J.M. Arber, A.C. Flood, C.D. Garner, C.A. Gormal, S.S. Hasnain and B.E. Smith, *Biochem. J.* **252** (1988) 421.
38. S.D. Conradson, B.K. Burgess, S.A. Vaughan, A.L. Roe, B. Hedman, K.O. Hodgson and R.H. Holm, *J. Biol. Chem.* **264** (1989) 15967.
39. H. Bortels, *Zentralb. Bakteriol. Parasitenkd. Abt. 2* **95** (1936) 193.
40. J.M. Arber, B.R. Dobson, R.R. Eady, P. Stevens, S.S. Hasnain, C.D. Garner and B.E. Smith, *Nature* **325** (1987) 372; J.M. Arber, B.R. Dobson, R.R. Eady, S.S. Hasnain, C.D. Garner, T. Matsushita, M. Nomura and B.E. Smith, *Biochem. J.* **258** (1989) 733.

41. G.N. George, C.L. Coyle, B.J. Hales and S.P. Cramer, *J. Am. Chem. Soc.* **110** (1988) 4057.
42. C.D. Garner, *J. Phys. C.* **8** (1986) 1111.
43. D.E. Sayers, E.A. Stern and J.B. Heriott, *J. Chem. Phys.* **64** (1976) 427; R.G. Shulman, P. Eisenberger, B.-K. Teo, B.M. Kincaid, and G.S. Brown, *J. Mol. Biol.* **124** (1978) 305.
44. N.J. Blackburn, S.S. Hasnain, N. Binsted, G.P. Diakun, C.D. Garner and P.F. Knowles, *Biochem. J.* **219** (1984) 985; C.D. Garner and J.R. Helliwell, *Chem. Br.* **22** (1986) 835.
45. R.C. Garratt, R.W. Evans, S.S. Hasnain and P.F. Lindley, *Biochem. J.* **233** (1986) 479.

12 X-ray microscopy

P.J. DUKE

12.1 Introduction

Until very recently, the imaging of materials using X-rays has been confined to X-ray radiography in which a detector such as a photographic plate is used to register a shadow image of a specimen illuminated by the radiation. Using this method, industrial and medical radiography has been developed to a high degree of sophistication. The X-rays are produced by the impact of high energy electrons on a cooled metal target and the energy and intensity of the X-rays can be optimised by a suitable choice of target material and electron voltage. Fine focus electron sources can be used to produce X-ray beams which radiate outwards from an almost point source to produce shadow images with a low degree of penumbral blurring. Resolution typically in the region of a millimetre can be obtained which is quite adequate for a large number of applications in industrial radiography.

However, X-rays are electromagnetic radiation and as such will be refracted and diffracted by the sample and in principle it should be possible to use methods analogous to those used in optical microscopy to produce images with much higher resolution than is possible using radiographic methods. Indeed, because the wavelength of the X-radiation is at least 100 times shorter than visible light, the resolution obtainable from an X-ray microscope should be very much better than that obtainable from the light microscope. That this potential has not so far been realised is because of two major considerations which prevent the methods of light microscopy being simply transferred over to the X-ray region. These are as follows:

(a) the refractive index of all materials is almost equal to unity for X-rays. This means it is impossible to construct an X-ray lens to focus the radiation and
(b) even if an X-ray lens could be constructed, the electron impact X-ray source (or X-ray tube) cannot be made sufficiently intense to produce a high resolution image in a reasonable length of time.

On the other hand there are definite incentives to seek solutions to these problems. The potentially superior spatial resolution possible with X-rays has already been mentioned. X-rays also have a greater penetrating power compared with visible radiation and their different interaction mechanisms with matter can be used to highlight different features of the specimen.

A further reason why X-ray microscopy has not been developed becomes clear when we compare X-rays and electrons. High energy electrons not only have a wavelength shorter even than X-rays which gives them a further resolution advantage compared with visible light but electrons carry an electric charge which means that they can be focussed using electromagnetic lenses. This key property has led to the development of the electron microscope for the biological and materials sciences. The present day commercial availability and wide use of the electron microscope in both its imaging and scanning form has meant that a wide variety of techniques has been developed for specimen preparation and staining. Familiarity with these, combined now with advanced methods of computerised image processing has led to the widespread use of the electron microscope and to a comparative neglect of all other methods of examining materials with sub-micrometre resolution.

Nevertheless, their lower penetrating power and higher scattering probability compared to X-rays have confined the use of the standard 1 keV transmission electron microscope to the study of thin specimens only or to specimens made artificialy thin by slicing. The use of heavy metal staining to improve contrast is essential for many examinations although the use of frozen hydrated specimens is becoming a valuable alternative to staining. At the same time the advent of synchrotron radiation sources of intense X-ray beams and the development of sub-micrometre structures for the diffractive focussing of these beams have given a new incentive to the development and application of X-ray microscopy. Although it is unlikely that X-ray microscopy will ever achieve the high resolution obtainable, under ideal conditions, with thin specimens in the electron microscope, it now seems possible that the different properties of X-rays compared with both visible light and electrons can contribute to the development of microscopy.

12.2 X-ray optical systems

Although it is widely believed, and enshrined in many text books, that X-rays cannot be focussed, this is correct only if focussing by refraction is in view. X-rays can be focussed using both reflection and diffraction and both methods are now in use, particularly at synchrotron radiation centres (see Chapter 1).

The passage of a beam of X-rays through a medium can be expressed in terms of the complex refractive index of the medium, n, given by

$$n = 1 - \delta - i\beta \tag{12.1}$$

The amplitude of the incident X-radiation will be attenuated by thickness t of the medium by a factor

$$\exp(-2\pi\beta t/\lambda) \tag{12.2}$$

where λ is the wavelength of the radiation, and will be advanced in phase by an amount

$$\phi = 2\pi\delta t/\lambda \qquad (12.3)$$

In fact δ is a small positive quantity (of order 10^{-2} to 10^{-3}) so that the real part of the refractive index is less than unity. It follows that X-rays can be totally externally reflected when their grazing angle relative to the surface of reflection is small enough. Specifically, since δ is small compared with unity, Snell's law gives

$$\theta_c = \sqrt{2\delta} \qquad (12.4)$$

for the grazing angle below which total external reflection will take place. This grazing angle is typically a few degrees for X-rays incident on a gold surface and this property is being used for the construction of X-ray focussing systems for microscopy [1] and also for X-ray astronomy [2]. Specially figured surfaces (hyperboloid and ellipsoid or paraboloid) must be used, as proposed, for example by Wolter [3]. These are difficult to construct to the required sub-wavelength accuracy. Likewise the surfaces must be smooth to obtain good reflectivity. These problems have not prevented grazing incidence optical systems being used successfully for the concentration of X-ray beams from synchrotrons as well as from conventional laboratory X-ray sources but have prevented their use as imaging optics for the X-ray microscope. Nevertheless considerable effort is being made to produce improved optical systems of this type [4, 5] and spatial resolution of test objects in the region of $1 \mu m$ has been obtained (see Chapter 13).

Reflective optics with multilayer coatings (Section 1.3.4) to improve the reflectivity at near-normal incidence have also been applied to the construction of microscope optics [6]. The fabrication tolerances are more stringent than for grazing incidence reflectors but the optical surfaces can be spherical or almost spherical which makes them easier to construct. A microscope with this type of optic is being constructed by Spiller [7].

The only X-ray focussing optic which has been used successfully to give sub-micrometre resolution is the Fresnel zone plate. Figure 12.1 shows a plane wave front which is producing an electromagnetic disturbance at a distant point P. Each point on the wave front can be viewed as a source of secondary wavelets, all of which contribute, by vector addition, to the X-ray amplitude at P. The wave front has been divided into concentric half-period zones such that the wavelets from each point in one zone have an additional path difference $\lambda/2$ (are phase retarded by π radians) compared with the next innermost zone. Each adjacent zone then contributes an equal amplitude at P but π radians out of phase with the contribution from the adjacent one. If now each alternate zone is rendered opaque then the contributions at P from the remaining zones all add in phase. A Fresnel zone plate is drawn as a series of concentric rings with just the property described above. The radius of the nth zone, r_n is given

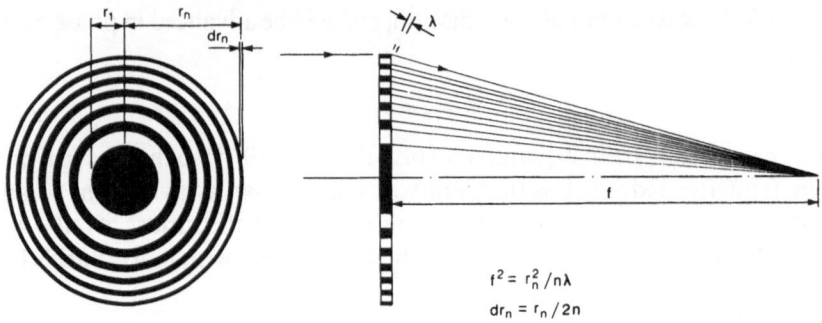

Figure 12.1. Schematic illustrating the point focussing properties of a circular diffraction grating or zone plate.

by

$$r_n^2 = nf\lambda + \frac{n^2\lambda^2}{4} \tag{12.5}$$

and the width of the nth zone by

$$\delta r_n = \frac{f\lambda}{2r_n} + \frac{n\lambda}{2} \tag{12.6}$$

so that a parallel beam of X-radiation, intercepted by such a zone plate, will be focussed (at least in part) to the point P which is at a distance f from the plate. The zone plate will behave as a lens with a focal length f.

The point spread distribution of the focussed radiation will have the usual Airy distribution generated by an aperture of diameter, D, equal to that of the zone plate. This distribution will have a half-width, R_D, given by

$$R_D = 1.22\frac{f\lambda}{2r_n} \tag{12.7}$$

It follows from Eqns. 12.6 and 12.7 that the spatial resolution provided by such a focussing zone plate is given simply by the minimum zone width (MZW), δr_n, so that

$$R_D = 1.22\,\delta r_n \tag{12.8}$$

The second (and higher order) terms in Eqns. 12.5 and 12.6 are negligible and have been neglected.

An X-ray focussing zone plate must be a very small object. If the spatial resolution achieved by the X-ray optical systems is to be of the same order as the wavelength of the radiation, it follows from Eqns. 12.5 and 12.6 that the MZW must also be of this order. In turn, for a reasonable number of zones (e.g. 100–1000) the diameter of the zone plate can only be a few micrometres. This puts very stringent conditions on the zone plate fabrication techniques.

12.3   Methods of zone plate fabrication

Zone plates with an MZW down to about 1 μm have been available for a
number of years and can be bought commercially. However, in order to
construct an X-ray optical system which is capable of imaging specimens to a
resolution better than 1/10th of a micrometre, zone plates of much better
MZW are required. These can only be made either by optical interferometric
techniques or by lithography using a scanning electron microscope. A third
method, that of sputtering layers onto a rotating wire is also being tried and
produces zone cylinders—a high aspect ratio version of the zone plate.

   The first successful X-ray focussing zone plates were constructed using laser

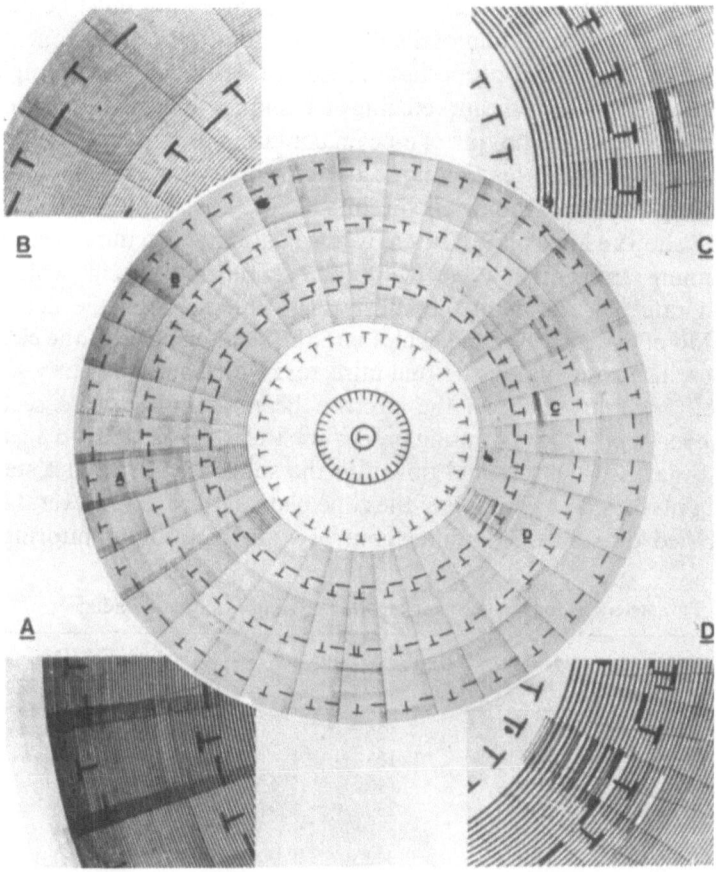

Figure 12.2.   Photomicrograph of a zone plate prepared using the carbon contamination writing
method described in Section 12.3 and [13]. This zone plate has a diameter of 47 μm and a
minimum zone width of 70 nm. The registration marks (T) are used to define each writing sector.
Enlarged views of four sectors are shown in insets A–D.

interferometry by Rudolph and Schmahl [8] at the University of Göttingen, Federal Republic of Germany. They were used at the ACO electron storage ring to produce the first X-ray optical system with a resolution better than 1 μm. Two types of zone plate were produced, condenser zone plates which were used for specimen illumination and microzone plates which were used as objective lenses in the microscope. Detailed descriptions of the methods used are given by Schmahl et al. [9], Guttmann [10] and Thieme [11].

Electron beam lithographic techniques, developed for the production of high feature density microcircuits in the electronics industry, can be used for the production of microzone plates suitable for the focussing of X-ray beams. The feature sizes required for high resolution zone plates are much less than are available routinely in microcircuits produced by electron beam lithography so special methods must be used to produce structures whose minimum feature size is less than 0.1 μm. Careful attention must be paid to calibration and the elimination of distortion in order to produce zone plates with the necessary optical properties. These are described, for example by Bögli et al. [12]. The e-beam etching techniques described above are complementary to the technique of e-beam contamination writing developed by Buckley et al. [13]. In this technique, the zone plate pattern is deposited as carbon atoms on a carbon, silicon nitride or boron nitride substrate. Figure 12.2 shows a zone plate written by this method. A Vacuum Generators HB5 scanning transmission electron microscope is used in which the deposition rate of carbon on the substrate is enhanced by the deliberate introduction of hydrocarbon molecules into the vacuum close to the electron beam focus. The zone plate is written in sectors which can be clearly seen in Figure 12.2 and movement of the electron beam from sector to sector is obtained by a stepping and rotating procedure. Reference is made to a pattern of fiducial marks which are laid down on the substrate as the first stage in the writing process. The creation of the zone plate pattern takes several hours and is carried out under computer control with operator monitoring and

Table 12.1 Parameters of zone plates manufactured at King's College, London.

r_n (μm)	dr_n (Å)	f(mm) at $\lambda = 30$ Å	n	m	$\lambda/d\lambda$	df(FWHM) (μm)
25	1000	1.6	126	1	210	12.4
23.5	750	1.13	158	1	260	7.2
21.5	750	1.03	145	1	240	7.2
22.5	500	0.71	230	1	377	3.2
28.5	300	0.55	476	1	780	1.2
18	300	0.35	303	1	500	1.2
18	180	0.21	498	1	816	0.4
25[a]	700	0.56	120	2	390	3.2

Source: Burge et al. [14].
[a]This zone plate has every other absorbing zone missing to enhance the diffraction efficiency in the second order.

intervention where necessary. Carbon patterns whose MZW is less than 200 Å have been produced by this method. These are the smallest feature size zone plates produced up to the present time. They can be replicated into gold patterns [13], but this has been successful only down to a MZW of about 500 Å. Table 12.1 lists the properties of zone plates constructed recently at King's College, London.

Sputtering techniques for the production of high aspect ratio zone plates are very promising as a method of producing zone plates which can be used to focus X-rays with wave lengths shorter than about 20 Å. The requirement for good contrast between the opaque and transparent zones becomes harder to achieve because of the reduced absorption coefficient of the material used for the construction of the zone plate as the wavelength becomes shorter. The method, which was first described by Rudolph and Schmahl [8] relies on the deposition of alternate layers of materials such as carbon and tungsten carbide which are, respectively semi-transparent and semi-opaque to X-rays in the 1–100 Å wavelength range. The materials are deposited on a thin rotating wire with monitoring of the thickness of the deposited layer. Zone plates produced by this method have been described, for example, by Bionta et al. [15] and by Saitoh et al. [16].

12.4   X-ray microscope designs

12.4.1   Source brilliance

The design of an X-ray microscope is constrained by the physical and optical properties of the zone plate. The small size of the zone plate means that it can only be used with a well collimated source of X-radiation unless it can be placed very close to the source itself. This may be possible with laser plasma source of X-rays but, in practice, X-ray microscopes using zone plate focussing optics have been confined to synchrotron radiation sources and are likely to remain so for some time to come. The high degree of collimation of the synchrotron source, particularly the radiation from an undulator magnet (see Section 1.2.3), which is collimated in both the vertical and the horizontal directions makes it particularly well matched to the physical dimensions of the zone plate. The key property of the source in this respect is its emittance which is the product of the spatial extent of the source and its divergence. A source will have both a horizontal and a vertical emittance (see Section 1.2.2): these will not, in general, be the same. The emittance is measured in millimetres × milliradians and is an invariant property of the beam of radiation along the path between the source and the microscope. An associated quantity is the source brilliance, which is the number of photons radiated by the source into unit bandwidth divided by the product of the horizontal and vertical emittance (Eqn. 1.5). Table 12.2 gives a comparison of the brilliance of a

Table 12.2 Brilliance comparison of synchrotron radiation sources.

Source	Location	Status	Energy	Brilliance
BESSY, Dipole	Berlin, FRG	Operating	800 MeV	2×10^{13}
SRS, Undulator	Daresbury, UK	Operating	2 GeV	2×10^{14}
NSLS, Undulator	Brookhaven, USA	Operating	2.5 GeV	3×10^{16}
PF, Undulator	Tsukuba, Japan	Operating	2.5 GeV	1×10^{16}
ALS, Undulator	Berkeley, USA	Calculated	1.5 GeV	5×10^{18}
ESRF, Undulator	Grenoble, France	Calculated	6 GeV	1×10^{18}

Note: Brilliance is defined as photons/(mm^2 mrad^2 s) into 0.1% band width at an X-ray wavelength of 44 Å and a circulating electron beam current of 100 mA.

number of selected synchrotron sources and includes those in use for X-ray microscope development as well as some sources under construction where X-ray microscopes are being planned for installation. It is clear from the table that there is a strong tendency to design and construct sources with higher and higher brilliance. This can only be achieved by careful design of the storage ring magnet systems although two sources, the SRS in the UK [17] and the Photon Factory in Japan [18] have recently been upgraded to improve their source brilliance.

12.4.2 *Monochromators*

In order to be useful for the microscope the beam from the storage ring must be conditioned using a monochromator. This is because the zone plate is a highly chromatic device whose focal length varies inversely with the wavelength of the radiation (see Eqn. 12.5). A high resolution monochromator is not required; it is sufficient that the resolving power be of the same order as the number of zones in the zone plate. Table 12.1 gives some numerical values of the monochromator resolving powers which are required to provide a bandwidth match [19]. In practice a compromise is needed because it is impractical to change the monochromator grating whenever a different zone plate is used. It may be possible in the future to use a grating monochromator in which the grating can be moved to present regions of different groove density to the incident beam but that degree of sophistication has not yet been reached. Good matching is important however because otherwise there is either loss of useful radiation if the resolving power of the grating is too high or the microscope performance is degraded by chromatic aberration.

12.4.3 *Scanning versus fixed beam arrangements*

An important decision is whether to build a scanning or a fixed beam microscope. A scanning instrument has a number of definite advantages over a fixed beam instrument. This arises principally from the very low efficiency of the zone plate as a focussing element. An amplitude zone plate, which relies on the attenuation of the radiation in the opaque zones to provide the necessary modulation of the incident wave front to generate focussed radiation at the sample has an efficiency of about 10% for radiation reaching the first order focus. This is the theoretical maximum and the practical achievement is usually only about 5% or even less than that. The microzone plate in a fixed beam instrument has the same function as a microscope objective lens so its low efficiency means that most of the radiation passing through the specimen cannot be focussed by the zone plate and is therefore wasted. This unused radiation may even degrade the image by contributing to the background in the image plane. If, as is often the case, the specimen is radiation sensitive, the fixed beam geometry leads to radiation damage without contributing to the observed image. This post-specimen optical system is not needed in the scanning instrument. The microzone plate is used to form a beam of focussed radiation, through which the specimen can be scanned. The radiation transmitted by the specimen is detected in some type of photon counting device with a high X-ray detection efficiency.

A schematic layout of the scanning transmission X-ray microscope (STXM) under development at the SRS [20] is shown in Figure 12.3. Radiation from

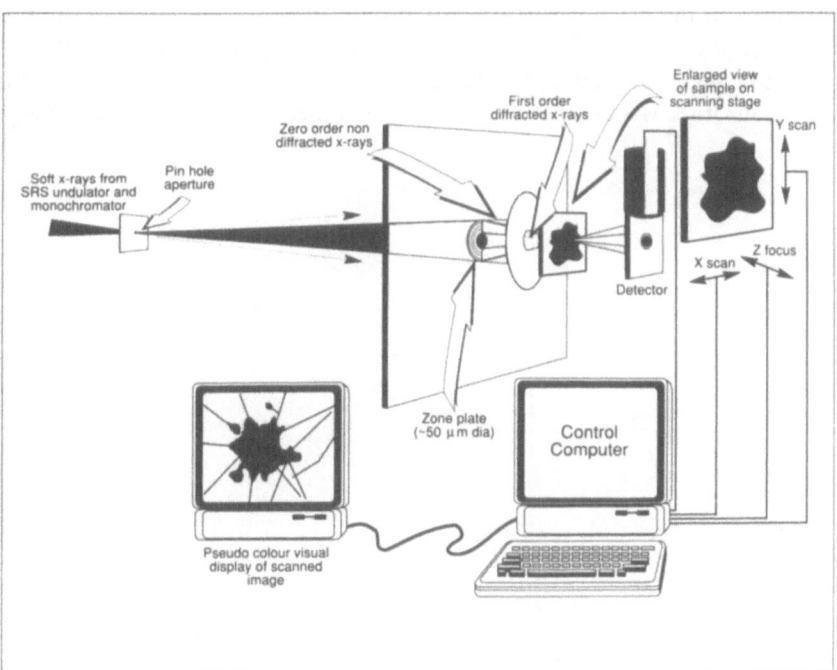

Figure 12.3.   Schematic showing the operation of scanning transmission X-ray microscope (STXM) described in Section 12.4.3 and [20].

the magnetic undulator passes through a monochromator and a zone plate is used to focus the exit slit of the monochromator on to the specimen. The X-radiation passing through the specimen is counted in a simple gas-flow proportional counter (Section 1.5.1) whose output is processed and displayed visually during the period of acquisition of the image. The motion of the scanning stage is provided by two piezoelectric crystals with linear feedback which are mounted on a motor driven coarse stage so that a wide range of scanning steps can be achieved with the ability to take rapid scans of a large area with low resolution in order to locate regions of interest in the specimen which can then be examined at high resolution. A circular aperture, not shown in the figure, is placed between the zone plate and the specimen to eliminate the zero order undiffracted radiation which would otherwise generate an unwanted background in the proportional counter. A small central region of the zone plate is rendered opaque by means of a gold layer so that the focal spot is formed within the geometric shadow.

Although the scanning X-ray microscope has the advantage of lower radiation damage to the specimen compared with the fixed beam instrument, a significant disadvantage is the time taken to collect the image which precludes the collection of data from a time-varying sample unless the time taken for a significant change in the sample is long compared with the imaging time. In fact actual imaging times obtained in practice are between minutes and hours, depending on the brilliance of the radiation source, the efficiency of the pre-specimen optics and the characteristics of the scanning stage. The advantage of the fixed beam instrument in this regard is that the entire image is obtained at once or, in other words, the pixel data is obtained in parallel rather than in

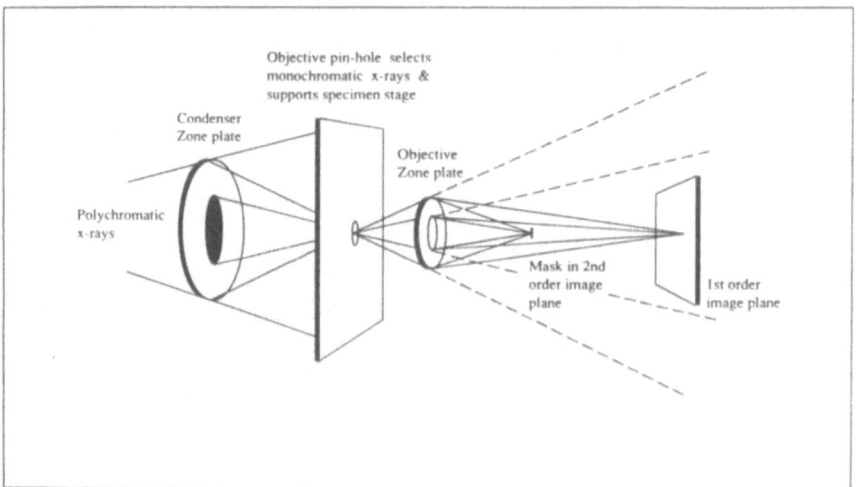

Figure 12.4. Schematic showing the principles of the fixed beam microscope described in Section 12.4.3 and [21].

series. Figure 12.4 shows a schematic arrangement of a fixed beam instrument [21] which uses two zone plates as the pre-specimen optical system. The basic arrangement is the same as in a conventional optical microscope; the condenser zone plate concentrates the X-radiation onto the specimen and the microzone plate has the function of the objective lens. Because the synchrotron source supplies polychromatic radiation to the microscope the condenser zone plate also acts as a monochromator whose bandwidth must be matched to the acceptance bandwidth of the microzone plate. Referring to Figure 12.4, the specimen is mounted on a circular aperture which is placed at the first order focus of the condenser. The aperture acts also to prevent zero and higher order radiation from reaching the region of the objective lens and the size of the aperture defines the bandwidth of the monochromator. The central region of the condenser zone plate itself is obscured in order to prevent zero order radiation reaching the specimen. The entire aperture of the microzone plate is then used to produce a magnified image of the specimen on an X-ray sensitive film located in the image plane. Using this arrangement, Schmahl et al. [22] were the first to demonstrate X-ray microscope operation with focussing optics. These first experiments were made at DESY in Hamburg and later at Orsay. More recently the microscope, operated by the same group but now installed at BESSY, in Berlin, has been used to obtain images of test specimens in a few seconds, using X-ray film as the detector [21].

12.5 Applications of X-ray focussing microscopes

There is at least one specific area of work where the X-ray microscope can claim to have a distinct advantage over its electron counterpart, namely the imaging of thick, wet specimens as close as possible to their natural state. Standard electron microscopy procedures include the slicing of dried specimens, their staining with heavy elements and their observation in ultra high vacuum. The success of this technique cannot be doubted and it might be claimed that at best all that X-ray microsopy could accomplish would be to confirm results already obtained. It is, of course easy to point out that specimens prepared in this way are very far removed from their natural state, that what is actually being observed is the distribution of the staining material and not the distribution of the tissue itself. On the other hand, extrapolation from what is seen in the light microscope would indicate that there is no strong discontinuity in what is being observed by the two techniques. Furthermore confocal optical microscopy [23] and the use of fluorescent staining [24] are encroaching on the region over which X-ray microscopy can claim exclusive rights. The parallel development of rapid freezing techniques in electron microscopy [25] and the use of amorphous ice [26] as a embedding agent for specimen preparation are also important new developments in electron microscopy which make it possible to observe unstained specimens in the electron microscope. Nevertheless, the X-ray microscope can continue to claim an important niche between optical and electron microscopy and to provide a region of overlap with both of these. X-ray microscopy can further claim the use of the characteristic X-ray absorption edges as a method of elemental mapping on a sub-micrometre scale. This applies to light elements such as calcium which are present naturally in many materials and also to other elements, such as gold, which can be introduced by the use of labelled, protein specific, antibodies. Subsequent sections will show how these techniques are starting to be used in X-ray microscopy on an experimental basis.

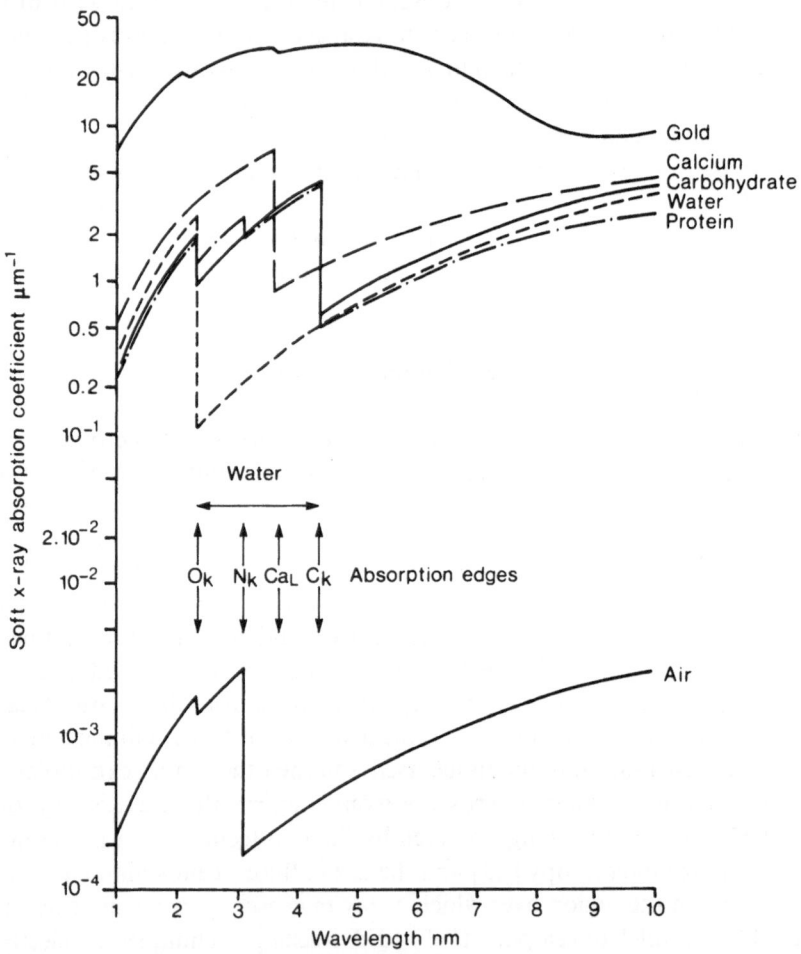

Figure 12.5. Linear absorption coefficients of X-rays in various materials as a function of wavelength. The behaviour of the coefficients in the vicinity of an absorption edge is illustrative only.

12.5.1 X-ray contrast

Contrast in X-ray microscopy is produced mainly by the photo-electric absorption of X-rays in the specimen. The absorption cofficient, μ, is related to the complex refractive index of Eqn. 12.1 by

$$\mu = 2\pi\beta/\lambda \tag{12.9}$$

and is a monotonic function of X-ray wavelength except when the X-ray energy, E, matches the binding energy of a particular core level (E(keV) $= 12.4/\lambda$(Å)). At this energy resonant absorption of the photon takes place and the value of μ shows a sharp increase. This behaviour is shown in Figure 12.5 for certain elements and compounds of particular biological interest. The sharp discontinuities in the behaviour of the absorption coefficients for

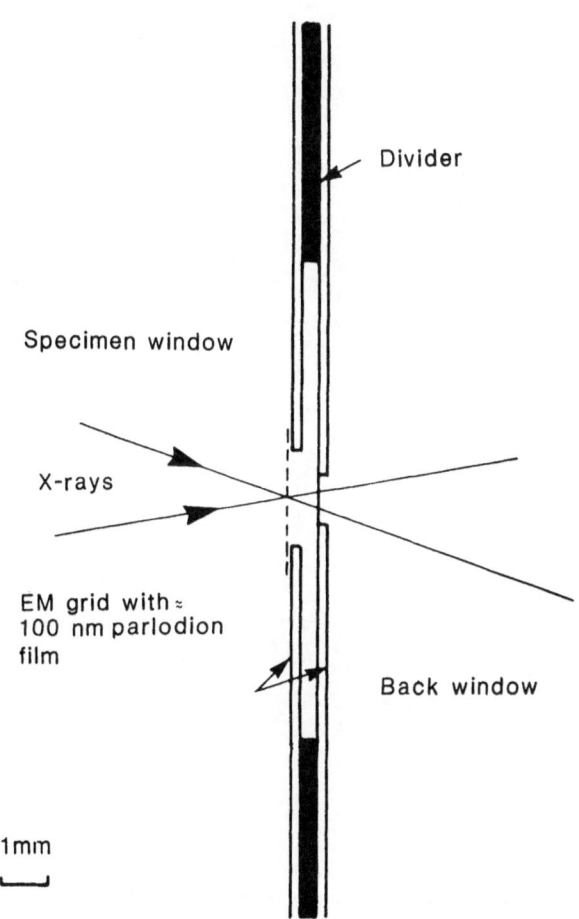

Figure 12.6. Schematic of the wet cell referred to in Section 12.5.1.

carbohydrate and protein occur at the carbon, nitrogen (in the case of protein only) and oxygen K absorption edges. This variation should be compared with that of the absorption coefficient for water which is smooth except for the oxygen K edge discontinuity. The X-ray absorption law follows a simple exponential behaviour, as a function of the thickness, t, of the specimen, so that

$$I = I_0 e^{-\mu t} \tag{12.10}$$

where I_0 and I are the incident and transmitted X-ray intensities. This means that in the region of the X-ray spectrum between the C and O edges, at 282 eV and 564 eV respectively, the so-called water window, there is strong natural

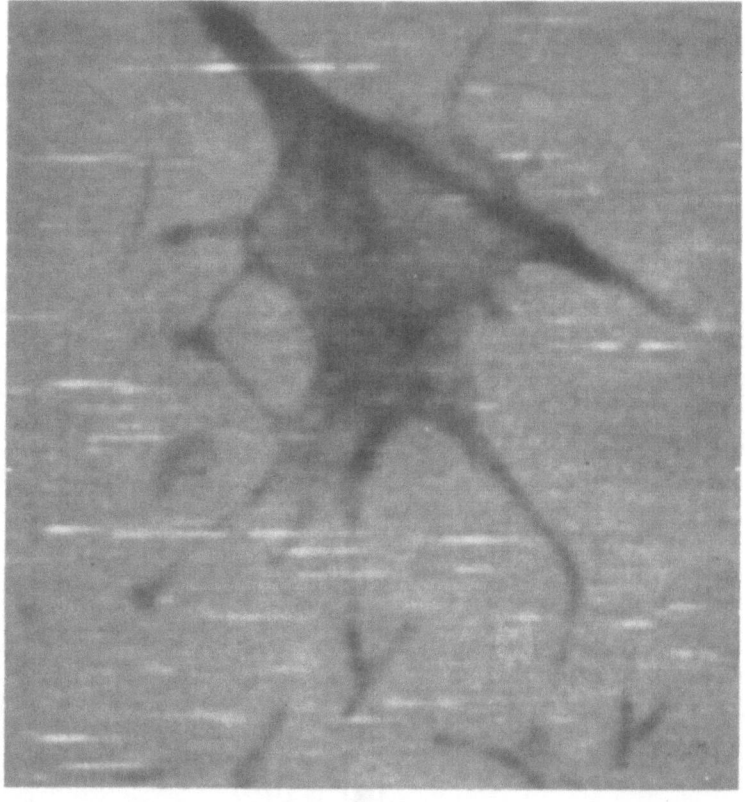

5μm

Figure 12.7. Enlarged reproduction of an image of an hydrated chick embryo neuronal growth cone, stained with gold-labelled antibodies to the nerve-specific protein 3D5. The image was obtained using the STXM of [20] at an X-ray energy of 379 eV.

contrast between the carbon-containing biological material and the surrounding aqueous medium. In order to make use of this contrast, the specimen must be held in a wet state throughout the exposure to the X-radiation. A specially designed cell can be mounted on the scanning stage and placed between the collimator and detector in the scanning microscope or across the condenser zone plate aperture in the case of the fixed beam instrument. A simple cell, used at the SRS, is shown in Figure 12.6. The sample is contained between two parlodion films which are separated by an electron microscope grid. Figures 12.7 and 12.8 show images of a neuronal growth cone and a fibroblast. The images shown in the figures are of preparations which have been stained with gold-labelled antibodies specific to certain proteins of the cellular surface so that the pictures represent the first attempt to obtain a protein map of a biological cell using focussing X-ray microscopy. A wet cell has also been constructed for use in the BESSY fixed beam microscope [21]. In this design, the biological cells were grown on a polimide foil mounted on a quartz ring. After the cells had been cultivated on the foil in a culture medium, the foil, held

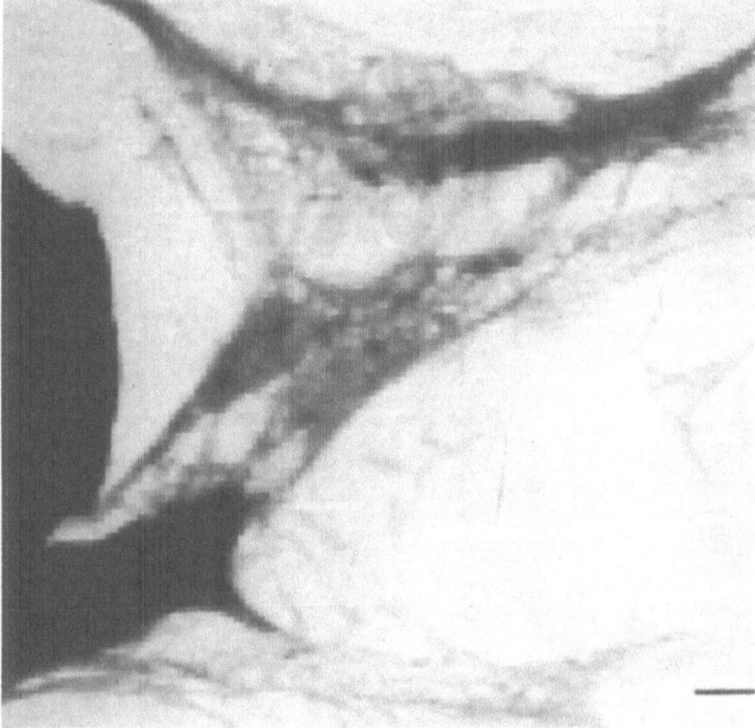

Figure 12.8.   Enlarged reproduction of an image of chick embryo fibroblasts, stained with gold-labelled antibodies to the protein vimentin. The image was obtained using the STXM of [20] at an X-ray energy of 379 eV. The scale bar is 5 μm.

(a)

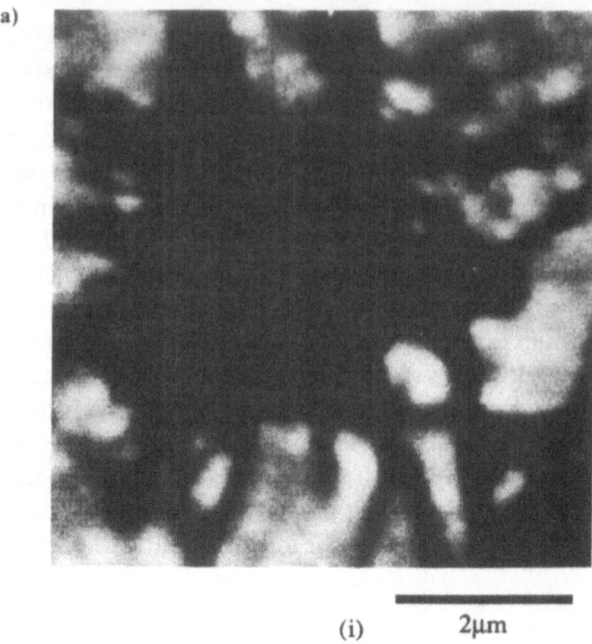

(i)   2μm

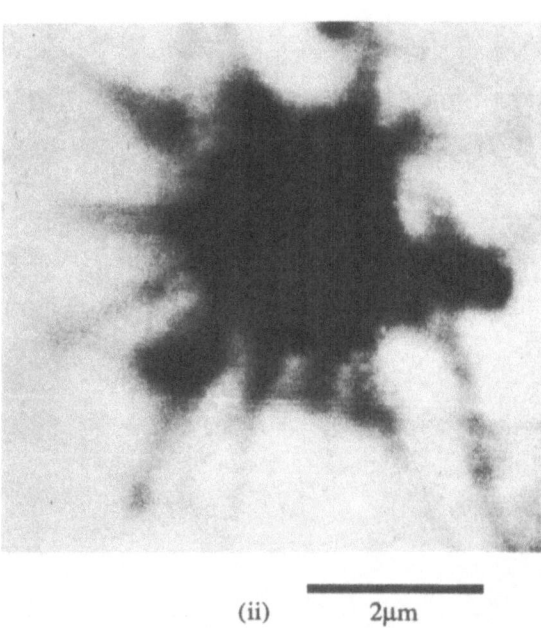

(ii)   2μm

Figure 12.9. (a) Enlarged reproduction of a pair of images of a single grain of Portland cement, hydrated for 48 h. The images were obtained using X-rays differing by only 2 eV in energy using the STXM of [20], (i) low energy image, (ii) high energy image. The striking difference in the appearance of the images is caused by the greatly increased absorption of X-rays by calcium at 351 eV [20]. (b) Difference images obtained from (a): (i) direct difference; (ii) logarithmic difference.

(b)

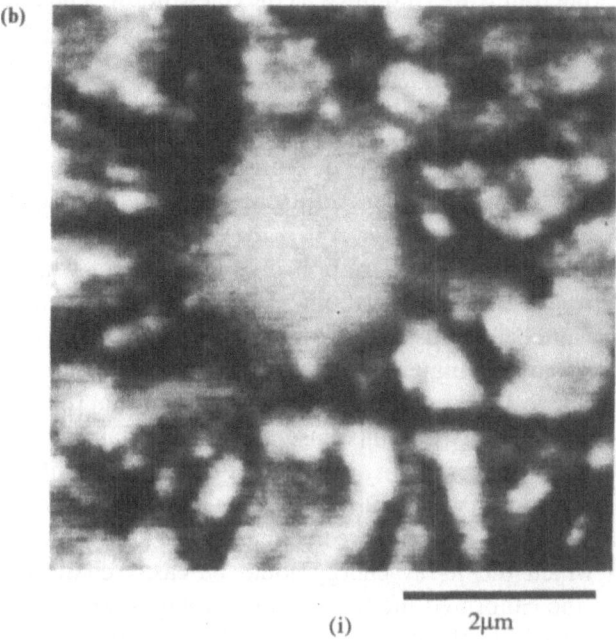

(i)   2μm

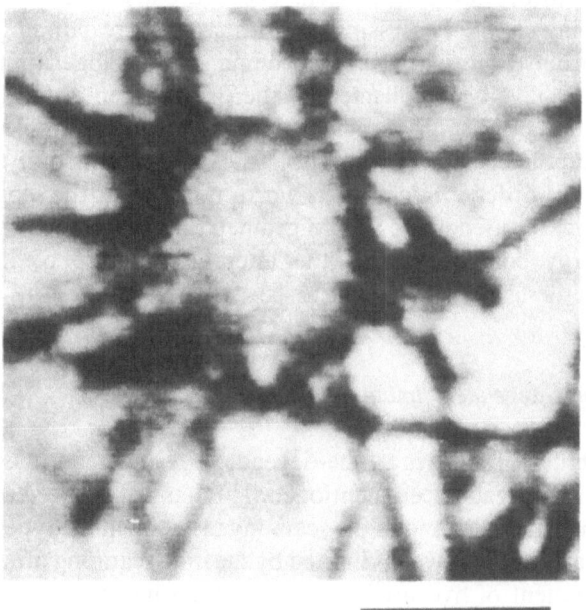

(ii)   2μm

in the quartz ring frame, was mounted as the cell window. The thin films used for specimen containment may not be completely water-tight so that a water reservoir needs to be provided. Furthermore it will be necessary to monitor the water content of the cell during the irradiation to be sure that the specimen remains fully hydrated. Images of wet specimens have also been reported by the Brookhaven group [27] and, more recently by the BESSY group using their fixed beam instrument in a phase contrast mode of operation [28].

Figure 12.5 also shows the variation of μ for calcium which presents a discontinuity (the L III edge) at about $E = 355$ eV. Images taken at X-ray energies above and below this energy can be used differentially to map the Ca distribution in a suitable material. The variation of μ is idealised in the figure and a more accurate representation [29] makes it clear that the X-ray energies must be chosen carefully to maximise the differential absorption effect. Figure 12.9 shows images of a particle of Portland cement after 48 h of hydration. The particle had, by this time, grown tentacles of a form of calcium silicate, Ca_3SiO_5. These images were taken using the X-ray microscope at the SRS [20] in which the X-ray energy could be easily changed by a small adjustment of the calibrated monochromator. Images were taken at 351 eV and 353 eV, well within the energy bandwidth of the microzone plate, and exhibit a distinct difference caused by the change in Ca absorption close to the L III edge. This is apparent especially in the region around the particle, in the centre of which the X-ray absorption is total in both images so that no differential absorption can be seen. The figure also shows images obtained after applying simple processing algorithms to corresponding pixels of each image in order to present a difference image and a logarithmic ratio image. Similar data have been collected by the Brookhaven group [30, 31], in this case using thin bone samples. These represent the first example of Ca mapping at sub-micrometer spatial resolution and with a sensitivity down to concentrations of order $2 \mu g/cm^2$. It is worth noting that the separated function monochromator used in the STXM geometry makes it much easier to vary the X-ray energy compared with the integral structure of the fixed beam microscope described in Section 12.4.3.

12.5.2   *Time-dependent imaging*

The possibility of studying time-dependent phenomena at sub-micrometre resolution has already been mentioned. This has not yet been attempted in the fixed beam instrument and in the scanning instrument the type of samples that can be observed in this way is limited by the long scanning times. Nevertheless the development of hydrated cement granules has been observed using the scanning instrument [20] and some results are shown in Figure 12.10. The

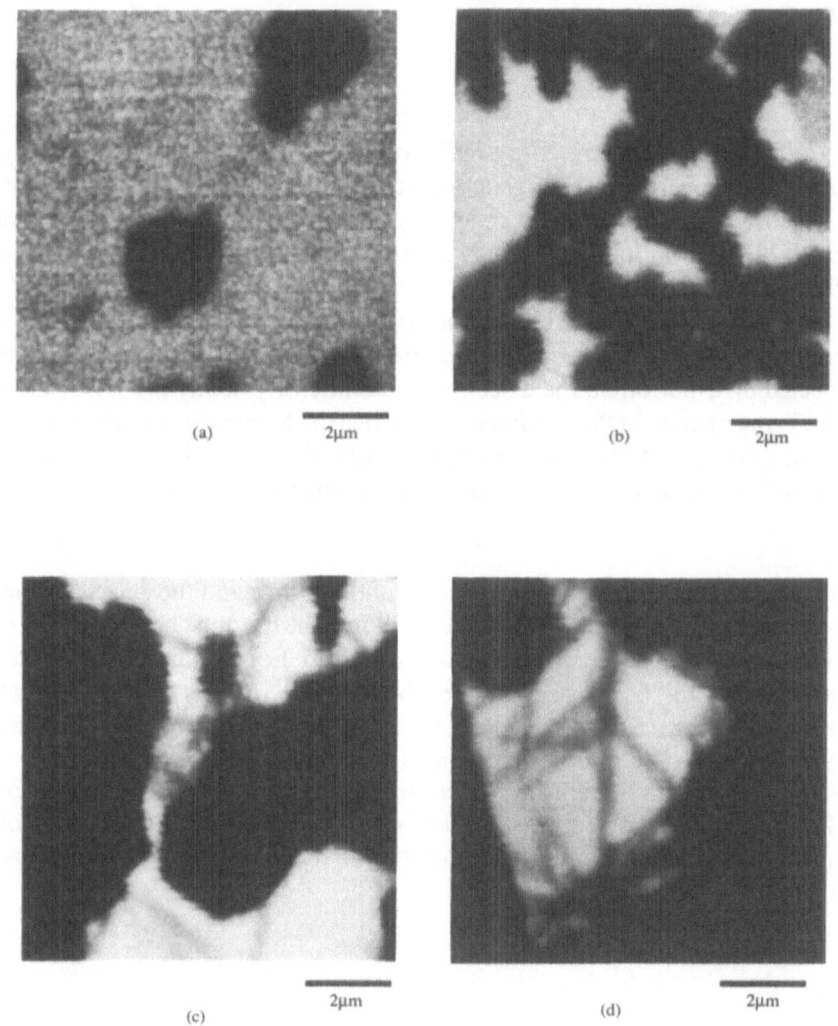

(a)   2μm   (b)   2μm

(c)   2μm   (d)   2μm

Figure 12.10. Enlarged reproductions of images of Portland cement taken at various stages of hydration as a function of time and at an X-ray energy of 380 eV using the STXM of [20]: (a) after 12 h; (b) after 22 h; (c) after 36 h; (d) after 48 h. The cement specimen shown here and in Figure 12.9 was supplied by Dr. RJ Oldman, ICI plc. (d) has been published previously by Morrison *et al.*, *Rev. Sci. Instrum.* **60** (1989) 2464.

granules were observed at approximately 12-h intervals over a 48 h period of hydration in a 5:1 cement/water mixture. The figure shows the granules in their initial state, the clustering of the granules after about 12 h, the start of tubule development after a further 12 h and the development of interlocking tubules 48 h after the start of the process. Each of these observations, though not of identical granules, was made with the cement in a wet state.

12.6 Contact X-ray microscopy

12.6.1 *Geometrical considerations*

The difficulty in focussing X-ray radiation and the advent of intense X-ray sources have generated a resurgence of interest in contact X-ray radiography. This method of recording X-ray images was described in detail by Coslett and Nixon [32] but was limited by the X-ray sources then available. The principle of the method is shown in Figure 12.11. A specimen, in close proximity to an X-ray detector such as film or photoresist, is irradiated with X-radiation from a point source. X-ray transmitted through the specimen (Eqn. 12.10) form a projected distribution on the detector, whose intensity at any point is related to the X-ray opacity of the specimen. In practice the X-rays do not radiate from a point source and the linear dimension (d) of a point in the specimen, projected onto the detector, is related to the source size (s) by

$$d = sb/a \tag{12.11}$$

where a and b are the respective distances of the source and the detector from the specimen. If the source size s is of order 1 mm, the b/a must be about 10^{-5} if a spatial resolution of order 100 Å is to be obtained. For a synchrotron source, whose horizontal beam size can be about 1 mm (or less in the vertical direction), the source to specimen distance, a, will often be about 10–15 m (see Chapter 1), so that b must be less than 100 μm under these conditions.

As well as generating absorption contrast, discontinuities in the specimen will produce diffraction contrast, which, in the straightforward application of contact microscopy also limits the resolution that can be obtained. A measure of the effect is given by the quantity.

$$r_d = \sqrt{b\lambda} \tag{12.12}$$

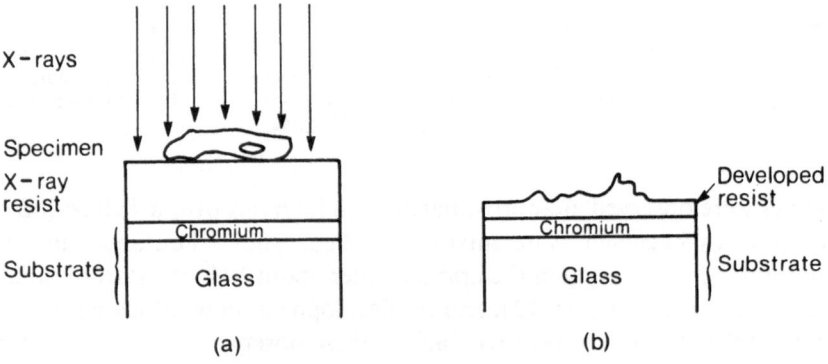

Figure 12.11.   X-ray lithography geometry showing irradiation of the resist through the sample and the resist after development.

which, for the conditions outlined above, and for radiation in the water window, means that b must be about $200\,\text{Å}$ if diffraction is not going to seriously degrade the image. This not only means that the specimen must be brought into intimate contact with the detector, but also that only thin specimens can be successfully imaged without serious diffraction effects.

12.6.2 *Photoresist*

The clarity of the image is also affected by the properties of the detector. High resolution X-ray film has a grain size of about $1\,\mu$m, and photographic plates prepared for nuclear particle detection can have an even smaller grain size. However, if the detector is to be matched to the resolution potential of the method, a much finer detector such as photoresist must be used. This material has a polymeric structure in which the incident X-rays induce bond-breaking in the polymer chain. The resist can then be developed using an etching agent which dissolves the damaged regions at a rate dependent on the local radiation damage so that, at the end of the development process, a contour image of the X-ray intensity on the resist surface is obtained. This image can be examined using an optical or an electron microscope and it is at this stage that the magnification takes place. Although the resolution limit of the detector is determined ultimately by the dimensions of the polymer molecules, in practice great care must be taken in the development process if artifacts are not to be introduced. These problems have been discussed in detail by Cheng et al. [33] and by Shinozaki and Robertson [34]. Their findings may be summarised as follows:

(a) after exposure the specimen must be removed from the resist surface but small fragments of the specimen may remain and, in addition, the surface may be damaged,

(b) the resist surface may collect an electrical charge during examination under the electron microscope unless the developed resist has been given a thin metallic coating,

(c) expert interpretation of the electron microscope image is required. In particular the possibility of diffraction effects at boundaries must be recognized and allowed for,

(d) highly irradiated regions may undergo a self-development process before etching,

(e) the dissolution rate of the exposed resist in the etching bath depends on the X-ray exposure so that contrast and the preservation of fine detail depend on the local exposure level,

(f) photoresist has a high sensitivity to UV radiation so that care must be taken to prevent UV light from the synchrotron source reaching the surface of the resist,

(g) photoresist is known to be inhomogenous and to have a position-dependent dissolution rate which can introduce random noise into the developed image.

Although there is a broad experience of the use of photoresist material for the fabrication of X-ray diffraction gratings and zone plates [13, 35], these applications do not place such stringent requirements on the properties of the resist material and are not hindered by a steep, non-linear response function. The non-linearity of the dissolution rate dependence on X-ray exposure creates serious difficulties for its use in contact microscopy and makes it hard to obtain clear quantitative information from the contact images. Nevertheless, the medium is the only X-ray detector with spatial resolution at the angstrom unit level, and as such demands serious attention in order to define the optimum regime for its use.

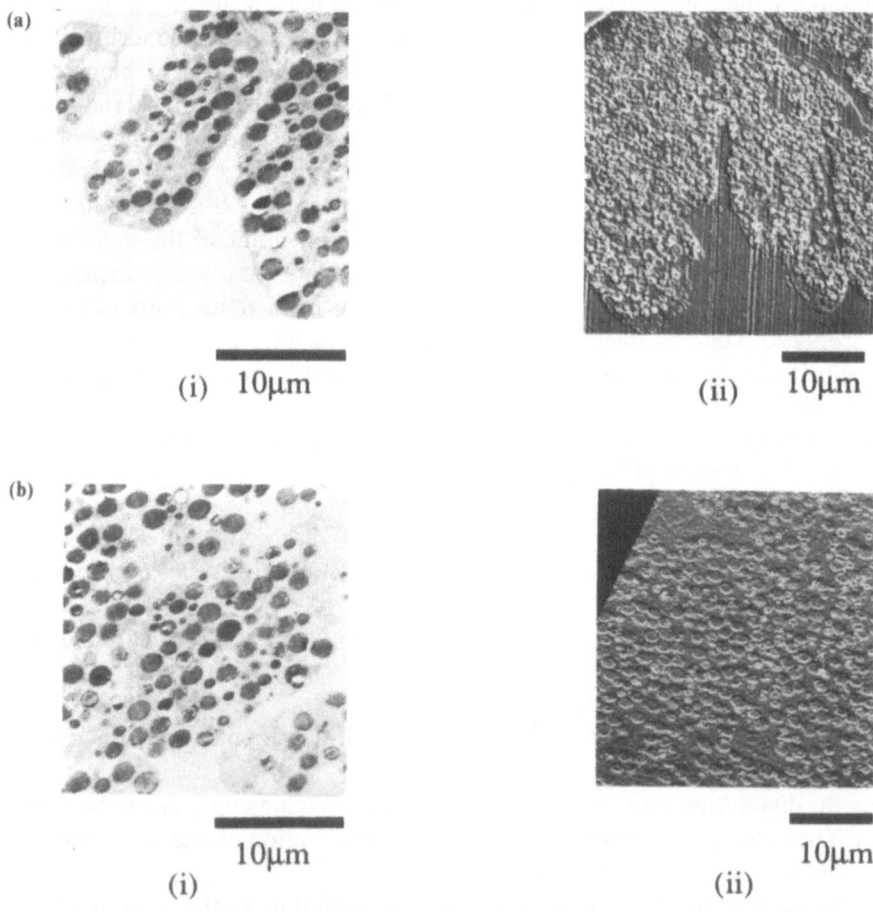

Figure 12.12.   (a) (i) TEM of an unstained section of lead-contaminated earthworm cells and (ii) X-ray contact image of a similar area. (b) as for (a) but (i) TEM of control tissue and (ii) X-ray contact image of nearby section [36].

12.6.3   Applications

Compared with X-ray film, photoresist is approximately 10^4 times less sensitive which means that the low detective quantum efficiency must be compensated for by the high photon flux which can be obtained from the synchrotron and which is needed to prove the value of the technique. An example of this kind of study is that of Richards et al. [36] in which the distribution of heavy metal contaminants in earthworms was investigated. Figures 12.12(a) and (b) show a transmission electron micrograph of a sample of chlorogogenous tissue from an earthworm which had lived in a high metallic environment and a contact X-ray image of the same tissue. The distribution of heavy metal is imaged clearly in both cases, and in the case of the X-ray contact image, the authors claim a resolution of around 700 Å.

The work described in reference [36] was carried out using broad-band radiation from the Daresbury SRS. Panessa-Warren et al. [37], on the other hand, used monochromatic radiation from the NSLS to image bacterial spores treated with vanadium-containing germicide. Images were taken at X-ray energies both above and below the nitrogen K edge (see Figure 12.5) and the vanadium L III edge. The nitrogen edge maps revealed details of the spore coat and exosporium, and, in the case of Bacillus thuringiensis, disruption of the spore coat following sporocidal treatment. The images taken close to the V L III edge revealed the distribution of vanadium after sporocidal action. This work is particularly important because it demonstrates the possibility of determining elemental distributions in partially hydrated, unsliced materials on a submicron scale.

12.7   Future developments

Future improvements to the X-ray microscope are likely to concentrate on the optimisation of the X-ray source and its associated optical system and more careful attention to the environment of the sample than has been the case up to the present time. Major developments include the development of phase contrast microscopy and X-ray holography.

12.7.1   Sources and optics

The aim of both source development and the improvement of the X-ray optical system must be to concentrate more photons onto a smaller area of sample, thus improving the product of imaging time and spatial resolution. At present, imaging times of a few seconds can only be obtained in the fixed beam microscope at the expense of a high dose rate to the specimen. Under such conditions, spatial resolution in the region of 500 Å can be obtained. Reduction of specimen dosage is possible with the scanning instrument but at the expense of long scanning times for a complete image and with poorer spatial resolution. However, the comparison of resolution values achieved using these two contrasting methods of microscopy has not been rigorously carried out so far.

The development of X-ray undulators on high brilliance sources of X-radiation will probably be the most significant future development. Table 12.2 indicates the present and future trends in this direction and shows that increases of several orders of magnitude in terms of photon brilliance can be expected in the future at the ALS and ESRF facilities. However, X-ray microscopy has not yet reached the degree of development which would enable it to command the construction of a totally optimised X-ray source and for the forseeable future the X-ray microscopist will have to be content with sources which have been built primarily for other purposes. This makes the development of improved X-ray optical systems particularly important and this is being pursued at the Daresbury SRS [19] at the present time. It is sometimes difficult, at multipurpose SR sources, to ensure that individual instruments get sufficient development time to reach their full potential. The X-ray microscope is no different to any other instrument in this respect. For example, the BESSY storage ring is usually operated in a low brilliance mode for X-ray lithographic applications. This priority reduces significantly the time available in the high brilliance mode [21, 38] of operation, as is required for microscopy. High stored beam current, a long stored beam life-time, good beam stability and operational reliability are also important source considerations for a successful X-ray microscopy programme.

Turning now to the microscope itself, the most significant development would be the construction of improved efficiency zone plate focussing elements. The efficiency, ε, of a semi-transparent zone plate is given by Simpson [39] as

$$\varepsilon = F/(m\pi)^2 \qquad (12.13)$$

where

$$F = (e^{-\mu_1 t} - e^{-\mu_2 t})^2 \qquad (12.14)$$

where m is the diffractive order of the zone plate focus (for a zone plate with perfectly positioned zones, $m = 1, 3, 5$ etc. and no light is diffracted into even orders). The exponential terms in the factor F simply describe the tranmission through the semi-open and semi-opaque zones, respectively. For a perfect free standing zone plate, $F = 1$, and $\varepsilon = 1/\pi^2$, or about 10% in first order. In practice about 10% of the radiation may be absorbed in the substrate and only about 60% in the opaque zones so that $F = 0.5$ and $\varepsilon = 2.5\%$ in this case. The figure quoted is for a carbon zone plate written on a carbon substrate. Replication of this zone plate in gold [13] will increase the F value to about 0.9 and ε to about 8%. Even then the zone plate will be unlikely to have the ideal geometry of equal area zones covering the whole area of the zone plate. Such defects will reduce the zone plate efficiency in an unquantifiable way. Zone plates which rely on the different amounts of radiation transmitted through the open and opaque zones are often known as amplitude zone plates. It is possible to increase the zone plate efficiency by making use of the phase shift of the transmitted radiation so that radiation transmitted through the semi-opaque zone interferes constructively instead of destructively at the zone plate focus. In order to do this, the correct thickness of the semi-opaque zones must be chosen, using Eqn. 12.3 together with the relation between the phase shift parameter δ and f_1, the real part of the atomic scattering factor, which, for a pure element is given by

$$\delta = \frac{r_0}{2\pi}\lambda^2 N f_1 \qquad (12.15)$$

in which r_0 is the classical electron radius and N is the number of atoms in unit volume. In order to calculate the efficiency of such a zone plate both the phase shift and the transmission amplitude of the radiation must be taken into account. The latter can be obtained from Eqn. 12.2 and

$$\beta = \frac{r_0}{2\pi}\lambda^2 N f_2 \qquad (12.16)$$

In these equations f_1 and f_2 are the atomic scattering factors which have been tabulated by Henke [40]. The zone plate efficiency can then be calculated from Eqn. 12.13 with

$$F = 1 + e^{-2\gamma\phi} - 2e^{-\gamma\phi}\cos\phi \qquad (12.17)$$

where

$$\gamma = \frac{\beta}{\delta} = \frac{f_2}{f_1} \qquad (12.18)$$

The first successful use of a phase zone plate is that of Hilkenbach et al. [41] who constructed a germanium zone plate with a minimum zone width of 1 μm and a zone thickness of about 0.4 μm. This zone plate, with a theoretical efficiency of 20% and a measured efficiency of 15% (in first order at λ23.6 Å), was used as the condenser in the scanning X-ray microscope developed by Niemann [42].

12.7.2 Phase contrast and holographic microscopy

The use of phase contrast in the formation of the X-ray image has been explored by Schmahl et al. [43, 44, 45], who have demonstrated that phase contrast (as opposed to transmission contrast) can be particularly sensitive to elemental differences in biological materials, particularly close to absorption edges. The introduction of a thin phase plate, into the back focal plane of the fixed beam microscope can produce a partially phase contrasted image. The first phase contrast images produced by X-rays have been published by Schmahl *et al.* [44] and by Nyakatura *et al.* [28].

X-rays were first used for the holographic recording of X-ray images by Aoki and Kikuta [46] using a conventional electron beam impact X-ray source. More recently the use of holographic methods has been demonstrated by Howells *et al.* [47] and by Joyeux *et al.* [48]. The principal difficulties preventing the development of X-ray holography at the present time are the availability of sufficiently intense coherent X-ray sources and a high resolution X-ray recording medium with good linear characteristics. A partially coherent X-ray source can be prepared by illuminating a pinhole with radiation from an X-ray undulator and X-ray holograms can be recorded in the Gabor geometry in which the reference beam is the incident X-ray beam itself and the holographic image is produced by interference between the reference beam and that part of the X-ray beam phase-shifted (and also attenuated) during its passage through the object. The method succeeds because no X-ray focussing elements are needed but image reconstuction is difficult because part of the reference beam wave-front is distorted by its passage through the object so that the central region of the holographic image must be rejected at the reconstruction stage. There is also a practical difficulty in that photoresist is the only recording medium with sufficiently high spatial resolution so that the recording of the image is subject to the same problems noted in the discussion of contact microscopy. Nevertheless, successful image reconstuction has been claimed [47, 48].

12.8   Conclusion

X-ray microscopy is at an early stage of development and the first images are being published. This chapter has concentrated on the work of the three well-established groups in zone plate X-ray microscopy but groups in Japan [49] and in the Peoples' Republic of China [50] are now developing instruments which are similar to those described here and which will add to the pool of expertise and experience in this field. The basic foundations of both focussing and contact microscopy have now been laid, the instruments are technically reliable and image interpretation techniques are being developed and characterised. Rapid progress is now being made and the application of the method to real problems in biology and materials science is now beginning.

References

1. A. Franks, B. Gale, K. Lindsey, D.J. Pugh, C.J. Robbie and M. Stedman, *Annu. Rev. NY Acad. Sci.* **342** (1980) 167
2. B. Aschenbach, *Rep. Prog. Phys.* **48** (1985) 579.
3. H. Wolter, *Ann. Phys Series 6* **10** (1952) 94.

4. A. Franks and B. Gale, in *X-Ray Microscopy*, eds. G. Schmahl and D. Rudolph, Springer-Verlag, Berlin (1984) p. 124.
5. S. Aoki, Y. Gohshi and A. Ida, In *X-Ray Microscopy Instrumentation and Biological Applications*, eds. P.C. Cheng and G.J. Jan, Springer-Verlag, Berlin (1987) p. 254.
6. T.W. Barbee Jr., in *X-Ray Microscopy*, eds. G. Schmahl and D. Rudolph, Springer-Verlag, Berlin (1984) p. 144.
7. E. Spiller, in *X-Ray Microscopy*, eds. G. Schmahl and D. Rudolph, Springer-Verlag, Berlin (1984) p. 226.
8. D. Rudolph and G. Schmahl, *Annu. Rev. NY Acad. Sci.* **342** (1980) 94.
9. G. Schmahl, D. Rudolph, P. Guttmann and O. Christ, in *X-Ray Microscopy*, eds. G. Schmahl and D. Rudolph, Springer-Verlag, Berlin (1984) p. 63.
10. P. Guttmann, in *X-Ray Microscopy*, eds. G. Schmahl and D. Rudolph, Springer-Verlag, Berlin (1984) p. 75.
11. J. Thieme, in *X-Ray Microscopy*, eds. G. Schmahl and D. Rudolph, Springer-Verlag, Berlin (1984) p. 91.
12. V. Bögli, P. Unger, H. Beneking, B. Greinke, P. Guttmann, B. Niemann, D. Rudolph and G. Schmahl, in *X-Ray Microscopy II*, eds. D. Sayre, M. Howells, J. Kirz and H. Rarback, Springer-Verlag, Berlin (1988) p. 80.
13. C.J. Buckley, M.T. Browne, R.E. Burge, P. Charalambous, K. Ogawa and T. Takeyoshi, in *X-Ray Microscopy II*, eds. D. Sayre, M. Howells, J. Kirz and H. Rarback, Springer-Verlag, Berlin (1988) p. 88.
14. R.E. Burge, A.G. Michette and P.J. Duke, in *Scanning X-Ray Microscopy* 1, Scanning Microscopy Intl. Chicago (AMF O'Hare), IL 60666, USA (1987) p. 891.
15. R.M. Bionta, A.F. Jankowski and D.M. Makowiecki, in *X-Ray Microscopy II*, eds. D. Sayre, M. Howells, J. Kirz and H. Rarback, Springer-Verlag, Berlin (1988) p. 142.
16. K. Saitoh, K. Inagawa, K. Kohra, C. Hayashi, A Iida and N. Kato, in *Tochigi* (1988) to be published.
17. D.J. Thompson and V.P. Suller, in *Proc. Int. Conf. on Synchrotron Radiation Instrumentation* (SRI88), Tsukuba, Japan (1988).
18. Photon Factory Activity Report, National Laboratory for High Energy Physics, Tsukuba, Japan (1987) p. 55.
19. H.A. Padmore, P.J. Duke, R.E. Burge and A.G. Michette, in *X-Ray Microscopy II*, eds. D. Sayre, M. Howells, J. Kirz and H. Rarback, Springer-Verlag, Berlin (1988) p. 63.
20. G.R. Morrison, S. Bridgewater, M.T. Browne, R.E. Burge, R.C. Cave, P.S. Charalambous, P.J. Duke, G.F. Foster, A.R. Hare, A.G. Michette, D. Morris and T. Taguchi, in *Tochigi* (1988), to be published.
21. D. Rudolph, B. Neimann, G. Schmahl and O. Christ, in *X-Ray Microscopy*, eds. G. Schmahl and D. Rudolph, Springer-Verlag, Berlin (1984) p. 192.
22. G. Schmahl, D. Rudolph, B. Niemann and O. Christ, *Annu. Rev. NY Acad. Sci.* **342** (1980) 368.
23. A.J. Boyde, in *Techniques and Applications of Modern Microscopy*, eds. P.J. Duke and A.G. Michette, Plenum Press, London (1989).
24. D. Shotton, *Proc. R. Microsc. Soc.* **23** (1988) 289.
25. D. Shotton, *Nature* **283** (1980) 12.
26. M. Stewart, in *Techniques and Applications of Modern Microscopy*, eds. P.J. Duke and A.G. Michette, Plenum Press, London (1989).
27. H. Rarback, J.M. Kenney, J. Kirz, M.R. Howells, P. Chang, P.J. Coane, R. Feder, P.J. Houzego, D.P. Kern and D. Sayre, in *X-Ray Microscopy*, eds. G. Schmahl and D. Rudolph, Springer-Verlag, Berlin (1984) p. 203.
28. G. Nyakatura, W. Meyer-Ilse, P. Guttmann, B. Niemann, D. Rudolph, G. Schmahl, V. Sarafis, N. Hertal, E. Uggerhøj, E. Skriver, J.O.R. Nørgaard and A.B. Maunsbach, in *X-Ray Microscopy II*, eds D. Sayre, M. Howells, J. Kirz and H. Rarback, Springer-Verlag, Berlin (1988) p. 365.
29. J.P. Connerade, *Contemp. Phys.* **19** (1978) 415.
30. F. Cinotti, M.C. Voisin, C. Jacobsen, J.M. Kenney, J. Kirz, I. McNulty, H. Rarback, R. Rosser and D. Shu, in *X-Ray Microscopy Instrumentation and Biological Applications*, eds. P.C. Cheng and G.J. Jan, Springer-Verlag, Berlin (1987) p. 311.
31. J.M. Kenney, C. Jacobsen, J. Kirz, H. Rarback, F. Cinotti, W. Thomlinson, R. Rosser and G. Schidlovsky, *J. Microsc.* **138** (1985) 321.

32. V.E. Cosslet and W.C. Nixon, *X-Ray Microscopy*, University Press, Cambridge, (1960).

33. P.C. Cheng, D.M. Shinozaki and K.H. Tan, in *X-Ray Microscopy Instrumentation and Biological Applications*, eds. P.C. Cheng and G.J. Jan, Springer-Verlag, Berlin (1987) p. 65.

34. D.M. Shinozaki and B.W. Robertson, in *X-Ray Microscopy Instrumentation and Biological Applications*, eds. P.C. Cheng and G.J. Jan, Springer-Verlag, Berlin (1987) p. 105.

35. H.I. Smith, E.H. Anderson, A.M. Hawryluk and M.L. Schattenburg, *X-Ray Microscopy*, eds. G. Schmahl and D. Rudolph, Springer-Verlag, Berlin (1984) p. 51.

36. K.S. Richards, A.D. Rush, D.T. Clarke and W.J. Myring, *J. Microsc.* **142** (1986) 1.

37. B.J. Panessa-Warren, G.T. Tortora and J.B. Warren, in *X-Ray Microscopy II*, eds. D. Sayre, M. Howells, J. Kirz and H. Rarback, Springer-Verlag, Berlin (1988) p. 421.

38. G. Mulhaupt, in *X-Ray Microscopy*, eds. G. Schmahl and D. Rudolph, Springer-Verlag, Berlin (1984) p. 4.

39. M.J. Simpson, Ph D Thesis, London University (1984).

40. B.L. Henke, P. Lee, T.J. Tanaka and R.L. Shimabukuro, *Atomic Data Nucl. Data Tables* **27** (1982) 1.

41. R. Hilkenbach, J. Thieme, P. Guttmann and B. Niemann, in *X-Ray Microscopy II*, eds. D. Sayre, M. Howells, J. Kirz and H. Rarback, Springer-Verlag, Berlin (1988) p. 95.

42. B. Niemann, P. Guttmann, R. Hilkenbach, J. Thieme and W. Meyer-Ilse, in *X-Ray Microscopy II*, eds. D. Sayre, M. Howells, J. Kirz and H. Rarback, Springer-Verlag, Berlin (1988) p. 209.

43. G. Schmahl and D. Rudolph, in *X-Ray Microscopy Instrumentation and Biological Applications*, eds. P.C. Cheng and G.J. Jan, Springer-Verlag, Berlin (1987) p. 231.

44. G. Schmahl, D. Rudolph and P. Guttmann, in *X-Ray Microscopy II*, eds. D. Sayre, M. Howells, J. Kirz and H. Rarback, Springer-Verlag, Berlin (1988) p. 228.

45. D. Rudolph, G. Schmahl and B. Niemann, in *Techniques and Applications of Modern Microscopy*, eds. P.J. Duke and A.G. Michette, Plenum Press, London (1989).

46. S. Aoki and S. Kikuta, *Jpn. J. Appl. Phys.* **13** (1974) 1385.

47. C. Jacobsen, J. Kirz, M. Howells, K. McQuaid, S. Rothman, R. Feder and D. Sayre, in *X-Ray Microscopy II*, eds. D. Sayre, M. Howells, J. Kirz and H. Rarback, Springer-Verlag, Berlin (1988) p. 253.

48. D. Joyeux, S. Lowenthal, F. Polack and A. Bernstein, in *X-Ray Microscopy II*, eds. D. Sayre, M. Howells, J. Kirz and H. Rarback, Springer-Verlag, Berlin (1988) p. 246.

49. Y. Kagoshima, S. Aoki, M. Kakuchi, M. Sekimoto, H. Maezawa, K. Hyodo and M. Ando, in *X-Ray Microscopy II*, eds. D. Sayre, M. Howells, J. Kirz and H. Rarback, Springer-Verlag, Berlin (1988) p. 296.

50. X.S. Xie, S.X. Kang, C.Z. Jia and T. Jin, *Nucl. Instrum. Methods A* **246** (1986) 698.

13 Synchrotron radiation trace element analysis

R.D. VIS

13.1 Introduction

The need for techniques able to determine trace element concentrations in a variety of specimens is growing rapidly. Various reasons can be given for this growth. Moreover, the arguments for developing advanced techniques for trace element analysis are different for the different branches of science using these techniques.

In biology, it has been demonstrated that low concentrations of a group of elements play a crucial role in living material. This group is called the essential elements and the number of members of this group is growing by performing experiments demonstrating that life itself needs a particular element in a particular range of concentrations, the so-called range of adequacy. The more recently observed essential elements have lower ranges of adequacy (examples are Al, As and the longer known Mo) increasing the demands set on the analytical techniques. Also the awareness that toxic elements, even in very low concentration, negatively influence the biosphere emphasises the importance of conclusive analytical procedures.

In material science or in the semi-conductor industry, there is a growing need for very high purity materials. To monitor the production process and to judge the final product we require sensitive analyses with precise quantitative results.

In geology, the abundance of trace elements in rocks and sediments often records their history in terms of temperatures and pressures present during formation or in terms of cooling rates. Also in this branch of science, the number of elements to be analysed is increasing and the average concentration level of interest is decreasing.

Environmental studies very often rely on accurate determination of trace element concentrations in air, water or soil. The increasing use of a greater variety of metals in the chemical industry with the associated risk of diffusion of these metals calls for adequate analytical methods.

For a number of applications in these fields, major progress is possible as soon as it becomes possible to measure the distribution of trace elements over a sample, often referred to as element mapping. The analytical techniques are in such cases called microtechniques, since the information is extracted from a very small sample volume. In biology and medicine for instance, knowledge of the elemental distribution over a histological section together with knowledge of the protein distribution may elucidate the role of a particular essential element. In geology, the gradient of a concentration may give information on the cooling rate or metamorphic processes that took place after the initial solidification. In environmental studies, sometimes minute amounts of sample need to be analysed, as for example, aerosols, also requiring micro-analytical techniques.

Of course the progress towards sensitive techniques for element analysis is too extensive to describe in this chapter. In spite of major progress made in a great number of analytical chemical methods we will restrict overselves to the application of synchrotron radiation to excite atoms in the sample. In order to introduce trace element X-ray fluorescence analysis, comparisons with techniques using an ion beam or an electron beam are included.

13.2   The development of accelerator-based techniques

Neutron activation analysis preceded accelerator-based techniques for the detection of low concentrations of elements. Using this technique, the gamma emission induced by (normally) thermalised neutrons is counted and, if suitable standards are irradiated simultaneously, quantitative results of elemental concentrations can be obtained. Very often, chemical separation procedures are performed after neutron-irradiation to clean up the gamma-spectra, in this way increasing the sensitivity of the method considerably. This sensitivity is obviously strongly dependent on the cross-section for neutron capture of the nucleus of the element of interest but low detection limits are within reach in favourable cases (0.01–1 μg/g).

The first real accelerator-based techniques were based on the analogy of neutron activation. Ions of sufficient energy were used to induce a suitable nuclear reaction and the radioactivity was counted with or without a preceding chemical separation. The technique of charged particle activation analysis (CPAA) requires ions of 5–20 MeV to overcome the Coulomb barrier of the nucleus. Cyclotrons are used in almost all cases.

In 1970, for the first time, protons of a few MeV energy were used to induce X-rays for analytical purposes [1]. Of course, due to the short life-time of vacancies in the inner shells of the atom, this is an in-beam technique. The method is referred to as particle induced X-ray emission (PIXE) and has grown rapidly to a well-established technique at present. Reviews of this technique may be found in [2–9]. The PIXE technique has been given a sound theoretical basis, and since it is an X-ray technique it will be useful to compare with synchrotron radiation applications.

It should be emphasised that X-ray fluorescence (XRF) itself is much older than the accelerator-based techniques. XRF started on a commercial basis as early as about 1950. A turning point in XRF was reached by the mid-1960s when automation in measurements and analytical calculations, including corrections, became feasible thanks to increased spectrometer capabilities and growing data handling facilities. These conditions favoured the development and re-evaluation of mathematical procedures, namely the fundamental parameter method based on Sherman's equation and the numerical or influence coefficient methods. Both approaches are still under development [10]. The reason that PIXE became viable in the mid-1970s was largely due to the development of solid state Li-drifted Si detectors with sufficient energy resolution to detect all elements separately (see Section 1.5.2). Previously crystal spectrometers had been used for analysing different emission lines in conjunction with proportional counter detectors (Section 1.5.1) an inefficient combination, considering the X-ray flux densities available from PIXE.

To complete our survey it is worth mentioning the use of electrons to excite X-rays. This application is almost exclusively found in electron microscopes as an additional feature for element mapping and is discussed later.

13.3 The use of synchrotron radiation

The availability of dedicated electron storage rings, with a few GeV electrons, radiating intense electromagnetic radiation triggered a number of research groups to use the X-ray region of the synchrotron spectrum for trace element analysis. To distinguish this method from the conventional XRF, it will be referred to as SXRF throughout this chapter. The motivation for the development of SXRF is found in the following advantages of this technique over the techniques mentioned above.

(i)   The detection power of SXRF is higher than for conventional XRF due to the high incident X-ray flux available.

(ii) The tunability of synchrotron radiation by the use of a suitable monochromator, offers the possibility of obtaining the highest sensitivity throughout the whole range of elements of interest by tuning the excitation energy just above the binding energy of the electrons in a particular shell of the element of interest.

(iii) The linear polarisation of the synchrotron beam (See Chapter 1) enables a lower background to be achieved. If the X-rays are detected in the plane of the storage ring at 90° to the incident exciting radiation no Compton radiation is emitted from the sample into the detector.

(iv) Compared with ion-bombardment, the radiation damage induced in the specimen under investigation is considerably less. Especially for biological applications this is a major advantage. Moreover, in principle one can perform SXRF in air or under a protective atmosphere instead of the vacuum necessary for ion-bombardment techniques. Even the analysis of living material is feasible with SXRF.

(v) The high intensity, in combination with the low divergence of the synchrotron beam permits the development of microbeams of synchrotron radiation for spatially resolved element analysis. Here, the lower radiation damage is an especially attractive feature, because ion microprobes with their inherent high flux density are limited in their applicability in biology by the deterioration of the samples under the beam.

In the following sections these advantages will be examined in a more quantitative way.

13.3.1   The sensitivity of SXRF

The intensity of the analytical signal of a given trace element is in general given by:

$$I = \sigma_z N_x n_z \Omega \varepsilon C \qquad (13.1)$$

where I is the intensity of the fluorescent peak and σ_z is the cross-section for the production of the measured X-ray. Note that in this factor the fluorescence yield and the branching ratio of the X-rays are included. N_x is the number of incident photons, n_z is the number of trace element atoms seen by the beam, Ω is the solid angle subtended by the detector, ε is the detector efficiency and C is the absorption of the X-rays on their way to the detector. Since sensitivities are usually expressed in counts/g, n_z should be converted to the mass of the trace element using

$$n_z = (a_z/A_z)N_0 \qquad (13.2)$$

with a_z and A_z the mass and the atomic weight of the element, and N_0 Avogadro's number.

Note that Eqn. 13.1 only holds for thin samples, that is in cases where absorption of incident flux, self absorption of the sample and secondary fluorescence by matrix elements are ignored. For thick samples and heavy matrices, one should apply adequate corrections, commonly known as ZAF corrections and available in the form of various software packages. For details about these corrections, which are in principle the same as for conventional XRF, see [10].

The sensitivity of any analytical technique is defined as:

$$S = (dI/da_z) = \sigma_z N_x N_0 \Omega \varepsilon C/A_z \qquad (13.3)$$

In Figure 13.1 cross-sections are given for a number of photon energies. For comparison, also K-production cross-sections are given for proton and electron bombardment. Note that, especially for well-tuned incident energies (the maxima of the photon curves), the cross-section for production with photons is considerably higher than for the other modes of excitation, leading

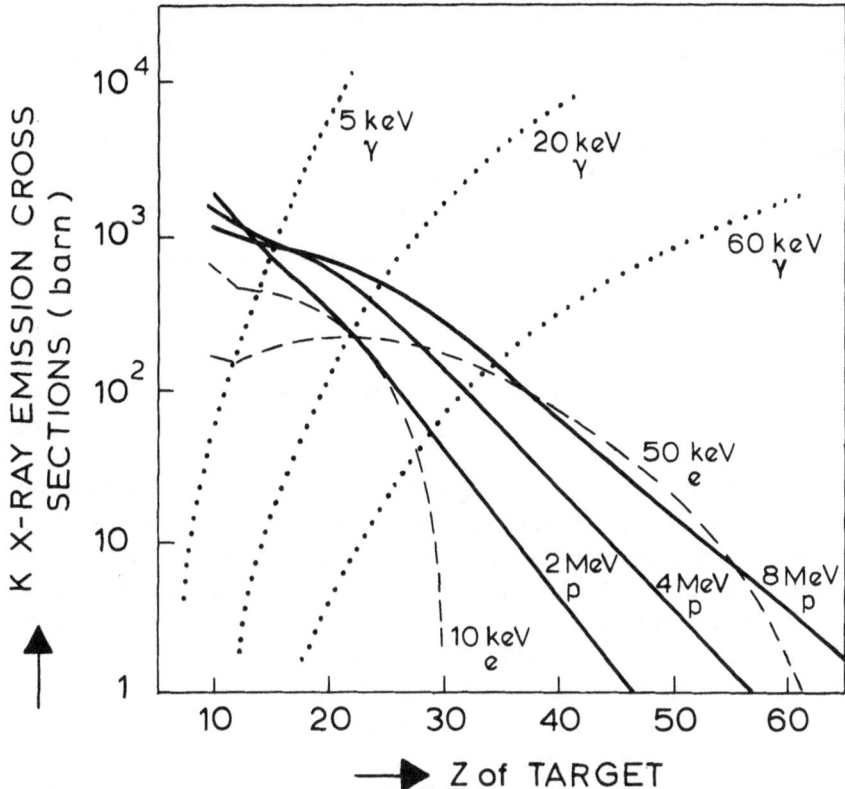

Figure 13.1.   The cross-section for X-ray production with photons, protons and electrons with energies regularly used for trace element analysis.

to improved sensitivity. To give an impression of the order of magnitude of this sensitivity we assume that we have a sample slice of 10 μm thickness, having a density of 1 g/cm^3 and containing 1 μg/g Zn.

Assuming further Ω to be 0.01 and εC to be 1 for the characteristic X-rays of Zn and choosing an excitation energy of 15 keV to have a favourable cross-section for Zn (Figure 13.1), then the only parameter left is N_x. The available flux of course depends on the machine used and on the distance between storage ring and experimental station. Moreover, the flux on target depends on the bandwidth one selects with the monochromator, and since for trace element work a high flux is more important than a high energy resolution in the beam, very often mosaic type crystals are used. Assuming here a bandwidth of $\delta E/E = 10^{-3}$ and assuming that one is able to select 0.2 mrad horizontally given by the size of the crystal and the location of the station, then for the SRS at Daresbury (U.K.) with 100 mA stored beam at 2 GeV and using a dipole magnet (1.2 T), N_x will be about 6×10^{10} photons/s.

Substituting these values in Eqn. 13.3 yields $S = 12$ counts cm^2/ng per s. This example illustrates that analysis at a level of μg/g is readily achievable.

With some extra effort analysis at a level of 10 ng/g should be possible, provided that one can keep the background at a sufficiently low level.

13.3.2   The detection limits of SXRF

Normally, the detection limits are defined by demanding that the analytical signal, I, has to be greater than three times the standard deviation (τ_B) of the background [11], i.e.

$$I = 3\tau_B \qquad (13.4)$$

The lowest concentration detectable (a_z) is thus given by

$$a_z = I/S = 3\tau_B/S = (3\tau_B A_z)/(\sigma_z N_x N_0 \Omega \varepsilon C) \qquad (13.5)$$

To calculate these limits, information is needed on the background processes in order to estimate τ_B. Although the main contribution consists of Rayleigh and Compton scattering in the sample and can be estimated, results of such calculations tend to be too optimistic. Other sources of background such as higher order reflections in the monochromator, multiple scattering in sample and detector shielding and the difficulty of describing the influence of the linear polarisation of the beam can cause discrepancies between calculated and experimental detection limits. Moreover, Compton scattering in the detector itself and also incomplete charge collection will produce background at lower energies than the scattered radiation from the sample. If the overall production cross-section of background under the fluorescent peak of interest is σ_B then, by analogy with Eqns. 13.1 and 13.2, one may write

$$I_B = \sigma_B a_M N_x N_0 \Omega \varepsilon C/A_M \qquad (13.6)$$

with a_M and A_M the mass and atomic weight of the matrix. Since $\tau_B = \sqrt{I_B}$ one obtains from Eqn. 13.5 for the detection limit

$$a_z = (3A_z/\sigma_z)\{(\sigma_B a_M)/(N_x N_0 \Omega \varepsilon C A_M)\}^{1/2} \qquad (13.7)$$

Note that the intensity of background under the peak is dependent on the detector resolution. This has triggered attempts to implement wavelength dispersive systems with much lower efficiency but better resolution in SXRF systems [12]. Gordon [13] calculated quantitative limits in a thin biological sample of 2 mg/cm^2 and found values less than 1 ppm for low Z elements and 1–10 ppm for high Z elements using a common solid state detector exposed for 1 min only. For the wavelength dispersive system the author calculated 3–30 ppb for biological and 30–300 ppb for geological samples. Jaklevic et al. [14] claim detection limits of 40–120 ppb in a 30 mg/cm^2 cellulose sample. These authors also investigated the influence of the excitation energy on a_z. Figure 13.2 shows the detection limits of Mn, Zn and Se in these cellulose samples as a function of incident energy. It is seen that the incident energy has

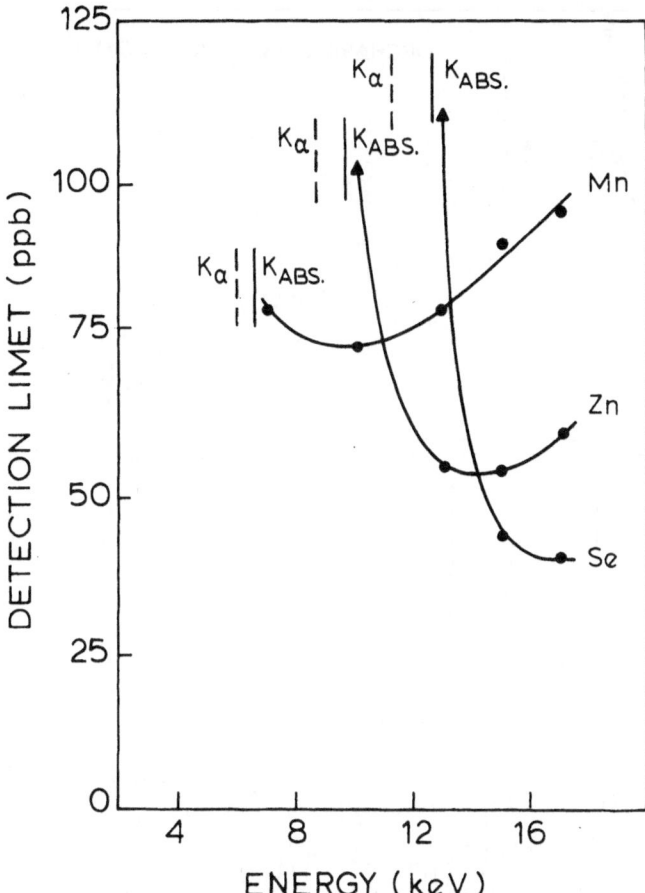

Figure 13.2.   Detection limits as a function of incident photon energy for Mn, Zn and Se. Both the absorption edge and the K_α X-ray energy are indicated (After Jaklevic *et al.* [14]).

to be chosen well above the absorption edge of the K-shell in order to avoid overlap in the spectra between the fluorescent peak of interest and the Compton scatter peak lying just below the incident energy. This limitation is more severe for conventional XRF by the lack of polarisation. Knöchel *et al.* [15, 16] measured detection limits using multi-element standards and white radiation for which they claim a smaller dependence of the detection limit as a function of Z. Their calculations are in good agreement with the experiments, reaching values down to $0.1 \ \mu g/g$ in $1 \ mg/cm^2$ samples.

Bos *et al.* [17] compared different techniques of excitation of trace elements with respect to detection limits. PIXE, XRF and SXRF were used to analyse orchard leaves (a NBS standard) and human hair (an IAEA standard). Figure 13.3 shows spectra of the measurements on orchard leaves with an indication of the experimental condition. Note again the effect of the polarisation when comparing XRF and SXRF. Figure 13.4 shows the limits of

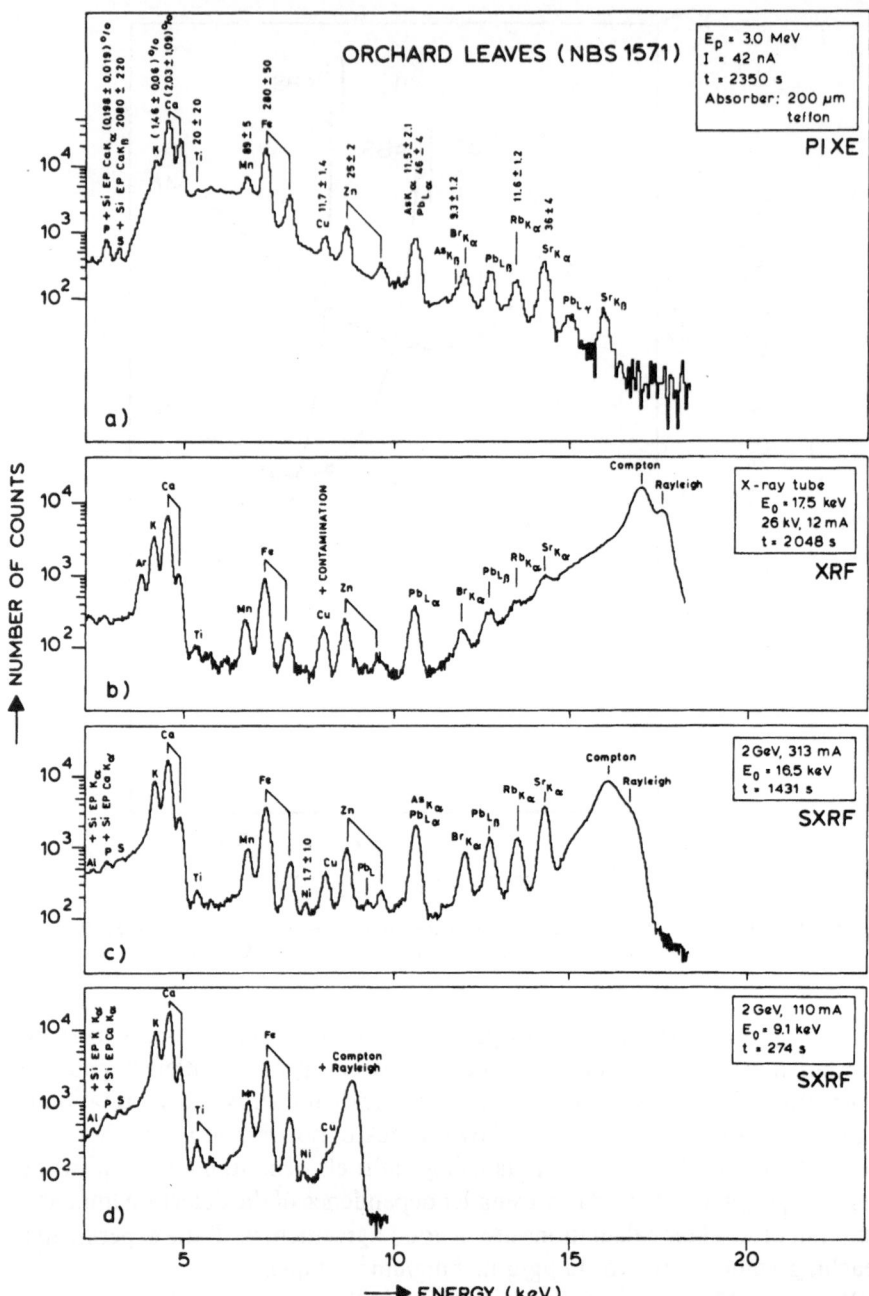

Figure 13.3. X-ray spectra of standard reference material 'Orchard leaves' obtained by excitation with protons, an X-ray tube and synchrotron radiation respectively (after Bos *et al.* [17]).

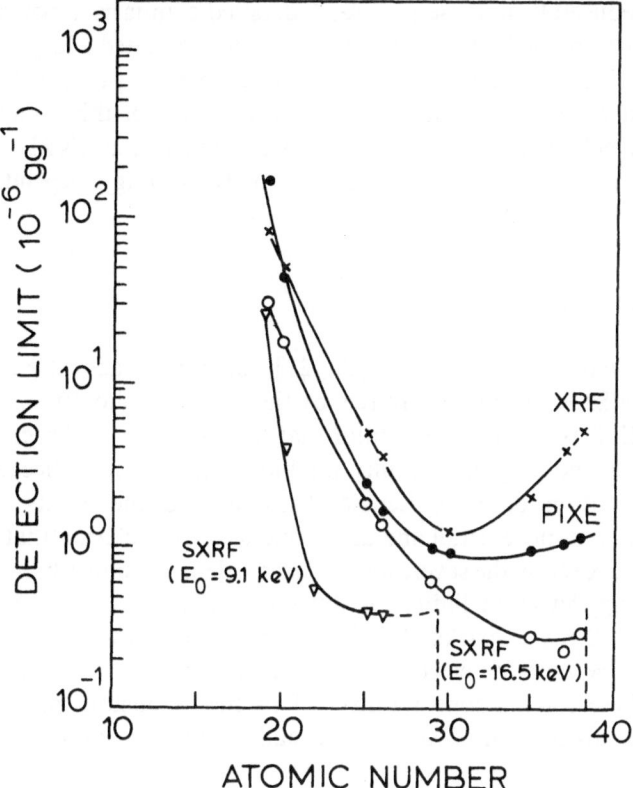

Figure 13.4. Minimum detectable limits obtained by different modes of excitation (after Bos *et al.* [17]).

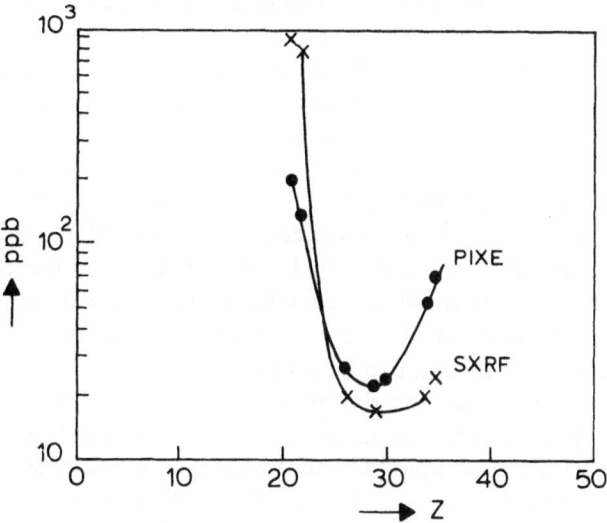

Figure 13.5. A comparison of detection limits between PIXE and SXRF obtained on a set of serum samples (after Lenglet *et al.* [18]).

detection deduced from these spectra. A detailed comparison between PIXE and SXRF was made by Lenglet et al. [18] who irradiated a large number of freeze-dried blood serum samples both with 2.5 MeV protons and 15 keV synchrotron radiation. Figure 13.5 gives the results of this comparison in terms of detection limits only. As a general conclusion SXRF should be capable of analysing trace elements at levels of 0.05 to 1 μg/g depending on the degree of optimisation [19].

13.3.3 Radiation damage

As was pointed out by Grodzins [20, 21], for most practical purposes the sensitivity of SXRF is limited entirely by the available photon flux, while for micro-PIXE, it seems to be doubtful that flux densities much greater than 1 nA/μm^2 can be of general utility for biomedical samples because of the inherent radiation damage. Lenglet [22] studied the effects of proton bombardment in more detail and also came to the conclusion that in micro-PIXE limits are set by the severe loss of matrix elements under bombardment. In comparing detection limits for different modes of excitation (protons, electrons and X-rays), Sparks [23] also considered the power dissipated in the specimen in W/μm^2 for electrons, protons and X-rays. In Figure 13.6 the results of these comparisons are given. This clearly demonstrates the much lower power dissipation of X-rays, especially monochromatic X-rays.

13.3.4 Concentration assignment using SXRF

The aim of SXRF is to perform trace element analysis in a quantitative way. Various methods are available to do this. As is often the case in PIXE work, one can 'spike' the sample with an internal standard and assign concentrations relative to this standard by appropriate corrections for differences in photo-electric cross-sections, X-ray absorption and detector efficiencies. The obvious advantage is that the incident flux, I, as well as the geometry, $\Omega \in C$, of the experimental arrangement cancel out, as can easily be seen from Eqn. 13.1. Another option is to use reference standards with roughly the same overall matrix composition and with known amounts of the elements of interest. These standards are irradiated in the same run as the samples. In this case the incident flux has to be monitored with ionisation chambers (Section 1.5.1) so that spectra can be normalised.

Alternatively, one can use the photons scattered from the matrix which are, of course, proportional to the incident flux but also (at least to a good approximation for thin samples) to the thickness of the sample. This means that, after suitable calibration with standards, the fluorescent peak to scatter peak ratio is proportional to the concentration of the trace element. Lenglet et al. [18, 22] used this procedure for the analysis of blood serum using for the

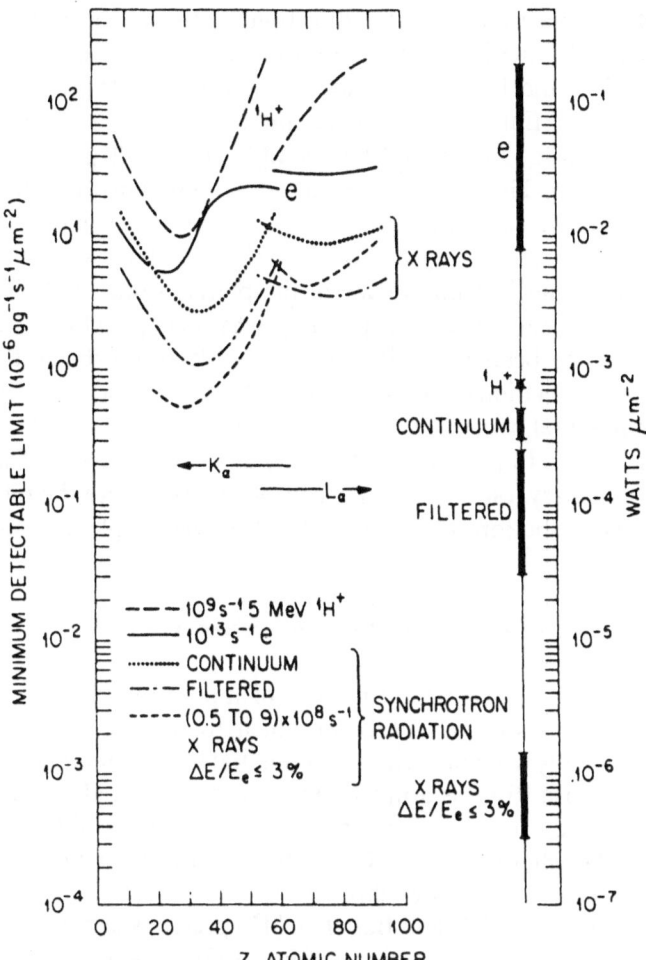

Figure 13.6. A comparison of detection limits for thick samples (left vertical scale) and dissipated power (right vertical scale) for charged particle and synchrotron radiation microprobes. The beam intensities used are 10^9 protons s^{-1} at 5 MeV; 10^{13} electrons s^{-1} at an energy of three times the absorption edge of the element involved and $(0.5-9) \times 10^8$ photons s^{-1} at energies of the various absorption edges. Also filtered and white synchrotron radiation beams are compared. (after Sparks [23]).

coherent scattering cross-section

$$d\sigma_R/d\Omega = r_e^2 \left\{ 1 - \tfrac{1}{2}\sin^2\theta(1 + P\cos 2\phi) \right\} \qquad (13.8)$$

and for the incoherent scattering cross-section, the Klein-Nishina formula

$$d\sigma_c/d\Omega = r_e(k/k_0)^2 \left\{ (1 - \tfrac{1}{2}\sin^2\theta(1 + P\cos 2\phi)) + [(k_0 - k)^2/(4k_0 k)] \right\} \qquad (13.9)$$

where r_e is the classical electron radius, P the degree of polarisation of the synchrotron radiation X-rays (See Section 1.2.1), ϕ the angle between the direction of the polarisation and the direction of observation, θ the scattering angle and k_0 and k the incoming and outgoing wavenumbers.

Equations 13.8 and 13.9 hold for scattering from free electrons. To correct for binding effects, Eqn. 13.8 should be multiplied by $F[Z, E \sin(\theta/2)]$, the atomic form factor, and Eqn. 13.9 by $S[Z, E \sin(\theta/2)]$, the incoherent scattering function. These correction factors are tabulated by Hubbel et al. [24]. Using this formalism, accuracies of 10% are achievable after suitable system calibration. According to Lenglet [22], the major source of error is introduced by insufficient knowledge of the precise value of P. This can be described with the basic equations given by Schwinger [25] but deviates from this description by betatron oscillations of the electron orbit in the storage ring or by slight deviation of the alignment of the trace element analysis station out of the horizontal plane. It is advisable to run standards regularly to check possible variations. In principle, in this way absolute analysis is feasible and sometimes necessary, especially in microprobe work, where the interest is in trace element distributions and where integrity of the sample is a pre-requisite.

13.4   Applications of synchroton X-ray fluorescence

In contrast to PIXE, where a large number of applications are reported in the literature, SXRF is still at a stage in which there have been only a few applications, mainly to demonstrate the potential of the technique. The reason is not only because PIXE is a much older technique but also in the limited availability of dedicated electron storage rings. A review of SXRF work is presented by Baryshev et al. [26]. The authors formulate requirements for the synchrotron source, the monochromator, the sample chamber and the detector. Hanson et al. [27] and Bos et al. [17] present spectra obtained from the reference material, orchard leaves (SRM 1521). A sea-water sample was measured by Knöchel et al. [16], while Sparks et al. [28] show results of the irradiation of human hair. Iida et al. [40, 41] underline the importance of having a wide band pass monochromator, for which they use a Pt-coated mirror in total reflection mode in combination with an absorber, to increase the incident flux and decrease the limit of detection as compared with white radiation. They demonstrated these improvements by measuring trace elements absorbed on ion-exchange resins. NBS bovine liver (SRM 1577) as well as a whole blood sample were measured by Giauque et al. [29] using a crystal monochromator. They also measured NBS (SRM 1632) coal and obtained a good agreement with the reference values of trace element concentrations for most of the elements. Lenglet et al. [18] measured blood serum from nephretic patients and observed that the haemodialysis negatively influenced the Br concentration leading to depletion of Br in the serum of these patients. The authors used a carbon mosaic monochromator for this study. Chen et al. [30] did measurements on vitrinites, a particular class of coal. The results were compared with micro-PIXE and the Fe and S distribution in the depth of the coal bed was determined. In another paper, Chen *et al.* [31] determined the occurrence of Au in an unoxidised Carlin-type ore sample. In both applications, collimated beams were used obtaining spot sizes ranging from $20 \times 20\,\mu m^2$ to $75 \times 75\,\mu m^2$. The light platinoids Ru, Rh and Pd were analysed by Khvostova *et al.* [32] and Baryshev *et al.* [33]; 25 and 75 keV synchrotron radiation was used and comparisons were made with spectra obtained by excitation with radioactive sources (^{109}Cd and ^{241}Am). Tarasov *et al.* [34] presented a valuable study of the application of SXRF on extraterrestrial matter. The samples consisted of lunar rock material, chondrules and Ca–Al rich inclusions from various chondrites and matrix material of these chondrites. The Y/Zr ratio was emphasised because this ratio seems to be an indication of the time of crystallisation. Also, the difference in trace element distribution in chondrules, matrix and Ca–Al rich inclusions was found indicating a varied genesis for these materials. Extraterrestrial material was also analysed by Sutton *et al.* [35] who presented trace element data of the Orgueil chondrite and data of Greenland stony particles. Finally, Chen *et al.* [36] demonstrated that an attractive combination may be found in SXRF and EXAFS in the fluorescent mode, in order to obtain relatively complete chemical characterisation of complex materials.

13.5   Total reflection

It was pointed out in 1922 by Compton [37], that total external reflection from a smooth surface was possible at X-ray wavelengths. Simple expressions for the critical angle, θ_c, and the penetration depth, Z_{min}, of X-rays are given in Section 1.3. For a complete treatment consult Parratt [38]. Recently, total

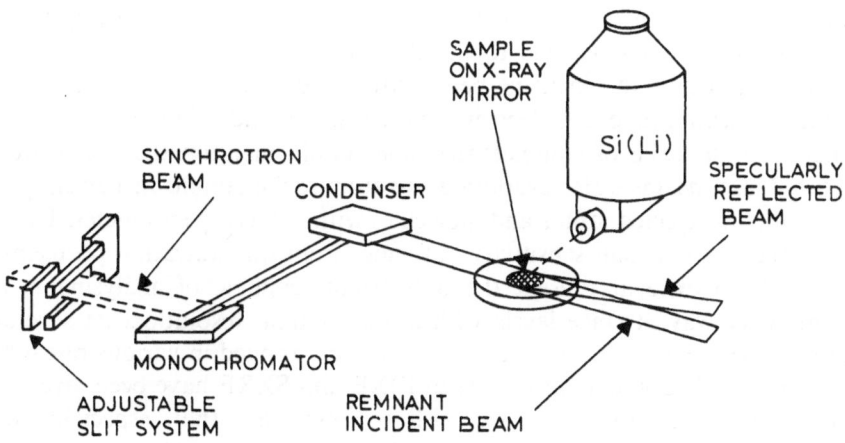

Figure 13.7.   A schematic diagram of the facility at NSLS for total reflection SXRF (from Pella *et al.* [39]).

reflection has extended the applicability of SXRF for trace analysis. If the synchrotron beam is shining on the sample under angles $\theta < \theta_c$, total reflection completely suppresses the scattering background in the XRF spectra. Of course, the penetration depth is very small, thus the technique becomes a real surface analysis method. A schematic view of the equipment used at the NSLS is given in Figure 13.7. Pella *et al.* [39] describe the capabilities of the method, including minimum detection limits for Se as a function of the glancing angle. They obtain values as low as 10 ppb in human blood serum down to 2 ppb Se in aqueous solution. Iida *et al.* [40, 41] also measured aqueous solutions in the total reflection mode. It should be kept in mind, that total reflection XRF was described in the 1970s, using an X-ray tube as excitation source [42]. Nevertheless, up to now the full potential of synchrotron radiation in combination with this surface sensitive technique has not been fully explored.

13.6 Spatial resolved SXRF

13.6.1 *The focussing of X-rays*

Although SXRF has a number of very attractive features for the analysis of trace elements in samples, as indicated in Section 13.4, the technique is in competition with quite a number of other analytical methods capable of measuring at a comparable concentration level, usually with more effort but with less cost. The situation is completely different as soon as a microprobe of synchrotron radiation becomes available, since the capability of measuring trace element concentrations at the sub-ppm level with lateral resolutions at the micrometre level is unique.

Microprobes of charged particles have already existed for some time but have some disadvantages. In the first place, although electrons can be focussed down to nanometre resolutions, the use of these beams for detecting trace elements is not possible, due to the high level of background caused by bremsstrahlung of the incident beam. Moreover, the lateral resolution is spoiled by scattering of the electrons in the samples and is in the micrometre range. For the sake of completeness, one should mention the use of well focussed laser beams which evaporate and atomise the sample on a given spot and detect the ejected atoms and molecules with a mass spectrometer. These techniques have a high sensitivity but suffer from quantification problems. Proton beams are at present the only beams capable of analysing in a quantitative way at trace levels with a good lateral resolution. At several laboratories, beams of $1-5\,\mu m$ are available and applied in various research programs [43]. Comparisons between PIXE and SXRF have been given in Sections 13.3 and 13.4; here we wish to emphasise that, especially for microbeams, the difference in radiation damage induced in the sample gives micro-SXRF a major advantage over micro-PIXE. In conclusion, microbeams of synchrotron radiation offer unique possibilities for trace element research combining good spatial resolution with quantitative analysis and relatively little radiation damage. Research groups at several synchrotron facilities are in the process of developing these beams.

The focussing of X-rays is not a new phenomenon and is discussed in Sections 1.3 and 1.4. Following early work by Gouy [44], Hámos [45, 46] and others the first operational focussing monochromators and spectrographs using bent crystals were designed and built by Johann [47] and Cauchois [48–51] in the 1930s. A detailed review on bent crystal monochromators has been given by Roberts et al. [52]. The possibility of obtaining X-ray images using curved mirrors also has an early history [53–56]. Subsequent theoretical and experimental work was devoted to the development of X-ray microscopes [57] and most recently X-ray telescopes for satellite observations. In the last 17 years the availability of synchrotron radiation has stimulated renewed interest in X-ray optics [19], not least in the realisation of X-ray microprobes. The first reported SXRF microprobe for obtaining elemental distributions, was constructed in 1972 by Horowitz and Howell [61, 62] at the Cambridge Electron Accelerator. A condensing mirror intercepted the beam and focussed it down to a spot of $1 \times 2 \, mm^2$. To improve the resolution further, a real microbeam was obtained by using a pinhole collimator made in a thin Au-substrate. A $2 \, \mu m$ resolution was achieved, albeit with low beam intensity. The X-ray beam constructed by Sparks et al. [28, 63, 64] used a curved pyrolytic graphite monochromator to reduce $37 \, keV$ X-rays into a $0.45 \, mm^2$ spot. The energy spread in the focussed beam was $460 \, eV$ FWHM and the reported flux was 1.5×10^{11} photons $s^{-1} mm^{-2}$. The motivation for developing this equipment was provided by reports based on PIXE work of the possible detection of superheavy elements $(Z = 126)$ in giant-halo inclusions in monazite.

From 1980 the number of publications dealing with the construction of micro-SXRF systems has increased. In general terms, Deslattes [65] comments on the use of monochromators and mirrors as well as on the effect of curvature and asymmetry of crystal planes. Sparks et al. [66] mention the relative merits of crystals versus mirrors for focussing synchrotron radiation at energies above $10 \, keV$, which is still on the low side to excite most of the trace elements of interest.

As mirrors have critical angles of reflection, typically 1/20 of the Bragg scattering angles for crystals, large mirror surface areas are required. Sagittal focussing is weak for mirrors and limits the amount of radiation intercepted. Ray tracing calculations can be used to predict the size of focus one might expect. Heald [67] reports on the limitations in applying bent cylindrical mirrors to X-ray beam lines. Howells et al. [68] propose a multilayer coated Kirkpatrick-Baez mirror (see also [56]) $(d = 20 \, Å, \, f = 0.2 \, m)$ to focus and monochromate radiation from $2–16 \, keV$ in the horizontal plane. This is followed by a vertically focussing mirror to achieve a beam spot of $3 \, \mu m$. Ice *et al.* [69, 70] examine mirror, multilayer and crystal combinations from the point of view of producing the most intense image of the source, tunability, energy range and energy resolution. The authors compare at 15 keV the ray tracing results from a non-dispersive multilayer system, a Kirkpatrick-Baez system and a doubly curved Ge(111) crystal. These Monte-Carlo simulations show that intensities over 10^8 photons s^{-1} μm^{-2} might be possible.

Sagittally bent Si(111) crystals have been tested by Batterman *et al.* [71] and Mills *et al.* [72]. They use slotted bent crystals in order to prevent anticlastic bending as described by Sparks *et al.* [73]. The entire crystal and bending device is so arranged that, coupled with the first crystal, a fixed exit beam height is maintained for all energies. Bilderback *et al.* [74] made a design of a doubly focussing tunable (5–30 keV), wide band pass optical system, made from layered synthetic microstructures (LSM). The first element is water cooled because it intercepts the direct synchrotron beam. It can be bent for vertical focussing and a set of LSMs is used to cover the energy range of interest. The second element consists of a LSM structure with a 10% band pass built onto a flexible substrate bent for sagittal focussing. Prins *et al.* [75–77] report the design of an X-ray microprobe at the SRS, Daresbury (U.K.). As focussing element a doubly bent Si(111) crystal is proposed which is expected to produce a $50 \times 50\,\mu$m^2 spot with a calculated intensity of $\sim 10^8$ photons s^{-1} μm^{-2}.

13.6.2 *Experimental work with synchrotron microprobes*

13.6.2.1 *Collimated systems*   As mentioned in the previous paragraph, the simplest way to obtain a microprobe is to place a pinhole in the white beam. The two main disadvantages are obvious. First, no increase in flux density takes place due to the lack of any focussing action and secondly, there is no energy definition in the beam. There is the possibility of filtering out low energy radiation, but this will worsen the detection limits obtainable. Nevertheless, very useful work with a collimated system is possible as is illustrated by Petersen *et al.* [78, 79]. The authors collimated the beam in two steps, namely with a first aperture ($>200\,\mu$m) made of lead and a second collimator consisting of one of a set of pinholes ranging from 10 to 150 μm. Results are reported of measurements on carbon samples and on aerosols. In a later stage Bavdaz *et al.* [80] report the use of an 1:1 mirror to remove higher energies placed in front of the collimating slits. The authors claim resolutions better than 10 μm and show an example of a measurement on a sample of the 42 line bible by Gutenberg. As was already mentioned in Section 13.4 Chen *et al.* [30, 31] also use collimated beams, sometimes in combination with a channel cut Si-monochromotor, the reported spot sizes varying from $20 \times 20\,\mu$m^2 to $75 \times 75\,\mu$m^2.

13.6.2.2 *Focussed systems*   One of the first descriptions of focussing devices to produce a synchrotron microbeam is given by Howell *et al.* [62] and was used in the first microprobe at the Cambridge Electron Accelerator [61]. Besides a mathematical formalism given to calculate the necessary mirror dimensions, experimental results are shown of the spot obtained at that time (1972–1973). The bent cylindrical mirror was made out of fused quartz; the cylinder had a radius of 10 cm and the surface area was $60 \times 10 \, cm^2$.

Bilderback *et al.* [81] show results of a focussing system consisting of a $10 \times 60 \, cm^2$ Pt-coated reflecting surface that can be elastically bent from a radius of 400 m to infinity. This mirror receives horizontally focussed X-rays after they are diffracted from a bent asymmetrically cut Ge(111) monochromator. In this way, they focussed the beam in the vertical plane from 5 mm to $300 \, \mu m$ and they measured a flux of 3×10^{12} photons/s in a $1.5 \times 0.3 \, mm^2$ area. The X-ray energy was 8 keV. (The storage ring operated at 5.3 GeV, 70 mA and the station was on a wiggler line intercepting 2 mrad.)

Van Langevelde *et al.* [82] and Lenglet *et al.* [18] describe the use of doubly bent Si(111) in a toroidal shape in order to focus X-rays of 20 keV. For such a system, the magnification (M) from Eqn. 1.10 is given by

$$M = u/v = 2u \sin \theta / Rs - 1$$

with u and v the object and image distance, θ the Bragg angle and R_s the sagittal radius. The authors used $100 \, \mu m$ thick Si wafers with a diameter of 50 mm. (The minimum value of R_s is about 10 cm to prevent breaking of the crystal.) As described in Section 1.3.2 in order to obtain an astigmatic optical system, the meridian radius, R_m is given by

$$R_m = R_s / \sin^2 \theta$$

For the chosen energy, the meridian curvature needed to be 10 m. The crystal was attached to an A1 mould of the appropriate shape using oxidised polyethylene foil as adhesive. The bent crystal was directly placed in the white beam, where it acted both as monochromator and focussing unit simultaneously. It was tested at Daresbury on a wiggler station, with the ring running at 2 GeV and ~ 100 mA. The demagnification of the object was about a factor

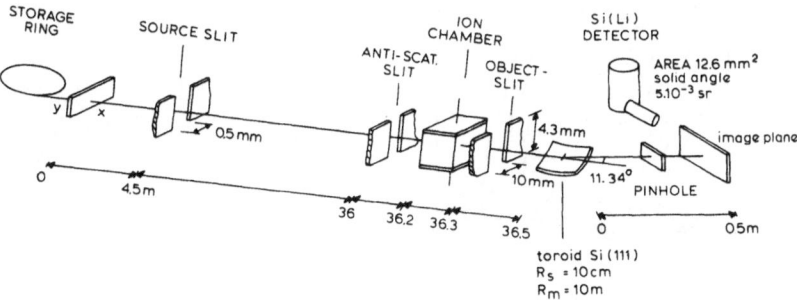

Figure 13.8. A schematic view of the synchrotron microprobe in the use at the SRS, Daresbury (UK) (after van Langevelde *et al.* [82]).

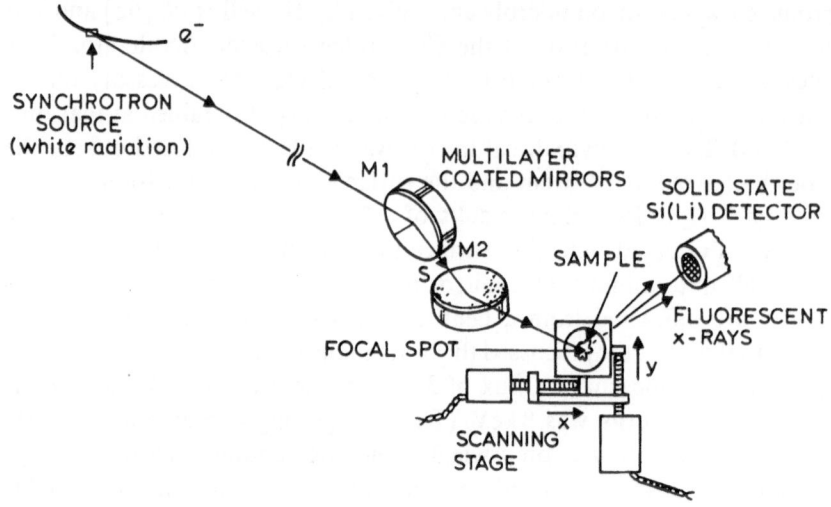

Figure 13.9. A schematic diagram of the synchrotron microprobe using multilayer coated mirrors (from Underwood *et al.* [86]).

of 500 and the measured flux was 200 photons s^{-1} mA^{-1} μm^{-2}. The set up was used to scan over the cross-section of a coronary artery. Recently, the authors replaced the toroidally shaped crystal by an ellipsoid, using a computerised lathe to fabricate the mould. Van Langevelde [83] reports better focussing properties and, also important, flux densities at least 2 orders of magnitude better than the values obtained with the toroid. A schematic view of the equipment is given in Figure 13.8.

Another approach to focus the beam is described by Thompson *et al.* [84, 85] and Underwood *et al.* [86]. They use two spherical mirrors in a

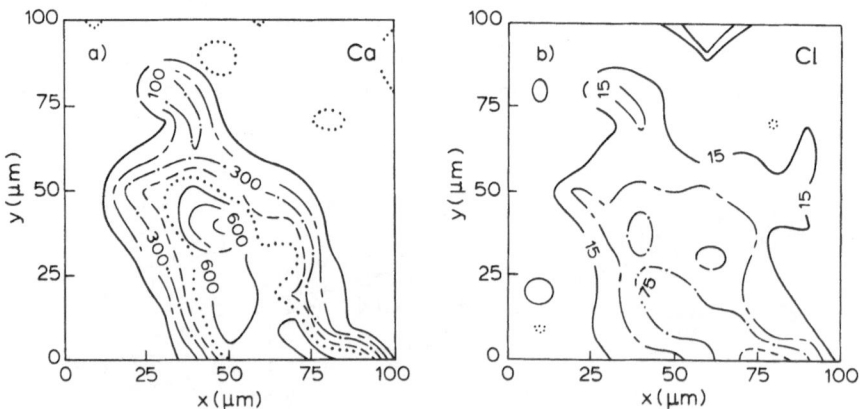

Figure 13.10. Results of a two-dimensional scan of a fluid inclusion containing an aqueous CaCl$_2$ solution in'a thinned quartz sample, obtained with the microprobe shown in Figure 13.9 (after Thompson *et al.* [85]).

Kirkpatrick-Baez geometry [56] which considerably eliminates the astigmatism inherent to spherical mirrors for images of off-axis points. A schematic diagram of such an X-ray microprobe is given in Figure 13.9. In order to have a high flux beam, the authors use multilayer reflectors selecting a relatively wide band pass of about 1 keV at 10 keV. Two different mirrors were used. The upstream mirror was coated with 200 layer pairs of W and C with a 2d spacing of 58 Å; the second mirror with 100 layer pairs with a spacing of 87 Å. The beamspot obtained is less than $10 \times 10 \, \mu m^2$ and the available intensities $10^8 - 10^9$ photons/s depending on the experimental station used. The first applications of this elegant equipment include the measurement of the Fe distribution along a filament of blue-green algae, a number of spot measurements on a tissue section of rat melanoma and measurements on fluid inclusions containing an aqueous calcium chloride solution. As illustration from the latter application, the Ca and Cl elemental maps are given in Figure 13.10. The resistivity of the mirrors against very intense beams is still under investigation.

13.7 Conclusions

Synchrotron radiation offers very attractive possibilities for trace element analysis. Especially if microprobes become available for elemental mapping, micro-SXRF should be capable of analysing at the sub-ppm level in an accurate, quantitative way with resolutions down to a few micrometers. Besides the fact that usually the photon beams have higher stability than ion beams, the photons are superior in terms of signal to induced damage ratio which is a major advantage when analysing vulnerable samples. Moreover, sample preparation procedures are normally less demanding for micro-SXRF. Measurements are made in air, so wet specimens can be examined. For the production of microprobes, the high intensity of synchrotron radiation is essential and brings the technique out of range of conventional XRF, not to mention the advantage of the linear polarisation of the synchrotron radiation.

Due to the limited availability of dedicated storage rings, not very many applications of (micro-) SXRF are described to date. Moreover, most applications are still at the stage of demonstration experiments rather than tackling an existing scientific problem related to trace elements. This may also be partially due to the fact that the development of the technique is mainly performed by physicists, not familiar with the problems existing in geology and biology. In order to realise the full potential of the technique it is necessary to form interdisciplinary research groups with sufficient access to the synchrotron sources to bring forward the branches of science where trace elements play crucial roles.

References

1. T.B. Johansson, R. Akselsson and S.A.E. Johansson. *Nucl. Instrum. Methods* **84** (1970) 141.
2. V. Valkovic, *Nuclear Microanalysis*, Garland Publications, New York (1977).
3. M.J. Owens and H.I. Shalgosky, *J. Phys.* **37** (1974) 593.
4. F. Folkmann, *J. Phys. E* **8** (1975) 429.
5. S.A.E. Johansson and T.B. Johansson, *Nucl. Instrum. Methods* **173** (1976) 473.
6. T.A. Cahill, *Annu. Rev. Nucl. Particle Sci.* **30** (1980) 211.
7. V. Valkovic, *Analysis of Biological Material for Trace Elements Using X-Ray Spectroscopy*, CRC Press, Boca Raton, FL, (1980).
8. R.D. Vis, *The Proton Microprobe: Applications in the Biomedical Field*, CRC Press, Boca Raton, FL (1985).
9. W. Maenhaut, in *Proc. VII Ion Beam Conference*, Johannesburg, South Africa (1988).
10. R. Tertian and F. Claisse, *Principles of Quantitative X-Ray Analysis*, Heyde, London (1982).
11. C.A. Currie, *Anal. Chem.* **40** (1968) 586.
12. M. Prins, W. Dries, W.J.M. Lenglet, S.T. Davies and D.K. Bowen, *Nucl. Instrum. Methods Phys. Res. B* **10/11** (1985) 299.
13. B.M. Gordon, *Nucl. Instrum. Methods.* **204** (1982) 223.
14. J.M. Jaklevic, R.D Giauque and A.C. Thompson, *Nucl. Instrum. Methods Phys. Res. B* **10/11** (1985) 303.
15. A. Knöchel, W. Petersen, G. Tolkiehn, *Nucl. Instrum. Methods* **208** (1983) 659.
16. A. Knöchel, W. Petersen and G. Tolkiehn *Anal. Chim. Acta* **173** (1985) 105.
17. A.J.J. Bos, R.D. Vis, H. Verheul, M. Prins, S.T. Davies, D.K. Bowen, J. Makjanic and V. Valkovic, *Nucl. Instrum. Methods Phys. Res. B* **3** (1984) 232.
18. W.J.M. Lenglet, R.D. Vis, F. van Langevelde and H. Verheul, *Anal. Chim. Acta* **195** (1987) 153.
19. V.P. Khvostova and V.A. Trunova *Nucl. Instrum Methods in Phys. Res. A* **261** (1987) 295.
20. L. Grodzins, *Neurotoxicology* **4**(3) (1983) 23.
21. L. Grodzins, *Nucl. Instrum Methods Phys. Res.* **218** (1983) 203.
22. W.J.M. Lenglet, Ph.D. Thesis, Free university, Amsterdam (1988).
23. C.J. Sparks, In *Synchrotron Radiation Research*, eds. H. Winick and S. Doniach, Plenum Press, New York (1980) p. 459.
24. J.M. Hubbel, W.J.M. Veigele, E.A. Briggs, R.T. Brown, D.T. Cramer and R.J. Howerton, *J. Phys. Chem. Rev. Data* **4** (1975) 471.
25. J. Schwinger, *Phys. Rev.* **70** (1946) 798.
26. V.B. Baryshev, G.N. Kulipanov and A.N. Skrinsky, *Nucl. Instrum. Methods Phys. Res. A* **246** (1986) 739.
27. H.L. Hanson, H.W. Kraner, K.W. Jones, B.M. Gordon, R.E. Mills and J.R. Chen, *IEEE Trans. Nucl. Sci.* **30**(2) (1983) 1339.
28. C.J. Sparks, S. Raman, E. Ricci, M.O. Krause and R.V. Gentry, *IEEE Trans. Nucl. Sci.* **26**(1) (1977) 1368.
29. R.D. Giauque, J.M. Jaklivic and A.C. Thompson, *Adv. X-Ray Anal.* **27** (1985) 53.
30. J.R. Chen, N. Martijs, E.C.T. Chao, J.A. Minkin, C.L. Thompson, A.L. Hanson, H.W. Kraner, K.W. Jones, B.M. Gordon and R.E. Mills, *Nucl. Instrum. Methods Phys. Res. B* **3** (1984) 241.
31. J.R. Chen, E.C.T. Chao, J.A. Minkin, J.M. Back, W.C. Bayby, M.L. Rivers, S.R. Sutton, B.M. Gordon, A.L. Hanson and K.W. Jones, *Nucl. Instrum. Methods Phys. Res. B* **22** (1987) 394.
32. V.P. Khvostova, V.N. Maximov, A.A. Yaroshevsky, V.B. Baryshev and G.N. Kulipanov, *Nucl. Instrum. Methods. Phys. Res. A* **261** (1987) 283.
33. V.B. Baryshev, G.N. Kulipanov, E.I. Zaytsev, Y.V. Terekhov and V.I. Kalyuzny, *Nucl. Instrum. Methods Phys. Res. A* **261** (1987) 279.
34. L.C. Tarasov, A.F. Kudryashova, A.V. Ivanov, A.A. Ulyanov, V.B. Baryshev, G.N. Kulipanov and A.N. Skrinsky, *Nucl. Instrum. Methods Phys. Res. A* **261** (1987) 263.
35. S.R. Sutton, M.L. Rivers and J.V. Smith, *Proc. 9th Int. Conf. on the Applications of Accelerators in Research and Industry*, Denton, USA (1986).
36. J.R. Chen, B.M. Gordon, A.L. Hanson, K.W. Jones, H.W. Kraner, E.C.T. Chao and J.A. Minkin, *Scanning Electron Microsc.* **IV** (1984) 1483.

37. A.H. Compton, *Philos. Mag.* **45** (1923) 1121.
38. L.G. Parratt, *Phys. Rev.* **95** (1954) 359.
39. P.A. Pella and R.C. Dobbyn, *Anal. Chem.* **60** (1988) 684.
40. A. Iida and Y. Gohshi, *Adv. X-Ray Anal.* **27** (1985) 61.
41. A. Iida, K. Sakurai, A. Yoshinaga and Y. Gohshi, *Nucl. Instrum. Methods Phys. Res. A* **246** (1986) 736.
42. W.C. Marra, P. Eisenberger and A.Y. Cho, *J. Appl. Phys.* **50**(11) (1979) 6927.
43. G.W. Grime and F. Watt, eds. *Proc. 2nd Int. Conf. on Nucl. Microprobe Technology and Applications*, Oxford, UK (1987); *Nucl. Instrum. Methods Phys. Res. B* **30** (1988) 227.
44. G. Gouy, *Ann. Phys.* **5** (1916) 241.
45. L. Hámos, *Naturwiss enschaften* **20** (1932) 705.
46. L. Hámos, *Nature* **134** (1933) 181.
47. H.H. Johann, *Z. Phys.* **69** (1931) 185.
48. Y. Cauchois, *C.R. Acad. Sci. Paris* **194** (1932) 362.
49. Y. Cauchois, *C.R. Acad. Sci. Paris* **195** (1932) 228.
50. Y. Cauchois, *J. Phys. Radium* **4** (1933) 61.
51. Y. Cauchois, *Ann. Phys. Paris* **1** (1934) 215.
52. B.W. Roberts and W. Parrisch, *Int. Tables for X-ray Crystallography*, Birmingham (1968).
53. F. Jentzsch, *Phys. Zeit.* **30** (1929) 268.
54. W. Ehrenberg, *Nature* **160** (1947) 330.
55. W. Ehrenberg, *J. Opt. Soc. Am.* **39** (1949) 741.
56. P. Kirkpatrick and A.V. Baez, *J. Opt. Soc. Am.* **38** (1948) 766.
57. V.E. Cosslett and W.L. Nixon, *X-Ray Microscopy*, Cambridge (1960).
58. Proc. Of the Int. Conf. on Synchrotron Radiation, Instrumentation and New Developments, *Nucl. Instrum. Methods* **152** (1978).
59. G. Schmahl and D. Rudolph *X-Ray Microscopy*, Springer-Verlag, Berlin (1984).
60. V.V. Aristov, Y.A. Basov, S.V. Redkin, A.A. Suigirev and V.A. Yunkin, *Nucl. Instrum. Methods A* **261** (1987) 72.
61. P. Horowitz and J.A. Howell, *Science* **178** (1972) 608.
62. J.A. Howell and P. Horowitz, *Nucl. Instrum. Methods* **225** (1975) 235.
63. C.J. Sparks, S. Raman, H.L. Yakel, R.V. Gentry and M.O. Krause, *Phys. Rev. Lett.* **38** (1977) 205.
64. C.J. Sparks, S. Raman, E. Ricci, R.V. Gentry and M.O. Krause, *Phys. Rev. Lett.* **40** (1979) 507.
65. R.D. Deslattes, *Nucl. Instrum. Methods* **172** (1980) 201.
66. C.J. Sparks, B.S. Borie and J.B. Hastings, *Nucl. Instrum. Methods.* **172** (1980) 237.
67. S.M. Heald, *Nucl. Instrum. Methods* **195** (1982) 59,
68. M.R. Howells and J.B. Hastings, *Nucl. Instrum. Methods* **208** (1983) 379.
69. G.E. Ice and C.J. Sparks, *Nucl. Instrum. Methods Phys. Res.* **232** (1984) 121.
70. G.E. Ice and C.J. Sparks, *Nucl. Instrum. Methods Phys. Res. A* **266** (1988) 394.
71. B.W. Batterman and L. Berman, *Nucl. Instrum. Methods* **208** (1983) 327.
72. D.M. Mills, C. Henderson and B.W. Batterman, *Nucl. Instrum. Methods. Phys. Res. A* **246** (1986) 356.
73. C.J. Sparks, G.E. Ice, J. Wong and B.W. Batterman, *Nucl. Instrum. Methods* **194** (1982) 73.
74. D.H. Bilderback, B.W. Lairson, T.W. Barbee, G.E. Ice and C.J. Sparks, *Nucl. Instrum. Methods* **208** (1983) 251.
75. M. Prins, S.T. Davies and D.K. Bowen, *Nucl. Instrum. Methods* **222** (1984) 324.
76. M. Prins, J.M. Kuiper and M.P.A. Viegers, *Nucl. Instrum. Methods Phys. Res. B* **3** (1984) 246.
77. M. Prins, J.A. van de Heide, A.J.J. Bos, D.K. Bowen and S.T. Davies, *IEEE Trans. Nucl. Sci.* **30**(2) (1983) 1243.
78. W. Petersen, Roentgenfluoreszensanalyse mit Hilfe des Synchrotronstrahlung, Internal Report DESY F41, Hasylab 84–02 (1984).
79. W. Petersen, P. Ketelsen, A. Knöchel, R. Pausch, *Nucl. Instrum. Methods Phys. Res. A* **246** (1986) 731.
80. M. Bavdaz, A. Knöchel, P. Ketelsen, W. Petersen, N. Gurker, M.H. Salehi and T. Dietrich, *Nucl. Instrum. Methods Phys. Res. A* **266** (1988) 308.
81. D. Bilderback, C. Henderson and C. Prior, *Nucl. Instrum. Methods Phys. Res. A* **246** (1986) 194.
82. F. van Langevelde, W.J.M. Lenglet, R.M.W. Overwater, R.D. Vis, A. Huizing, M.P.A.

Viegers, C.P.G.M. Zegers and J.A. van de Heide, *Nucl. Instrum. Methods Phys. Res. A* **257** (1987) 436.

83. F. van Langevelde, Private communication.
84. A.C. Thompson, Y. Wu, J.H. Underwood and T.W. Barbee, *Nucl. Instrum. Methods. Phys. Res. A* **255** (1987) 603.
85. A.C. Thompson, W.H. Underwood, Y. Wu, R.D. Giauque, K.W. Jones and M.L. Rivers, *Nucl. Instrum. Methods Phys. Res. A* **266** (1988) 318.
86. J.H. Underwood, A.C. Thompson, Y. Wu and R.D. Giauque, *Nucl. Instrum. Methods Phys. Res. A* **266** (1988) 296.

14 Atomic and molecular science

J.B. WEST

14.1 Introduction

Experiments in this area of research are generally oriented more towards physical processes than chemical ones, but the success in the development of new experimental techniques, particularly using synchrotron radiation or high power lasers has encouraged new applications, principally in the area of molecular chemistry. The methods developed by physicists interested in the ways in which electrons interact in atoms, and correlations between different electron orbitals, have begun to be used successfully by chemists. Their interests are, of course, a little different; their purpose is to identify the symmetry of orbitals in polyatomic molecules, and what part these play in molecular binding. The overall aim is to give a description of chemical interactions at the molecular orbital level. The techniques of photo-ion spectroscopy, photo-electron spectroscopy and fluorescence spectroscopy, when applied over the broad wavelength range available from a synchrotron radiation source, yield much new and specific information in this respect, and are a considerable improvement on the fixed excitation energy methods available previously.

The purpose of this chapter is to give an account of the experimental techniques used for this kind of work, and show how they are applied to problems which can generally be described as belonging to the area of molecular chemistry. Thus the data obtained are fundamental in nature, but are complementary to data obtained using other techniques and are essential to our understanding of chemical reactions at the molecular level. It is beyond the scope of this chapter to provide an exhaustive review of all synchrotron radiation based research in this area, but the examples presented here show how techniques used in atomic and molecular spectroscopy are being applied to chemistry. Modern synchrotron radiation sources provide an improvement in intensity of about two orders of magnitude over laboratory sources, which has opened up new possibilities. Differential processes can now be studied in detail and the behaviour of individual ionisation channels mapped over a wide range of excitation energies. Good spatial and energy resolution are both available, making it possible to study small samples and resolve narrow resonant structure. Most of the work covered here is concerned with the soft

X-ray, VUV and UV ranges, i.e. photon energies from $\sim 2000\,eV$ down to 5 eV. Those interested in the optical techniques and beam line technology used to cover this range, where diffraction gratings are employed primarily, are referred to the literature (see, e.g. [1]); the experimental details given in this chapter concentrate on the techniques used to obtain the data.

14.2 Photo-electron spectroscopy

This technique has been used for some time, by both physicists and chemists, for the identification of molecular orbitals, and the acronyms ESCA (electron-spectroscopy for chemical analysis), UPS and XPS (UV and X-ray photo-electron spectroscopy, respectively) are routinely used. The standard equipment used in the laboratory employs line sources such as a helium discharge lamp, generating radiation energies of 21.2 eV and 40.8 eV, or X-ray characteristic line sources. Data obtained in this way have been extensively tabulated [2, 3]. Although a lot has been learned and the usefulness of the technique established, it is realised that measurements made at one or two wavelengths cannot be definitive in assigning band origins. The measurement of partial cross-sections and angular distributions over a wide range of photo-ionisation energies has made it possible to apportion ionisation bands into their separate components, and it is in this area that synchrotron radiation has made a major contribution in the last few years.

As an example of an experimental system for measurements of this kind, Figure 14.1 shows the layout of an electron spectrometer used extensively at the Daresbury SRS to study organic halides and metallo-organic complexes. This instrument has the facility to rotate about the light beam direction and is thus able to measure photo-electron angular distributions as well as photo-electron spectra. Monochromatic light is provided by a toroidal grating monochromator (TGM) optically coupled to the SRS by mirrors used at grazing angles of incidence (a brief description of TGMs is given in Section 1.4.2; full details can be found in [1]). Photon energies available from this monochromator lie in the range 10–120 eV, with a minimum photon resolution of 20 meV; thus many molecular bands, extending down to core levels, are accessible. The radiation leaving the monochromator exit slit is polarised, primarily, but not completely, in the horizontal direction perpendicular to the light beam (see Section 1.2.1); photo-electron angular distributions are measured relative to the major, horizontal E-vector component, in the plane perpendicular to the light beam.

As will be seen later, electron angular distributions are also of value in assigning molecular bands. This system is capable, however, of measuring partial cross-sections, i.e. the cross-section for ionisation into a particular channel, on a relative basis. This data can be put on an absolute basis by normalisation to total cross-section values, where available, from a separate

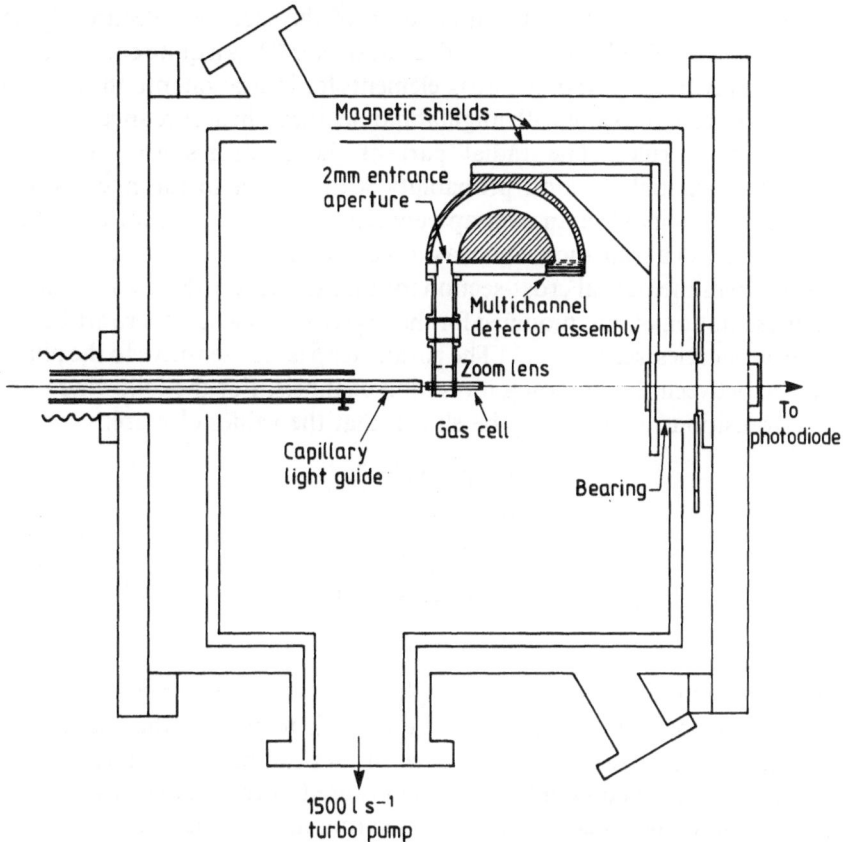

Figure 14.1.   Angle resolving photo-electron spectrometer.

experiment. The expression

$$d\sigma/d\Omega = \sigma_i/4\pi \left\{ 1 + \beta/4(1 + 3P\cos 2\theta) \right\} \qquad (14.1)$$

gives the differential cross-section for the ith ionisation channel, measured at
an angle θ with respect to the main polarisation axis; σ_i is the photo-ionisation
cross-section for this channel. P is the polarisation of the incident light, defined
as $(I_\parallel - I_\perp)/(I_\parallel + I_\perp)$ where $I_\parallel$ and $I_\perp$ are the parallel and perpendicular
components of the polarisation respectively, and β is the angular asymmetry
parameter. It can be seen readily that if P is known, β can be calculated from
measurements at two angles, using Eqn. 14.1; furthermore, once P is known, a
value of θ can be chosen such that the term containing β comes to zero. Thus
the intensities measured at this angle, the so called 'pseudo-magic' angle, are
directly proportional to the cross-section of the ith ionisation channel.
Measurements at this angle and one other angle are therefore sufficient to give
both the partial cross-section and angular distribution for the band being
studied, at a particular photon energy.

The form of Eqn. 14.1. is a consequence of the fact that electric dipole transitions are involved; see, e.g. [4]. The parameter β has a quite complicated form which involves the radial matrix elements for ionisation into the l_{+1} and l_{-1} channels, from the state with angular momentum l, in accordance with the dipole selection rules. The radial part of the electron's wave function represents its spatial position probability with respect to the nucleus; the matrix element is in effect an overlap between the probabilities for the initial and final states of a transition, and if a dipole operator is used in calculating it, the result gives the partial cross-section for that transition. (For a mathematical representation of this, modified for the case of solids where there are bands of electron energies, see Eqn. 6.2.) The parameter β is also sensitive to the phase difference between the outgoing ionisation channels. For the following pure atomic transitions, it can easily be shown that the values of β are:

$$s - p \longrightarrow \beta = 2;$$
$$p - s \longrightarrow \beta = 0;$$
$$p - d \longrightarrow \beta = 1;$$
$$d - p \longrightarrow \beta = 0.2;$$
$$d - f \longrightarrow \beta = 0.8.$$

These expressions hold when $L - S$ coupling is valid; values of β can fall to -1 when the spin orbit interaction is large enough to require the use of $j - j$ coupling. The fact that $\beta = 2$ for $s - p$ transitions can be used to measure the polarisation of the incoming light. Helium is used as the sample gas; making measurements of the ionisation of the 1s electrons into the p-continuum at two angles and using the fact that $\beta = 2$ in Eqn. 14.1 yields the value of P, for photon energies beyond the IP of helium at 24.2 eV. Below this energy, argon is used for which β is well known from accurate measurements [5].

The above experimental information on atomic angle resolved photoelectron spectroscopy is included here for background purposes; for a more complete description the reader is referred to the extensive literature available on this subject (see, e.g. [6] and references therein). Recently, substantial improvements have been made to increase the sensitivity of electron analysers by using multichannel detection at the exit plane [7, 8]; this is of particular value for gaseous samples where the pressure has to be kept low to be compatible with the ultra high vacuum required in the optical beam line, and also where the chemical samples themselves may not be available in large quantities. Much of the work described here would not have been possible without this enhancement.

Before giving examples of the use of such equipment in identifying molecular orbital structure, a brief description of 'Cooper minima' and resonant phenomena is appropriate. Cooper minima occur when the radial matrix element for one of the photo-ionisation channels changes sign, and if this is the dominant channel at that particular photon energy, the cross-section σ_i goes through a minimum, and may even reach zero. This phenomenon, first predicted by Seaton [9] and fully developed and explained by Cooper [10], occurs for example in argon at photon energies around 50 eV. In this particular case there are three channels contributing to the total photo-ionisation cross-section, $3p - \varepsilon s$, $3p - \varepsilon d$ and a very small contribution from $3s - \varepsilon p$. It is the $3p - \varepsilon d$ channel whose matrix element goes negative in this region, and since this channel forms the dominant contribution to σ_t here a minimum in the cross-section results [11]. Shape resonances, so-called because of the shape of the electronic potential curves, are closely related to the centrifugal barrier effect in atoms. In this case electrons excited or ionised to high angular momentum states ($l > 2$) are trapped until there is sufficient surplus energy from the incoming photon to overcome the centrifugal barrier. For example, in the case of xenon, this causes a delayed onset of the 4d-εf continuum transitions expected at the 4d threshold. Experimentally, this is seen as a large broad peak in the absorption cross-section displaced to higher energy [12, 13]. Auto-ionisation is a general phenomenon for both atoms and molecules, and arises because there are discrete, though short-lived, bound states which lie above the first ionisation threshold. There is then interference between the direct ionisation process and excitation to the bound neutral state, which causes in some cases strongly asymmetric absorption profiles or even so-called 'window' resonances where the absorption decreases. The underlying theoretical basis of these phenomena is explained in a review article by Fano and Cooper [4].

These ideas in atomic spectroscopy find many applications in molecular spectroscopy. Angular momentum l is now no longer a good quantum number, being replaced by Λ which represents the sum of the atomic orbitals which constitute the molecular orbital. Partial ionisation channels with high angular momentum are thus more accessible; e.g. the outgoing wave resulting from the ionisation of σ electrons can contain high angular momentum components, and shape resonances are frequently encountered in molecular spectra. Their presence, however, is not always apparent from straightforward absorption; it is usually necessary to examine the ionisation channels individually. This is why the measurement of partial cross-sections is important; in regions of resonance structure (e.g. shape or auto-ionisation) phase changes in the outgoing wave occur, and the β parameter is sensitive to this. Thus measuremenmts of β help to identify the channel in which a resonance occurs and can provide information on the individual components of molecular orbitals.

For molecules, such effects manifest themselves in a breakdown of the Franck-Condon principle, which states that during an absorption or emission process, the distances between the atomic nuclei which make up the molecule do not change. This implies that there is no interaction between the molecular electronic states and the vibrational motion of the nuclei. Figure 14.2 shows a typical configurational coordinate diagram for the simple case of a diatomic

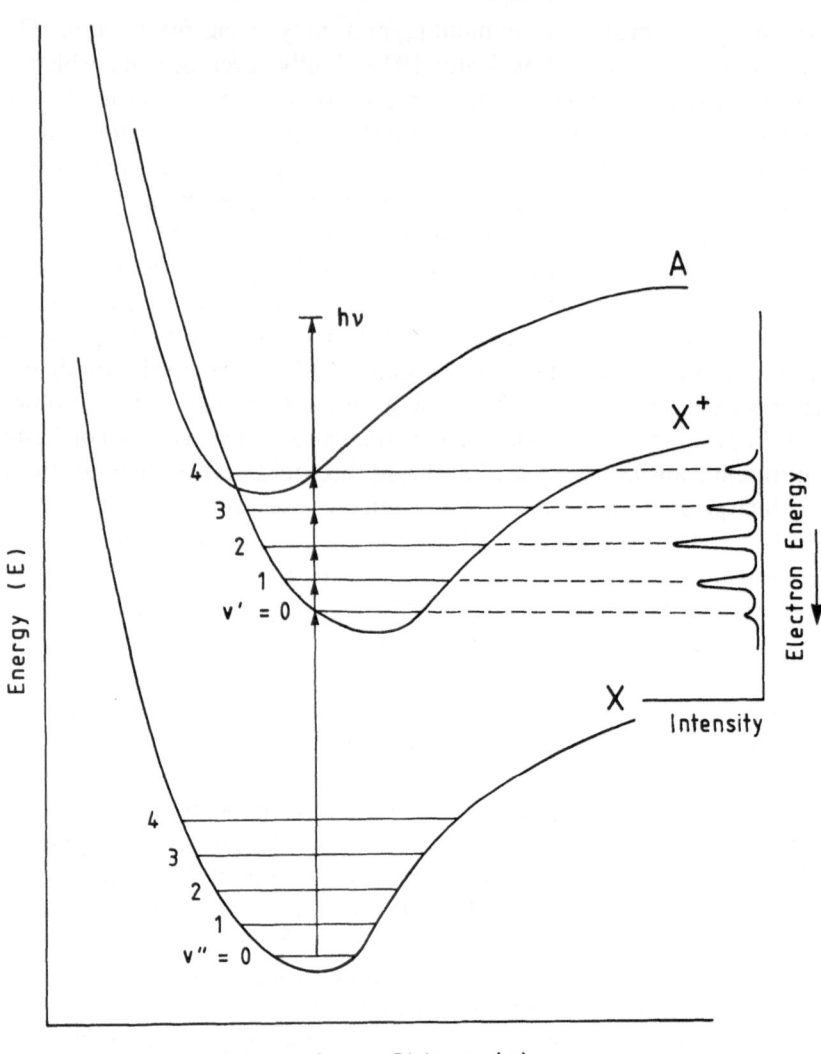

Figure 14.2. Configurational coordinate diagram for a diatomic molecule.

molecule, where r is the internuclear distance, E the system energy. The vibrational levels are shown as the horizontal lines within the potential wells, shown with their quantum numbers v'' and v'; the ground state of the neutral molecule is labelled X and the state labelled X^+ is the ground state of the molecular ion. The Franck-Condon principle states in effect that all transitions on this diagram are vertical. If the mean inter-nuclear distances of the two states involved is the same, this implies that most of the absorption intensity would be contained in the transition $v'' = 0$ to $v' = 0$, since at room temperature only the lowest vibrational level of the neutral ground state has any substantial population. Transitions to higher v' values are less probable because the nuclei spend more time at the extremities of the vibration and the 'overlap' with the ground state vibrational wave function is small. However, if as a result of removing an electron the mean internuclear distance changes, a vertical transition on the diagram now accesses the higher v' levels. Figure 14.2 shows this case, where the mean value of r is shown to increase indicating that the electron removed was a *bonding* electron. The resulting photo-electron spectrum is shown on the right of the figure; the incoming photon energy is represented by the line from $v'' = 0$ to the bar labelled hv. Then the electron energies are represented by the lengths of the lines from the v' values to hv, and their intensities are proportional to the 'Franck-Condon factors' associated with each of them. These are calculated from the overlap integrals between the ground stage $v'' = 0$ and the v' levels of the molecular ionic state.

The third level labelled A in Figure 14.2 is a state of the neutral molecule above the ionisation potential of the ground state and is thus auto-ionising. Transitions to its vibrational levels are possible from the neutral ground state when the incoming photon has the correct or resonant energy, and these will be dominated by their Franck-Condon factors. This state can then auto-ionise, ejecting an electron, to the vibrational levels of the ground state of the ion, and the intensities of these transitions will in their turn depend on a second set of Franck-Condon factors. Thus the photo-electron spectrum seen will depend on the convolution of the two sets of Franck-Condon factors, and in general will be very different from the case for direct ionisation. It may also be that there is interaction between the molecular electronic state and the vibrational motion, depending somewhat on the lifetime of the auto-ionising level. This further complication yields a spectrum very different from that expected, and to unravel effects such as these measurements must be made both 'off' and 'on' resonance, and preferably continuously over the width of the resonance. For this a continuum source of high intensity is essential.

Some experiments will now be described which demonstrate the use of these ideas. Potts et al. [15] made a series of measurements on the chlorofluoro-methanes over the photon energy range 18–80 eV. Figure 14.3 shows photo-electron spectra taken at a photon energy of 80 eV and an analyser angle of 0°, i.e. along the main component of polarisation. The band assignments shown were derived from a one electron model of ionisation. The purpose of this work was to measure β for these bands, and the result of this is shown in Figure 14.4. In Figures 14.4(a), (b) and (c), the oscillatory behaviour of β is reminiscent of similar behaviour in argon [16], which is isoelectronic with the chloride ion. It can thus be associated with a Cooper minimum, and is attributed to chlorine lone pair orbitals. The fact that the effect is less prominent for the $2a_2 + 6a_1$ band of CF_2Cl_2 and the 5e band of CF_3Cl indicates that mixing is occurring with other molecular orbitals of the same symmetry and reduced Cl 3p character. In contrast, in Figure 14.4(d) the fluorine 2p lone pair orbitals show behaviour reminiscent of earlier measurements on neon [17], despite the fact that the $5a_1$ orbital of CF_3Cl is assumed to have marked C–Cl bonding

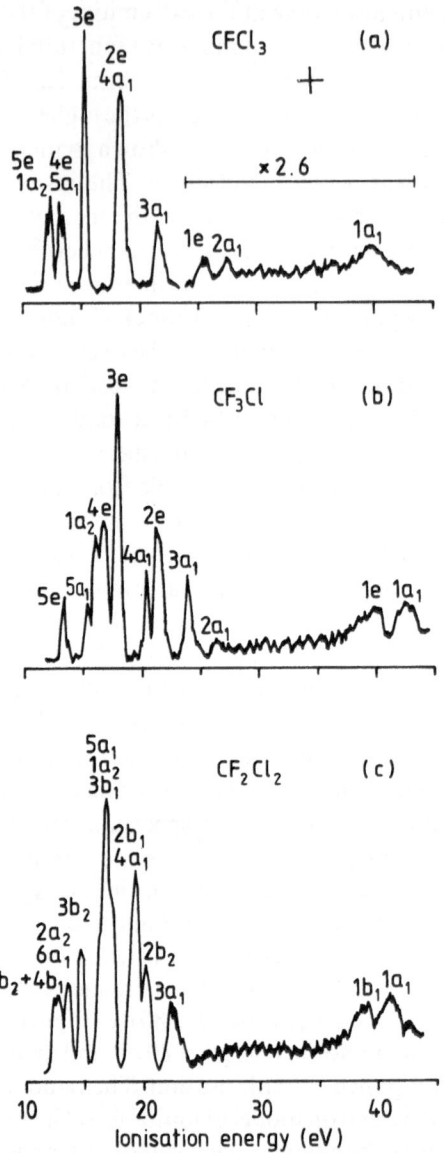

Figure 14.3. Photo-electron spectra for the chlorofluoromethanes taken at $\theta = 0°$, for a photon energy of 80 eV (after Potts *et al.* [15]).

character, indicating that the chlorine 3p character cannot be large in this case. For a complete analysis of all these data the reader is referred to the original paper by Potts *et al.* [15]; the aim here is to indicate how such measurements can be used to give information on mixing of molecular orbitals and their relationship to pure atomic behaviour. A substantial study of small inorganic molecules has been made by Carlson *et al.* [18] who have made by far the

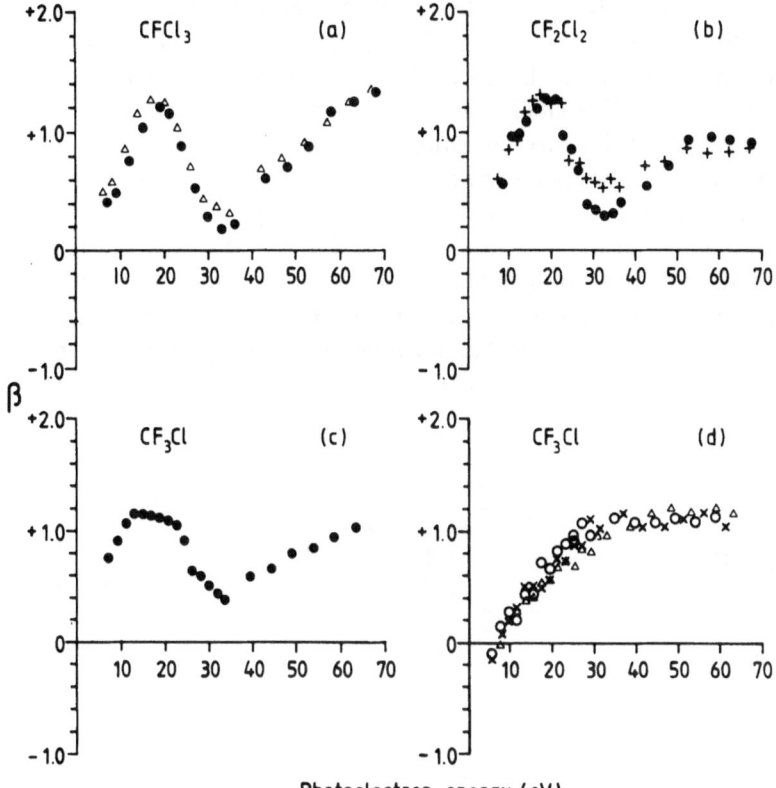

Figure 14.4. The angular distribution parameter β as a function of photo-electron energy (redrawn and adapted from Potts *et al.* [15]) for ionisation from the (a) $1a_2 + 5e$ ($\bullet$) and $4e + 5a_1$ ($\triangle$) orbitals of $CFCl_3$; (b) $4b_2 + 4b_1$ ($\bullet$) and $2a_2 + 6a_1$ (+) orbitals of CF_2Cl_2; (c) $5e$ ($\bullet$) orbital of CF_3Cl; (d) $5a_1 + 1a_2 + 4e$ ($\triangle$), $3e$ ($\times$) and $4a_1$ ($\bigcirc$) orbitals of CF_3Cl.

major contribution in this area. Effects due to the Cooper minimum are discussed in a useful summary by Carlson *et al.* [18]; shape resonance effects are highlighted for a series of tetrachlorides by Piancastelli *et al.* [19].

The measurement of partial cross-sections, σ_i gives further information on interactions between molecular orbitals. In this example a series of organometallic complexes has been studied in this way using the electron spectrometer illustrated in Figure 14.1, including the chromium, molybdenum and tungsten carbonyls [20] and the metallocenes of iron, ruthenium and osmium [21] all of which show strongly mixed metal-ligand character. The electron spectometer was used at the 'pseudo-magic' angle, which it may be recalled from the discussion following Eqn. 14.1 removes angular distribution effects. Photoelectron spectra were then taken at a range of photon energies between 16 and 115 eV, and the relative partial cross-sections for the individual molecular orbitals obtained. Before describing these, experimentally it is now important to establish the detection efficiency of the electron spectrometer over the range of electron energies being detected, so that intensities at one energy can be quantitatively related to those at another. The absolute efficiency is almost impossible to determine because of uncertainties in sample density and geometry, but relative efficiencies can be obtained by making a series of measurements on helium to generate electrons with the same range of energies. The electron intensities can then be compared with the precisely known photo-ionisation cross-section of helium for those photon energies [22]. Thus this provides relative partial cross-sections, which can be put on an absolute basis by normalisation to absolute total cross-section values where these are known.

Figure 14.5 is an energy level diagram for a metallocene with D_{5h} symmetry, showing the order of the orbitals identified in the photo-electron spectra in Figure 14.6. The spectra shown for the three compounds were taken at a

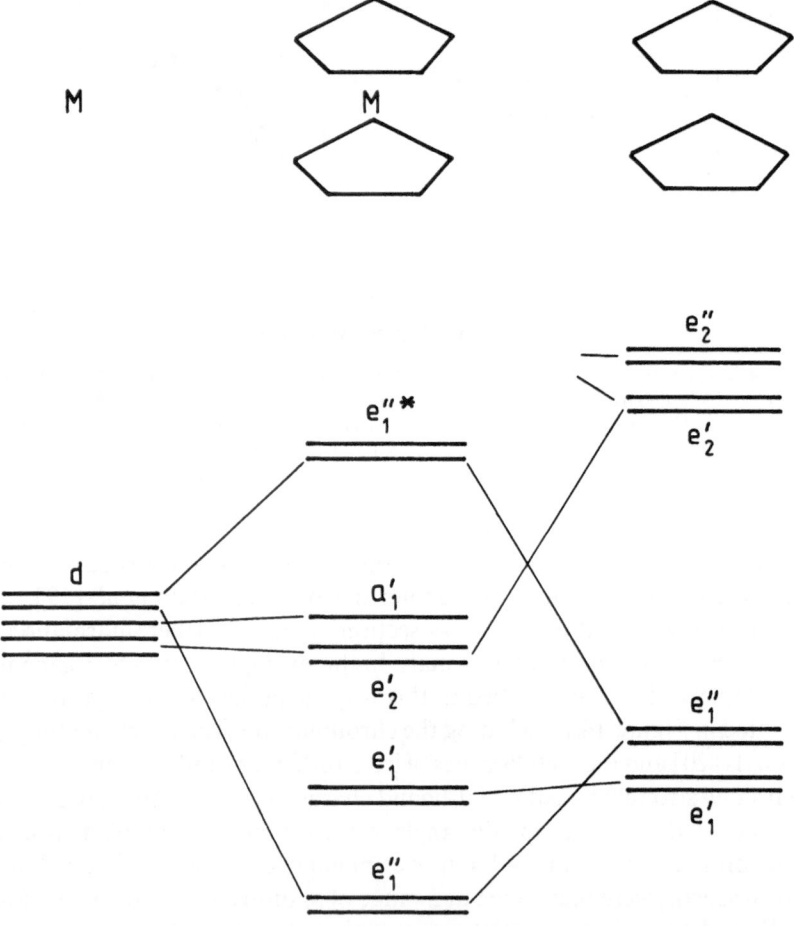

Figure 14.5. Schematic diagram for a metallocene with D_{5h} symmetry; the metal d-levels are shown on the left, the C_5H_5 levels on the right of the figure; the resulting combined levels are shown in the centre (after Cooper *et al.* [20]).

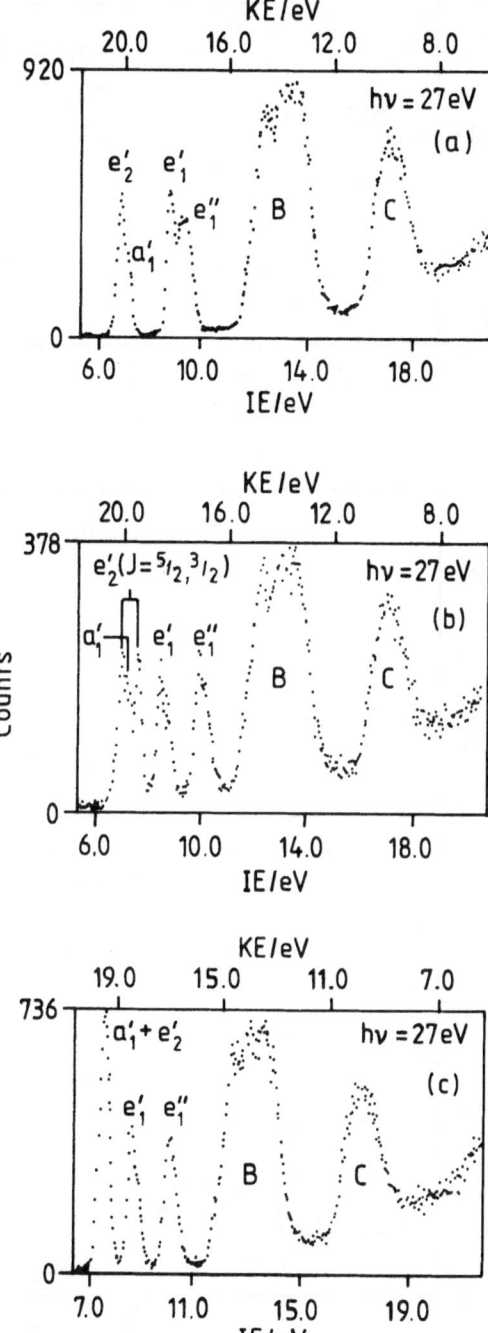

Figure 14.6. Photo-electron spectra of (a) ferrocene, (b) ruthenocene and (c) osmocene taken at a photon energy of 27 eV (after Cooper *et al.* [20]).

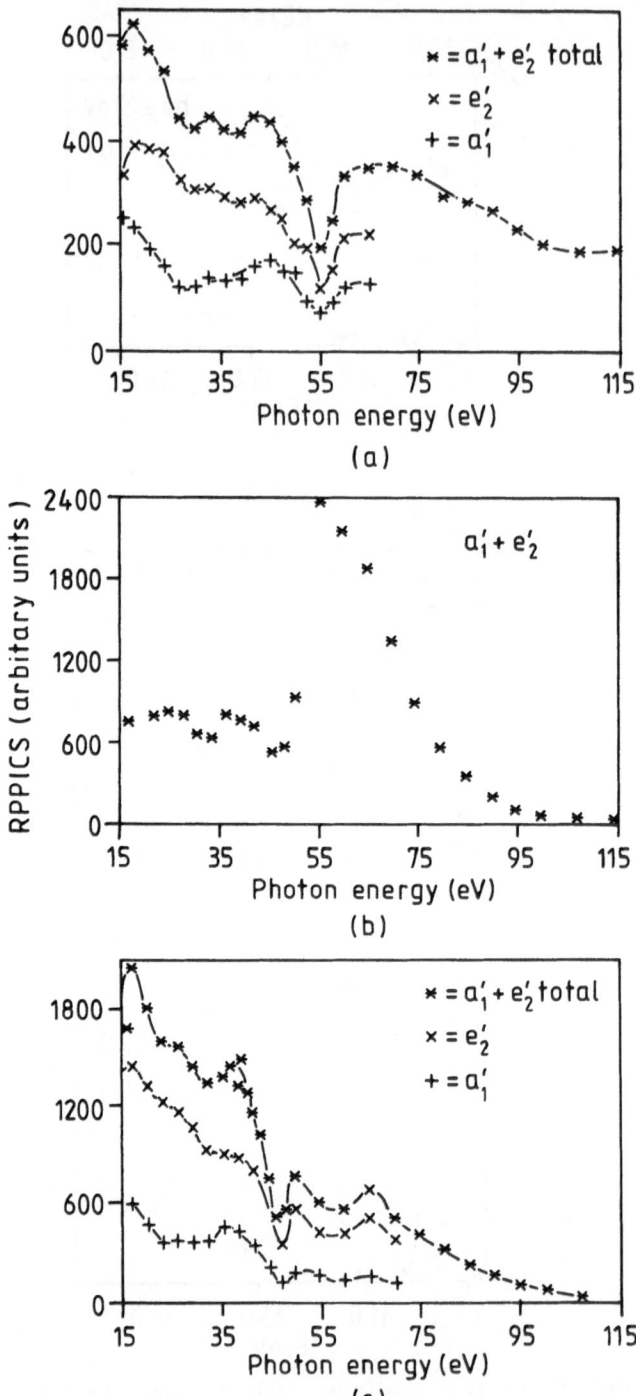

Figure 14.7. Relative partial photo-ionisation cross-sections as a function of photon energy for (a) ferrocene, (b) ruthenocene and (c) osmocene (after Cooper *et al.* [20]).

photon energy of 27 eV; the assignments of the bands are well established from earlier line source work reviewed by Green [23]. They can be separated into three regions: the a'_1 and e'_2 bands which arise predominantly from metal d orbitals; the e'_1 and e''_1 bands associated with ligand valence orbital bands; and the unassigned regions labelled B and C in the spectra arising from ligand σ ionisations. Figure 14.7 shows the relative partial cross-sections for the first region for the three metallocenes. At around 60 eV a peak is seen in all three curves, but is much more pronounced for ruthenium. It is associated with excitation of the metal p levels to d states, and is directly analogous to so called 'giant' resonances in atoms [24]. These are caused by collapse of the d wave functions into the atomic core, giving large oscillator strength to np-nd excitations which then auto-ionise. Clearly the degree of collapse in these compounds is different for the three metal atoms and probably reflects the degree of overlap of these orbitals with the other molecular orbitals, although Cooper et al. [21] point out that the cross-section of ruthenocene in this region is much larger anyway; detailed theoretical calculations will be needed to ascertain this. Similar differences are seen in the partial cross-sections for the second region, where only ruthenium shows a substantial peak. Thus metal character is appearing in the ligand e'_1 and e''_1 orbitals despite the fact that mixing with metal d bands is symmetry forbidden. The third region also shows this effect, particularly for band C. Since band C has carbon character, this suggests that mixing is occurring between the e''_1 orbitals and carbon 2p levels; alternatively, inter-channel coupling in the final states between the $a'_1 + e'_2$ ionisation channels and the $e'_1 + e''_1$ channels is a possibility; Cooper et al. describe these considerations in more detail. They also find evidence for a centrifugal barrier delayed maximum for ruthenocene, which they attribute to the peak at 24 eV in Figure 14.7. The wealth of features seen in these partial cross-sections yields considerable information on the interaction between the metal atoms and the organic part of the molecule. The use of continuum sources in this manner could well change ideas on the ways in which these molecules are bonded.

14.3 Photo-ion spectroscopy

The fragmentation pathways in a molecule following ionisation or excitation can be detected using photo-ion spectroscopy. This technique is particularly powerful when combined with photo-electron spectroscopy since the excited state of the residual ion can then also be identified. This implies the use of coincidence techniques, and experiments of this kind are not, as yet, widespread on synchrotron radiation sources. Straightforward mass spectrometry, in which the interest has centred on the charge states and nature of the photofragments, has been extensively used. The thresholds for fragment production can be related to thermochemical thresholds, and partial cross-sections for the production of particular fragments measured. Work of this nature has been thoroughly reviewed by Berkowitz [25]; since that review, synchrotron radiation work has begun to make a contribution, pioneered primarily by Nenner et al. [26] at the ACO storage ring at Orsay, France, and Samson and co-workers (see, e.g. Masuoka and Samson [27]) using the Tantalus storage ring at the University of Wisconsin, USA. The purpose behind much of this work was to study electron correlations in atoms and diatomic, or small, molecules, by measuring multiple ionisation branching ratios and identify the effects on these of resonances in the photo-ionisation continuum. Thus effort has been devoted more towards solving problems of a physical rather than a chemical nature, as far as synchrotron radiation work is concerned. The exception is when fluorescence from excited fragments is detected; this has opened up many new possibilities in chemistry, and will be discussed later.

More recently, the development of *triple* coincidence experiments has also provided opportunities to study molecules from a chemical point of view. In this case, two charged fragments are detected from one fragmentation event, and in coincidence with the photo-ejected electron; the acronym PEPIPICO,

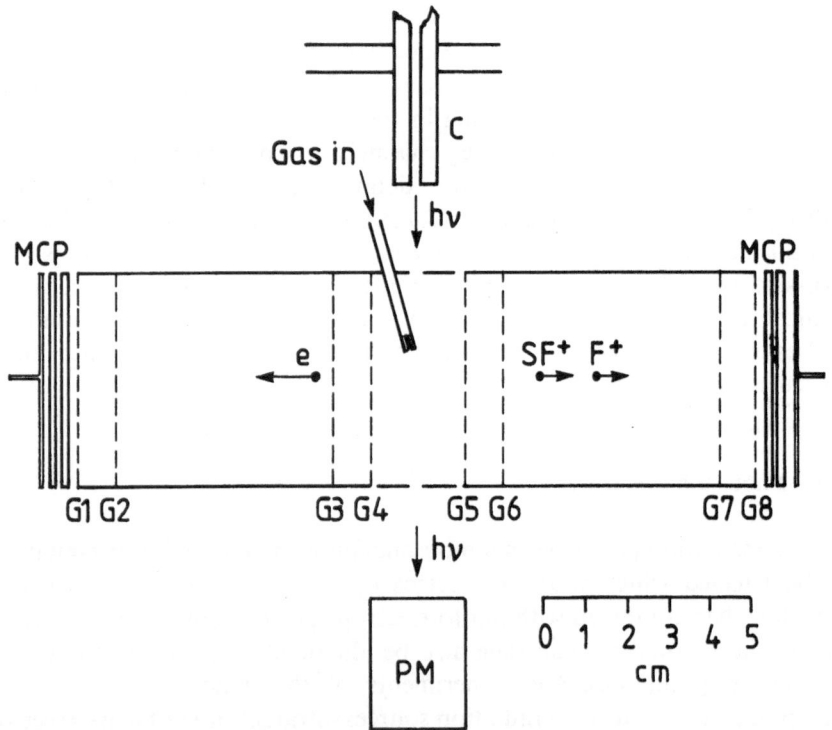

Figure 14.8. Layout of the electron-ion-ion coincidence experiment, from Frasinski et al. [28]. C, capillary; MCP, microchannel plates; PM, sodium salicylate coated photomultiplier tube; G1–G8, highly transmitting grids. For further details, see text.

Photo Electron Photo Ion COincidence is frequently used when referring to experiments of this kind, the first of which was performed by Frasinski *et al.* on the Daresbury SRS [28]. The apparatus they used, a time-of-flight mass spectrometer, is shown in Figure 14.8. Samples are introduced in gaseous form as an effusive jet, and interact with the photon beam from a normal incidence monochromator emerging from the glass capillary as shown. Use of glass capillaries in this way has two advantages: first, the angles of incidence of the light on the walls are sufficiently small that even VUV radiation is reflected efficiently; second, they provide a very effective vacuum differential between the monochromator and the experiment. An electrostatic field of ~ 50 V/cm is placed across the central two grids G4 and G5 so that electrons are driven towards one of the channel plate detectors, and ions towards the other. For the ion channel, grid G6 allows spatial focussing to be optimised, and G1 and G2, G7 and G8 shield the flight tubes electrostatic field penetration from the high voltages on the multichannel plates.

The electronic counting system is CAMAC based; the electrons provide the start pulse for a multi-hit time-to-digital converter, and the ions provide the stop pulses after drifting down the time-of-flight tube. The two digitised ion flight times are fed to a memory, and Figure 14.9, upper half shows data for SF_6 obtained in this way at a photon energy of 56.4 eV. The data are displayed as a false colour map, with the two ion flight times plotted along the two perpendicular axes. In the lower half of Figure 14.9 the structures of interest are labelled. The elliptical features along the 45° axis are caused by after-pulsing in the detector when an ion is detected, and from their times of flight the ions responsible can be identified. The densities of these features reflect the relative intensities with which these ions are produced. The true coincidences are labelled 1–4 in Figure 14.9 and by using the features identified along the diagonal axis it can be seen that this molecule fragments preferentially in the following ways:

(1) $SF_5^+ + F^+$,
(2) $SF_3^+ + F^+$
(3) $SF_2^+ + F^+$
(4) $SF^+ + F^+$

The length of these lines represents the kinetic energy release in the dissociation process, and their intensities indicate the preferred dissociation process; clearly the molecule breaks up primarily into SF_3^+ and F^+ ions at this photon energy.

The fact that these coincidence lines are at 45° to the horizontal axis is significant; it implies that a two body fragmentation took place, since any increase in one fragment's momentum is compensated by a corresponding loss in the other's. Since the momentum of a fragment is proportional to its time of flight, this leads to the features shown, and furthermore suggests that in the case of feature (4), a third, neutral particle may be involved in the

SF5+
SF4+
SF3+
SF2+
SF2 2+ + SF+
S+
F+
F2+
SF+ + F+
SF2+ + F+
SF3+ + F+
SF5+ + F+
Ion (2) time of flight
Ion (1) time of flight

8.16
0
0
8.16
SF5+
SF4+
SF3+
SF2+
SF2++ +SF+
S+
4
3
2
1
F+
F++
Ion (2) time of flight (μs)
Ion (1) time of flight (μs)

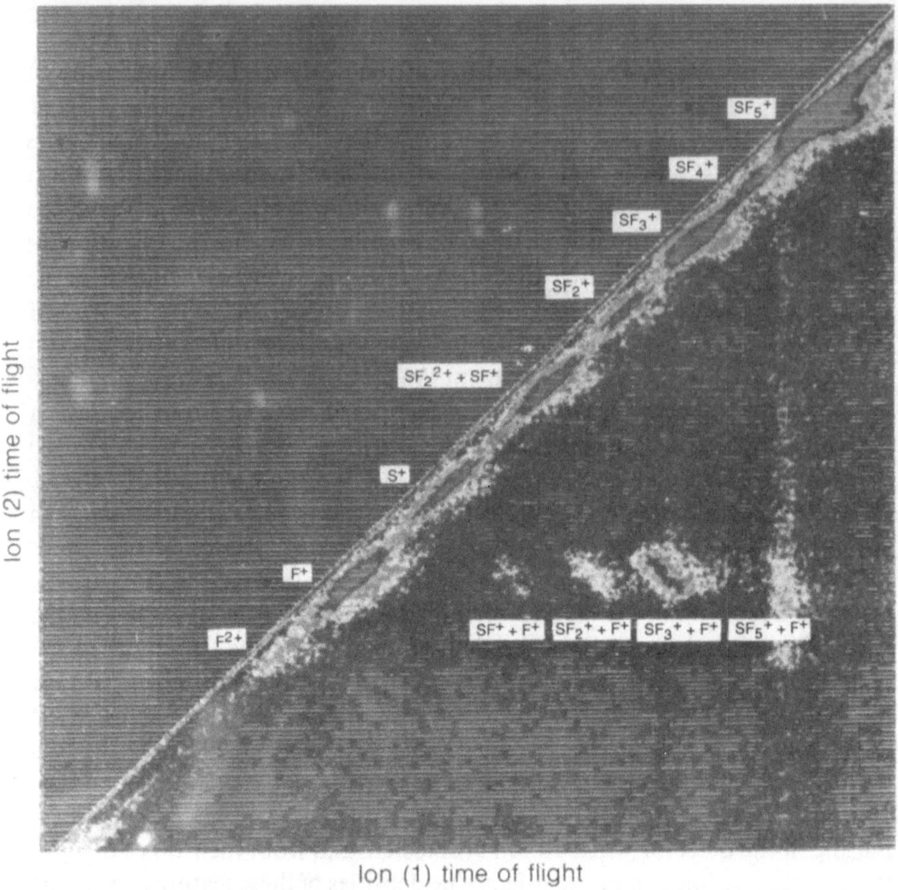

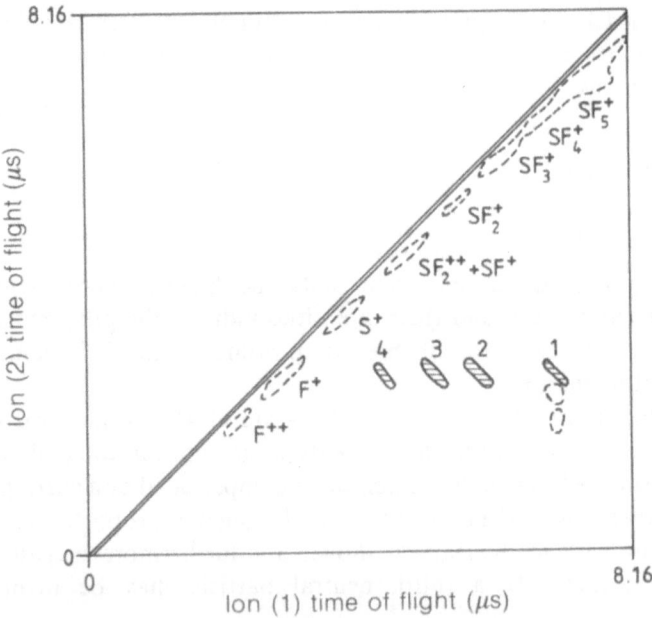

fragmentation process since this line is not quite at $45°$ to the axis. Alternatively, Frasinski et al. [28] suggest that a further fragmentation takes place following the one detected.

By taking a series of such spectra over a range of photon energies, it has proved possible to determine thresholds for the production of particular pairs of ions, and map the way this molecule fragments as the ionisation potentials of its molecular bands are exceeded. Thus this method is providing a wealth of information on fragmentation pathways in polyatomic molecules, and is being actively pursued for several other molecules both at Daresbury and at Orsay, where a similar technique has been developed by Eland [29]; in his paper a full discussion of the dynamics of these processes is given for a series of doubly charged molecular ions.

An interesting development of mass spectrometry was pioneered by Koyano and Tanaka [30], in which state selected ion-molecule reactions are being studied. In their case a laboratory light source was used for the work, the Hopfield continuum in helium which covers the photon energy range between 12 and 20 eV. The aim was to measure how reaction probabilities in an elementary process

$$A + B \longrightarrow C + D$$

depend on the quantum states of the reactants A and B and the collision geometry. Additionally, in what final states are the products C and D? The technique is described as TESICO, $\underline{T}$hreshold $\underline{E}$lectron $\underline{S}$econdary $\underline{I}$on $\underline{C}$oincidence, and the experimental layout is shown in Figure 14.10. Monochromatic photons interacting with the sample gas in chamber (I) produce species 'A' in the above equation. The photon energy is tuned to produce this species in a known state, by detecting a threshold electron in the hemispherical electron analyser (EA). The potential across (I) is arranged so that the ions travel in the opposite direction, and are focussed by the lenses L_1 and L_2 into the reaction chamber (R). Lens L_1 draws the ions out of the photo-ionisation region; lens L_2 serves the purpose of decelerating the primary ion beam down to thermal energies and focussing it into the reaction chamber. The second reactant is introduced into this chamber and the products drawn out by lens L_3 and analysed by the quadrupole mass spectrometer (Q).

This equipment has been used very successfully by Tanaka et al. [31] to study isotope effects in the charge exchange reactions

$$O_2^+ + HD \longrightarrow O_2H^+(O_2D^+) + D(H)$$

The cross-sections for both reactions showed large variations both with the electronic state and vibrational level of O_2^+, but were very similar to each

---

Figure 14.9. Ion-ion coincidences in the double ionisation of SF_6 at an incident photon energy of 56.4 eV, from Frasinski et al. [28]. The major features in the upper half of the figure are identified in the lower half of the figure (hatched areas 1–4 denote 1 count, dotted outline > 30 counts) and in the text.

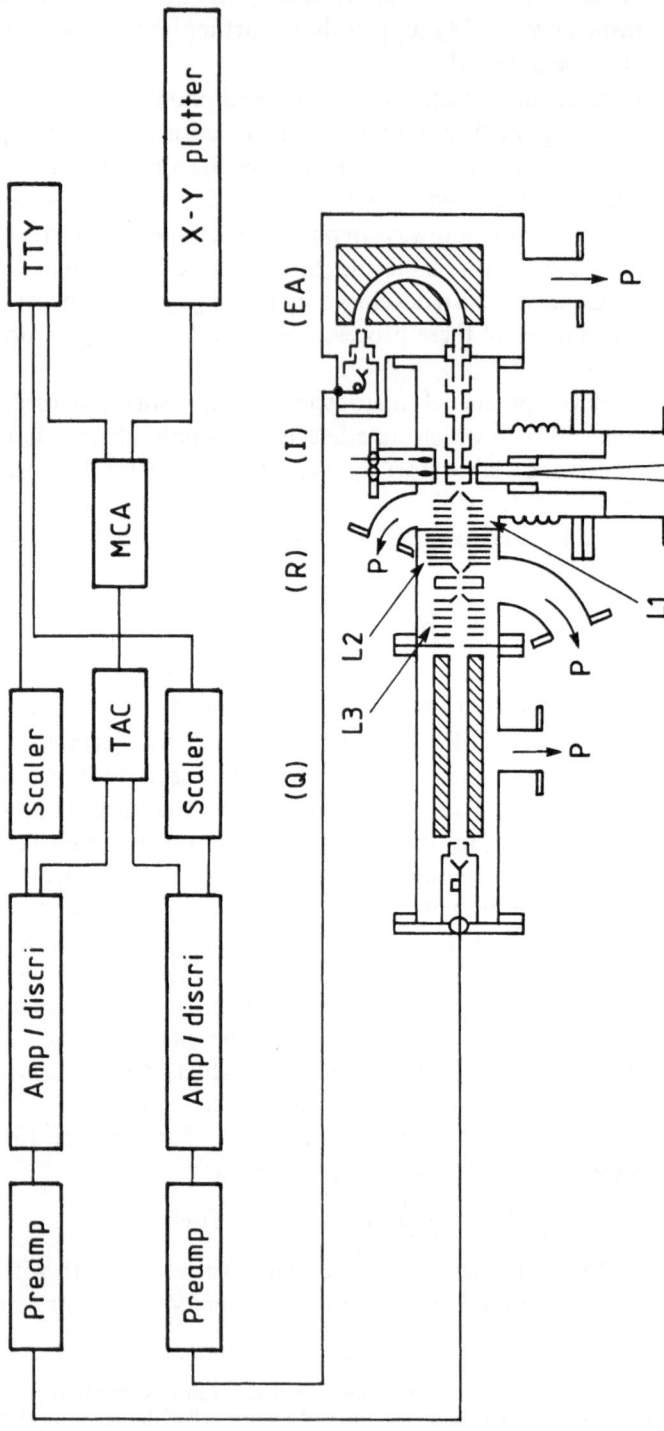

Figure 14.10. The TESICO apparatus, redrawn and adapted from Koyano and Tanaka [30]. Light enters from the bottom of the diagram; the exit slit of the monochromator is just below the interaction chamber (I), and the photodiode monitoring it just above. Pumping is provided at the points marked P; for further description, see the text.

other. Kato *et al.* [32] have used the method to study the reaction

$$N_2{}^+(v) + Ar \longrightarrow Ar^+ + N_2$$

To enhance their signal they employed Threshold Electron Spectroscopy (TES) in which only those electrons whose energies were $\leqslant 14$ meV were

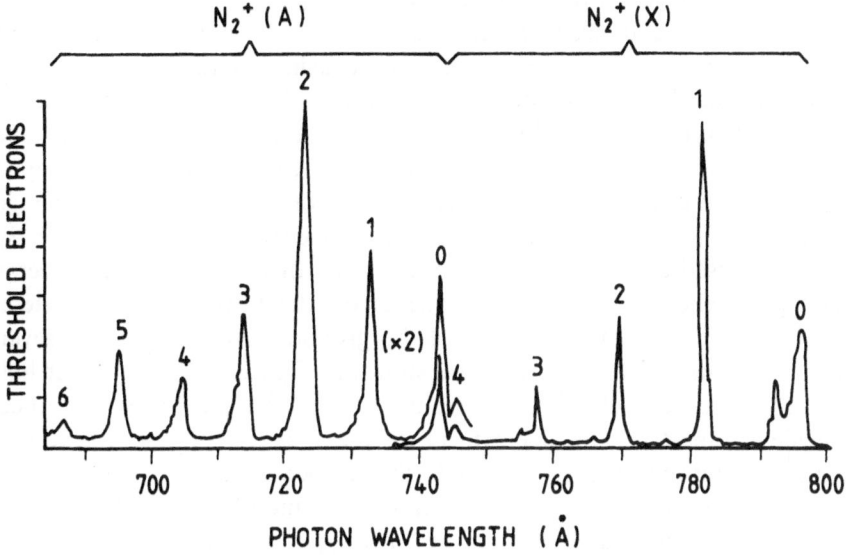

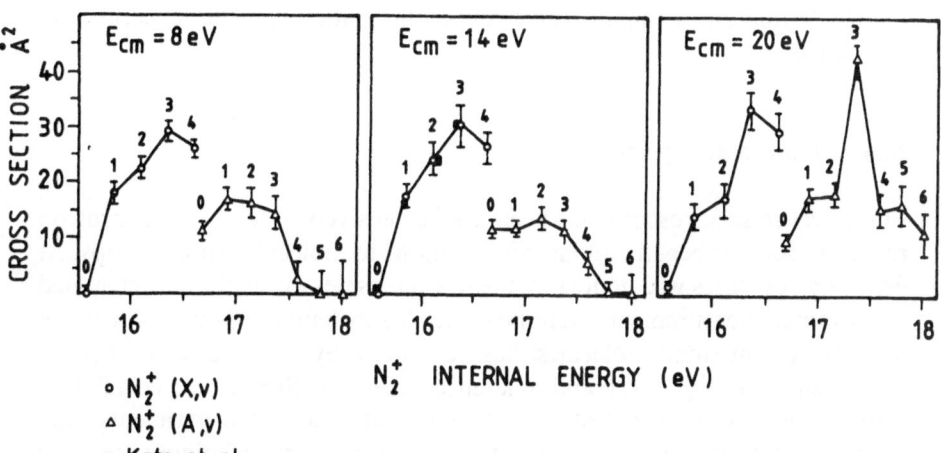

Figure 14.11. Upper half, threshold electron spectrum of nitrogen; lower half, dependence of the charge exchange reaction $N_2{}^+(v) + Ar \rightarrow Ar^+ + N_2$ on the vibrational quantum number of $N_2{}^+$ (after Govers *et al.* [33]).

detected, in coincidence with the ions. This method gives access to higher vibrational levels through auto-ionising states of the neutral molecule, as described earlier, states which must lie close to each of the vibrational thresholds of the ground state of the N_2^+ ion, but requires a continuum source of excitation. A line source would not in general populate the higher vibrational levels, with their low Franck-Condon factors, unless, quite fortuitously, it happens to excite an auto-ionising level.

Even with this enhancement, however, Kato *et al.* were unable to extend their measurements to the $v = 4$ level because, using the rather weak helium continuum, they had to degrade their resolution to 25 meV to obtain sufficient ion signal. This demonstrated the need for the photon source not only to have continuous wavelength coverage but high intensity, and left the field wide open to exploitation using the more intense synchrotron radiation. This challenge was taken up by Govers *et al.* [33] working at the Orsay storage ring ACO. By using the pulsed time structure of the ring, their threshold electron detector was highly efficient. Their measurements on the nitrogen-argon system, covering five vibrational members of the ground state of N_2^+ and seven members of the first excited state, are shown in Figure 14.11; the upper half gives the threshold electron spectrum for N_2, and the lower half their coincidence spectra. The dependence of the reaction on the vibrational quantum number is clearly seen, and different behaviour for three collision energies is also evident. In a later paper, Govers and Guyon [34] gave a full account of their measurements on the 'prototype' system

$$H_2^+(v) + He \longrightarrow HeH^+ + H \quad \text{and} \quad He + H^+ + H$$

and showed how it is possible to understand the dynamics of an endo-thermic chemical reaction. Clearly the potential of the method is enormous, though requiring a high degree of sophistication in the experimental technique.

14.4 Fluorescence spectroscopy

Fluorescence spectroscopy has been used extensively ever since synchrotron radiation sources began making contributions to molecular spectroscopy. At first, work on gases was limited by the low intensities available, but integrated fluorescence measurements were possible. A substantial amount of work in this respect, on small molecules, has been done by Lee and co-workers at Wisconsin (see, e.g. [35] and references therein). Synchrotron radiation sources have also been used widely for time resolved experiments, and examples of the application of these two complementary techniques are given below. For the convenience of the reader, it is worth noting that photon wavelengths, usually given in nanometres(nm), are often used in work of this kind; electron energies are not often used. Thus it may be useful to recall the conversion relationships.

$$(nm) = 1239.8/E\,(eV) \quad \text{and} \quad (nm) = 10^7/(\text{wave number})$$

Fluorescence analysis has been used with great effect to identify pair states in halogen and inter-halogen molecules, following the discovery that these states are highly reactive. To understand this, much attention has been focussed in recent years on the spectroscopy of these states, using both laser optical-double-resonance and synchrotron radiation techniques. In an experiment which combined data taken on the UVSOR storage ring at the Institute for Molecular Science, Japan and the SRS at Daresbury, Hiraya et al. [36] measured the fluorescence excitation spectrum of I_2, at the same time measuring the absorption spectrum. Normal incidence spectrometers were used at each facility, that at Daresbury having higher resolution. The intention was to examine the formation of ion pair states in this molecule; the apparatus required was simple, since the fluorescence wavelengths of interest were in the near UV and visible region, and the exciting wavelengths were all in the region beyond the LiF window cut-off at ~ 1150 Å. Therefore the samples could be contained in a sample cell sealed by LiF windows, and the fluorescence observed at right angles to the incident light. On the end window a scintillation counter (see Section 1.5.1) served to detect the transmitted VUV light; absorption spectra were measured using this facility.

In Figure 14.12 the fluorescence excitation and absorption spectra of I_2 are shown. The sharp structure is due to Rydberg transitions, and the quantum fluorescence yield from these is expected to be low since they are known to be extensively pre-dissociated. Also, any fluorescence from them is expected to occur in the VUV, near to the corresponding absorption system, apart from any cascade processes. Thus, by detecting only long wavelength fluorescence Hiraya et al. discriminated in favour of ion pair states. This is evident from the fluorescence cross-section measurements in Figure 14.12, where the broad structure between 178 and 200 nm is known to be due to ion pair emission; even the sharp structure seen in this figure may be due to population of ion pair states through interaction with Rydberg states, although at the present time this remains a hypothesis. A higher resolution scan of this peak is shown in Figure 14.13, where vibrational structure is evident, $v' = 150$ occurring at ~ 193 nm. From this it is clear that the ion pair states have large equilibrium internuclear distances. They give rise to oscillatory continua, decaying into a dissociative ground state, and are observed by wavelength analysis of their fluorescent emission [37].

The quenching behaviour of these ion pair states, particularly by the rare gases, provides the opportunity of studying chemical reactions in detail. O'Grady and Donovan [38] show that the $I_2(D\,0_u{}^+)$ state, selectively populated by excitation with UV light at 192 nm, reacts rapidly with Xe, yielding the electronically excited product $XeI(B\,^2\Sigma_{1/2})$. They explain how this is in distinct contrast to the behaviour with Ar, where a collisional cascade is

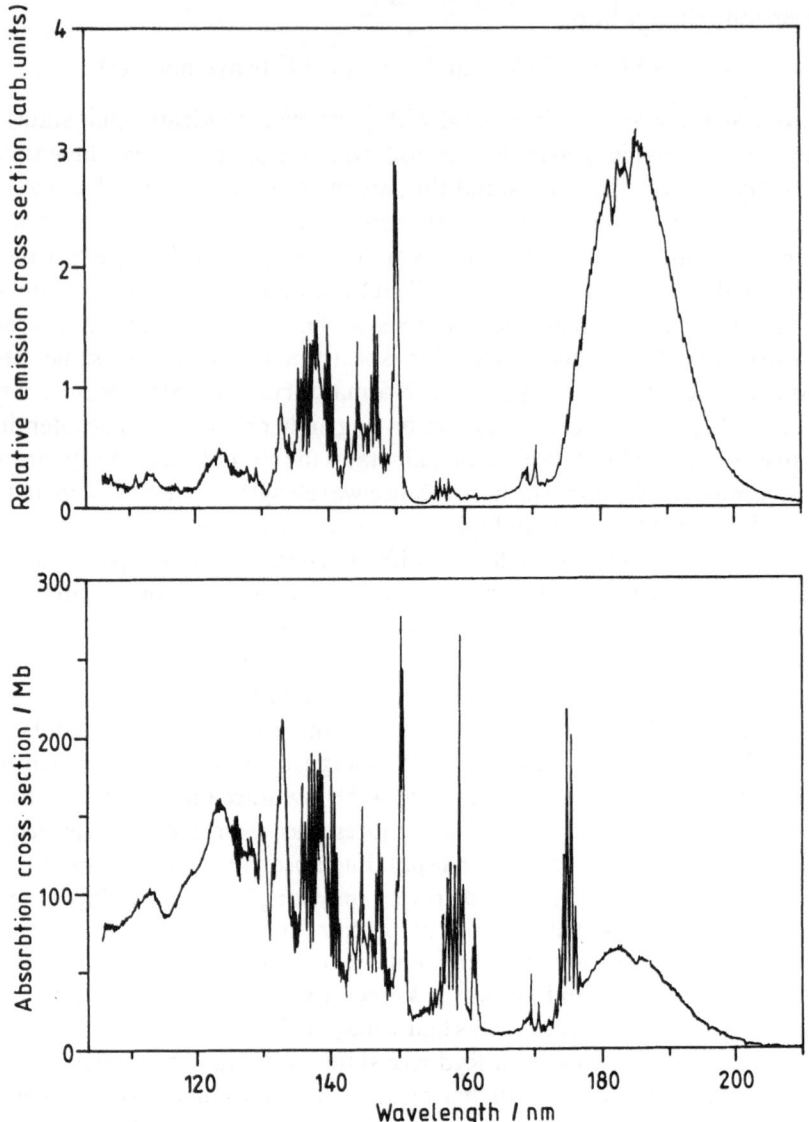

Figure 14.12. Upper half, fluorescence excitation spectrum and lower half: absorption spectrum of I_2 (from Hiraya *et al.* [36]).

known to take place ending up with the iodine in the lowest ion pair state, $I_2(D'^3\Pi_{2g})$. This state is responsible for intense narrow band emission at 340 nm, through the transition to $A'^3\Pi_{2u}$. Since the D' state is extremely resistant to quenching by the rare gases except Xe, this process finds an application in the optically pumped I_2 laser; high pressure argon is used to assist in populating the D' state, and intense emission at 340 nm occurs from the transition $D' \rightarrow A'^3\Pi_{2u}$ [39].

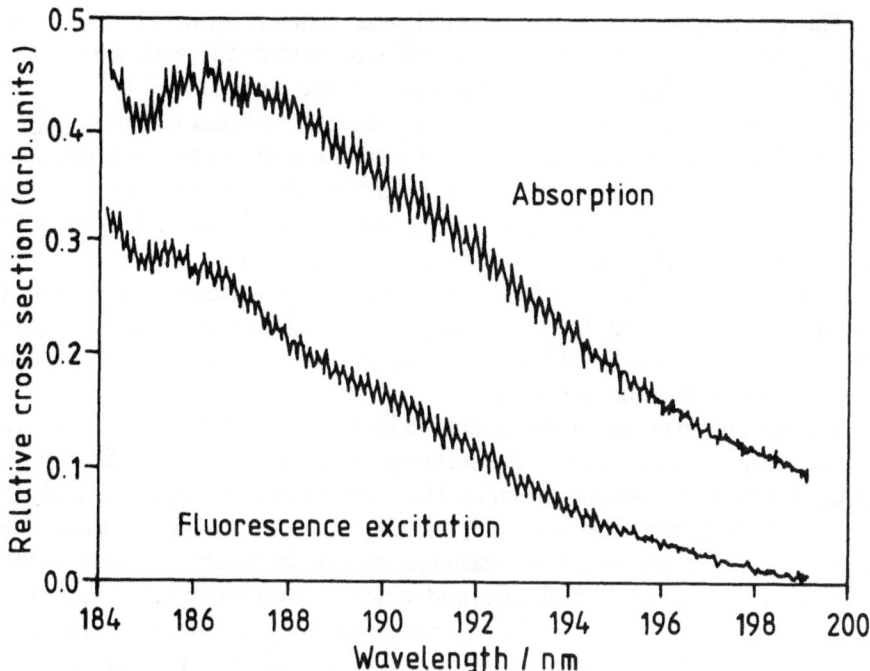

Figure 14.13. As figure 14.12; high resolution data in the peak region centred at 188 nm.

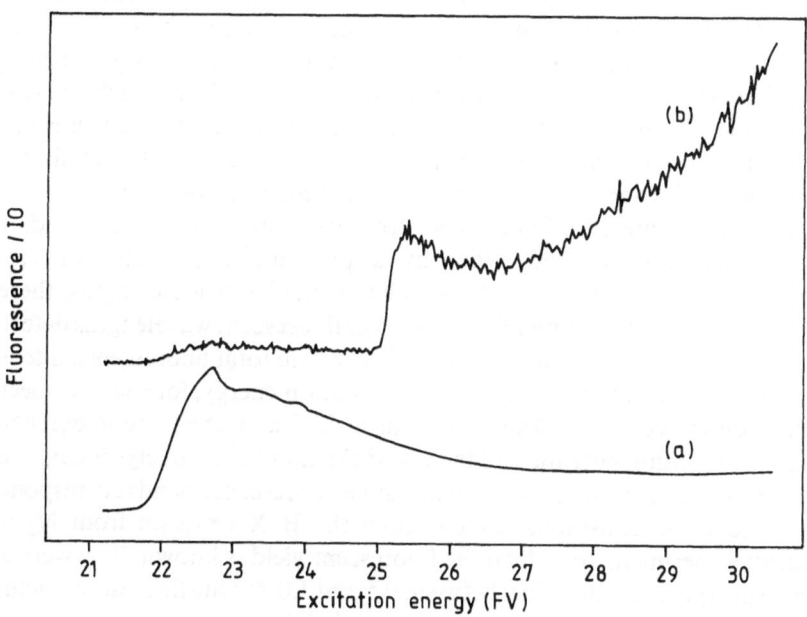

Figure 14.14. Fluorescence measurements on CF_4 from Lambert *et al.* [42]. (a) UV fluorescence (250–390 nm) results; (b) VUV fluorescence (120–200 nm) results.

The reactivity of the ion pair states of I_2 was further confirmed by Donovan *et al.* [40]. Their measurements on collision interactions with a series of hydrocarbon molecules, monitored through the intensity of the 340 nm emission band, confirmed the ability of these states to break carbon-halogen bonds. The photochemistry of the halogens and inter-halogens, where it is now apparent that the excited ion pair states are much more reactive than the ground state, has also been studied using time resolved spectroscopy [41]. In this case fluorescence lifetimes and quenching rates were measured giving quantitative information on the chemical interactions described above. Much remains to be done since the spectroscopy of these molecules has only recently begun to be explored; it holds the potential that photochemical reactions can be understood at a fundamental level.

When combined with fluorescence lifetime measurements, fluorescence spectroscopy can be a very useful technique assisting in the identification of decay processes in ionised molecules. The method has been applied recently to molecules of industrial interest, CCl_4 and CF_4. These are frequently used in the plasma etching process, the discharge nature of which involves many collisions. This leaves the molecules and molecular ions involved in a variety of excited states, and to model such systems photophysical data are required. Lambert *et al.* [42] have made fluorescent yield and lifetime measurements on molecules of the type MX_4, where M = C, Si or Ge and X = F or Cl. The purpose was to identify the modes of decay of the excited electronic states of these molecules. Experimentally, the arrangement is similar to that used for the halogen work described above, except that the photon wavelengths used to ionise the molecules are below the LiF window cut-off so the system must be windowless. Thus differential pumping is required, and this is provided by a 2 mm bore capillary, which also channels the light from the monochromator to the sample efficiently. In this case the sample is in the form of an effusive jet, intersected by the light beam from the end of the capillary. The details of the experimental layout used at the SRS are contained in [43] where a similar system used to measure fluorescence from calcium atoms is described. The fluorescence was detected either by a photomultiplier tube for visible wavelengths or by a CsI solar blind tube for the VUV wavelengths, thereby providing some wavelength selectivity in the fluorescent wavelengths detected. Results for CF_4 are shown in Figure 14.14, where total fluorescence intensity was measured as a function of photon excitation energy, for the two spectral regions described above. The thresholds at 21.7 and 25 eV are in agreement with the ionisation potentials of the C and D states, respectively, of CF_4^+, and Lambert *et al.* confirm that the broad visible fluorescence bands correspond to C–X and C–A transitions. By detecting the B–X emission from N_2 in a separate experiment, for which the fluorescent yield is known, they were able to give an approximate value, between 0.5 and 1.0, for the fluorescent yield for the C state, and conclude that this is a major decay pathway.

A refinement of fluorescence spectroscopy is to measure the lifetime of the radiation, and this is done with pulsed discharge light sources or pulsed lasers in the laboratory. However, as discussed in Section 1.1.2, synchrotron radiation is actually a very high frequency pulsed source, because the electrons circulate in bunches. It is therefore possible to use it for time resolved spectroscopy over a broad range of exciting wavelengths. This is generally done by injecting one bunch into the storage ring, rather than filling the ring completely with many bunches. The result, for example at the Daresbury SRS, is a source giving a pulse of light with a duration of 200 ps every 320 ns, and other synchrotron radiation sources are similar.

Lambert *et al.* exploited this property of synchrotron radiation to measure the radiative lifetime of the C state in CF_4, using a Schott filter to isolate the emission from the C–X, A transitions. They discovered that the decay was a single exponential over a range of excitation energies between the threshold at 21.7 eV and 30 eV. The lifetime varied between 9.7 and 8.6 ns, and this was attributed to a change in the non-radiative decay rate with exciting photon energy. In fact this non-radiative decay rate is small near threshold, since the fluorescence quantum yield is approximately unity there, but in the other tetrafluorides studied this was not the case. Jahn-Teller distortion from tetrahedral symmetry does not occur for CF_4, but does for the others, and this may be related to the fact the C states in these other molecules do decay non-radiatively.

Experiments of this kind show that detailed information on decay processes can be obtained using synchrotron radiation. However, to complete the picture it would be desirable: (a) to disperse the fluorescence; this will enable the separation of different fluorescence channels and should help in the analysis of multi-exponential decays, e.g. the bi-exponential decay seen for the $GeF_4{}^+D$ state [42]; and (b) to study the non-radiative decay channel by detecting the fragment ions, preferably in coincidence with the fluorescent photon. Experiments of this kind are in preparation at the Daresbury SRS.

Fluorescence lifetime measurements have been applied to the photochemistry of large molecules, particularly those of biological interest [44, 45], and details of the measurement and analysis of membrane systems using synchrotron radiation, for example, are given in Chapter 15. The fluorescence lifetime method can also be used to detect dynamic changes in an excited state, by examining the fluorescence spectrum of an excited molecule on a time scale equivalent to relaxation processes taking place in the molecule. Excimer molecules, such as the rare gas dimers, are important in UV and VUV lasers, and have been studied in this way. Using the very high intensity experimental station SUPERLUMI installed at the DORIS storage ring [46], Moller and Zimmerer [47] show recent measurements on the Xe_2 molecule. These are reproduced in Figure 14.15, which shows the fluorescence spectrum measured at a series of time delays (Δt) following the excitation pulse. The first spectrum. was taken within a time window (δt) of 500 ps, the remainder with a time window of 1 ns. The last spectrum shown is a time integrated one. It can be

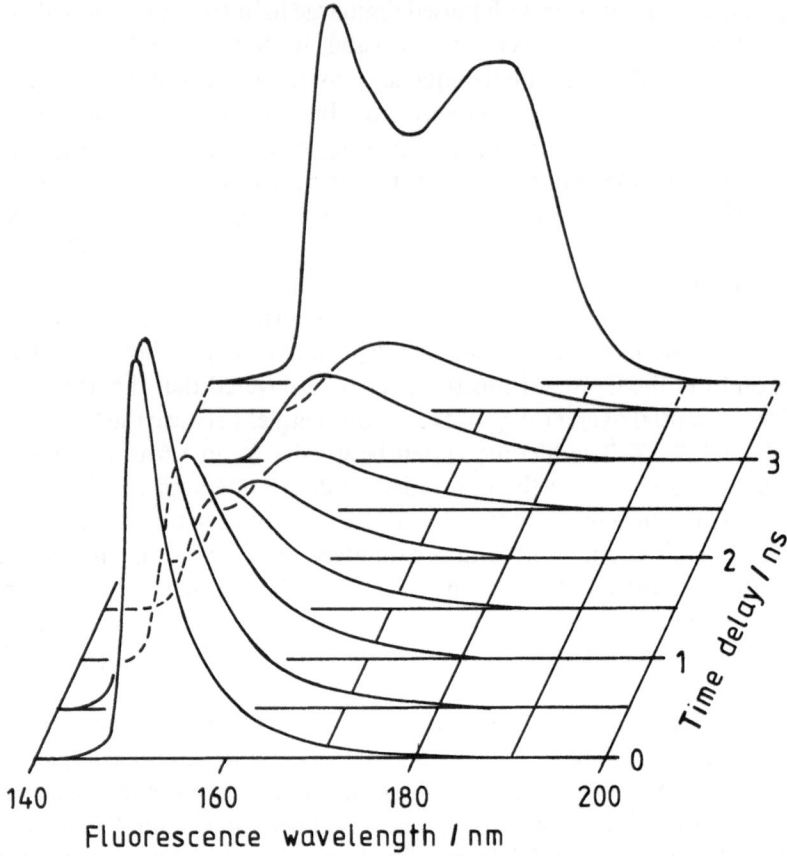

Figure 14.15.   Evolution of the fluorescent decay of the Xe$_2$ molecule on a nanosecond time scale
(from Moller and Zimmerer [47]).

seen that dramatic changes take place on a sub-nanosecond time scale, which
in this case are attributable to vibrational relaxation of the dimers as they
collide with xenon atoms. Rigler *et al.* [48] show how the time resolution can
be improved to 100 ps with the use of multichannel plates as detectors;
deconvolution techniques provide a further improvement of a factor of 10.
Thus, exploiting both the wide spectral range and pulsed nature of the
synchrotron radiation source, it is possible to follow the decay of selected
molecular states, and to examine the decay mechanism whether by collisions
or other non-radiative processes.

14.5   Conclusion

A selection of experiments has been presented here to demonstrate how atomic
and molecular spectroscopy techniques commonly used in physics and

chemistry have been adapted for use on synchrotron radiation sources. This is by no means an exhaustive selection, the majority of this chapter being devoted to the photochemistry of 'small' molecules. Large molecules are beyond the scope of this chapter, but the references given in the text will guide the reader in that direction. Synchrotron radiation sources are now increasingly used by chemists and biologists to study the photochemistry of molecules. This trend seems likely to continue and is helped by the fact that the techniques of optical instrumentation and detector technology used in this research area are becoming well established and more reliable.

References

1. J.B. West and H.A. Padmore, *Handbook on Synchrotron Radiation*, Vol. II, ed. G.V. Marr, North-Holland, Amsterdam (1987) Chap. 2.
2. D.W. Turner, C. Baker, A.D. Baker and C.R. Brundle, *Handbook of Molecular Photoelectron Spectroscopy*, Wiley-Interscience, New York (1970).
3. K. Kimura, S. Katsumata, Y. Achiba, T. Yamazaki and S. Iwata, *Handbook of HeI Photoelectron Spectra of Fundamental Organic Molecules*, Japan Scientific Societies Press, Tokyo and Halstead Press, New York (1981).
4. J.W. Cooper and R.N. Zare, in *Lectures in Theoretical Physics* 11c Gordon and Breach, New York, (1970) p. 317.
5. D.M.P. Holland, A.C. Parr, D.L. Ederer, J.L. Dehmer and J.B. West, *Nucl. Instrum. Methods.* **195** (1982) 331.
6. A.C. Parr, S.H. Southworth, J.L. Dehmer and D.M.P. Holland, *Nucl. Instrum. Methods Phys. Res.* **222** (1984) 221.
7. P.J. Hicks, S. Daviel, B. Wallbank and J. Comer, *J. Phys. E* **13** (1980) 713.
8. A.A. MacDowell, I.H. Hillier and J.B. West, *J. Phys. E* **16** (1983) 487.
9. M.J. Seaton, *Philos Trans R. Soc (London) A* **208** (1951) 418.
10. J.W. Cooper, *Phys. Rev.* **128** (1962) 681.
11. J.B. West and G.V. Marr, *Proc. R. Soc. London A* **349** (1976) 397.
12. D.L. Ederer, *Phys. Rev. Lett.* **13** (1964) 760.
13. A.P. Lukirskii, I.A. Brytov and T.M. Zimkina, *Opt. Spectrosc.* **17** (1964) 234.
14. U. Fano and J.W. Cooper, *Rev. Mod. Phys.* **40** (1968) 441.
15. A.W. Potts, I. Novak, F. Quinn, G.V. Marr, B. Dobson, I.H. Hillier and J.B. West, *J. Phys. B* **18** (1985) 3177.
16. R.G. Houlgate, J.B. West, K. Codling and G.V. Marr, *J. Electron Spectrosc* **9** (1976) 205.
17. K. Codling, R.G. Houlgate, J.B. West and P.R. Woodruff, *J. Phys. B* **9** (1976) L83.
18. T.A. Carlson, M.O. Krause, W.A. Svensson, P. Gerard, F.A. Grimm, T.A. Whitley and B.P. Pullen, *Z. Phys. D* **2** (1986) 309.
19. M.N. Piancastelli, P.R. Keller, J.W. Taylor, F.A. Grimm, T.A. Carlson, M.O. Krause and D. Lichtenberger, *J. Electron Spectrosc Related Phenom.* **34** (1984) 205.
20. G. Cooper, J.C. Green, M.P. Payne, B.R. Dobson and I.H. Hillier, *J. Am. Chem. Soc.* **109** (1987) 3836.
21. G. Cooper, J.C. Green and M.P. Payne, *Mol. Phys.* **63** (1988) 1031.
22. G.V. Marr and J.B. West, *Atomic Data and Nuclear Data Tables* **18** (1976) 497.
23. J.C. Green, *Struct. Bond* **43** (1981) 37.
24. M.W.D. Mansfield and G.H. Newsom, *Proc. R. Soc. London A* **357** (1977) 77.
25. J. Berkowitz, *Photoabsorption, Photoionisation and Photoelectron Spectroscopy*, Academic Press, New York (1979).
26. I. Nenner, P.M. Guyon, T. Baer and T.R. Govers, *J. Chem. Phys.* **72** (1980) 6587.
27. T. Masuoka and J.A.R. Samson, *J. Chim. Phys.* **77** (1980) 623.
28. L.J. Frasinski, M. Stankiewicz, K.J. Randall, P.A. Hatherly and K. Codling, *J. Phys. B* **19** (1986) L819.

29. J.H.D. Eland, *Mol. Phys.* **61** (1987) 725.
30. I. Koyano and K. Tanaka, *J. Chem. Phys.* **72** (1980) 4858.
31. K. Tanaka, T. Kato, P.M. Guyon and I. Koyano, *J. Chem. Phys.* **79** (1983) 4382.
32. T. Kato, K. Tanaka and I. Koyano, *J. Chem. Phys.* **77** (1982) 834.
33. T.R. Govers, P.M. Guyon, T. Baer, K. Cole, H. Frohlich and M. Lavollee, *Chem. Phys.* **87** (1984) 373.
34. T.R. Govers and P.M. Guyon, *Chem. Phys.* **113** (1987) 425.
35. L.C. Lee, X. Wang and M. Suto, *J. Chem. Phys* **85** (1986) 6294.
36. A. Hiraya, K. Shobatake, R.J. Donovan and A. Hopkirk, *J. Chem. Phys.* **88** (1988) 52.
37. R.J. Donovan, M.A. MacDonald, K.P. Lawley, A.J. Yencha and A. Hopkirk, *Chem. Phys. Lett.* **138** (1987) 571.
38. B.O. O'Grady and R.J. Donovan, *Chem. Phys. Lett.* **122** (1985) 503.
39. M.J. Shaw, C.B. Edwards, F. O'Neill C, Fotakis and R.J. Donovan, *Appl. Phys. Lett.* **37** (1980) 346.
40. R.J. Donovan, B.V. O'Grady, L. Lain and C. Fotakis, *J. Chem. Phy.* **78** (1983) 3727.
41. R.J. Donovan, G. Gilbert, M.A. MacDonald, I.H. Munro, D. Shaw and G.R. Mant, *Chem. Phys. Lett.* **109** (1984) 379.
42. I.R. Lambert, S.M. Mason, R.P. Tuckett and A. Hopkirk, *J. Chem. Phys.* **89** (1988) 2675.
43. J.B. West, *Z. Phys. D* **5** (1987) 265.
44. I.H. Munro and A.P. Sabersky, *Synchrotron Radiation Research*, eds, H. Winwick and S. Doniach, Plenum Press, New York (1980) Chapt. 6.
45. R.B. Cundall and I.H. Munro, eds. *Applications of Synchrotron Radiation to the Study of Large Molecules of Chemical and Biological Interest*, Proc. Study Weekend at Daresbury Laboratory, Warrington, UK (1979).
46. P. Gurtler, E. Roick, G. Zimmerer and M. Pouey, *Nucl. Instrum. Methods.* **208** (1983) 835.
47. T. Moller and G. Zimmerer, *Phys. Scripta T* **17** (1987) 177.
48. R. Rigler, O. Kristensen, J. Roslund, P.Y. Thyberg, K. Oba and M. Eriksson, *Phys. Scripta T* **17** (1987) 204.

15 Time-resolved spectroscopy

Y.K. LEVINE and H. VAN LANGEN

15.1 Introduction

Time-resolved spectroscopy has long been a valuable tool for the study of the kinetics of chemical reactions. The technique utilises a short light pulse which generates the desired transient molecular species. These pulses are commonly obtained from sources such as flash-lamps or lasers. The decay of the concentration is then followed by monitoring the time dependence of the absorption or emission spectrum of the molecules under study.

In these experiments, intense light pulses are needed to create a sufficiently high concentration of the transient species to be observed by the detection system. Furthermore, the reaction time of the detection equipment must be faster than the kinetics of the reaction, for otherwise the change in the concentration cannot be followed. This places a stringent constraint on the timescale of the reactions which can be studied, as the electronic equipment available at reasonable cost has a response time of a few nanoseconds. Consequently, one is often restricted to the study of reactions whose time course is microseconds or slower. Of course, specialised equipment with a response time on the picosecond timescale can be obtained at a cost. Unfortunately, this is beyond the means of most research groups.

The limitations imposed by the response time of the detection system can be circumvented in one of two ways. The most common approach is the application of photon correlation techniques [1–4] which can extend the study of reaction kinetics to the sub-microsecond timescale. This method is particularly useful when the emission of fluorescent light is used to follow the time-course of the process. Alternatively, the changes in the optical density of the sample can be monitored through the application of a weak interrogation light pulse following the intense pulse which creates the transient species [6, 7]. The time interval between the two pulses is produced by the simple expedient of an optical delay line. In this way the interrogating pulse traverses a longer, but well-defined, path before incidence on the sample. The time delay is then simply defined by the additional optical path length of the interrogation pulse. Both these detection schemes yield equivalent information [6, 7]. However, the advantage of the former is that it does not require the initial intense pulse and is convenient for use with weakly emitting samples.

In the foregoing it was implicitly assumed that the initiating pulse is very short compared to the decay of the system response, so that no significant changes in the concentration of the transient species take place during the pulse. Most sources, however, produce light pulses of widths between 2 ns and 100 ps, so that the assumption can be reasonably justified if the half-time of the reaction is a microsecond or longer. Nevertheless, systems with even faster kinetics can still be investigated. In this case the experimental decay curves yield the true system response convolved with the pulse profile [2–4]. The two signals may in principle be disentangled numerically, as the pulse shape can be determined experimentally by monitoring elastic light-scattering from a somewhat turbid suspension of ludox particles. A complicating factor in this procedure is the circumstance that the sample is excited with a light pulse at one wavelength, while the response is monitored at a different, usually longer wavelength. In the analysis of the experiments it then becomes necessary to take account of the fact that the time structure of the pulses generated by sources such as lasers and flash-lamps varies with wavelength [4] and that furthermore, photomultiplier tubes also exhibit a colour effect in their response [8]. Consequently, the determination of the pulse shape at the wavelength of observation is not straightforward. To a reasonably good approximation, however, the effect of the spectral dependence of the instrumental response, the colour effect, can be described as a simple time-shift between the true system response and the pulse shape determined at the exciting wavelength [8–10]. Nevertheless, this problem introduces uncertainties into the numerical deconvolution procedures. Ideally, one would like to get rid of these colour effects altogether, but the use of a light source with a stable and spectrally independent pulse shape alone helps to eliminate many of the difficulties.

It is a quirk of life that the really interesting systems are just beyond the grasp of the researcher. Usually the signals are just too weak to be observed with the apparatus at hand. Any modern experimental set-up possesses an in-built capability for signal-averaging, so that the desired response can be raised above the noise level by the repetitive accumulation of the experimental signals. The repetition rate of the excitation pulse must of course be so adjusted that the system can return to equilibrium in the interval between two successive pulses. At the same time, the repetition frequency should be sufficiently high, so that an acceptable signal-to-noise ratio can be obtained in a reasonable time and certainly before the sample deteriorates. The repetition rate of the apparatus is, however, dictated by technical factors so that it usually becomes a matter of fitting the quest to the instrument at hand.

How does all this relate to a synchrotron radiation source? The foremost advantage of synchrotron radiation over a conventional laboratory light source is its wide spectral emission which ranges from the X-ray region to the infrared. This means that the chemist is now in principle free to choose the problem to investigate on scientific merits. Moreover, as discussed in Section 1.2.2, when the ring operates in the single-bunch mode, a sub-nanosecond light pulse reappears every fraction of a microsecond. This means that kinetic studies on the nanosecond timescale can now be carried out. Of course, the price paid is reduction of approximately a decade in the beam current, I_b, and hence the average total intensity I of the radiation produced by the ring, so that most of the activity on other experiments usually halts.

The intensity of the synchrotron pulse, while significantly higher than that obtained from a flash-lamp, is not sufficiently high to create a large population of excited state species; certainly not on a par with a powerful laser system. Thus, the synchrotron should preferably be used in studies where the emission from the sample is used for monitoring the course of the reaction.

The synchrotron light pulse possesses the remarkable property that its time profile is independent of the emitted wavelength [11] and is therefore particularly attractive to experimentalists. As the pulse width is commensurate with the timescale of the system response, the observed signal must be deconvolved in order to extract the true system response. But, as we can determine the system response function at the wavelength of observation, the deconvolution process can be carried out with confidence.

It was noted above that a synchrotron radiation source is really only suited to time-resolved studies in which the emission of light is used to monitor the processes in the sample. We can also utilise the polarisation properties of the exciting and emitted light to obtain more detailed information about the sample [5, 12–14]. The synchrotron radiation is virtually perfectly polarised in a horizontal direction (see Section 1.2.1) that is to say the electric field vector of the light is horizontal. The light absorption process for a molecular system is inherently anisotropic, so that a particular population of molecules is selectively excited along the polarisation direction of the incident light. This is known as the photoselection principle [12, 13]. In this way we have effectively imposed an axis on a previously isotropic sample. The same photoselection also applies to the emission process. We can therefore select sub-populations with particular spatial orientations for our study on observing the intensity of the emitted light with different directions of polarisation. It is a happy circumstance that the excited molecules can change their orientation in space relative to the imposed axis, before decaying to the ground state. These rotations will of course change the spatial distribution of the molecules and consequently modify the intensities of the polarised emission expected from a static system. The resulting depolarisation therefore can be used to character- ise the rotational dynamics of the molecules in the sample.

This technique of time-resolved depolarisation has in fact become a powerful tool in the study the dynamics of molecules in general and in particular has found many useful biochemical applications [4, 5, 14, 15]. For example, the changes in the rates of rotation can be used to monitor association/dissociation processes of macromolecular complexes. Here we make use of the simple fact that larger molecules have a higher moment of inertia and therefore will undergo slower rotational motions. In a similar vein, one can follow the changes in the shapes of molecules brought about by a variety of physical or chemical perturbations. Furthermore, the internal flexibility of protein and DNA molecules as well as membrane systems can also be brought to light. In most of this work the fluorescence emission from the molecules is used to monitor the rotational motions. The fluorescence lifetime, that is the lifetime of the excited electronic state, usually in the $1-50$ ns range, effectively defines the time-window during which the rotations can be followed. A storage ring operating in the single-bunch mode is thus a particularly suitable light source for studying molecular rotations on the timescale of the fluorescence lifetime.

It is in fact in the biochemical applications of time-resolved fluorescence spectroscopy that the synchrotron scores so highly because of its intensity and wavelength tunability. The broad spectral range of synchrotron radiation can be used to excite specific chromophores in the complex biological macro-molecules. In this way local structures may be probed without interfering signals from other species. Unfortunately not every biochemical macro-molecule possesses fluorescent chemical groups. Thus the biological systems must all too often be labelled at specific sites with chromophores exhibiting a suitable photophysical behaviour. Most of the useful chromophores, intrinsic or extrinsic, contain conjugated electron systems and in general absorb in the $250-400$ nm region ($3-5$ eV). This spectral range is not easily covered with conventional laboratory laser systems, although flash-lamps can be used. These lamps, however, cannot match the intensity and repetition rate of a synchrotron source.

The reader will have understood by now that this chapter concentrates on the biochemical application of time-resolved fluorescence spectroscopy. This reflects not only the interests of the authors but also the growing importance of the technique in current research [15]. Time-resolved fluorescence spec-troscopy was introduced briefly in Chapter 14 in the context of the photo-chemistry of small molecules. In this chapter we consider the technique in greater detail, particularly in relation to the study of organic molecules tethered to biological membranes. First theoretical aspects are discussed and particular emphasis is placed on the physical significance of the fluorescence anisotropy which is measured experimentally. We then deal with the experimental protocol. Finally, a discussion of the numerical analysis of the experimental data is presented. This is illustrated with the results of experiments on membrane systems obtained on the high aperture time-resolved spectroscopy station, 12.1, at the SRS.

15.2 Principles of fluorescence polarisation

15.2.1 *General expressions*

Fluorescence is a two-step process involving the absorption of a photon at time $t = 0$ and the emission of a photon of a different wavelength a time t later

[12–14]. If we assume that these processes are independent, we can write the emitted fluorescence intensity $I(t)$ from a molecule at time t after excitation as the product

$$I(t) \propto P_{abs}(0) P_{em}(t) F(t) \qquad (15.1)$$

where $P_{abs}(0)$, $P_{em}(t)$ and $F(t)$ are, respectively, the probability that the molecule absorbs a photon at time $t = 0$, that it emits a photon and is still excited at time t. The normalised function $F(t)$ describes the fluorescence decay of the molecule from the excited stated and we shall now assume that the decay is isotropic and in no way influenced by the orientation of the molecule in the sample. This assumption is reasonable, but difficult to test in an experiment.

The probabilities of absorption and emission, $P_{abs}(0)$ and $P_{em}(t)$, respectively, can be expressed in terms of Fermi's golden rule, which is a standard result of perturbation theory [16]. In doing this, we shall implicitly make the following assumptions:

(1) The exciting light intensity is low and no photochemical processes are initiated. Thus the fluorescence intensity varies linearly with the intensity of the excitation beam.
(2) The absorption and emission transitions are dipole-allowed and are characterised by unique absorption μ and emission ν moments in the molecular frame.
(3) All the internal photophysical processes induced by excitation decay fast on the timescale of the fluorescence lifetime.
(4) The inter-molecular distances between the fluorescence molecules are large, so that their electronic interactions are negligible.
(5) The emission spectrum is not affected by changes in the micro-environment of the molecule on the timescale of the fluorescence decay.
(6) The molecular shape and symmetry are the same in the ground and excited electronic states.

The last assumption lies at the heart of the interpretation. A little reflection will reveal that with this experiment we are only monitoring the behaviour of the molecule in its excited electronic state. Assumption (6) above allows us to equate the rotational motions of the molecule in the excited state with those in the ground state. It is then important to note that we are probing the molecular behaviour in a somewhat indirect manner.

Consider now an idealised experiment in which an instantaneous light pulse, plane-polarised in a direction e_i, is incident on the sample. The fluorescent emission is observed along another direction through a polariser, set at a direction of polarisation e_f. According to Fermi's rule [16], Eqn. 15.1 can now be written in the form

$$I_{if}(t) = \langle [e_i \cdot \mu(0)]^2 [e_f \cdot \nu(t)]^2 \rangle F(t) \qquad (15.2)$$

where $\mu(0)$ and $v(t)$ are unit vectors, respectively, along the absorption moment at time $t = 0$ and the emission moment at time t. It is important to note that we have here set all the unknown molecular constants and concentrations equal to unity.

The scattering geometry most commonly used in time-resolved experiments is shown in Figure 15.1. The exciting light is incident along the laboratory X-axis and is polarised along the vertical Z-axis. The fluorescence intensity is observed along the Y-axis, with the polariser set either along the Z-axis, $I_{VV}(t)$, or along the X-axis, $I_{VH}(t)$. The cartesian coordinates of the unit vectors e_i, e_f, $\mu(0)$ and $v(t)$ in the laboratory frame can now be expressed in terms of their spherical coordinates (Figure 15.1) as

$$e_i \equiv \{0, 0, 1\}, \quad e_{fVV} \equiv \{0, 0, 1\}, \quad e_{fVH} \equiv \{1, 0, 0\} \quad (15.3a)$$

$$\mu(0) \equiv \{\sin\theta_\mu \cos\phi_\mu, \sin\theta_\mu \sin\phi_\mu, \cos\theta_\mu\} \quad (15.3b)$$

$$v(t) \equiv \{\sin\theta_v \cos\phi_v, \sin\theta_v \sin\phi_v, \cos\theta_v\} \quad (15.3c)$$

On substituting Eqns. 15.3 into Eqn. 15.2 we can easily calculate the scalar products to obtain

$$I_{VV}(t) \propto F(t)\langle \cos^2\theta_\mu \cos^2\theta_v \rangle$$
$$I_{VH}(t) \propto F(t)\langle \cos^2\theta_\mu \sin^2\theta_v \cos^2\phi_v \rangle \quad (15.4)$$

However, because we observe the fluorescence emission from a large number of molecules, we need to average our expressions over all the possible orientations of their emission and absorption moments. This averaging is implied by the angled brackets $\langle \ldots \rangle$. It is interesting to note that the intensities (Eqn. 15.4) are insensitive to a change of sign of either of the

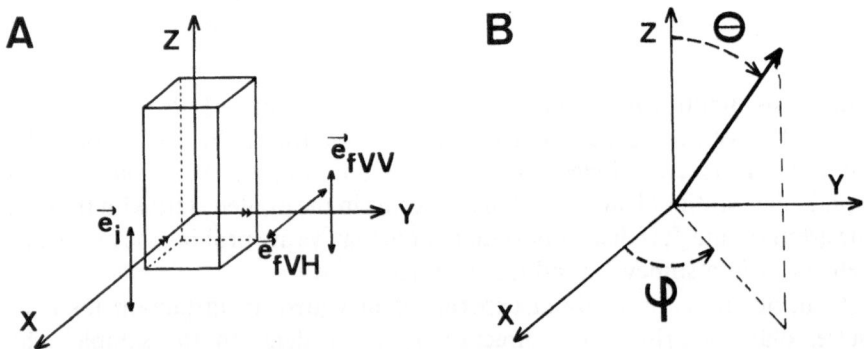

Figure 15.1. (A) the scattering geometry of a fluorescence anisotropy experiment. XYZ denote the laboratory-fixed axes and e denotes the polarisation direction of the light beam. The incident beam (along the X-axis) is vertically (V) polarised (along the Z-axis) and the emission observed along the Y-axis is either X-axis (VH) or Z-axis (VV) polarised. (B) The angles $\{\theta \varphi\}$ specifying the orientation of a vector or a cylindrically symmetric molecule in spherical coordinates.

transition moments. Thus, our experiment cannot distiguish for example between the vector μ and $-\mu$.

We shall now define a quantity which plays a central role in fluorescence depolarisation experiments. The time-resolved fluorescence anisotropy $r(t)$, is defined as

$$r(t) = \frac{I_{VV}(t) - I_{VH}(t)}{I_{VV}(t) + 2I_{VH}(t)} \tag{15.5}$$

The usefulness of this quantity lies in the fact that in macroscopically isotropic systems the denominator is proportional to the fluorescence decay function $F(t)$: $I_{VV}(t) + 2I_{VH}(t) \propto F(t)$. The anisotropy is then solely determined by the time behaviour of the transition moments.

It is furthermore clear from Eqn. 15.4, that the observed intensities and their decays are determined by the orientations of the dipole moments. Thus the rotational motions of the molecules are monitored indirectly via those dipoles. Of course, the changes in the orientations of the dipoles are a direct consequence of the rotations of the molecules to which they are fixed. However, it is now important to realise that both the absorption and emission moments are fixed in the molecular frame and that their relative orientation in space is the same at all times. Thus in order to proceed further, we need to disentangle the molecular motions of the molecules and the orientations of the dipoles in the molecular frame. We shall do this with particular reference to membrane systems as the results for molecular solutions can be obtained simply as a limiting case.

15.2.2   Membrane systems

15.2.2.1   *Aligned samples*   We shall start by considering the case of elongated (cylindrically symmetric) molecules which are preferentially aligned along the Z-axis, their orientations in the laboratory frame are now defined by the polar angle θ_m and the azimuthal angle φ_m (Figure 15.1). Let us suppose that the molecules are distributed randomly about that axis, so that every angle φ_m is equally probable. This situation is encountered when membrane fragments or macromolecules are aligned macroscopically in squeezed gels [12].

The probability of encountering a molecule with its long axis at angles between θ_m and $\theta_m + d\theta_m$ is now given by $f(\theta_m)$. The distribution of the molecules above the XY-plane is furthermore identical to that below that plane, so that effectively a reflection of the coordinate system through the origin leaves the probability distribution function $f(\theta_m)$ unchanged. This symmetry has the important consequence that the mathematical form of $f(\theta_m)$ requires the property, $f(\theta_m) = f(\pi - \theta_m)$. The next step is the evaluation of the various averages appearing in Eqn. 15.4. As we have no explicit expression for

$f(\theta_m)$, we shall now show that we can get a long way by simply considering the averages themselves.

It is known from the quantum mechanical theory of the hydrogen atom [16] that those orbitals (s,p,d,....) whose form is invariant to rotations about the quantization Z-axis, are described mathematically by the Legendre polynomials $P_L(\cos \beta)$. Here $L = 0$ for an s-orbital, $L = 1$ for a p-orbital, $L = 2$ for a d-orbital etc. The first four polynomials are given by

$$P_0(\cos \beta) = 1, \quad P_1(\cos \beta) = \cos \beta$$

$$P_2(\cos \beta) = 1/2(3 \cos^2 \beta - 1), \quad P_3(\cos \beta) = 1/2(5 \cos^3 \beta - 3 \cos \beta)$$

$$P_4(\cos \beta) = 1/8(35 \cos^4 \beta - 30 \cos^2 \beta + 3). \tag{15.6}$$

They satisfy the orthonormality relation [16]

$$\int_0^\pi P_L(\cos \beta) P_{L'}(\cos \beta) \sin \beta \, d\beta = 2/(2L + 1) \delta_{LL'} \tag{15.7}$$

where $\delta_{LL'} = 1$ if $L = L'$ and $\delta_{LL'} = 0$ otherwise. Inspection of Eqn. 15.6 reveals that the polynomials of *odd* order L change sign on the substitution $\beta \to \pi\text{-}\beta$, so that they do not fulfil the symmetry requirement of our distribution function. It can be shown that this is a general property of all the odd polynomials [16]. We thus conclude that the distribution function can be written as a sum over all the *even* Legendre polynomials as

$$f(\theta_m) = 1/2 \sum_{L=0}^\infty (2L + 1) \langle P_L \rangle P_L(\cos \theta_m), \quad L \text{ even} \tag{15.8}$$

Equation 15.8 can be simply verified using Eqn. 15.7.

Here $\langle P_L \rangle$ denotes the average of the corresponding Legendre polynomial over the distribution function and has the special name of 'order parameter'. The name follows from the simple fact that $\langle P_L \rangle = 1$ if all the molecules are perfectly aligned along the Z-axis ($\theta_m = 0$), and $\langle P_L \rangle = 0$ for a random alignment of the molecules. In studies of membrane systems and liquid crystalline materials it is usual to characterise the orientational order in terms of these order parameters [12, 17]. Unfortunately, only the first two non-trivial parameters $\langle P_2 \rangle$ and $\langle P_4 \rangle$ are accessible experimentally as we shall now show. The normalised orientational probability function $f(\theta_m)$ can be nevertheless reconstructed from this limited knowledge by the use of the maximum entropy formalism [18–20]

$$f(\theta_m) = N_0 \exp\{ \lambda_2 P_2(\cos \theta_m) + \lambda_4 P_4(\cos \theta_m) \} \tag{15.9}$$

Statistical considerations, as well as Eqn. 15.9, show that knowledge of $\langle P_2 \rangle$ effectively yields only the width of the distribution, while $\langle P_4 \rangle$ provides information on its overall shape.

The expressions for the intensities of the experimental signals given in Eqn. 15.4 now become

$$I_{VV}(t) \alpha F(t) \langle \cos^2 \theta_\mu \cos^2 \theta_v \rangle$$

$$I_{VH}(t) \alpha \, 1/2 \, F(t) \left\{ \langle \cos^2 \theta_\mu \rangle - \langle \cos^2 \theta_\mu \cos^2 \theta_v \rangle \right\} \qquad (15.10)$$

where we have used the rotational symmetry of the distribution around the Z-axis. Consequently, the expression for the fluorescence anisotropy, Eqn. 15.5, is

$$r(t) = \frac{3 \langle \cos^2 \theta_\mu \cos^2 \theta_v \rangle}{2 \langle \cos^2 \theta_\mu \rangle} - 1/2 \qquad (15.11)$$

It will be instructive to look at the limiting values of the anisotropy at time $t = 0$ and $t \to \infty$. In the former limit emission takes place before the molecule has undergone a rotation. The latter limit simply implies that emission takes place after the molecule has undergone many rotational steps, so that its orientation in space is no longer correlated with its orientation at time $t = 0$ [21, 22]. Thus if τ_c is the average time taken by a molecule to rotate by an angle of 1 radian, then effectively a time $5\tau_c$ can be taken to represent infinity. By the same token τ_c is the effective decay time of $r(t)$. Now the correlation function appearing in Eqn. 15.11 simplifies to the product of the individual averages yielding

$$r(\infty) = 3/2 \, \langle \cos^2 \theta_v \rangle - 1/2$$

$$= S_v \qquad (15.12)$$

where we have recognised that the expression for $r(\infty)$ is the average $\langle P_2 \rangle$ for the emission transition moment, henceforth denoted as S_v. The limit at $t = 0$ can only be obtained in a general way, making use of the cosine rule

$$\cos \theta_\mu = \cos \theta_m \cos \beta_\mu + \sin \theta_m \sin \beta_\mu \cos(\varphi_m - \alpha_\mu)$$
$$\cos \theta_v = \cos \theta_m \cos \beta_v + \sin \theta_m \sin \beta_v \cos(\varphi_m - \alpha_v) \qquad (15.13)$$

where now $\{\theta_m, \varphi_m\}$ and $\{\alpha, \beta\}$ are, respectively, the orientations of the molecule and of the transition moments. After some laborious but straightforward algebraic manipulations we obtain [12]

$$r(0) = \frac{S_v + 2H_0(0)}{2S_\mu + 1} \qquad (15.14)$$

where

$$H_0(0) = 1/4 \left\{ G_{00} + \langle P_2 \rangle^2 \right\} (3 \cos^2 \beta_\mu - 1)(3 \cos^2 \beta_v - 1)$$
$$+ 3/4 \, G_{01} \sin 2\beta_\mu \sin 2\beta_v \cos(\alpha_\mu - \alpha_v)$$
$$+ 3/4 \, G_{02} \sin^2 \beta_\mu \sin^2 \beta_v \cos 2(\alpha_\mu - \alpha_v) \qquad (15.15)$$

and

$$G_{00} = 1/5 + 2/7 \langle P_2 \rangle + 18/35 \langle P_4 \rangle - \langle P_2 \rangle^2$$
$$G_{01} = 1/5 + 1/7 \langle P_2 \rangle - 12/35 \langle P_4 \rangle$$
$$G_{02} = 1/5 - 2/7 \langle P_2 \rangle + 3/35 \langle P_4 \rangle \qquad (15.16)$$

For the sake of completeness we note that

$$S_\mu = 1/2 (3 \cos^2 \beta_\mu - 1) \langle P_2 \rangle = P_2(\cos \beta_\mu) \langle P_2 \rangle$$
$$S_v = 1/2 (3 \cos^2 \beta_v - 1) \langle P_2 \rangle = P_2(\cos \beta_v) \langle P_2 \rangle \qquad (15.17)$$

On setting $\langle P_2 \rangle = \langle P_4 \rangle = 0$, we recover the expressions for an isotropic molecular distribution for both time limits [5, 14]

$$r(0) = 1/5 \{ 3 \cos^2 (\beta_\mu - \beta_v) - 1 \}$$
$$r(\infty) = 0 \qquad (15.18)$$

However, we now see that $r(\infty)$ contains information about $\langle P_2 \rangle$ and $P_2(\cos \beta_v)$, whereas $r(0)$ yields the order parameters $\langle P_2 \rangle$, $\langle P_4 \rangle$ as well as $P_2(\cos \beta_\mu)$ (Eqn. 15.17). The important lesson to draw from this example is that the directions of the transition moments must be known in order to obtain the order parameters from the limiting anisotropy values obtained from experiments on systems with microscopic orientational order.

Expressions for the fluorescence anisotropy in aligned systems, such as pressed gels, in which the molecules are preferentially aligned along the X- or Y-axis, can be found in [12]. However, it is important to note that now the denominator in Eqn. 15.15 is no longer proportional to $F(t)$.

15.2.2.2 *Isotropic dispersions* The situation discussed above is idealised in the sense that experimentally the membrane fragments in aqueous suspensions are not perfectly aligned in space, but are randomly dispersed in the laboratory frame. It is important to realise at this point, that the microscopic alignment of the molecules in each individual fragment is still sensed experimentally, since the preferential molecular orientations are dictated by intermolecular interactions on a local scale. The anisotropy of the microscopic intermolecular interactions cannot be washed out by a macroscopically isotropic distribution of the fragments.

The evaluation of the fluorescence anisotropy can now be carried out either the hard way or the even harder way. The hard way [22, 23] involves the use of advanced mathematical methods which are beyond the scope of this chapter. The even harder way, is to carry on with the approach set out here utilising cartesian coordinates. To summarise we shall quote the approximate, but very useful result derived from the Brownian rotational diffusion model [21, 22]. In this model the molecules are assumed to undergo small-step rotational motions in a potential well of the form $U(\theta_m) = - kT \{ \lambda_2 P_2(\cos \theta_m) + \lambda_4 P_4(\cos \theta_m) \}$. The orientational probability function $f(\theta_m)$ is now given by $f(\theta_m) = \exp \{ - U(\theta_m) \}$. This potential has been chosen for the simple reason that the pair of coefficients $\{\lambda_2, \lambda_4\}$ are uniquely related to the pair of order parameters $\{\langle P_2 \rangle, \langle P_4 \rangle\}$ (Eqn. 15.9). The anisotropy decay is now given by [23]

$$
\begin{aligned}
r(t) = 2/5 \{ & g_0[G_{00}\exp(-t\varphi_{00}) \\
& + 2G_{01}\exp(-t\varphi_{01}) + 2G_{02}\exp(-t\varphi_{02}) + \langle P_2 \rangle^2] \\
& + g_1[G_{10}\exp(-t\varphi_{10}) + 2G_{11}\exp(-t\varphi_{11}) + 2G_{12}\exp(-t\varphi_{12})] \\
& + g_2[G_{20}\exp(-t\varphi_{20}) + 2G_{21}\exp(-t\varphi_{21}) + 2G_{22}\exp(-t\varphi_{22})] \}
\end{aligned}
$$

$$(15.19)$$

with

$$
\begin{aligned}
g_0 &= P_2(\cos \beta_\mu) P_2(\cos \beta_\nu) \\
g_1 &= 3/4 \sin 2\beta_\mu \sin 2\beta_\nu \cos(\alpha_\mu - \alpha_\nu) \\
g_2 &= 3/4 \sin^2 \beta_\mu \sin^2 \beta_\nu \cos 2(\alpha_\mu - \alpha_\nu)
\end{aligned}
$$

$$(15.20)$$

and

$$
\begin{aligned}
G_{01} &= G_{10}, G_{02} = G_{20} \\
G_{12} &= G_{21} = 1/5 - 1/7 \langle P_2 \rangle - 2/35 \langle P_4 \rangle \\
G_{11} &= 1/5 + 1/14 \langle P_2 \rangle + 8/35 \langle P_4 \rangle \\
G_{22} &= 1/5 + 2/7 \langle P_2 \rangle + 1/70 \langle P_4 \rangle
\end{aligned}
$$

$$
\begin{aligned}
\varphi_{00} &= 6D_\perp G_{10}/G_{00} \\
\varphi_{01} &= 6D_\perp G_{11}/G_{01} \\
\varphi_{02} &= 6D_\perp G_{12}/G_{02} \\
\varphi_{10} &= D_\perp(1 + 2/7 \langle P_2 \rangle + 12/7 \langle P_4 \rangle /G_{01} + D_\parallel \\
\varphi_{11} &= D_\perp(1 + 1/7 \langle P_2 \rangle - 8/7 < P_4 \rangle)/G_{11} + D_\parallel \\
\varphi_{12} &= D_\perp(1 - 2/7 \langle P_2 \rangle + 2/7 \langle P_4 \rangle)/G_{21} + D_\parallel \\
\varphi_{20} &= 2D_\perp G_{10}/G_{20} + 4D_\parallel \\
\varphi_{21} &= 2D_\perp G_{11}/G_{21} + 4D_\parallel \\
\varphi_{22} &= 2D_\perp G_{12}/G_{22} + 4D_\parallel
\end{aligned}
$$

$$(15.21)$$

Here $D_\perp$ and $D_\parallel$ are the coefficients of diffusion about axes parallel and perpendicular, respectively, to the long molecular axis.

Equation 15.19 reveals a fundamental difference between the anisotropy decay in ordered systems and molecular solutions. In microscopically ordered systems, the decay coefficients $\{\varphi_{mn}\}$ (Eqn. 15.20) are determined by both the order parameters and the diffusion coefficients. Consequently the decay times obtained from a multi-exponential analysis of the anisotropy decay cannot be directly related to the rotational dynamics of the molecules. On the other hand, in molecular solutions for which $\langle P_2 \rangle = \langle P_4 \rangle = 0$, the dependence on the order vanishes. Now, since $D_\parallel > D_\perp$, we expect the correlation functions to decay at significantly different rates so that a simple multi-exponential fit to $r(t)$ will yield the diffusion constants.

Interestingly we find for the membrane systems the model-independent limits of the fluorescence anisotropy at time $t = 0$ and $t \to \infty$ are

$$r(0) = 1/5\left\{3\cos^2(\beta_\mu - \beta_\nu) - 1\right\}$$
$$r(\infty) = 2/5\,P_2(\cos\beta_\mu)P_2(\cos\beta_\nu)\langle P_2\rangle^2 \qquad (15.22)$$
$$= 2/5\,S_\mu S_\nu$$

Consequently we need to know either β_μ or β_ν in order to extract the order parameter $\langle P_2\rangle$ from the limiting values. Moreover, the value of the order parameter $\langle P_4\rangle$ can only be determined from a fit of the experimental decay to Eqn. 15.19.

15.2.3   Excitation with horizontally polarised light

The experiments described above utilise an excitation light beam polarised vertically, that is in a direction perpendicular to the scattering plane. The question arises now as to why the use of a horizontally polarised beam has been excluded in the study of macroscopically isotropic systems. The reason for this can be understood without resorting to complex mathematical formulations. By the principle of photoselection, a beam polarised along the X-axis, creates a population of excited molecules along the polarisation direction. This population, however, has no preferential orientation in the ZY-plane. Thus the intensity of the fluorescence emission polarised along the Z-axis must be identical to that polarised along the Y-axis for reasons of symmetry. We make good use of this scattering geometry in setting up the experiment, for any deviations from this condition observed in practice, must be due to a differential sensitivity of the detection equipment to the polarisation of the emitted light. The ratio of the two intensities thus yield the correction factor, G, which must be applied to the intensities observed in the anisotropy experiment.

15.2.4   Determination of $F(t)$

We have seen above that in macroscopically isotropic systems or ordered systems possessing rotational symmetry around the Z-axis, the intrinsic fluorescence decay function can be obtained from the sum of the experimental intensities, $F(t) \propto I_{VV}(t) + 2I_{VH}(t)$ [2]. This can be done by the numerical addition of the two signals, each of which measured for the same length of time. On the other hand, the decay can also be determined in a single experiment if we now set the polariser on the emission side with its axis at an angle of $54.7°$ to the vertical, the transmitted intensity will be proportional to

$$I(t) \propto I_{VV}(t)\cos^2(54.7°) + I_{VH}(t)\sin^2(54.7°) \qquad (15.23)$$
$$\propto 1/3\left\{I_{VV}(t) + 2I_{VH}(t)\right\}$$

which is proportional to $F(t)$. The relevant procedures required in the determination of $F(t)$ in macroscopically aligned systems have been discussed in [12, 24–27].

15.3 Experimental methods and instrumentation

15.3.1 *The single photon counting method*

At the heart of every time-resolved anisotropy instrument lie the electronic components which confer the time resolution on the observations. Experiments utilising pulsed sources invariably make use of single photon counting methods [1–3]. The essential component is the time-to-amplitude converter (TAC), the rise time of which (typically 0–10 V in 100 ns) defines the time-resolution of the experiment. The voltage generated by the TAC between the start and stop pulses is transferred to a multi-channel analyser (MCA) where it is digitised and registered in a memory address, the number of the channel being determined by the height of the pulse. In this way a histogram of the stop pulses is built up, sorted according to their delay relative to the start pulse.

The combination of TAC and MCA is conventionally triggered by the excitation pulse from the source and stopped by the first fluorescence photon emitted by the sample. However, at the SRS this order is reversed for the practical reason that the synchrotron pulses arrive at the sample at a frequency of about 3 MHz, while the fluorescence photons are detected at a maximal frequency of 30 kHz. In this way the apparatus is subjected to fewer switching operations and this reduces the pile-up problem [1–4, 9, 10].

15.3.2 *Experimental set-up and methods*

The experimental setup for time-resolved spectroscopy at Daresbury is shown schematically in Figure 15.2. The experimental geometry utilises an L-format, that is the two polarised intensities are measured alternately by rotating the polariser on the emission side by 90° every 25 s. In the other experimental geometry, T-format [14], the two signals are monitored simultaneously on either side of the sample.

The synchrotron radiation arriving from the ring is essentially white and the excitation wavelength needs to be selected with a monochromator. The excitation beam passes through a number of optical components and becomes depolarised to some extent. The polarisation state of the beam is then redefined with a Glann-Thompson (G-T) prism before incidence on the sample. The advantage of this prism is its high throughput in the UV region, but it is important to ensure that the light beam passing through it is perfectly parallel. This is achieved by inserting a lens between the prism and the monochromator.

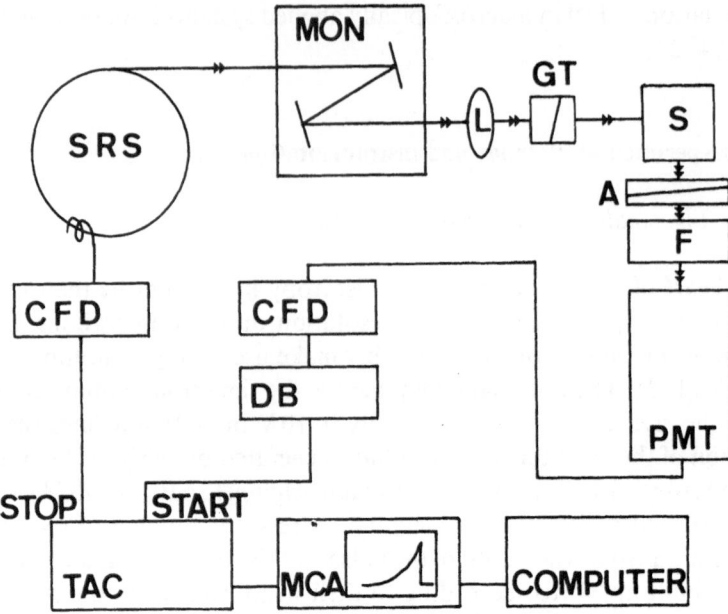

Figure 15.2. A schematic diagram of the experimental setup at station 12.1 of the SRS for fluorescence anisotropy experiments. Symbols used: SRS, Synchrotron Radiation Source; CFD, constant fraction discriminator which can also be used in the leading edge mode; DB, delay box; TAC, time-to-amplitude converter; MCA, multichannel analyser; MON, monochromator: L, lens; GT, Glann-Thompson prism; S, sample; A, analyser polaroid sheet; F, filter combination; PMT, photomultiplier tube.

The polarisation of the fluorescence emission is defined by a polaroid sheet, A, which is larger than the G-T prism, placed immediately after the sample. The wavelength of observation is chosen next, usually by means of a cut-off filter followed by an interference filter, F. The object of the cut-off filter is the rejection of any stray excitation light which may well be more intense than the emission. The interference filter lets through light within a well-defined wavelength region. The cut-off filter may often be replaced by a salt solution in a 1-cm cuvette. For work in the near UV a $NaNO_2$ solution can be used. This combination of filters is considerably cheaper than an analyser monochromator and has the added advantage that it can be placed just before the photomultiplier tube (PMT) and does not affect the polarisation direction. The PMT has a high efficiency, so that a large number of electrons is emitted per incident photon. In timing experiments this leads to the broadening of the signal and sets a lower limit on the signal decays which can be observed reliably. It can, however, be compensated for with a microchannel plate. This device limits the spread of the electron paths through the PMT, reducing the broadening effects and helping to extend the time window to the sub-nanosecond region.

The PMT provides the start pulse for the TAC which is defined by a constant fraction discriminator, CFD. The stop pulse is provided by a ring pulse generated by the electronics of the storage ring and is defined with a leading edge discriminator. In order to position the signal in the MCA, one of the pulses is delayed electronically.

The detection system used in the experiments is optimised to work with low-level light signals. An unwelcome side-effect of this, is its sensitivity to stray rf radiation and rf shielding is a necessary precaution. In this and other respects, it is useful to utilise the first 50 channels of so of the MCA to measure the background signal of the PMT.

In fluorescence anisotropy experiments the signals are commonly collected until the averaged noise is reduced to an acceptable level. Typically this occurs when more than 10 000 counts have been accumulated in the signal maximum. This means a total measurement time of around 20 min, for a synchrotron source which is far superior to experiments utilising other pulsed light sources.

15.4 Numerical analysis

The experimental decay curves $I_{VV}(t)$ and $I_{VH}(t)$ and the lamp profile are obtained in a digital form from a readout of the MCA memory and are stored on disc or magnetic tape. Ordinarily this means the collection of three times 1024 data points and their analysis. This can only be carried out with the help of a computer and some sophisticated software [3,4]. In the numerical analysis we shall consider, the data will be compared with model-based calculations, such as a multi-exponential curve fitting. This approach has the distinct advantage that the estimates of the errors in the measurements are not compromised. Thus on fitting the data to one model or another by least squares techniques, the residuals between the calculated and experimental points can be weighted in an unambiguous way. Single photon counting is essentially a Poisson process [1–4, 10, 28], so that the standard deviation in the observed intensity I at each experimental point is simply $\sqrt{I}$.

There are two problems to be tackled in analysing the data. In the first place, the data do not represent the true signals as they are convolved with the finite response profile of the system; the width of the pulse, the broadening effects in the PMT, etc. This profile is determined experimentally from elastic scatter from a somewhat turbid sample at the emission wavelength. Of course it is important that this is done under identical optical conditions to the real measurements. We now consider this profile as a sum of instantaneous excitation pulses, each of which leads to the intensity responses calculated in Section 15.2. If $L(t)$ is the given profile, then the measured response $R(t)$ is given by the convolution integral [3, 10, 28]

$$R(t) = \int_0^t L(t')\,I(t-t')\,dt' \qquad (15.24)$$

The way to proceed now is to take a particular expression for the intensity $I(t)$

and add to it the background signal of the PMT. The calculated curve is now convolved with the experimental lamp profile $L(t)$ and the result is fitted to the data. We shall here follow common practice and recommend the use of non-linear least squares procedures utilising the Marquardt algorithm [29]. It is important to note that least squares techniques fit the data in such a way as to conserve the area under the experimental curve [30]. This often introduces strong correlations between the adjustable parameters, as for example is the case for the amplitudes and decay times in a multi-exponential analysis. Elsewhere in this volume, the analogous situation of fitting interference functions to model structures has already been discussed in powder diffraction analysis (Chapter 2) and EXAFS analysis (Chapter 6). The data analysis of time-resolved fluorescence consists of the simultaneous fit of the two intensities using the relations derived from Eqn. 15.4

$$I_{VV}(t) = 1/3 \; F(t)\{1 + 2r(t)\}$$
$$I_{VH}(t) = 1/3 \; F(t)\{1 - r(t)\} \tag{15.25}$$

This implies the manipulation of 2048 data points. The goodness of fit criteria are now purely statistical. They include the minimum reduced χ^2 values which should ideally be unity [29], the random distribution of the weighted residuals, H_i, about zero and the lack of correlation between the residuals as can be monitored from their autocorrelation [2–4, 10, 28]. The quantities themselves are defined as [29]

$$H_i = \{R_{calc}(t_i) - R_{exp}(t_i)\}/\{R_{exp}(t_i)\}^{1/2} \tag{15.26}$$

$$\chi^2 r = \frac{\sum_{i=1}^{n} H_i^2}{(n - p)}$$

where n is the number of data points and p the number of model parameters optimised during the fit. The results from a typical analysis are presented graphically in Figure 15.3. It is important to note that the experimental data and the lamp profile are given on a logarithmic scale. This tends to accentuate the low intensity regions such as the after pulse.

The reader will have noticed the sudden switch from the continuous variables implied in the theoretical relations to the discrete nature of the experimental data. Equation 15.24 can cast into the discrete form [3, 10, 28]

$$R(k) = \sum_{j=0}^{k-1} L(k - j)I(j), \quad k = [1, n] \tag{15.27}$$

where $L(k - j)$ is the response profile in channel $(k - j)$, $I(j)$ is the true intensity in channel j and n is the number of channels in the MCA. As the time behaviour of $I(t)$ is described as a sum of exponentials, $I(t) = \sum \alpha_i \exp(-t\beta_i)$, Eqn. 15.27 can be

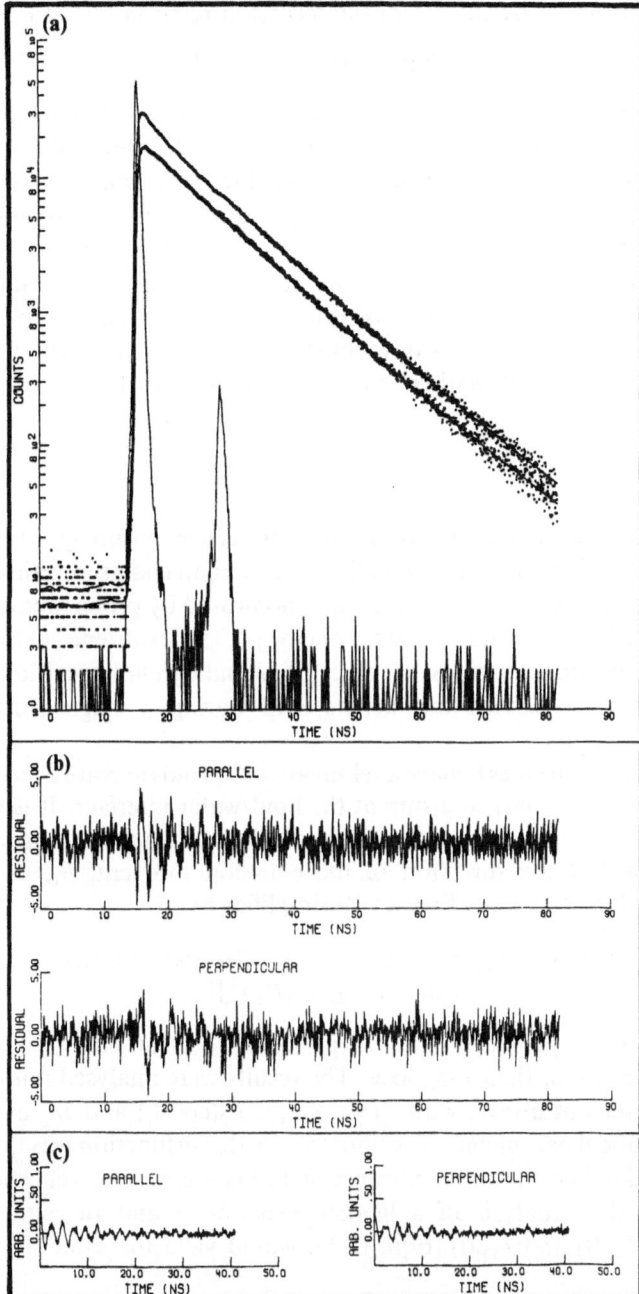

Figure 15.3. The simultaneous fit of the observed time-dependent intensities $I_{VV}(t)$ and $I_{VH}(t)$ for DPH molecules in vesicles of the plant lipid SQDG containing 25% by weight of cholesterol. (A) the excitation pulse profile, $L(t)$, and the experimental points. The continuous lines represent the best fit to Eqns. 15.25 and 15.29. (B) The residuals of the fit and (C) their autocorrelation functions; note the oscillations arising from rf pickup of 500 MHz components from the storage ring. Each channel is equivalent to 80 ps.

evaluated efficiently by the use of the recursion relation [10, 28]

$$R^i(k) = R^i(k-1)\exp(-\beta_i) + L(k)\alpha_i \qquad (15.28)$$

where $R^i(k)$ is the contribution of the ith component to channel k.

An important question which must now be addressed is the number of adjustable parameters which can be extracted reliably from the fits. We have found that in general the fluorescence decay $F(t)$ can be expressed satisfactorily as a bi-exponential process requiring four parameters in its description. With this recipe for $F(t)$, the fitting procedure becomes numerically unstable if the decay of the anisotropy $r(t)$ is described by more than six parameters. Interestingly, the parameters describing $r(t)$ extracted from the fit are not affected by the detailed mathematical description of $F(t)$.

15.5 Results

We shall now consider a time-resolved fluorescence anisotropy study of the effects of increasing unsaturation of the hydrocarbon chains in vesicle systems on the orientational order and dynamics monitored by probe molecules. The molecules 1-[4-(trimethylammonio)-phenyl]-1,3,5-hexatriene (TMA-DPH) were used to probe vesicles of diacylphosphatidylcholines [26, 27]. The experiments were carried out at the SRS during a single-bunch mode operation.

These probe molecules behave as elongated ellipsoids and are expected to be anchored with the charged group at the lipid/water interface. It is generally accepted that the absorption transition moments lies along the long molecular axis ($\beta_\mu = 0$), but the direction of the emission moment, β_ν, is not well characterised. In this case, Eqn. 15.19 simplifies to

$$r(t) = 2/5\{g_0[G_{00}\exp(-t\varphi_{00}) + 2G_{01}\exp(-t\varphi_{01}) \\ + 2G_{02}\exp(-t\varphi_{02}) + \langle P_2\rangle^2]\} \qquad (15.29)$$

and we see from Eqn. 15.21 that the experiment does not sense the rotation of the molecules about their long axes. The results were analysed following the procedures set out above, with $\langle P_2\rangle$, $\langle P_4\rangle$, $P_2(\cos\beta_\nu)$ and $D_\perp$ as the four adjustable model parameters. The fluorescence decay function was taken to be bi-exponential. The physial significance of the fits was now also checked on the expectation that analysis of a lifetime experiment and the simultaneous analysis of $I_{VV}(t)$ and $I_{VH}(t)$ (Eqn. 15.25) would yield the same fluorescence decay $F(t)$.

15.5.1 The initial anisotropy $r(0)$

In all the systems studied β_ν was found to lie in the range 13–21°, corresponding to $r(0)$ values of 0.37–0.32 (Eqn. 15.22). This is in good agreement with values for $r(0)$ obtained from a straightforward multi-exponential fit to the data. In view of the fact that previous studies reported in the literature assume at the outset that $r(0) = 0.4$, i.e. $\beta_v = 0$, we have analysed the data systematically by keeping $P_2(\cos \beta_v)$ constant during the fitting procedures. The χ_r^2 values obtained decreased from 1.49 for $P_2(\cos \beta_v) = 1$ to a minimum at 1.46 for $P_2(\cos \beta_v) = 0.81$, the value found in the unconstrained analysis, before increasing steeply as $P_2(\cos \beta_v)$ is reduced further. The small reduction in the value of χ_r^2 obtained in this way indicates either a small improvement in the quality of the fit over the entire decay or a marked change in the fit across a narrow time interval. It turns out that the decrease is due entirely to a significant improvement in the fit to a small number of channels directly following the excitation pulse. The reason for this effect is that for $P_2(\cos \beta_v) = 1$ the decay coefficients φ_{01} and φ_{02} are so large, that the second and third terms in Eqn. 15.29 only contribute to the first 10 channels (≈ 800 ps). In this way about 40% of the anisotropy amplitude disappears within the first nanosecond of the decay; at longer times the decay is dominated by the first term in Eqn. 15.29. Thus an extrapolation of the long time component to zero time will yield a value for $r(0)$ markedly lower than 0.4. This effect is alleviated considerably on lowering the value of $P_2(\cos \beta_v)$. Now, a slower decay of the middle terms in Eqn. 15.29 is obtained so that they contribute to the signal at longer times. The minimum in the χ_r^2 surface is the result of a trade-off between the two effects. It must be emphasised here that this sort of consideration has been made possible by the fact that the pulse profile can be measured at the excitation wavelength, so that the experimental signals at short times can be analysed reliably.

The analysis above is based on the explicit assumption that $\beta_\mu = 0$. The validity of this approach can be tested by fitting the experimental data to the general relation, Eqn. 15.19. We have found that this procedure is numerically unstable and that the solutions strongly depend on the initial values assigned to the model variables at the start of the non-linear least squares search. The problem here is that the optimisation procedure first increases the diffusion coefficient $D_\parallel$, before attempting to improve the fit by changing the directions of the transition moments. In this way the decay of the last six terms in Eqn. 15.19 is accelerated, so that their contribution to the experimental signal declines. Moreover, the procedure tends to equalise the two angles β_μ and β_v. We believe therefore that the anisotropy decay, certainly in membrane vesicle systems, can only be analysed if the direction of at least one transition moment is known.

15.5.2 *The orientational order*

The order parameter $\langle P_2 \rangle$ is found to decrease markedly on increasing the degree of unsaturation of the acyl chains. The same trend is observed in vesicle systems containing cholesterol, but now significantly higher values of $\langle P_2 \rangle$ are found. Surprisingly and in contrast to expectations the order parameter $\langle P_4 \rangle$ is found to be insensitive to the degree of unsaturation. Nevertheless, $\langle P_4 \rangle$ increases markedly on the addition of cholesterol, an observation consistent with the picture of increasing orientational order.

These effects are best illustrated by the reconstructed orientational distribution function $f(\theta_m)$ (Eqn. 15.9) or the number density $f(\theta_m) \sin \theta_m$. The number density simply denotes the number of molecules found at a given orientation. Typical distributions are shown in Figure 15.4. Interestingly, we find a bimodal distribution of TMA-DPH molecules; one population lying parallel to the bilayer surface and the other aligned preferentially along the normal to it. The parallel population grows with increasing unsaturation,

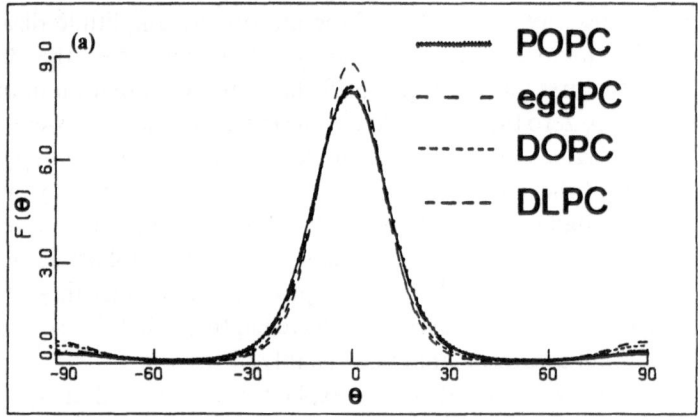

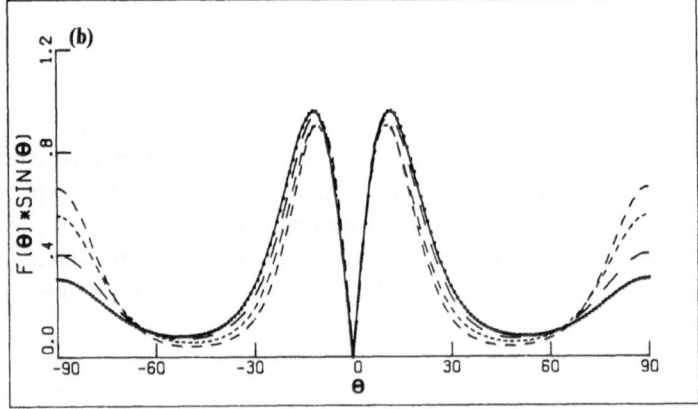

Figure 15.4. (a) The orientational distribution functions $f(\theta_m)$ and (b) the corresponding number densities $f(\theta_m) \sin \theta_m$ for TMA-DPH molecules in vesicles of diacylphosphatidylcholines (PC) with increasing degree of unsaturation in the acyl chains: palmitoyl-oleoylPC (POPC), eggPC, dioleoylPC (DOPC) and dilineoylPC (DLPC).

while the distribution of molecules lying along the normal to the surface narrows. This finding, is surprising to say the least. Nevertheless, cholesterol was found to suppress the parallel population and to enhance the normal one in line with our expectations.

The anisotropy experiment does not identify the location of the probe molecules in the bilayer structure. However, in the light of studies showing that TMA-DPH molecules can partition between the bilayer and aqueous phases (where they do not fluoresce), we believe that a fraction of the molecules lies at the water/lipid interface of the vesicle, parallel to its surface.

15.5.3 *Reorientational dynamics*

The value of the diffusion coefficient $D_\perp$ has been found to be very strongly correlated with the value of $r(0)$. This arises from the fact that $D_\perp$ is determined unambiguously from the initial slope of $r(t)$ [22, 23]. In fact we have found it to decrease by a factor of 5 and more on reducing $P_2(\cos \beta_v)$ from 1.0 to 0.81 in our analysis. The diffusion constants were found to increase significantly on increasing the unsaturation of the lipid chains. Surprisingly, the motional rates increased still further on the addition of cholesterol. This latter observation is at odds with the popular concept of membrane fluidity.

At this point we would like to emphasise that the diffusion coefficients $D_\perp$ and $D_\parallel$ appearing in Eqns. 15.19–15.21 have been defined in a general way in the spirit of the theory of Brownian motion [21]. It simply represents the mean square angular displacement of the molecule per unit time; it is the rotational analogue of the linear diffusion coefficient. The reader is invited to note that there is no a priori relation between diffusion constants and the microviscosity experienced by the probe. The concept of viscosity is only valid provided the sticky boundary condition holds. This condition implies that there is no relative motion between the probe and its environment, so that effectively we require the probe and its solvation shell to rotate as a single entity within the bilayer.

17.6 The future

Over the past few years the authors have become convinced of the many advantages offered for their research by synchrotron radiation. In fact their current work on angle- and time-resolved studies of macroscopically ordered membrane systems would not be feasible without this facility. The speed with which useful data are collected, particularly from small samples, opens up entirely new opportunities for experimentalists. Indeed, one can already foresee sub-nanosecond studies on protein and photosynthetic systems. An exciting possibility is the implementation of pulse-probe experiments in combination with laser systems. But perhaps the most challenging prospect is the combination of time-resolved spectroscopy with fluorescence microscopy techniques in order to probe biochemical processes on the cellular level.

Acknowledgements

The authors thank Dr David Shaw for his help with many of the experiments reported in this chapter. The work has been supported by the Dutch Foundation for Biophysics under the Netherland Organisations for Scientific Research (NWO).

References

1. W.R. Ware, in *Creation and Detection of the Excited State*, Dekker, New York (1971).
2. M.G. Badea and L. Brand, in *Methods in Enzymology*, eds. C.H.W. Hirs and S.N. Timasheff, Academic Press, New York (1979), Vol. 61H.
3. D.V. O'Conor and D. Phillips, *Time-Correlated Single Photon Counting*, Academic Press, London (1984).
4. R.B. Cundall and R.E. Dale, eds., *Time-Resolved Fluorescence Spectroscopy in Biochemistry and Biology*, Plenum Press, New York (1983).
5. R.E. Dale, in *Polarized Spectroscopy of Ordered Systems*, eds. B. Samori and E.W. Thulstrup, Kluwer, Dordrecht, The Netherlands (1988) p. 491.
6. D.H. Waldeck, A.J. Cross, D.B. McDonald and G.R. Fleming, *J. Chem. Phys.* **74** (1981) 3381.
7. A.J. Cross, D.H. Waldeck and G.R. Fleming, *J. Chem. Phys.* **78** (1983) 6455.
8. Ph. Wahl, J.C. Auchet and B. Donzel, *Rev. Sci. Instrum.* **45** (1974) 28.
9. A. Gafni, R.L. Modlin and L. Brand, *Biophys. J.* **15** (1975) 283.
10. A. Grinvald, *Anal. Biochem.* **75** (1976) 260.
11. I.H. Munro and N. Schwentner, *Nucl. Instrum. Methods* **208** (1983) 819.
12. M. van Gurp, G. van Ginkel and Y.K. Levine, *J. Theor. Biol.* **131** (1988) 333.
13. J.B. Birks, *Photophysics of Aromatic Molecules*, John Wiley, London (1970).
14. J.R. Lakowicz, *Principles of Fluorescence Spectroscopy*, Plenum Press, New York (1983).
15. J.R. Lakowicz, ed., *Time-Resolved Laser Spectroscopy in Biochemistry*, SPIE, Bellingham (1988) Vol. 909.
16. E. Merzbacher, *Quantum Mechanics*, John Wiley, New York (1961).
17. G.R. Luckhurst and G.W. Gray, eds., *The Molecular Physics of Liquid Crystals*, Academic Press, New York (1979).
18. D.I. Bower, *J. Polym. Sci.* **19** (1981) 93.
19. B.W. van der Meer, R.P.H. Kooyman and Y.K. Levine, *Chem. Phys.* **66** (1982) 39.
20. H. Pottel, W. Herreman, B.W. van der Meer and M. Ameloot, *Chem. Phys.* **102** (1986) 37.
21. U. Segre and P.L. Nordio, in ref [17], Chap. 18.
22. C. Zannoni, A. Arcioni and P. Cavatorta, *Chem. Phys. Lipids* **32** (1983) 179.
23. B.W. van der Meer, H. Pottel, W. Herreman, M. Ameloot, H. Hendrickx and H. Schröder, *Biophys. J.* **46** (1984) 515.
24. H. van Langen, G. van Ginkel and Y.K. Levine, in *Time-Resolved Laser Spectroscopy in Biochemistry*, ed. J.R. Lakowicz, SPIE, Bellingham (1988) p. 377.
25. G. Deinum, H. van Langen, G. van Ginkel and Y.K. Levine, *Biochemistry* **27** (1988) 852.
26. H. van Langen, G. van Ginkel and Y.K. Levine, *Liquid Crystals*, **3** (1988) 1301.
27. H. van Langen, G. van Ginkel, D. Shaw and Y.K. Levine, *Eur. Biophys. J.*, **17** (1989) 37.
28. A. Grinvald and I.Z. Steinberg, *Anal. Biochem.* **59** (1974) 583.
29. P.R. Bevington, *Data Reduction and Error Analysis for the Physical Sciences*, McGraw-Hill, New York (1969).
30. T. Awaya, *Nucl. Instrum. Methods* **165** (1979) 317.

Index